Advances in Intelligent Systems and Computing

Volume 757

Series editor

Janusz Kacprzyk, Polish Academy of Sciences, Warsaw, Poland
e-mail: kacprzyk@ibspan.waw.pl

The series "Advances in Intelligent Systems and Computing" contains publications on theory, applications, and design methods of Intelligent Systems and Intelligent Computing. Virtually all disciplines such as engineering, natural sciences, computer and information science, ICT, economics, business, e-commerce, environment, healthcare, life science are covered. The list of topics spans all the areas of modern intelligent systems and computing such as: computational intelligence, soft computing including neural networks, fuzzy systems, evolutionary computing and the fusion of these paradigms, social intelligence, ambient intelligence, computational neuroscience, artificial life, virtual worlds and society, cognitive science and systems, Perception and Vision, DNA and immune based systems, self-organizing and adaptive systems, e-Learning and teaching, human-centered and human-centric computing, recommender systems, intelligent control, robotics and mechatronics including human-machine teaming, knowledge-based paradigms, learning paradigms, machine ethics, intelligent data analysis, knowledge management, intelligent agents, intelligent decision making and support, intelligent network security, trust management, interactive entertainment, Web intelligence and multimedia.

The publications within "Advances in Intelligent Systems and Computing" are primarily proceedings of important conferences, symposia and congresses. They cover significant recent developments in the field, both of a foundational and applicable character. An important characteristic feature of the series is the short publication time and world-wide distribution. This permits a rapid and broad dissemination of research results.

More information about this series at http://www.springer.com/series/11156

Dipankar Deb · Valentina E. Balas
Rajeeb Dey
Editors

Innovations in Infrastructure

Proceedings of ICIIF 2018

Springer

Editors
Dipankar Deb
Department of Electrical Engineering
IITRAM
Ahmedabad, Gujarat, India

Rajeeb Dey
Department of Electrical Engineering
National Institute of Technology, Silchar
Silchar, Assam, India

Valentina E. Balas
Aurel Vlaicu University of Arad
Arad, Romania

ISSN 2194-5357 ISSN 2194-5365 (electronic)
Advances in Intelligent Systems and Computing
ISBN 978-981-13-1965-5 ISBN 978-981-13-1966-2 (eBook)
https://doi.org/10.1007/978-981-13-1966-2

Library of Congress Control Number: 2018950222

This Springer imprint is published by the registered company Springer Nature Singapore Pte Ltd.
The registered company address is: 152 Beach Road, #21-01/04 Gateway East, Singapore 189721, Singapore

Preface

This volume constitutes a part of the proceedings of first International Conference on Innovations in Infrastructure (ICIIF) 2018, which was held on 18–19 May, 2018, in Ahmedabad, India. The first International Conference on Innovations in Infrastructure (ICIIF) was organized by the Institute of Infrastructure Technology Research and Management (IITRAM), Ahmedabad, in collaboration with Gujarat Knowledge Society, Government of Gujarat. It was funded by the Student Start-up and Innovation Policy (SSIP) initiative of the Government of Gujarat; DST, Government of India; ISRO; and many other industrial partners who aligned with the view of disseminating the progress of innovative, scientific and technological inventions. The conference provided a forum for discussion on issues, concepts, skill development and possible innovations in the infrastructure sector. ICIIF aimed at bringing the best technical minds working in the field of infrastructure development to a common platform to share their knowledge of technical expertise, experience and forthcoming challenges in this area. The conference articles include latest research trends showcasing the important achievements and upcoming challenges in the development of the infrastructure sector. The conference papers included in the proceedings, published post-conference, were grouped into the following areas of research:

Innovation in Telecommunication Infrastructure
Innovation in Control Engineering
Innovation in Power System Infrastructure
Innovation in Smart Infrastructure
Innovation in Waste Management
Sustainable Solutions for Infrastructure Development
Innovation in Thermal Engineering
Innovation in Manufacturing
Innovation in Industrial Infrastructure
Innovation in Renewable Energy
Innovation in Hybrid Vehicles
New Optimization Techniques

Innovation in Mechatronics
Innovation in Machine Learning
Innovation in Indigenous and Rural technologies
Innovations in Infrastructure Management

In ICIIF 2018, we had three eminent keynote speakers: Professor Valentina Emilia Balas (Romania), Mr. Aninda Bose (Senior Editor, Springer), and Padma Shri Prof. Ajoy Ray (IIT Kharagpur).

We are thankful to all the authors who have submitted papers so as to keep the quality of ICIIF 2018 at high levels. The editors of this book would like to acknowledge all the authors for their contributions and also the reviewers. We have received invaluable help from the members of the International Programme Committee and the chairs responsible for different aspects of the workshop. We appreciate also the role of special sessions' organizers. Thanks to all of them, we had been able to collect many papers of interesting topics. We had very interesting presentations and stimulating discussions.

Our special thanks to Prof. Janusz Kacprzyk (Series Editor, Advances in Intelligent Systems and Computing Series, Springer) for giving us the opportunity to publish this edited volume. We are grateful to Springer, especially to Dr. Thomas Ditzinger (Executive Editor, Applied Sciences and Engineering, Springer-Verlag) and Mr. Aninda Bose (Senior Editor, Hard Sciences, Springer Nature India) for the excellent collaboration, patience and help during the evolvement of this volume.

We hope the volume will provide useful information to professors, researchers and graduated students in various fields of engineering to refer to the innovative techniques as a solution for their problems, and all will find this collection of papers inspiring, informative and useful. We also hope to see you in a future ICIIF event.

Ahmedabad, India Dr. Dipankar Deb

Contents

About the Editors

Dipankar Deb completed his Ph.D. from the University of Virginia, Charlottesville, with Gang Tao, IEEE Fellow and Professor in the Department of ECE in 2007. In 2017, he was elected to be IEEE Senior Member. He has served as Lead Engineer at GE Global Research, Bengaluru (2012–2015), and as Assistant Professor in electrical engineering, IIT Guwahati (2010–2012). Presently, he is Associate Professor in the Department of Electrical Engineering at the Institute of Infrastructure Technology Research and Management, Ahemedabad, India. He leads efforts towards innovation. He is Student Startup and Innovation Project Coordinator at IITRAM. He mentors students to build intellectual property rights (patents).

Valentina E. Balas is currently Full Professor in the Department of Automatics and Applied Software at the Faculty of Engineering, "Aurel Vlaicu" University of Arad, Romania. She holds a Ph.D. in applied electronics and telecommunications from the Polytechnic University of Timisoara, Romania. She is the author of more than 250 research papers in refereed journals and international conferences. Her research interests are in intelligent systems, fuzzy control, soft computing, smart sensors,information fusion, modeling and simulation. She is Editor-in-Chief of *International Journal of Advanced Intelligence Paradigms (IJAIP)* and *International Journal of Computational Systems Engineering (IJCSysE)*; editorial board member of several national and international journals; and is Director of Intelligent Systems Research Centre in Aurel Vlaicu University of Arad. She is Member of EUSFLAT and SIAM; Senior Member of IEEE; and TC Member of Fuzzy Systems (IEEE CIS), Emergent Technologies (IEEE CIS) and Soft Computing (IEEE SMCS).

Rajeeb Dey is presently working with National Institute of Technology, Silchar, Assam, India, as Assistant Professor in the Department of Electrical Engineering. Before joining NIT Silchar, he has served Sikkim Manipal University, Sikkim, for 12 years in various positions (Lecturer, Reader and Associate Professor). His research interests are time-delay systems and control, robust control, control of biomedical systems and application of wireless communication in control. He is presently a reviewer of many SCI(E) journals related to control engineering and applied mathematics. He is Senior Member of IEEE and CSS, Life Member of System Society of India and Member of Institution of Engineers (India).

Public Key Cryptography Using Harmony Search Algorithm

Suman Mitra, Gautam Mahapatra, Valentina E. Balas and Ranjan Chattaraj

Abstract Privacy is a very important requirement for viability of modern information sharing through cyberspace and the modern cryptology is ensuring success. Harmony Search Algorithm (HSA) is a new meta-heuristic computation technique inspired from musical improvisation techniques, where searching for a perfect harmony is the objective of this technique. Public Key Cryptography heavily relies on key pairs which are large prime numbers. Our adaptation of the HSA tries to provide a fast key generation mechanism with a feasible implementation. The keys are ranked based on their harmony and the best harmony is selected as the result of the search which in turn is used to generate the key pair of RSA, a Public Key Cryptography technique as a test of effectiveness and success.

Keywords Public Key Cryptography (PKC) · Harmony Search Algorithm (HSA) · Fast key generation · Random Number Generator (RNG) · RSA Keys management · Prime numbers

S. Mitra
Department of Computer Science, Asutosh College, University of Calcutta, 700026 Kolkata, India
e-mail: sumanmitra@outlook.com

G. Mahapatra (✉)
Department of Computer Science and Engineering, Birla Institute of Technology Mesra, Off-Campus Deoghar, Ranchi, Jharkhand, India
e-mail: gsp2ster@gmail.com

V. E. Balas
"Aurel Vlaicu" University of Arad, Arad, Romania
e-mail: balas@drbalas.ro; valentina.balas@uav.ro

R. Chattaraj
Department of Mathematics, Birla Institute of Technology Mesra, Off-Campus Deoghar, Ranchi, Jharkhand, India
e-mail: chattaraj6@gmail.com

D. Deb et al. (eds.), *Innovations in Infrastructure*, Advances in Intelligent Systems and Computing 757, https://doi.org/10.1007/978-981-13-1966-2_1

1 Introduction

Cryptology is a two-step system: the first step is related with making, generally called cryptography where original messages are transformed in unreadable form called ciphertext and the second step is breaking called cryptanalysis, taking the challenges to transform ciphertext into original messages only by the intended user. Cryptology guarantees properties like confidentiality, integrity, and authenticity of information which is shared using highly vulnerable public communications system. The Public Key Cryptography (PKC) [1] is an asymmetric cryptography technique where a key pair is used—one key known as Public Key is used for encryption (making) and another key known as Private Key is used for decryption (breaking). Generally, key is a prime integer number, and factorization of integer number into only two prime integer numbers is an intractable problem and this is because it is a one-way mathematical function [2]. The processes of finding keys are mostly dependent on identifying big prime integer numbers which are done by Random Number Generators (RNGs). RNGs are a very important building block in the techniques and protocols used in cryptography [3]. The quality of the key pair is ultimately dependent on finding pure random numbers which increase the security performance of the keys. There are several techniques in literature uses hardware, software, and their hybridized form for this key management [4, 5]. Nowadays, these key management for PKC have been devised as optimizing problem and meta-heuristic optimization frameworks are well suited than other deterministic techniques [6]. Finding prime integers as key pair is an optimized searching process, well suited for HSA, a newly developed meta-heuristic computation framework [7]. In the work, we are focusing on the randomness of keys using HSA which is inherently random, parallel, and robust and inspired from complex musical improvising system. Any meta-heuristic computation framework like evolutionary algorithms [8] and simulated annealing [9] are working by imitating natural phenomenon. Harmony Search Algorithm (HSA) [7] is a meta-heuristic algorithm which is made from the process of improvising music to search for a perfect state of harmony. Musical performances try to achieve harmony pleasing to an aesthetic standard and HSA tries to find a global solution, i.e., a perfect state using an objective function. Pitches of every instrument are combined to determine the quality of the harmony just as the value of the objective function is determined by the values of the decision variables. Key generation in cryptology has been dealt with many aspects but the use of HSA in the process is yet to be explored successfully. A unique or non-duplicate key guarantees better results in encryption process that is theoretically impossible to break in the decryption process. This work tries to explore the use of new nonconventional techniques in the key management process.

2 Literature Review

Key management in PKC is a most important issue for its viability and performance measures. In PKC management key pair, i.e., pair of prime integer numbers cannot be generated twice, therefore fast prime number generation is required. With the help of

BAN logic used in symmetric cryptography and combination optimization technique, Hao et al. [10] have proposed a framework for automatic protocol generation for key management. Bahadori et. al. [11] described a technique using Euclid's extended algorithm for large random prime number generation which is used in the secure and fast key generation for RSA. Spillman [6] has Genetic Algorithm (GA)-based meta-heuristics for cryptanalysis. Latter Particle Swarm Optimization (PSO) [12], Differential Evolutionary Algorithm (DEA) [8], Artificial Neural Network [13], Ant Colony Optimization (ACO) [14], Bee Colony Optimization (BCO) [15], and most recently Cuckoo Search Algorithm [16] have been tested successfully in the field of cryptanalysis after designing different kinds of attacks for unauthentic breaking of the encrypted message. Use of hybrid form like ANN-GA [17] of nature-inspired algorithms is also studied in the literature. Patidar et al. [18] exploited the interesting properties of chaotic systems to design a random bit generator, in which two chaotic systems are cross-coupled with each other. Statistical tests are applied to evaluate the randomness of the bit streams generated by the Cross-coupled Chaotic Tent Map Based Bit Generator (CCCBG).

3 Swarm Intelligence for Random Number Generator

3.1 Key Generation

In public key cryptography, strong random number generation is very important throughout every phase of processing. RNGs are suitable in any cryptographic application only when its outputs are strongly unpredictable in the absence of knowledge of inputs. The security performance of cryptographic systems totally depends on some secret data that is known only to the authorized persons, but unknown & unpredictable to others. This unpredictability is achievable in the strong randomized environments. In our proposed key management process, we have considered these randomized environments as an optimization problem and HSA is used for best possible unpredictable keys to provide secured cryptographic system. To test quality of our proposed RNG, we use Standard Statistical Test suits designed by NIST [19] for testing pseudo-RNG.

3.2 Harmony Search Algorithm

Like jazz music improvisation, each musician plays a pitch within possible range to make harmony together. If the all the pitches together make a good harmony, those individual pitches are stored in the memory of those players for a chance of making good harmony next time thus making the harmony more pleasing. For example, lets suppose a band is composed of three musicians: a guitarist, a pianist, and a cellist. The three musicians play the notes C3 (130.81 Hz), F3 (174.61 Hz), and G3 (196.00 Hz) accordingly thus making a harmony together known as the C-major chord. In engineering terms, these frequencies can be interpreted as the values of decision variables, stored in Harmony Memory (HM) and the value computed by the objective function for a harmony is also stored in memory. Now, if in the next iteration the cellist plays Bb3 (233.08 Hz), the harmony changes to C-7 chord which might be more aesthetically pleasing and if so then these pitches are stored in the HM thus increasing the chances of making a perfect harmony. Now, when a musician plays an improvised pitch he/she follows any of the three rules: (1) playing a pitch from memory, (2) playing a pitch adjacent to a pitch memory, and (3) playing a totally random pitch from the possible range of the instrument. The three rules are chosen using two parameters: Harmony Memory Consideration Rate (HMCR) is used to choose between rules (1), (2) together, and (3), the Pitch Adjutancy Rate (PAR) is used to decide between rules (1) and (2). The classical HSA for minimizing problems comprises of the following steps:

Step 1: Initialize algorithm parameters and objective function.
Initialize $NVAR, HMS, HMCR, PAR, BW, iterno, MAX\_ITR$
Step 2: Initialize Harmony Memory.
$HM = [X_i^j] \;\; i = 1 \ldots NVAR, j = 1 \ldots HMS, X_i^j = rand(L_i, U_i)$
$for \;\; i = 1 \; to \; HMS$
  $z^i = f(X^i)$ [Evaluate harmony value of a solution vector]
$sort(X^i, z^i)$ [Rank each solution vector with the harmony value]
Step 3: Perform Harmony Search
repeat Step 4 through Step 6 until $iterno < MAX\_ITR$
Step 4: Improvise new Harmony Vector
$for \;\; k = 1 \; to \; NVAR$
  $if \;\; rand(0, 1) < HMCR \;\; then$
    $if \;\; rand(0, 1) < PAR \;\; then$
      $X'_k = X_r \pm rand(L_k, U_k) \times BW$ [X_rRandom member of HM]
    $else$
      $X'_k = X_r$
  $else$
    $X'_k = X_r$
Step 5: Update Harmony Memory.
$z' = f(X')$
$if \;\; z' < z_i^w \;\; then$
    $X_i^w = X'$
    $x_i^w = f(X_i^w)$ [X_i^wworst solution vector]
$sort(X_i, x_i)$
$storeX_i^b$ [X_i^bbest harmony in ith iteration]
Step 6: Update iteration counter.
$iterno = iterno + 1$
Step 7: Finished.
return X_i^b

3.3 Public Key Cryptography

The PKC is an asymmetric cryptography technique where one key, known as Public Key is used for encryption and other known as Private Key is used for decryption. RSA is used as our case study of PKC. Finding this key pair is an optimization problem can be solved using meta-heuristic computational framework like HSA. The process that of finding keys is largely dependent on finding big prime numbers which are a case of finding random numbers. The challenge is finding large prime numbers which are generated in a purely random fashion and without repetition.

4 HSA Key Management for RSA

Here, we propose a technique for key pair generation for RSA algorithm using HSA. The private key requires two large prime numbers for which we first use HSA and then, the key pair for RSA after maintaining the properties of RSA key pair. The proposed procedure is described in the following step-by-step form:

Step 1: Initialize Algorithm Parameters and Setup Objective function.
Parameters $HMS, HMCR, PAR, BW, MAX\_ITR$ are initialized. Harmony Memory (HM) holds a set of solution vectors X^j, $j = 1 \ldots HMS$. These solution vectors are made up of a set of parameters which is denoted by $X_i, i = 1 \ldots NVAR$. Here, $NVAR = N_b/2$ and N_b is the number of bits initialized as the bit length for the private key. N_b is divided into two equal parts as each prime number and its multiplied value is used to get the N_b bit key.
Each solution vector is evaluated through the objective function to determine its harmony value. First, entropy is calculated using Shannons Entropy [20] formula: $H(X) = -(p \times \log_2 p) - ((1 - p) \times log_2(1 - p))$ where p denotes percentage of randomness. The chi-square value against Shannon Entropy value is calculated using the formula: $C^2 = \dfrac{(ObservedH(X) - ExpectedH(X))^2}{ExpectedH(X)}$. Lastly, Coefficient of Auto-correlation (Phi-coefficient) is calculated: $\phi^2 = \dfrac{C^2}{n}$, where $n = HMS$. The square root of coefficient of auto-correlation (ϕ) is used as the harmony value for ranking the solution vectors. The pseudo-code of this cost function given at the end of this the step- by-step method descriptions.

Step 2: Initialize Harmony Memory
The HM consists of solution vectors of numbers as bits. Each vector is the size of $NVAR$ and there are HMS number of solutions in the memory. The memory is $HM = X_i^j, i = 1 \ldots HMS, j = 1 \ldots NVAR$. Each solution is evaluated using the cost function and the respective harmony value is stored. Each vector is ranked after all the vectors are initialized and evaluated.

Step 3: Improvise a new Harmony Vector

A New Harmony vector is improvised using either one of the three rules for each parameter: (i) Find a random value of the parameter from HM and perform a slight improvisation using BW, (ii) Find a random value of the parameter from HM, and (iii) Assign a random value to the parameter between L_i and U_i. The rules are applied using $HMCR$ and PAR on the basis of chances. After improvising a vector, it is evaluated using cost function.

Step 4: Update Harmony Memory.

If the new improvised solution vector has a better harmony value than the worst harmony in the HM, then the worst vector is replaced with the new improvised vector. Also, the best harmony vector is found after sorting each solution.

Step 5: Check Termination Criteria and check for Primality.

$Step$ 3 and $Step$ 4 are repeated for MAX_ITR number of times. The best solution is then checked if it is prime or not. If not then 2 is added or subtracted to the number and again checked for primality until a prime number is found. Miller Rabin [21] primality test is used to determine if the number is non-prime or not.

Step 6: Generate Second Prime and check for Co-Prime.

$Step$ 2 through $Step$ 5 are repeated to generate a second prime number. After the two prime numbers are found, their co-primality are checked. If the numbers are not co-primes, then a new set of primes are generated using the above steps and is repeated until a pair of co-prime numbers are found

Step 7: Compute Private Key.

The public key *e* is a common key which is predetermined. The private key *d* is calculated using the formula $d = e^{-1} mod \phi(n)$; $\phi(n) = (p-1) \times (q-1)$. The key pair *e* and *d* are returned as solution of the algorithm.

Function **CostKeyPair**(X^{new}, X^{old})
begin
for i=1 **to** NVAR **do**
begin
    $DM = abs(X_i^{old} - X_i^{new})$
    $N1 = i \times DM$
    $N2 = i \times X_i^{new}$
end
$p = abs(N1 - N2)/N2$
$H = -(p \times log_2 p) - (1-p) \times log_2(1-p)$
$CHI = (H - E_H)^2 / E_H$
$V = \sqrt{CHI/HMS}$
return V
end

5 Experiments

5.1 Experimental Setup

The experiments have been conducted on a Windows 10 × 64 OS using Python 3.6 development environment running on an Intel Core i5 6400 @2.70 GHz and 8 GB DDR4 @2400 MHz RAM and a Gigabyte H110M-DS2-CF with a NVIDIA GeForce GTX 750 Ti. The parameters set for the algorithm are as follows: (Table 1).

5.2 Results

The algorithms runtime and the intermediate HM values in each iteration is recorded. Below are two graphs to show the performance of the proposed algorithm. The first graph shows the harmony value of the best harmony in each iteration. The two colors orange and blue represent the two numbers harmony value in each iteration respectively. The graph shows that the numbers have a similar harmony value towards the end of the iterations. Here, we shall clarify that this nature is observed for multiple variations of NVAR sizes and on multiple instances of same NVAR size. The graph in Fig. 1 is made with NVAR size of 1024 bits.

The next graph shows the average harmony of the HM in each iteration. Similarly, the orange and blue colors reflect the average harmony in two runs for the two numbers. In Fig. 2 the graph shows the decay in harmony value in progressive iterations. This implies that the randomness of the numbers in the solution vectors is increasing in the Harmony Memory. The values of the best harmonies after the two searches are complete are very small numbers which indicate that they have low auto correlation, that will result in a better randomness. This graph is also made with 1024 bits NVAR size. The algorithm was tested on different NVAR sizes.

Table 1 Algorithm parameters for the tests of the discussed results

Parameters	Values
N_b	1024, 1536 . . . 4096
$NVAR$	512, 768 . . . 2048
HMS	500
$HMCR$	0.85
PAR	0.2
BW	0.05
L	0
U	1

Fig. 1 Best harmony value on every iteration

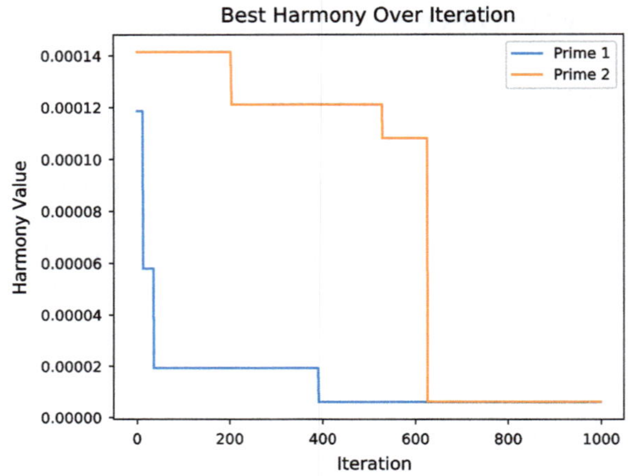

Fig. 2 Average harmony value on every iteration

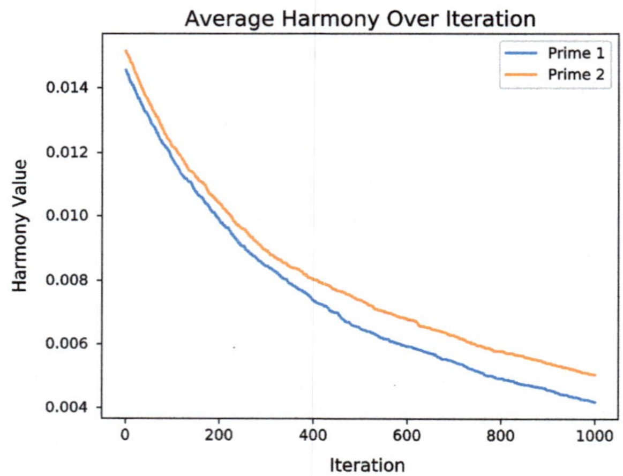

In Fig. 3, the graph is a runtime graph of the algorithm with NVAR sizes up to 4096. The points are taken as average of 25 individual runs. The intervals are of 512 bits starting from 512 bits. This following graph shows that the algorithms perform really well for different NVAR sizes, which indicates that this process can be implemented on high-security standards of encryption systems which requires a large number bit in its keys.

The algorithm was also tested for its quality of generating pseudo-random numbers with a standard set of tests designed by NIST. Here we present the result of the algorithm based on the tests in the NIST test suite standard, in the following Table 2.

Each of the tests generates a p-value as its result which indicates the score of the algorithm's data provided to that test. For all the tests done the expected conclusion is based on that if $p < 0.01$, then the result would be concluded nonrandom.

Table 2 NIST Test suite results on different NVAR sizes (MBF—Mono Bit Frequency, BF—Block Frequency, LC—Linear Complexity, AE-Approximate Entropy, CS—Cumulative Sums, RE—Random Excursions, REV—Random Excursions Variant, CSR—Cumulative Sums Reverse)

Tests	1024	1536	2048	2560	3072	3584	4096
MBF	0.36457	0.79847	0.79088	0.10919	0.15934	0.86734	0.06984
BF	0.34657	0.40988	0.40195	0.82220	0.22368	0.91884	0.06945
Runs	0.99556	0.18482	0.82632	0.99314	0.71778	0.00555	0.31144
Spectral	0.58074	0.30811	0.29168	0.31167	0.14512	0.80626	0.02770
NOTM	0.85407	0.73539	0.84414	0.05307	0.95080	0.72674	0.72577
LC	0.91969	0.06983	0.98560	0.00509	0.80881	0.03572	0.91966
Serial	0.99407	0.99999	1.00000	1.00000	1.00000	1.00000	1.00000
	0.99881	1.00000	1.00000	1.00000	1.00000	1.00000	1.00000
AE	0.99949	0.02503	0.00155	0.00001	0.00000	0.00000	0.00000
CS	0.35752	0.56454	0.25466	0.11016	0.20124	0.89617	0.13493
RE	0.79278	0.43776	0.99451	0.12885	0.95299	0.03053	0.15670
	0.22092	0.64381	0.98800	0.19887	0.83246	0.23943	0.20582
	0.46603	0.50461	0.53353	0.64072	0.75773	0.31606	0.78945
	0.23803	0.18603	0.80136	0.49178	0.72369	0.36632	0.58594
	0.72905	0.62339	0.50249	0.86779	0.66446	0.20747	0.38614
	0.54090	0.68286	0.00000	0.19336	0.76162	0.20406	0.39734
	0.05093	0.68072	0.00007	0.55543	0.50142	0.49047	0.26306
	0.23449	0.80836	0.00133	0.94220	0.57987	0.22047	0.55422
REV	0.21184	0.18381	0.76643	0.38730	0.22516	0.04235	0.67125
	0.18380	0.36131	0.75183	0.28293	0.20554	0.05254	0.69854
	0.16127	0.40457	0.73410	0.20650	0.21092	0.07039	0.94472
	0.13529	0.63144	0.71192	0.15186	0.30076	0.06350	0.88017
	0.12809	0.85968	0.68309	0.12896	0.57844	0.08484	0.86763
	0.16301	0.68850	0.64343	0.07859	0.71092	0.28459	0.63660
	0.20779	0.87437	0.58388	0.04166	0.48290	0.18645	0.21876
	0.26206	0.91871	0.63735	0.06735	0.42801	0.05501	0.38648
	0.38203	0.72367	0.68309	0.23476	0.37749	0.13965	0.80259
	0.84597	0.47950	0.68309	0.92113	0.92189	0.62246	0.08012
	0.91070	0.35833	0.47950	0.73160	0.57130	0.31977	0.02092
	0.60219	0.23568	0.85513	0.72315	0.75887	0.54483	0.05735
	0.50874	0.28505	0.75762	0.65337	0.76685	0.54530	0.05878
	0.62722	0.34578	1.00000	0.44778	0.34318	0.51151	0.08012
	0.88358	0.39377	0.71192	0.40320	0.35939	0.79502	0.11344
	0.91419	0.43277	0.57130	0.39459	0.76482	1.00000	0.07142
	0.90021	0.46521	0.24625	0.31874	0.83949	0.87374	0.17524
	0.83210	0.49272	0.11314	0.26930	0.47555	0.65429	0.16314
CSR	0.51924	0.79943	0.41536	0.13771	0.10709	0.98303	0.02170

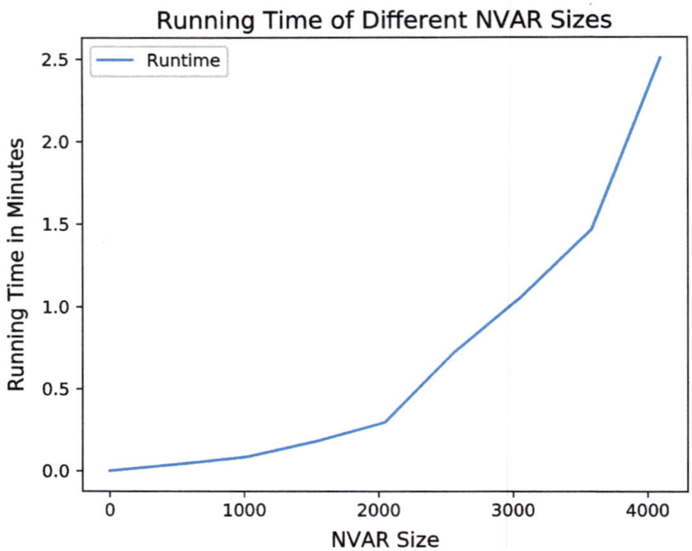

Fig. 3 Running time for different sizes of NVAR

6 Conclusion

The algorithm was tested rigorously for performance check on different conditions. Changing the NVAR size to higher numbers made sure that algorithm maintains its functionality despite having huge computational overhead. The performance of the algorithm is very satisfactory, but implementing parallel programming architecture will make a huge impact in the runtime of the algorithm. Also, reviewing the NIST test suite results gives the conclusion that the algorithms performance of delivering large keys are of a random nature, which implies that the algorithm is successful in maintaining the primitive security measure of the RSA algorithm.

Acknowledgements The authors wish to acknowledge the support of the Post Graduate Teaching and Research Council of Asutosh College.

References

1. Nechvatal, J.: Public-key cryptography. No. NIST-SP-800-2. NATIONAL COMPUTER SYSTEMS LAB GAITHERSBURG MD, (1991)
2. Rivest, R.L., Shamir, A., Adleman, L.: A method for ob-taining digital signatures and public-key cryptosystems. Commun. ACM **21**(2), 120–126 (1978)
3. Crow, J.: Prime Numbers in Public Key Cryptography. This is a paper from the SANS Institute Reading Room site (2003)

4. Hao, Chen, Clark, J.A., Jacob, J.L.: Automated design of security protocols. Comput. Intell. **20**(3), 503–516 (2004)
5. Mishra, M., et al.: A study on the limitations of evolutionary computation and other bio-inspired approaches for integer factorization. Procedia Comput. Sci. **62**, 603–610 (2015)
6. Spillman, R., Janssen, M., Nelson B., Kepner, M.: Use of a genetic algorithm in the cryptanalysis of simple substitution ciphers. Cryp-tologia 17(1):31–44 (1993)
7. Geem, Z.W., Kim, J.H.: A new heuristic optimization algorithm: harmony search. Simulation **76**(2), 60–68 (2001)
8. Storn, R., Price, K.: Differential evolutiona simple and efficient heuristic for global optimization over continuous spaces. J. Global Optim. **11**(4), 341–359 (1997)
9. Kirkpatrick, S., Gelatt, C.D., Vecchi, M.P.: Optimization by simulated annealing. Science **220**(4598), 671–680 (1983)
10. Clark, J., Jacob, J.: A survey of authentication protocol literature: Version 1.0. (1997)
11. Bahadori, M., Mali, M.R., Sarbishei, O., Atarodi, M., Sharifkhani, M.: A novel approach for secure and fast generation of RSA public and private keys on SmartCard. In: NEWCAS Conference (NEWCAS), 2010 8th IEEE International, pp. 265–268. IEEE (2010)
12. Eberhart, R., Kennedy J.: A new optimizer using particle swarm theory. In: Micro Machine and Human Science, 1995. MHS'95. In: Proceedings of the Sixth International Symposium on, pp. 39–43. IEEE (1995)
13. Bishop, C.M.: Neural Networks for Pattern Recognition. Oxford University Press (1995)
14. Dorigo, M., Di Caro, G.: Ant colony optimization: a new meta-heuristic. In: Evolutionary Computation, 1999. CEC 99. Proceedings of the 1999 Congress on, vol. 2, pp. 1470–1477. IEEE (1999)
15. Karaboga, D., Basturk, B.: A powerful and efficient algorithm for numerical function optimization: artificial bee colony (ABC) algorithm. J. Global Optim. **39**(3), 459–471 (2007)
16. Yang, X.-S., Deb, S.: Cuckoo search via Lvy flights. In: Nature & Biologically Inspired Computing, 2009. NaBIC 2009. World Congress on, pp. 210–214. IEEE (2009)
17. Jhajharia, S., Mishra, S., Bali S.: Public key cryptography using neural networks and genetic algorithms. In: Contemporary Computing (IC3), 2013 Sixth International Conference on, pp. 137–142. IEEE (2013)
18. Patidar, V., Sud, K.K., Pareek, N.K.: A pseudo random bit generator based on chaotic logistic map and its statistical testing. Informatica, **33**, 4 (2009)
19. Rukhin, A., Soto, J., Nechvatal, J., Smid, M., Barker, E.: A statistical test suite for random and pseudorandom number genera-tors for cryptographic applications. Booz-Allen and Hamilton Inc Mclean Va (2001)
20. Shannon, C.E.: Prediction and entropy of printed English. Bell Labs Technical J. **30**(1), 50–64 (1951)
21. Rabin, M.O.: Probabilistic algorithm for testing primality. J. Number Theory **12**(1), 128–138 (1980)

Performance Evaluation of Various Digital Modulation Techniques Using GNU Radio

B. Siva Kumar Reddy, Dhrumil Modi and Sumit Upadhyay

Abstract Recently, wireless technology became an emerging technology in commercial, medical and research domains. The communication between a transmitter and receiver are broadly divided into two such as wired and wireless. Modulation is required for any kind of communication to reduce the height of an antenna, avoid signal interference, etc. Therefore, this paper mainly focuses on the implementation of various digital modulation techniques such as amplitude-shift keying (ASK), frequency-shift keying (FSK), phase-shift keying (PSK), Gaussian frequency-shift keying (GFSK), quadrature amplitude modulation (QAM) and Gaussian minimum-shift keying (GMSK), which are currently being used in all wireless technologies. In this paper, an open-source free software, GNU Radio is employed to analyse the performance of modulation techniques and the necessary results are presented.

Keywords GNU radio · PSK · GMSK · GFSK · QAM

1 Introduction

In a basic communication system, the incoming signal from a source (audio, video, text, temperature, etc.) is converted into electrical signals by a transducer and then, the signal is processed through various stages such as coding, analog or modulation, power amplifier and transmitted by an antenna. The transmitted signal is received by the receiver antenna and processed through demodulation, decoder and transducer to get the original signal. However, in a digital communication system, the incoming signal is sampled and converted into a discrete signal before transmitting and then, a particular digital modulation technique is applied [1].

B. Siva Kumar Reddy (✉) · D. Modi · S. Upadhyay
Department of Electrical Engineering, Institute of Infrastructure Technology Research and Management, Ahmedabad 380026, Gujarat, India
e-mail: bsivakumar100@gmail.com

D. Deb et al. (eds.), *Innovations in Infrastructure*, Advances in Intelligent Systems and Computing 757, https://doi.org/10.1007/978-981-13-1966-2_2

1.1 Digital Modulation Techniques

A. *Amplitude-shift keying (ASK)*:
   In this technique, incoming digital samples are divided into different discrete amplitude levels based on numbers of message signals. When the amplitude level of message signal is high (logic 1), then the amplitude level of a carrier signal appears in the output. When the amplitude of message signal becomes low (logic 0), then the carrier signal will not be present in the output. Therefore, the presence of a carrier wave indicates binary 1 and its absence indicates the binary 0 [2].
B. *Frequency-shift keying (FSK)*:
   Presence of data in the message signal represents a high-frequency carrier signal (logic 1) and absence of data represents a low-frequency carrier signal (logic 0) in the modulated signal [3].
C. Phase-shift keying (PSK):
   Symbols are separated in terms of phase on the constellation mapping. The phase of the carrier signal is changed based on logic 1 and logic 0 of message signal [4].
D. Gaussian minimum-shift keying (GMSK):
   GMSK is a type of continuous-phase frequency-shift keying, in which a Gaussian filter is used in series of the MSK modulation [5].
E. Gaussian frequency-shift keying (GFSK):
   In this technique, a Gaussian filter is used to smoothen the data pulses by reducing interference and sideband power [6].
F. Quadrature amplitude modulation (QAM):
   In this technique, symbols are separated both in amplitude and phase in the constellation mapping [7]. For example, in 4-QAM, four symbols are transmitted in which each symbol has 2 bits of data. Thus, number of bits can be increased to achieve higher data rates [8].

2 GNU Radio

The implementation of software-defined radios (SDRs) takes a large amount of time, effort and cost as it has to process a tremendous amount of data in real time [9]. It is not possible to process such a huge amount of data in real time using MATLAB and other similar tools which have large expense in acquiring licenses. Therefore, in this paper, an open-source free software, GNU Radio is used to avoid this problem. GNU Radio is an open-source free software development toolkit for building any kind of real-time signal processing applications, and it offers more than 150 building blocks for signal processing [10]. In GNU Radio, all the blocks are written in C++ and connected using Python, and both can be interfaced with the use of Simplified Wrapper and Interface Generator (SWIG) as shown in Fig. 1. Users can modify existing blocks or and can

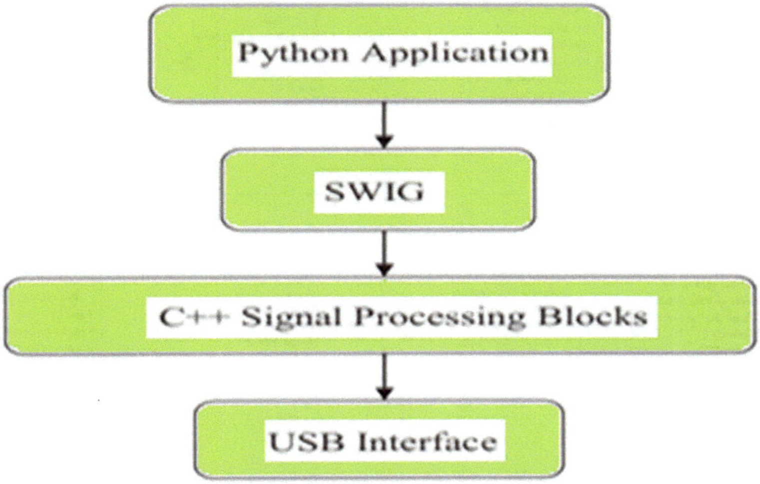

Fig. 1 GNU Radio architecture

also create new signal processing blocks. Some of the important blocks like Throttle provide protection to the system from damage due to high-speed reading/writing operations and high sample rate devices. Software-defined radios (SDRs) can be implemented by connecting Universal Software Radio Peripheral (USRP) hardware to a GNU installed computer. This universal platform converts high-frequency radio signals to low-frequency digital samples and sends them to the host computer, where signal processing takes place.

3 Simulations Model

As shown in Fig. 2, we are transmitting square signal at 126 k sample rate from the signal source. Packet encoder block converts the signal to digital form and transmits the data in the form of packets and sends it to the PSK modulation block, and then it is sent to receiver side through the channel model. FFT sink and scope sinks are used to observe the output in frequency domain and time domain, respectively. Figures 3, 4 and 5 show the GNU flow graphs to implement GMSK, GFSK and QAM, respectively. In the signal source block, waveform option allows us to select a particular type of signals such as square, trapezoidal, sinusoidal and cosine; frequency and amplitude are at user interest. It is allowed to test the behaviour of signal at each output just by connecting either scope sink or FFT sink. The notebook option in each block represents to observe the output in a single window. Number of constellations represents the type of modulation, for example four denotes 4PSK, 64 denotes

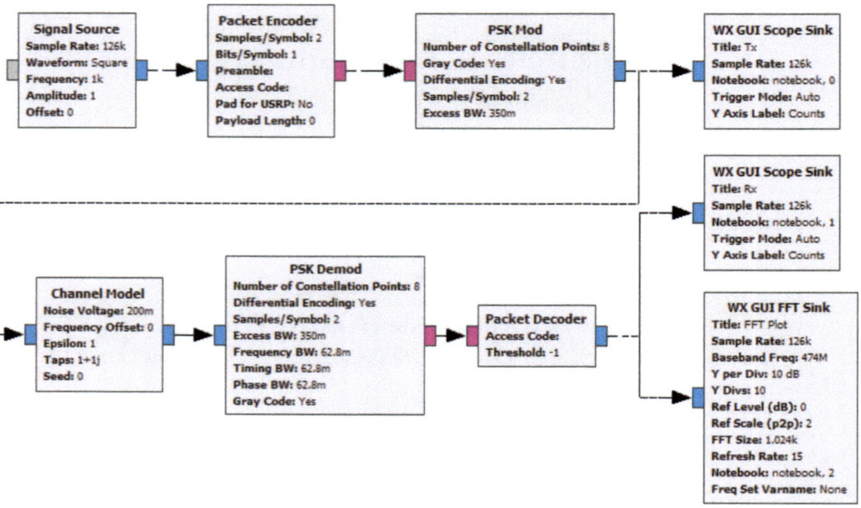

Fig. 2 GNU schematic to implement PSK transceiver

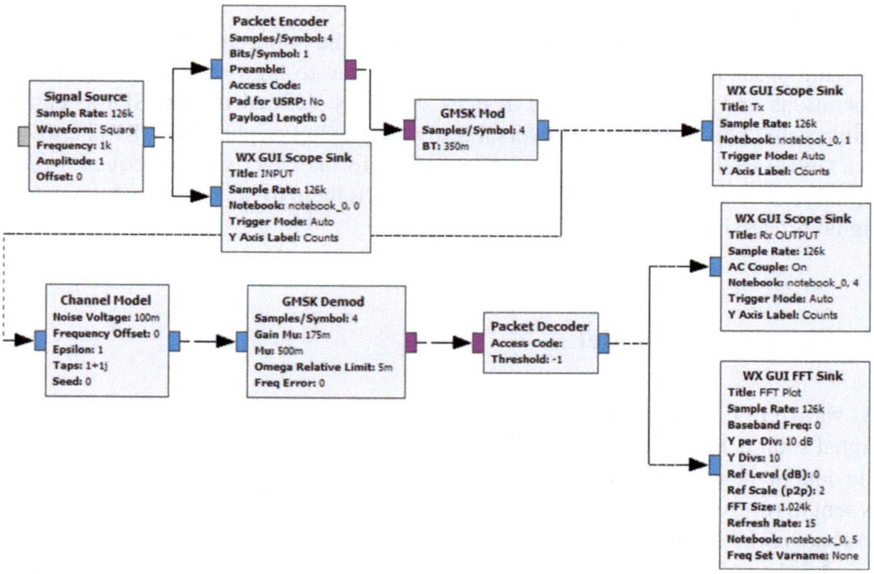

Fig. 3 GNU schematic to implement GMSK transceiver

64PSK. Thus, various wireless communication standards, systems and protocols can be implemented and tested easily using GNU Radio.

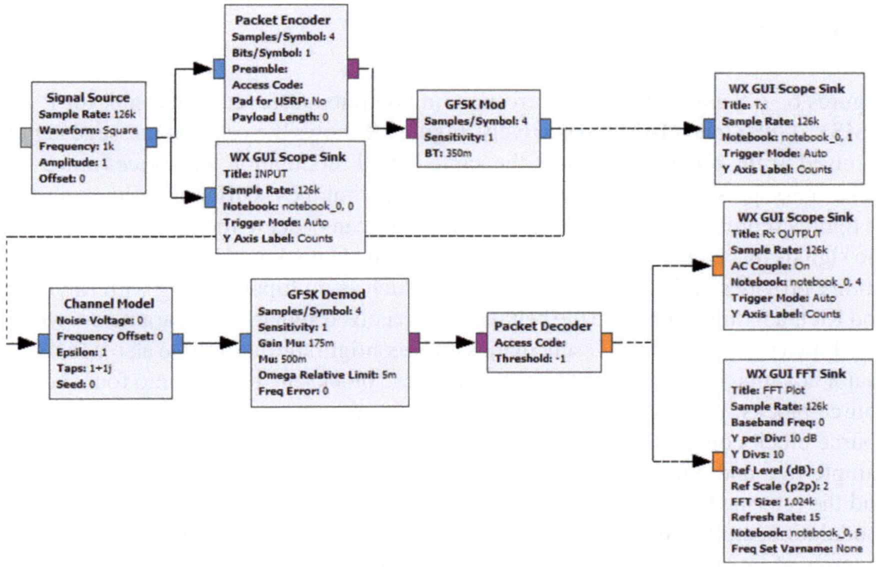

Fig. 4 GNU schematic to implement GFSK transceiver

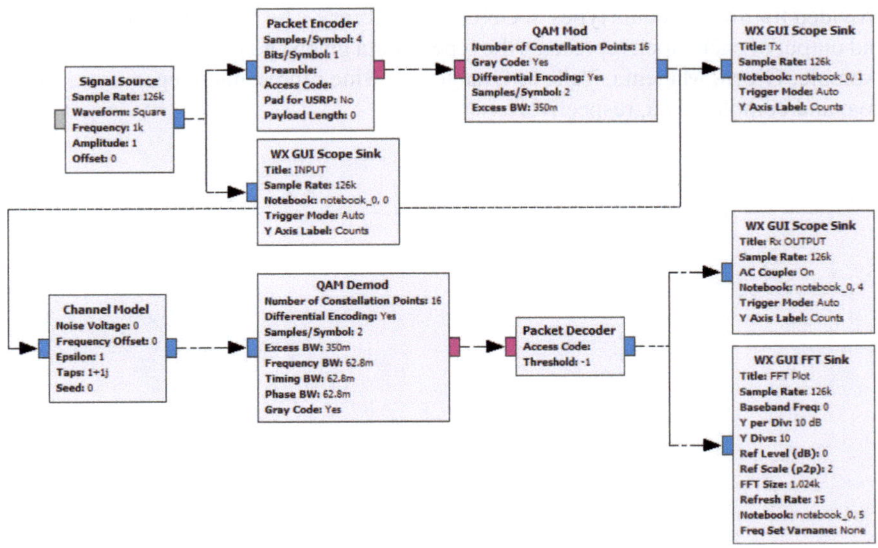

Fig. 5 GNU schematic to implement GFSK transceiver

4 Results

Figures 6, 7, 8 and 9 show the corresponding outputs observed scope sink for PSK, GMSK, GFSK and QAM, respectively. Figure 10 shows the original message signal received at the receiver, which is the same for all modulation techniques since the same message signal is considered for all the techniques. Different colours shown in figures represent real and complex values. We can observe the impact of noise on the signals by changing the noise voltage value in channel model block. GNU Radio library provides various simulation channels such as multipath fading with Rayleigh and Rician fading to observe the behaviour of received signals. The parameters which are shown on blocks in GNU schematics are reconfigurable at runtime also which is a major advantage of using GNU Radio. All these blocks are divided into four such as source blocks, intermediate blocks, sink blocks and type conversion blocks. Signal source block consists of a set of variables to produce an output, namely frequency, sample rate and amplitude. Source blocks are used to generate any kind of signal and the intermediate block is used to perform specific signal function (say filtering, modulation, etc.) on the generated signal and sink block is used to observe the output in terms of time domain, frequency domain, waterfall output and audio domain, etc. Each block has its own data type such as float, short, byte and integer, therefore, type conversion blocks are used to convert one data type to other and different colours are provided for different data types in GNU Radio library. The different colours at input and output of each block represents the type of data given and taken, respectively. For example, Green, Magenta, Yellow, Orange and Blue represents integer, byte, short, float and complex data, respectively.

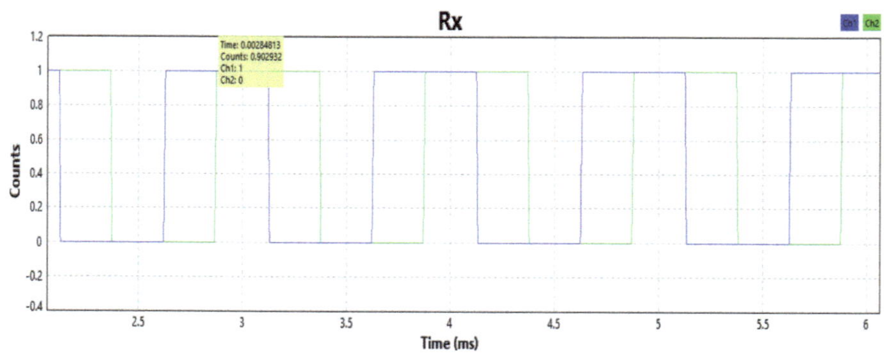

Fig. 6 Output waveform of PSK

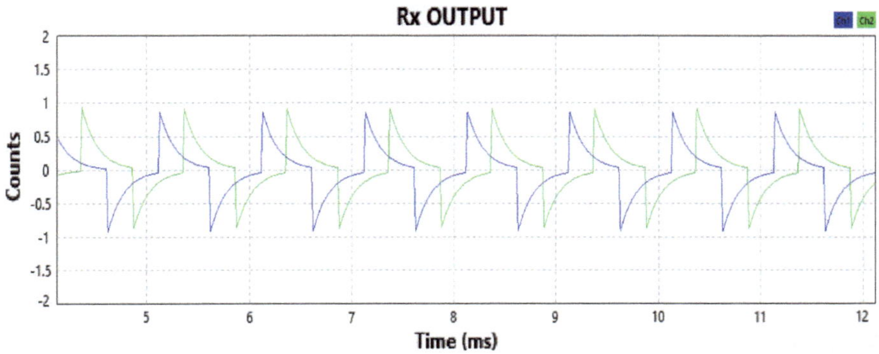

Fig. 7 Output waveform of GMSK

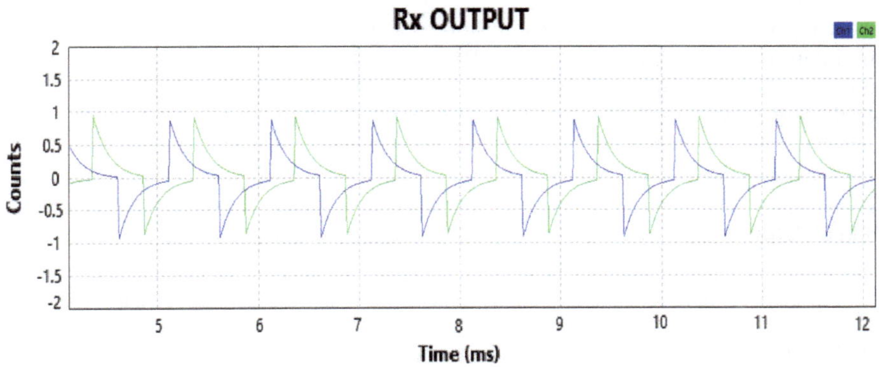

Fig. 8 Output waveform of GFSK

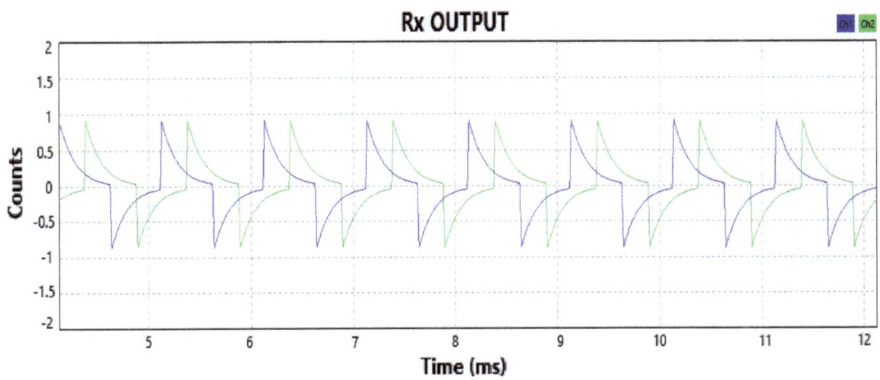

Fig. 9 Output waveform of QAM

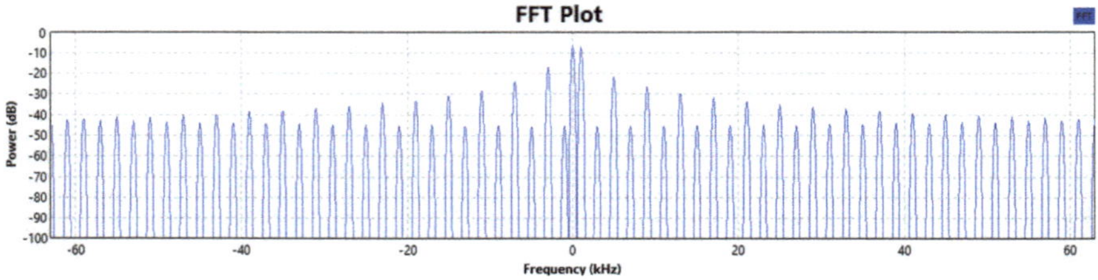

Fig. 10 FFT output waveform of all modulation methods

5 Conclusion

In this paper, various digital modulation techniques are analysed by creating schematics using GNU Radio. The transmission and reception of a signal with different modulation techniques such as ASK, PSK, GMSK, GFSK and QAM are observed in terms of time domain and frequency domain. Furthermore, the necessity of GNU Radio in the area of wireless communications is also presented.

References

1. Proakis, J.G., et al.: Communication Systems Engineering, vol. 2. Prentice Hall, New Jersey (1994)
2. Frenzel, L.: Understanding modern digital modulation techniques. Electron. Des. Technol. Commun. (2012)
3. Ho, P., Tan, H.: A coherent multi-user frequency shift keying receiver for wireless sensor networks based on clustering algorithms. IEEE Wireless Commun. Letters **5**(6), 580–583 (2016)
4. Faruque, S.: "Phase Shift Keying (PSK)". Radio Frequency Modulation Made Easy, pp. 69–83. Cham, Springer (2017)
5. Wu, C., et al.: Single-channel blind source separation of co-frequency overlapped GMSK signals under constant-modulus constraints. IEEE Commun. Lett. **20**(3), 486–489 (2016)
6. Cui, S., Xiong, F.: Compensation of frequency selectivity and antenna effect for the energy detection Gaussian frequency-shift keying ultra-wideband system. J. Eng. **1**(1) (2016)
7. Al-Saffar, D., Edwards, R.M.: A software defined radio comparison of received power with quadrature amplitude modulation and phase modulation schemes with and without a human. In: Antennas and Propagation (EuCAP), 2016 10th European Conference on. IEEE (2016)
8. Siva Kumar Reddy, B., Lakshmi, B.: Adaptive modulation and coding with channel estimation/equalization for WiMAX over multipath faded channels. Wireless Communications, Networking and Applications. Springer, New Delhi pp. 459–472 (2016)
9. Arslan, M.Y., Sundaresan, K., Rangarajan, S.: Software-defined networking in cellular radio access networks: potential and challenges. IEEE Commun. Mag. **53**(1), 150–156 (2015)
10. GNU Radio, Available at "www.GNUradio.org"

Parameter Validation to Ascertain Voltage of Li-ion Battery Through Adaptive Control

Bhavani Sankar Malepati, Deepak Vijay, Dipankar Deb and K. Manjunath

Abstract The objective of this paper is to ascertain the open- circuit voltage of a Lithium-ion battery in the presence of an uncertain parameter. The parameter relating to the state of charge (SoC) and open-circuit voltage are considered. An adaptive control law is developed to compensate for the uncertain parameter by considering its possible range of variation and estimating that parameter. Lyapunov stability analysis is carried out to ensure asymptotic stability and signal boundedness of the states associated with the system. The effectiveness of the methodology employed is verified by studying the dynamics of the battery in charging and discharging modes of operation. The robustness of the proposed adaptive controller is validated by considering the change in uncertain parameter to ensure asymptotic tracking.

Keywords State of charge (SoC) · Parameter uncertainty · Adaptive control
Model Reference Adaptive Control (MRAC) · Li-ion battery

1 Introduction

Nowadays, the applications of battery energy storage system are increasing at a rapid rate with its utilization in many aspects such as power bank for mobile phone [7], transportation through electric vehicle [18], energy backup for microgrid [16], etc. The power output of the battery system needs to be controlled as per system requirement. The main challenges associated while dealing with battery energy storage

B. S. Malepati (✉) · D. Vijay · D. Deb · K. Manjunath
Institute of Infrastructure Technology Research and Management,
Ahmedabad 380026, Gujarat, India
e-mail: bhavanishankar.m1996@gmail.com

D. Vijay
e-mail: dvijay1996@gmail.com

D. Deb
e-mail: dipankardeb@iitram.ac.in

K. Manjunath
e-mail: kmanjunath7@gmail.com

D. Deb et al. (eds.), *Innovations in Infrastructure*, Advances in Intelligent Systems and Computing 757, https://doi.org/10.1007/978-981-13-1966-2_3

system is estimation of SoC and determination of battery voltage for a particular charged condition [2, 12, 22]. The terminal voltage of the battery is usually determined by portraying the characteristics between no-load voltage (symbolized as v_{oc}) and SoC [22]. The particular characteristics of a typical Li-ion battery is shown with a polynomial expression [25].

The motivation behind the present work is to accurately define the no-load voltage versus SoC relationship such that the problem of voltage determination is alleviated. Specifically, ascertaining the coefficients of polynomial expression relating no-load voltage and SoC. The problem of uncertainty in the parameter is overcome through adaptive control. Adaptive control is the suitable option for designing the controller to achieve the best possible operation under uncertain and varying system conditions [5]. In general, the design of controller involves maximizing the performance index. However, the conventional control does not have the potential to adapt to system uncertainties and variations. In this aspect, adaptive controllers possess an edge over other techniques. It involves adaptive updation of control parameter in the presence of uncertainity and/or model variances and thereby enabling the system to achieve desired response [21]. Adaptive control can also be employed along with parameter estimation [1]. There exist adaptive techniques such as Model Reference Adaptive Control (MRAC), Inverse Adaptive Control (IAC), etc., in designing the controller [4, 6, 10, 19, 20]. As stated earlier, the considered model of Li-ion battery consists of an uncertain parameter which can be estimated by employing adaptive control. In particular, Model Reference Adaptive Control (MRAC) is employed to estimate the uncertain parameter. Here, the dynamics of the working battery model will be in accordance to the dynamics of the reference battery model.

The rest of the document is organized as follows. Modeling of lithium-ion battery is explained in Sect. 2. Section 3 deals with the adaptive control law developed to meet the objective of parameter validation. Simulation results are verified in Sect. 4. Finally, concluding remarks are drawn in Sect. 5.

2 Modeling of Lithium-Ion Battery

A typical lithium-ion battery can be modeled in many ways [3, 9, 15, 26, 27]. Modeling of battery storage system can be carried out in a simpler manner through equivalent circuit representation [17]. In the present work, equivalent circuit representation is followed up to second order [8] where a dependent voltage source v_{oc} is placed in series with a source resistance R_s and two RC combinations as shown in Fig. 1. Here, the dependent voltage source (i.e., open-circuit voltage of battery v_{oc}) is treated as the function of SoC. The mathematical relation between v_{oc} and SoC is usually shown with a second-order polynomial expression. However, the aforesaid relation is found to be linear in certain region [17]. For simplicity, only linear region is considered in the present work. The open-circuit voltage is expressed as

$$v_{oc} = a_1 x_{\mathrm{soc}} + a_0. \quad (1)$$

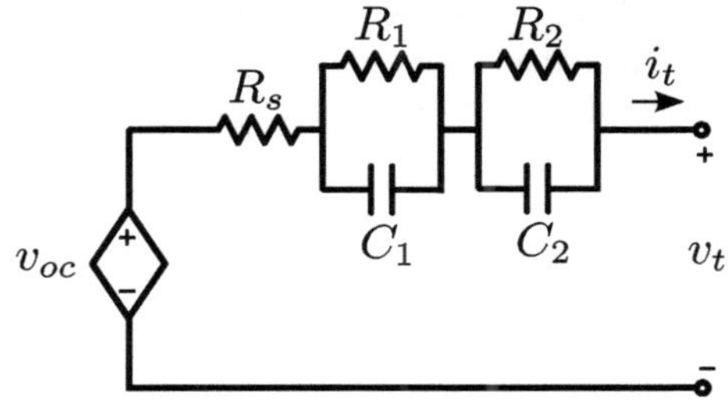

Fig. 1 Equivalent circuit representation of Li-ion battery

In other words, the characteristics between these variables is taken as a straight line with slope as 'a_1' and y-intercept as 'a_0'. The state variables of the considered battery are x_{soc} (SoC), voltage across capacitors C_1 and C_2 (symbolized as v_1 and v_2). The variation of SoC with time is governed by the following expression.

$$x_{\text{soc}}(t) = x_{\text{soc}}(0) - \frac{1}{Q}\int_0^t i_t \mathrm{d}t \tag{2}$$

where, 'Q' and '$x_{\text{soc}}(0)$' refers to rating of the battery in Ampere hour or Coulomb and initial state of charge, respectively.

The state equations of the battery are as follows:

$$\begin{aligned}\dot{x}_{\text{soc}} &= \frac{-1}{Q} i_t \\ \dot{v}_1 &= \frac{-1}{R_1 C_1} v_1 + \frac{1}{C_1} i_t \\ \dot{v}_2 &= \frac{-1}{R_2 C_2} v_2 + \frac{1}{C_2} i_t.\end{aligned}$$

The terminal voltage equation of the battery is shown below.

$$v_t = a_1 x_{\text{soc}} + a_0 - v_1 - v_2 - R_s i_t. \tag{3}$$

The above equation holds good in discharging mode of operation, i.e., battery acting as a source. The same equation in charging mode of operation can be obtained by reversing the direction of terminal current. The particular battery system can be modeled by taking terminal current and voltages as input and output variables, respectively. The described system consists of the uncertain parameters a_0 and a_1 in the output. In this paper, the adaptive control law is developed to address the uncertainty pertaining to a_1, assuming a_0 is a known parameter. The equations can be rearranged by choosing a new state x_1 given by

$$x_1 = a_1 x_{\text{soc}}. \tag{4}$$

In realizing the same system with new variables, the uncertainty can be seen in state x_1 rather than in the output. For simplicity, consider $v_1 = x_2$ and $v_2 = x_3$. For the terminal current of u(t), the state equations are as follows:

$$\begin{aligned}
\dot{x}_1 &= \frac{-a_1}{Q}u(t) \\
\dot{x}_2 &= \frac{-1}{R_1C_1}x_2 + \frac{1}{C_1}u(t) \\
\dot{x}_3 &= \frac{-1}{R_2C_2}x_3 + \frac{1}{C_2}u(t) \\
y(t) &= a_0 + x_1 - x_2 - x_3 - R_s u(t).
\end{aligned} \tag{5}$$

In many applications, each state variable may not be directly measurable [24]. In the working system, SoC of Li-ion battery cannot be measured directly. The other state variables are insignificant in the present context as the physical phenomena of battery is realized in the form of equivalent circuit. Hence, a state estimator like Kalman filter is used with rational assumptions [12, 26]. It is an algorithm which estimates the system states where few (or all) states are not directly measurable. It requires process noise co-variance data along with state-space model of the system [11].

3 Design of the Adaptive Feedback Control Law

A reference system is chosen such that the system has desired characteristics and it is defined as follows:

$$\begin{aligned}
\dot{\hat{x}}_1 &= \frac{-\hat{a}_1}{Q}r(t) \\
\dot{\hat{x}}_2 &= \frac{-1}{R_1C_1}\hat{x}_2 + \frac{1}{C_1}r(t) \\
\dot{\hat{x}}_3 &= \frac{-1}{R_2C_2}\hat{x}_3 + \frac{1}{C_2}r(t) \\
y_r &= a_0 + \hat{x}_1 - \hat{x}_2 - \hat{x}_3 - R_s r(t),
\end{aligned} \tag{6}$$

where $\hat{a}_1$ is the estimated parameter in the reference system, r(t) is the reference signal and the reference states are $\hat{x}_i$, $i = 1, 2, 3$.

Constraint over parameter estimation. The parameter estimate $\hat{a}_1$ is to be updated from an adaptive law and should stay in prespecified regions (that is, $\hat{a}_1 > 0$, since actual parameter a_1 is greater than zero) needed for implementation of the adaptive control. This constraint is built with the understanding that the terminal voltage available is more when SoC is more.

Subtracting (5) from (6) and defining $\tilde{a}_1(t) = \hat{a}_1(t) - a_1(t)$, we obtain the error signals $e_i = \hat{x}_i - x_i, i = 1, 2, 3$ as

$$\begin{aligned}
\dot{e}_1 &= \frac{\hat{a}_1}{Q}(u - r) - \frac{\tilde{a}_1}{Q}u \\
\dot{e}_2 &= \frac{-1}{R_1 C_1}e_2 - \frac{1}{C_1}(u - r) \\
\dot{e}_3 &= \frac{-1}{R_2 C_2}e_3 - \frac{1}{C_2}(u - r).
\end{aligned} \tag{7}$$

We define the adaptive update law for the parameter estimate as

$$\dot{\hat{a}}_1 = \frac{p_1}{\gamma Q}e_1 u, \tag{8}$$

and choose adaptive control law $u(t)$ given by

$$u(t) = \left(\frac{-p_1 e_1(t)}{Q}\hat{a}_1(t) + \frac{p_2 e_2(t)}{C_1} + \frac{p_3 e_3(t)}{C_2}\right) + r(t), \tag{9}$$

where $p_i, i = 1, 2, 3$ are the constants which could be adjusted in order to obtain the desired response.

Lemma 1 *The adaptive control scheme with control law u(t) in the (9) updated by the adaptive law in (8), when applied to linear error dynamics (7) ensures that the all closed loop signals within the system are bounded and the tracking error asymptotically approaches zero, that is,* $\lim_{t\to\infty} e_i(t) = 0, i = 1, 2, 3$ [13, 14, 23].

Proof We define a positive definite function V as follows :

$$V = \frac{1}{2}\sum_{i=1}^{3} p_i e_i^2 + \frac{1}{2}\gamma_1 \tilde{a}_1^2 \tag{10}$$

where $p_i > 0, i = 1, 2, 3$ and $\gamma > 0$, and we ascertain the dynamical system for asymptotic stability. After differentiating V in (10), we get

$$\dot{V} = p_1 e_1 \dot{e}_1 + p_2 e_2 \dot{e}_2 + p_3 e_3 \dot{e}_3 + \gamma_1 \tilde{a}_1 \dot{\tilde{a}}_1 \tag{11}$$

Since $\tilde{a}_1(t) = \hat{a}_1(t) - a_1(t)$, so by differentiating $\tilde{a}_1(t)$, we obtain

$$\dot{\tilde{a}}_1(t) = \dot{\hat{a}}_1(t) - \dot{a}_1(t)$$

where $\dot{a}_1$ vanishes as a_1 is constant term. Now, substituting the values of $\dot{e}_i(t)$ from (7) and $\dot{\tilde{a}}_1$ from above in Eq. (11), we have

$$\dot{V} = \frac{p_1\hat{a}_1}{Q}e_1(u-r) - \frac{p_1\tilde{a}_1}{Q}e_1u - \frac{p_2}{R_1C_1}e_2^2 + \frac{p_2}{C_1}e_2(r-u) - \frac{p_3}{R_2C_2}e_3^2 + \frac{p_3}{C_2}e_3(r-u) + \gamma_1\tilde{a}_1\dot{\hat{a}}_1. \tag{12}$$

On substituting $\dot{\hat{a}}_1$ from (8), the following expressions can be obtained:

$$\dot{V} = \frac{p_1\hat{a}_1}{Q}e_1(u-r) - \frac{p_2}{R_1C_1}e_2^2 + \frac{p_2}{C_1}e_2(r-u) - \frac{p_3}{R_2C_2}e_3^2 + \frac{p_3}{C_2}e_3(r-u). \tag{13}$$

In the above equation, consider V_1 given by

$$V_1 = -\frac{p_3}{R_2C_2}e_3^2 - \frac{p_2}{R_1C_1}e_2^2. \tag{14}$$

It can be seen that $V_1 \leq 0$, because V_1 carries negative definiteness in itself. Therefore,

$$\dot{V} = V_1 + \frac{p_1\hat{a}_1}{Q}e_1(u-r) - \frac{p_3}{C_2}e_3(u-r) - \frac{p_2}{C_1}e_2(u-r). \tag{15}$$

Substituting the adaptive control law given by (9) in the above equation, we get

$$\dot{V} \leq 0. \tag{16}$$

This implies that, asymptotically, the defined function V will become zero. Hence we conclude $\lim_{t\to\infty} e_i(t) = 0$, $i = 1, 2, 3$. It can be noted that the asymptotic stability is not lost though the constraint described earlier is not imposed in the proof.

4 Case Study

In this section, simulation studies are presented to show the efficacy of the employed methodology. The specifications corresponding to Li-ion battery are shown in Table 1. The gains p_1, p_2 and p_3 are assigned with 5×10^4, 1 and 1 respectively and γ as 10^{-4}. The optimal values of these parameters are selected by changing each of them in an iterative manner from a very low value to very high value. Table 2 shows the nominal values and range of the parameters a_0, a_1 and SoC [15, 17].

Table 1 Main specifications of the Li-ion Battery

Symbol	Description	Value	Units
Q	Capacity of the battery	2.5	Ah
R_s	Series resistance	0.01	Ohms
R_1, R_2	RC branch resistances	0.016	Ohms
C_1, C_2	RC branch capacitances	2200	Farads

Table 2 Nominal and range of parameters for a typical Li-ion battery

Parameter/state	Nominal value	Range
a_0	3	[2.5, 5]
a_1	0.5	[0.05, 0.6]
SoC	0.8	[0.05, 0.95]

In this simulation, the considered Li-ion battery model is subjected to both charging and discharging modes of operation. Initially, battery is charged with a current 1.2 A for 300 s and then discharged with 0.8 A for next subsequent 300 s as shown in Fig. 2. The chosen lower limit for the estimated parameter for a_1 is 0.001. The applicability of adaptive control is verified by introducing a sudden disturbance in the parameter a_1 at 450th sec, which changes from initial 0.5–0.7. Although such a disturbance may be unrealistic, the control technique is designed to trace the particular parameter. It can be seen in Fig. 3 that the change in the actual parameter leads to the change in estimated parameter. The tracking error e_1 is observed to settle at zero as shown in Fig. 4. Figures 5 and 6 shows the Terminal Voltage and the estimated SoC respectively. It is evident that both the quantities tend to increase during charging and decrease while discharging. The rate of change of voltage and SoC becomes negative after 300 s since the current in battery changed its direction. Also, it is noteworthy that there is no major change in the output voltage in the event of disturbance introduced in the parameter a_1 at 450th s. The reason behind this particular observation is due to the appropriate estimation of a_1 by adaptive controller.

It is to be noted that the task of ascertaining terminal voltage is accomplished through proper estimation of uncertain parameter a_1. In other words, the terminal voltage of the battery can be determined by assigning the final estimated value of the parameter in the voltage equation shown in (1).

5 Concluding Remarks

In this paper, an adaptive control technique is presented against the problem of parameter uncertainty associated with Li-ion battery. The task of determining the terminal voltage of the battery is accomplished through validation of uncertain parameter

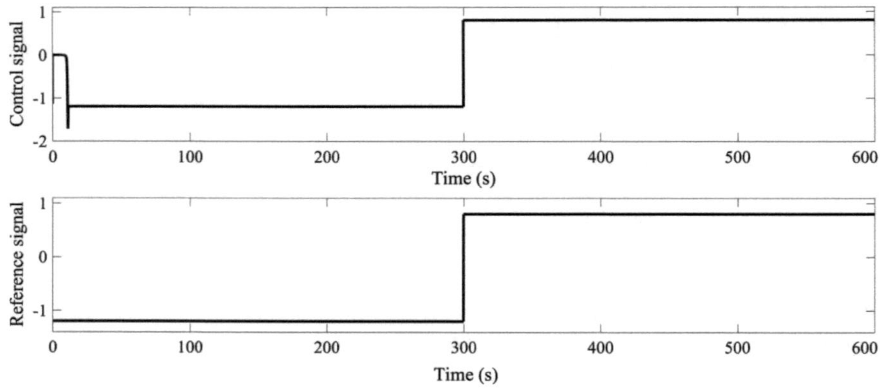

Fig. 2 Top: Control input signal; Bottom: Reference input signal

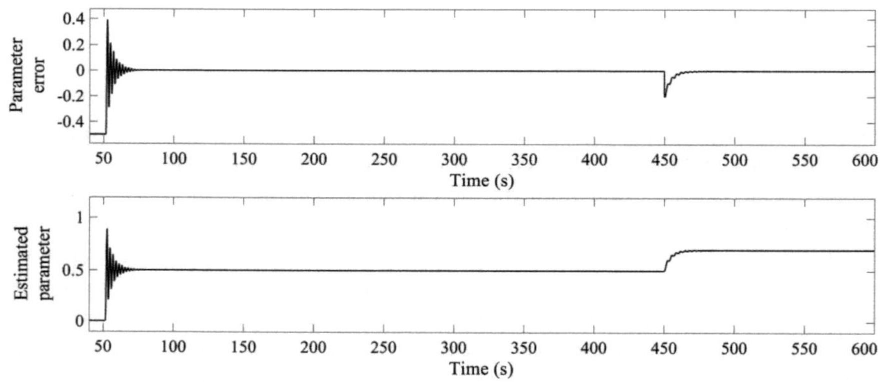

Fig. 3 Top: Parameter error, Bottom: Parameter estimate

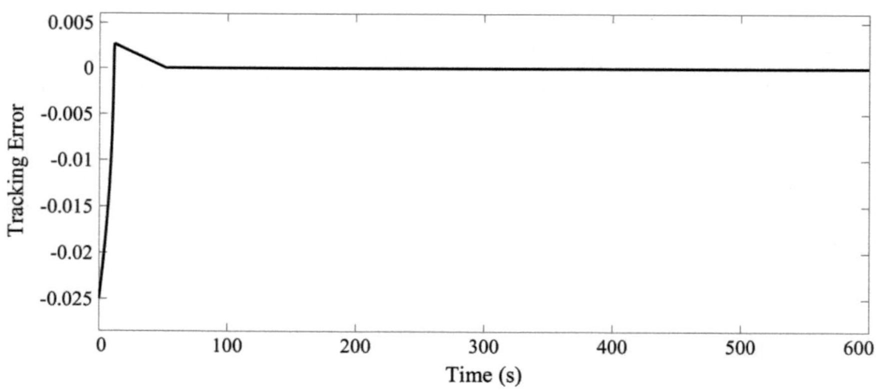

Fig. 4 Tracking error of the state x_1

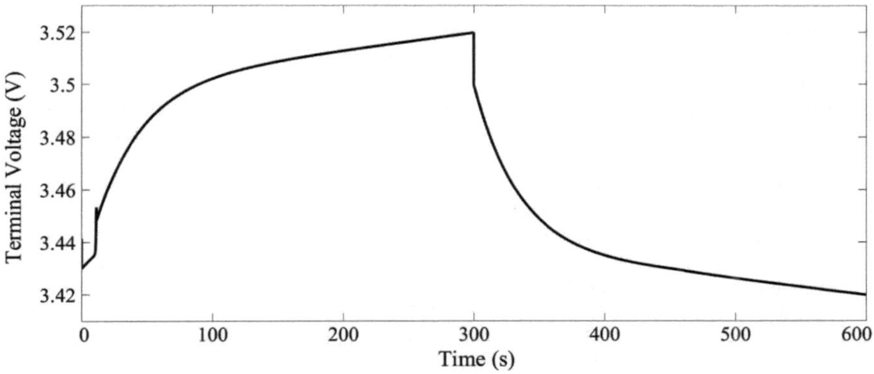

Fig. 5 Terminal voltage of the battery

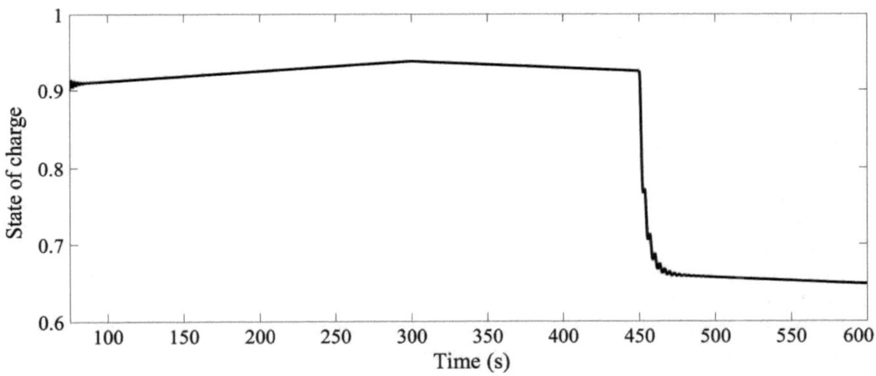

Fig. 6 SoC of the battery

with an assumption that SoC can be estimated. Moreover, the proposed technique is verified even under the variation of uncertain parameter in order to maintain the desired response. The methodology described may find a practical application in the testing and operation of Li-ion batteries.

References

1. Baloh, M., Tao, G., Allaire, P.: Adaptive estimation of magnetic bearing parameters. In: Control Applications, 1999. Proceedings of the 1999 IEEE International Conference on, vol. 2, pp. 1193–1198. IEEE (1999)
2. Cacciato, M., Nobile, G., Scarcella, G., Scelba, G.: Real-time model-based estimation of soc and soh for energy storage systems. IEEE Trans. Power Electron. **32**(1), 794–803 (2017)
3. Chen, M., Rincon-Mora, G.A.: Accurate electrical battery model capable of predicting runtime and iv performance. IEEE Trans. Energy Conversion **21**(2), 504–511 (2006)

4. Deb, D., Tao, G., Burkholder, J.O.: An adaptive inverse compensation scheme for signal-dependent actuator nonlinearities. In: Decision and Control, 2007 46th IEEE Conference on, pp. 4821–4826. IEEE (2007)
5. Deb, D., Tao, G., Burkholder, J.O., Smith, D.R.: Adaptive synthetic jet actuator compensation for a nonlinear tailless aircraft model at low angles of attack. In: American Control Conference, 2006, pp. 6–pp. IEEE (2006)
6. Deb, D., Tao, G., Burkholder, J.O., Smith, D.R.: Adaptive compensation control of synthetic jet actuator arrays for airfoil virtual shaping. J. Aircraft **44**(2), 616–626 (2007)
7. Giegerich, M., Koffel, S., Filimon, R., Grosch, J., Fuhner, T., Wenger, M., Gepp, M., Lorentz, V.: Electrothermal modeling and characterization of high capacity lithium-ion battery systems for mobile and stationary applications. In: Industrial Electronics Society, IECON 2013-39th Annual Conference of the IEEE, pp. 6721–6727. IEEE (2013)
8. Gong, X., Xiong, R., Mi, C.C.: A data-driven bias-correction-method-based lithium-ion battery modeling approach for electric vehicle applications. IEEE Trans. Ind. Appl. **52**(2), 1759–1765 (2016)
9. Gu, R., Malysz, P., Yang, H., Emadi, A.: On the suitability of electrochemical-based modeling for lithium-ion batteries. IEEE Trans. Trans. Electrification **2**(4), 417–431 (2016)
10. Joshi, S.M., Tao, G., Patre, P.: Direct adaptive control using an adaptive reference model. Int. J. Control **84**(1), 180–196 (2011)
11. Kovvali, N., Banavar, M., Spanias, A.: An introduction to kalman filtering with matlab examples. Synth. Lectures Signal Process. **6**(2), 1–81 (2013)
12. Li, X., Choe, S.Y.: State-of-charge (soc) estimation based on a reduced order electrochemical thermal model and extended kalman filter. In: American Control Conference (ACC), 2013, pp. 1100–1105. IEEE (2013)
13. Li, Z., Hovakimyan, N.: L 1 adaptive controller for mimo systems with unmatched uncertainties using modified piecewise constant adaptation law. In: Decision and Control (CDC), 2012 IEEE 51st Annual Conference on, pp. 7303–7308. IEEE (2012)
14. Li, Z., Soh, C.: Lyapunov stability of discontinuous dynamic systems. IMA J. Math. Control Inf. **16**(3), 261–274 (1999)
15. Lin, X., Perez, H.E., Mohan, S., Siegel, J.B., Stefanopoulou, A.G., Ding, Y., Castanier, M.P.: A lumped-parameter electro-thermal model for cylindrical batteries. J. Power Sources **257**, 1–11 (2014)
16. Manjunath, K., Sarkar, V.: System level parameter tuning of an islanded microgrid with improved computational efficiency. In: Power and Energy Systems: Towards Sustainable Energy (PESTSE), 2016 Biennial International Conference on, pp. 1–7. IEEE (2016)
17. Parvini, Y., Vahidi, A., Fayazi, S.A.: Heuristic versus optimal charging of supercapacitors, lithium-ion, and lead-acid batteries: an efficiency point of view. IEEE Trans. Control Syst. Technol. **26**(1), 167–180 (2018)
18. Saxena, N., Hussain, I., Singh, B., Vyas, A.: Implementation of grid integrated pv-battery system for residential and electrical vehicle applications. IEEE Trans. Ind. Electron. (2017)
19. Tan, C., Tao, G., Qi, R., Yang, H.: A direct mrac based multivariable multiple-model switching control scheme. Automatica **84**, 190–198 (2017)
20. Tao, G.: Inherent robustness of mrac schemes. Syst. Control Lett. **29**(3), 165–173 (1996)
21. Underwood, S.J., Husain, I.: Online parameter estimation and adaptive control of permanent-magnet synchronous machines. IEEE Trans. Ind. Electron. **57**(7), 2435–2443 (2010). https://doi.org/10.1109/TIE.2009.2036029
22. Wei, Z., Zou, C., Leng, F., Soong, B.H., Tseng, K.J.: Online model identification and state-of-charge estimate for lithium-ion battery with a recursive total least squares-based observer. IEEE Trans. Ind. Electron. **65**(2), 1336–1346 (2018)
23. Xu, G.Q., Yung, S.P.: Lyapunov stability of abstract nonlinear dynamic system in banach space. IMA J. Mathe. Control Inf. **20**(1), 105–127 (2003)
24. Yang, J., Chen, Y., Yu, K.: State and unknown information estimation for non-linear systems with both input uncertainty and output disturbance. IMA J. Math. Control Inf. **33**(2), 427–439 (2014)

25. Yao, L.W., Aziz, J., Kong, P.Y., Idris, N.: Modeling of lithium-ion battery using matlab/simulink. In: Industrial Electronics Society, IECON 2013-39th Annual Conference of the IEEE, pp. 1729–1734. IEEE (2013)
26. Yu, Z., Huai, R., Xiao, L.: State-of-charge estimation for lithium-ion batteries using a kalman filter based on local linearization. Energies **8**(8), 7854–7873 (2015)
27. Zheng, L., Zhu, J., Wang, G., Lu, D.D.C., He, T.: Lithium-ion battery instantaneous available power prediction using surface lithium concentration of solid particles in a simplified electrochemical model. IEEE Trans. Power Electron. (2018)

Fractional-Order Adaptive Sliding Mode Control for a Two-Link Flexible Manipulator

Sibbir Ahmed, Kshetrimayum Lochan and Binoy Krishna Roy

Abstract This work addresses some robust control strategies for controlling the tip position tracking of a flexible two-link robotic arm manipulator with the consideration that the system is having very less friction and less mechanical vibration at the joints. The flexible two-link flexible manipulator is modelled using Euler-Lagrangian mechanics. Further, a fractional-order modelling is also done for the same system. The parameters of the system are taken from a real-time system. The sliding mode control strategy which is one of the robust control techniques can be made more robust using adaptive sliding mode and fractional-order adaptive sliding mode. The use of an adaptive sliding mode makes the tuning of the control parameters easier with a better response. The simulation results reveal that the adaptive sliding mode strategy is more robust in terms of changes in the initial conditions as well as disturbances to the system.

Keywords Adaptive sliding mode control · Fractional-order · Sliding mode control · Two-link flexible manipulator

1 Introduction

The use of automation in industries is now reaching the pinnacle. The robotic arm manipulator is the dire need of automation. So, an exhaustive research is going on regarding the control of robotic arm manipulators. We have two types of manipulators. These are (i) rigid robotic arm manipulator and (ii) flexible arm robotic manipulator. The use of both types depends on the requirement of an industry, as

S. Ahmed (✉) · K. Lochan · B. K. Roy
National Institute of Technology Silchar, Silchar, India
e-mail: sibbir1993@gmail.com

K. Lochan
e-mail: lochan.nits@gmail.com

B. K. Roy
e-mail: bkrnits@googlemail.com

D. Deb et al. (eds.), *Innovations in Infrastructure*, Advances in Intelligent Systems and Computing 757, https://doi.org/10.1007/978-981-13-1966-2_4

both the manipulators have some advantages over the other [1]. In this paper, we are dealing with a two-link flexible arm robotic manipulator. We will control the tip position as well as the tip deflection of a robotic arm manipulator. The main challenge is that it is an underactuated problem and the modelling of a two-link flexible manipulator (TLFM) itself is another challenge. The modelling methods include Euler-Lagrangian mechanics, assumed mode method, finite element method, etc. Here, we are using a model derived from Euler-Lagrangian mechanics.

A lot of control strategies are there to control the tip position as well as the tip deflection of a robotic arm manipulator. But, as the subject of the application varies these control laws also vary. Control laws like trajectory tracking [2], PID control [3], adaptive control [4, 5] back-stepping control [6] were reported earlier. On the use sliding mode also there are lots of research like adaptive sliding mode [7], fractional-order adaptive sliding mode [8] and robust sliding mode based on MRAC [9], etc. But, for a two-link flexible arm robotic manipulator, control techniques applied are still less. People are still doing a good research to produce a robust control for a two-link flexible manipulator. The design of the controller also varies with the system modelling as well as the field of use, of the robotic arm manipulator.

The studies in the sector of the fractional-order differential equation are increasing in a large extent. Research scholars are finding that fractional-order differential equation can model a practical system more accurately than an integer-order differential equation. The addition of this fractional order to classical control is really enriching the field of control. The results are becoming more efficient, more realistic, and becoming more reliable.

The motivation behind using fractional order over integer order is its closeness to the real-time system dynamics. Flexible robotic arm shows the relaxation phenomenon of material. This relaxation phenomenon is due to the flexibility of material, i.e. elasticity. The robotic arm is not purely elastic, it can be somehow related to the viscoelasticity, which means it is neither pure elastic nor purely viscous. And this viscoelasticity is the field where there is a much use of fractional calculus.

In this paper, we will be designing a fractional-order system as well as fractional-order controller of commensurate order for a two-link flexible arm robotic manipulator. The control technique we have considered here is a fractional-order adaptive sliding mode control (FOASMC) which is desired to work better than adaptive SMC (ASMC).

This paper is indexed in the following ways, in Sect. 2, we have shown the dynamics of a two-link flexible manipulators. Section 3 deals with modelling of the system. Section 4 gives the controller design followed by Sect. 5 where the stability of the sliding surface is proved. Then, there is Sect. 6 with results and discussion. Finally, we have concluded the paper in Sect. 7.

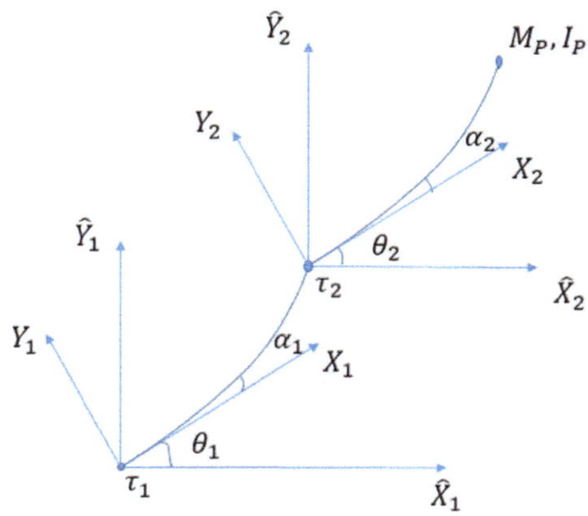

Fig. 1 Schematic diagram of FLTM

2 Dynamics of a TLFM

The system under consideration is a two-link flexible manipulator whose each of the two links is controlled by one independent motor connected to its link base. The motor is connected to the link base by a harmonic drive. The kinetic and potential energy are calculated to apply Euler-Lagrangian mechanics. System dynamics is obtained using E-L formulation [10].

The potential energy of the system is given as (Fig. 1)

$$V = P.E = \frac{1}{2}k_{\text{stiff}}\alpha^2$$

$(\hat{X}_i \hat{Y}_{i)}$ is the general coordinate at the hub of each link. (X_i, Y_i) is the coordinate tangential to the ith link at the hub of each link. θ_i is the angle made with the ith general coordinate by ith tangential coordinate. α_i is the angle made with the ith tangential coordinate by the ith flexible link where $i = 1, 2$.

The kinetic energy is

$$\begin{aligned} T &= K.E_{\text{total}} = KE_{\text{base1}} + KE_{\text{Link2}} + KE_{\text{base2}} + KE_{\text{link2}} \\ KE_{\text{base1}} &= \frac{1}{2}J_{\text{eqv}}\dot{\theta}_1^2 \; KE_{L\text{link1}} = \frac{1}{2}J_{\text{link1}}\left(\dot{\theta}_1 + \dot{\alpha}_1\right)^2 \\ KE_{\text{base2}} &= \frac{1}{2}J_{\text{eq2}}\dot{\theta}_2^2 \quad KE_{L\text{link2}} = \frac{1}{2}J_{\text{link2}}\left(\dot{\theta}_2 + \dot{\alpha}_2\right)^2 \end{aligned} \tag{1}$$

The Lagrangian equation is

$$L = K.E. - P.E. = T - V$$

$$L = \left[\frac{1}{2}J_{\mathrm{eqv}}\dot{\theta}_1^2 + \frac{1}{2}J_{\mathrm{link1}}\left(\dot{\theta}_1 + \dot{\alpha}_1\right)^2 + \frac{1}{2}J_{\mathrm{eq2}}\dot{\theta}_2^2 + \frac{1}{2}J_{\mathrm{eq2}}\dot{\theta}_2^2\right] - \left[\frac{1}{2}k_{\mathrm{stiff}}\left(\alpha_1^2 + \alpha_2^2\right)\right] \tag{2}$$

The Euler-Lagrangian equations are

$$\frac{\mathrm{d}}{\mathrm{d}t}\left(\frac{\partial L}{\partial \dot{\theta}}\right) - \frac{\partial L}{\partial \theta} = T - B_{\mathrm{eq}}\dot{\theta} \tag{3}$$

$$\frac{\mathrm{d}}{\mathrm{d}t}\left(\frac{\partial L}{\partial \dot{\alpha}}\right) - \frac{\partial L}{\partial \alpha} = 0 \tag{4}$$

Putting the value of L in (2), then solving, we get

$$\ddot{\theta}_1 = q_1\alpha_1 - q_2\dot{\theta}_1 + q_3\dot{\alpha}_1 + k_1u \tag{5}$$

$$\ddot{\theta}_2 = q_6\alpha_2 - q_7\dot{\theta}_2 + q_8\dot{\alpha}_2 + k_2u \tag{6}$$

$$\ddot{\alpha}_1 = q_2\dot{\theta}_1 - q_4\alpha_1 - q_5\dot{\alpha}_1 - k_1u \tag{7}$$

$$\ddot{\alpha}_2 = q_7\dot{\theta}_2 - q_9\alpha_2 - q_{10}\dot{\alpha}_2 - k_2u \tag{8}$$

where $q_1, q_2, q_3, q_4, q_5, q_6, q_7, q_{8,}q_9, q_{10}, k_1, k_2$ are the system parameters.

3 Modelling of the System

In the system modelling, the values of the system parameters that we have considered are taken from the real-time system that is present in our laboratory. These values are calculated by the instrument provider, i.e. QUANSER by various experiments. In the following part of this paper, these parameters are termed as b_i, where $i = 1$–10.

3.1 Link-1

The integer-order model of link-1 is

$$x_1 = \theta_1; x_2 = \alpha_1; x_3 = \dot{\theta}_1; x_4 = \dot{\alpha}_1;$$

$$\begin{aligned}
\dot{x}_1 &= x_3 \\
\dot{x}_2 &= x_4 \\
\dot{x}_3 &= b_2x_2 - b_1x_3 + b_3u \\
\dot{x}_4 &= -b_7x_2 + b_1x_3 - b_3u
\end{aligned} \tag{9}$$

where $b_1 = 62.9563$, $b_2 = 628.889$, $b_3 = 140.4712$ and $b_4 = 863.3324$.

A fractional-order model represents the system more accurately. A fractional commensurate order model of the system is considered here. The order of differentiation, α is taken as 0.98.

The FO model of link-1 is:

$$\begin{aligned} D^{\alpha}x_1 &= x_3 \\ D^{\alpha}x_2 &= x_4 \\ D^{\alpha}x_3 &= b_2x_2 - b_1x_3 + b_3u \\ D^{\alpha}x_4 &= -b_7x_2 + b_1x_3 - b_3u \end{aligned} \tag{10}$$

3.2 Link-2

The integer-order model of link-2 is

$$x_1' = \theta_2; x_2' = \alpha_2; x_3' = \dot{\theta}_2; x_4' = \dot{\alpha}_2;$$

$$\begin{aligned} \dot{x}_1' &= x_3' \\ \dot{x}_2' &= x_4' \\ \dot{x}_3' &= -b_4x_3' + b_5x_2' + b_9x_4' + b_6u \\ \dot{x}_4' &= b_4x_3' - b_8x_2' - b_{10}x_4' - b_6u \end{aligned} \tag{11}$$

where $b_4 = 496.8$, $b_5 = 2271.1$, $b_6 = 1.39$, $b_8 = 3336.2$, $b_9 = 28.4$, $b_{10} = 41.6$.

The FO model of second link of flexible manipulator

$$\begin{aligned} D^{\alpha}x_1' &= x_3' \\ D^{\alpha}x_2' &= x_4' \\ D^{\alpha}x_3' &= -b_4x_3' + b_5x_2' + b_9x_4' + b_6u \\ D^{\alpha}x_4' &= b_4x_3' - b_8x_2' - b_{10}x_4' - b_6u \end{aligned} \tag{12}$$

4 Design of Control Law

4.1 Design of Control Law for Link-1

Since we are dealing with the tracking problems, so the sliding surface is designed in terms of error dynamics. We will be making the sliding variable zero which in turn will minimize the error and will do better tracking for both the links. The constants in the sliding surfaces are tuned manually.

Using error dynamics, the FO sliding surfaces considered are

$$s_1 = D^{\alpha} e_1 + c_1 e_1 \tag{13}$$

$$s_2 = D^{\alpha} e_2 + c_1 e_2 \tag{14}$$

$$e_1 = x_1 - x_d \tag{15}$$

$$e_2 = x_2 - x_{d1} = x_2, \text{as } x_{d1} = 0 \tag{16}$$

$$D^{\alpha} e_1 = D^{\alpha} x_1 - D^{\alpha} x_d \tag{17}$$

$$D^{\alpha} e_2 = D^{\alpha} x_2 \tag{18}$$

The final sliding surface is

$$S_1 = a_1 s_1 + a_2 s_2 \tag{19}$$

The FO sliding surface is designed as

$$\begin{aligned} D^{\alpha} S_1 = {} & (a_1 b_7 - a_2 b_7) x_2 + x_3 (a_2 b_1 - a_1 b_1 + a_1 c_2 x_4 \\ & + a_1 c_1 D^{\alpha} e_1 - a_1 D^{\alpha} D^{\alpha} x_d + (a_1 b_3 - a_2 b_3) u \end{aligned} \tag{20}$$

From Eq. (20) making $D^{\alpha} S_1 = 0$, we calculate the equivalent control u_{eq}. For maintaining the sliding variable zero, the discontinuous control is taken as $[\rho \tan h(s)]$

$$\begin{aligned} A & = (a_1 b_7 - a_2 b_7) x_2 + x_3 (a_2 b_1 - a_1 b_1 + a_1 c_2 x_4 \\ & \quad + a_1 c_1 D^{\alpha} e_1 - a_1 D^{\alpha} D^{\alpha} x_d \\ B & = -(a_1 b_3 - a_2 b_3 \\ u & = -\frac{A}{B} - \rho \tanh(s) \end{aligned} \tag{21}$$

Using adaptive parameters, we get

$$\begin{aligned} A_1 & = (a_1 \hat{b}_7 - a_2 \hat{b}_7) x_2 + x_3 (a_2 \hat{b}_1 - a_1 \hat{b}_1) \\ & \quad + a_1 \hat{c}_2 x_4 + a_1 c_1 D^{\alpha} e_1 - a_1 D^{\alpha} D^{\alpha} x_d \\ u & = -\frac{A_1}{B} - \rho \tan h(s) \\ \tilde{b} & = b - \hat{b},\ \tilde{c} = c - \hat{c} \end{aligned} \tag{22}$$

The Lyapunov function candidate is considered as

$$V = \frac{1}{2}\left(S_1^2 + \tilde{b}_1^2 + \tilde{b}_2^2 + \tilde{b}_7^2 + \tilde{c}_1^2 + \tilde{c}_2^2\right) \tag{23}$$

Using this Lyapunov function candidate, adaptive parameters are obtained as

$$\begin{aligned} D^{\alpha} b_1 &= -S_1 a_1 x_3 + S_1 a_2 x_3 + k_2 \tilde{b}_1 \\ D^{\alpha} b_2 &= S_1 a_1 x_2 + k_1 \tilde{b}_2 \\ D^{\alpha} b_7 &= -S_1 a_2 x_2 + k_3 \tilde{b}_7 \\ D^{\alpha} c_1 &= S_1 a_1 \dot{e}_1 + k_4 \tilde{c}_1 \\ D^{\alpha} c_2 &= S_1 a_2 x_4 + k_5 \tilde{c}_2 \end{aligned} \tag{24}$$

4.2 Design of Control Law for Link-2

For link-2 also we have the same logic as link-1 for considering the sliding surfaces, i.e. to minimize the sliding variable thereby making error zero that finally leads to better tracking.

Using error dynamics, the FO sliding surfaces are considered as

$$s_1' = D^{\alpha} e_1' + c_1 e_1' \tag{25}$$

$$s_2' = D^{\alpha} e_2' + c_1 e_2' \tag{26}$$

$$e_1' = x_1' - x_d \tag{27}$$

$$e_2' = x_2' - x_{d1} = x_2', \, as \; x_{d1} = 0 \tag{28}$$

$$D^{\alpha} e_1' = D^{\alpha} x_1' - D^{\alpha} x_d \tag{29}$$

$$D^{\alpha} e_2' = D^{\alpha} x_2' \tag{30}$$

The final sliding surface are

$$S_2 = a_1 s_1' + a_2 s_2' \tag{31}$$

$$\begin{aligned} D^{\alpha} S_2 = {} & (a_1 b_5 - a_2 b_8) x_2' + x_3'(a_2 b_4 - a_1 b_4) \\ & + (a_1 b_9 + a_2 c_2 - a_2 b_{10}) x_4' \\ & + a_1 c_1 D^{\alpha} e_1' - a_1 \ddot{x}_d + (a_1 b_6 - a_2 b_6) u \end{aligned} \tag{32}$$

From Eq. (32) making $D^{\alpha} S_2 = 0$, we calculate the equivalent control u_{eq2}. For maintaining the sliding variable zero, the discontinuous control is taken as $[\rho \tan h(s_2)]$

$$\begin{aligned} A' &= (a_1 b_5 - a_2 b_8) x_2' + x_3'(a_2 b_4 - a_1 b_4) + (a_1 b_9 \\ &\quad + a_2 c_2 - a_2 b_{10}) x_4' + a_1 c_1 D^{\alpha} e_1' - a_1 D^{\alpha} D^{\alpha} x_d \\ B' &= -(a_1 b_6 - a_2 b_6) \\ u &= -\frac{A'}{B'} - \rho \tan h(s_2) \end{aligned} \tag{33}$$

Using adaptive parameters, we get

$$\begin{aligned}A_1^{'} &= (a_1\hat{b}_5 - a_2\hat{b}_8)x_2' + x_3'(a_2\hat{b}_4 - a_1\hat{b}_4)\\ &\quad + (a_1\hat{b}_9 + a_2\hat{c}_2 - a_2\hat{b}_{10})x_4' + a_1c_1D^{\alpha}e_1' - a_1D^{\alpha}D^{\alpha}x_d\\ u &= -\frac{A_1'}{B'} - \rho\ \tan h(s_2)\\ \tilde{b} &= b - \hat{b},\ \tilde{c} = c - \hat{c}\end{aligned} \tag{34}$$

Another Lyapunov function candidate is considered as

$$V = \frac{1}{2}\left(S_2^2 + \tilde{b}_4^2 + \tilde{b}_5^2 + \tilde{b}_8^2 + +\tilde{b}_9^2 + \tilde{b}_{10}^2 + \tilde{c}_1^2 + \tilde{c}_2^2\right) \tag{35}$$

Using this Lyapunov function candidate, adaptive parameters are obtained as

$$\begin{aligned}D^{\alpha}b_4 &= S_2a_2x_3^{'} - S_2a_1x_3^{'} + k_5\tilde{b}_4\\ D^{\alpha}b_5 &= S_2a_1x_2^{'} + k_6\tilde{b}_5\\ D^{\alpha}b_8 &= -S_2a_2x_2^{'} + k_7\tilde{b}_8\\ D^{\alpha}b_9 &= S_2a_1x_4^{'} + k_8\tilde{b}_9\\ D^{\alpha}b_{10} &= -S_2a_2x_4^{'} + k_9\tilde{b}_{10}\\ D^{\alpha}c_1 &= S_2a_1\dot{e}^{'} + k_{10}\tilde{c}_1\\ D^{\alpha}c_2 &= S_2a_2x_4^{'} + k_{11}\tilde{c}_2\end{aligned} \tag{36}$$

By designing the fractional-order adaptive sliding surfaces for both link-1 and link-2, we have derived the control input for both the link. The obtained control inputs are applied to the systems. The simulated results that we have obtained using its control inputs are listed in the next section of the paper.

5 Stability of the Sliding Surface

For link-1, the stability of the considered sliding surface (19) is analysed below. During sliding of the states through sliding surfaces, $S_1 = 0$ so putting in (19)

$$a_1(D^{\alpha}e_1 + c_1e_1) + a_2(D^{\alpha}e_2 + c_1e_2) = 0 \tag{37}$$

Taking the Laplace transformation, then solving, we get

$$e_2(s) = \frac{a_1s^{\alpha-1}e_1(0)}{a_2c_2 + a_2s^{\alpha}} - \frac{a_1c_1 + a_1s^{\alpha}}{a_2c_2 + a_2s^{\alpha}}e_1(s) \tag{38}$$

Taking inverse Laplace transformation as

$$e_2(t) = \left(\frac{a_1}{a_2}\right)E_\alpha(-c_2t^\alpha)e_1(0) + \left(\frac{a_1}{a_2}\right)c_1\left(t^{\alpha-1}E_\alpha(-c_2t^\alpha)\right)e_1(t) \tag{39}$$

here E_α is the Mittag-Leffler (ML) function.

The property of a Mittag-Leffler function is that it act as power law at $t = 0$ and at $t = \infty$ it act as inverse power law. It decays at a faster rate than an exponential one.

The condition for a ML function to be Convergent is that the coefficient of t has to be negative. So c_2 has to be positive. In [11], convergence of such system is verified. This condition is satisfied for our value of c_2 which, in turn, proves the stability of the fractional-order sliding surface. Here, e_2 converges to e_1 and $e_1(0)$ and again to solution of e_1 is another ML function which again converges to zero. So, $e_1\&e_2$ converges to zero making the system dynamically stable and produces better tracking responses. The same can be explained for the sliding surface for link-2 which again ensures the stability of S_2 (31).

6 Results and Discussions

The control strategy that is proposed in this paper for controlling the performance of a two-link flexible manipulator is working in a better sense. Here, we are showing the use of fractional-order sliding mode adaptive controller over our fractional-order system. The results produced are listed below. The constants of the link-1 are taken as:

$h = 0.002$; $\rho = 10$; $c_1 = 16.4$; $c_2 = 0.01$; $a_1 = 16$; $a_2 = 0.009$; $r_1 = 0.08$; $r_2 = 0.08$; $r_3 = 0.08$; $r_4 = 0.08$; $r_5 = 0.08$;

Constants for link-2 are taken as:

$h = 0.002$; $\rho = .1$; $c_1 = 10$; $c_2 = 0.05$; $a_1 = 20$; $a_2 = 0.01$; $r_5 = 9.02$; $r_6 = 9.02$; $r_7 = 9.02$; $r_8 = 9.02$; $r_9 = 9.02$; $r_{10} = 0.04$; $r_{11} = 0.04$

In Figs. 2 and 3, we can see the tracking of link-1(L-1) and Figs. 4 and 5 show the tracking of link-2(L-2) with two different sets of initial conditions (ICs) for both the links, i.e. $[0.002\ 0\ 0\ 0]^T$ and $[0.2\ 0.01\ 0\ 0]^T$ respectively. For initial condition $[0.002\ 0\ 0\ 0]^T$ the tracking for both the systems are nearly the same and it is tracking from the start. For initial condition $[0.2\ 0.01\ 0\ 0]^T$ with FOASMC link-1and -2 are tracking nearly within 0.2 s and 0.25 s while ASMC is tracking during steady state and from 0.86 s, respectively.

So, it explains that for link-1and -2 with the use of FOASMC, the tracking of the system becomes better in transient period of the system. While in the steady state both are producing nearly the same results.

Figures 6 and 7 are giving tip deflections of link-1 and Figs. 8 and 9 are giving tip deflection of link-2 with two different sets of initial conditions for both the links, i.e. $[0.002\ 0\ 0\ 0]^T$ and $[0.2\ 0.01\ 0\ 0]^T$. For the first initial condition in the link-1 the

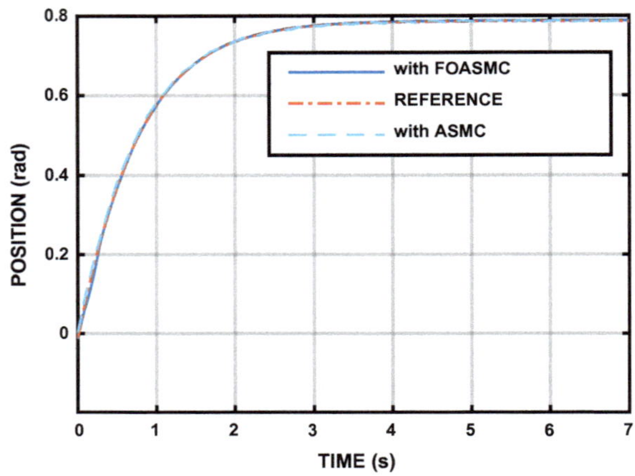

Fig. 2 Tracking of L-1, IC $[0.002\ 0\ 0\ 0]^T$

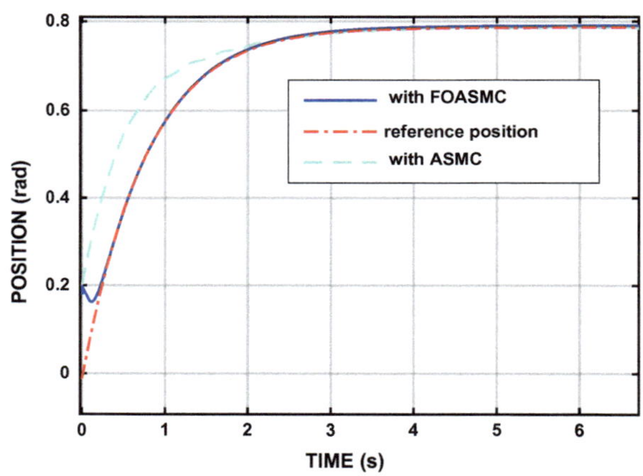

Fig. 3 Tracking of L-1, IC $[0.2\ 0.01\ 0\ 0]^T$

deflection is varying from −0.06 to 0.07 with ASMC and then dying out while with FOASMC it is in between −0.03 and 0.03 and nearly becomes stable in 7 s.

For the second initial condition with FOASMC, the tip deflection is within −0.03 to 0.03 and getting stable by nearly 6 s while in ASMC, it is still oscillating in −0.05 to 0.05. In link-2 for IC $[0.002\ 0\ 0\ 0]^T$, deflection is in order of 10^{-3} with FOASMC and it is from −0.25 to 0.1 with ASMC. For initial condition $[0.2\ 0.01\ 0\ 0]^T$ in link-2, it is seen that tip deflection is within −0.15 to 0.05 with ASMC and again in order of 10^{-3} using FOASMC. So, in terms of tip deflection also, we can see that the performance of FOASMC is better than ASMC.

Figures 10 and 11 give the sliding surfaces of link-1 and Figs. 12 and 13 show the sliding surfaces of link 2 with two different set of initial conditions. It is seen that

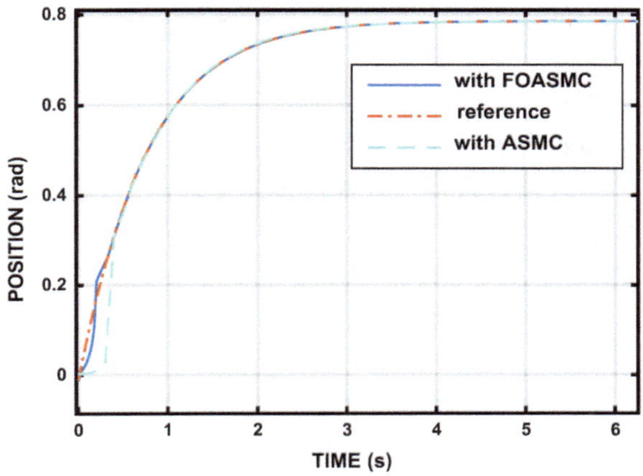

Fig. 4 Tracking of L-2, IC $[0.002\ 0\ 0\ 0]^T$

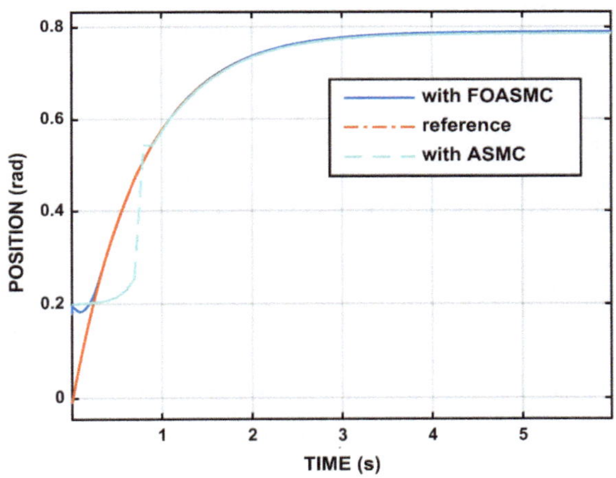

Fig. 5 Tracking of L-2, IC $[[0.02\ 0.01\ 0\ 0]^T$

both the sliding surfaces are converging to zero in a finite period. From Figs. 14, 15, 16, 17, 18, 19, 20 and 21, they show the control inputs to link-1 and link-2 for two different initial conditions. It is observed that the control input, in case, if FOASMC is high to make the system track from the initial positions.

Figures 22 and 23 show the tracking responses of link-1 and -2 with sinusoidal disturbances. These figures clearly show better tracking even with disturbances and thus, exhibit the robustness property of the proposed controller.

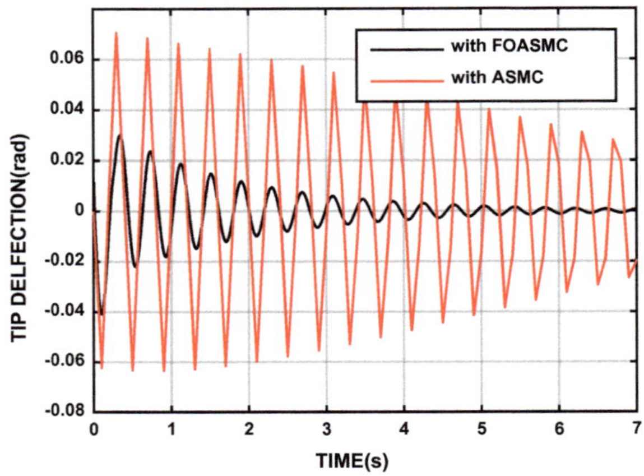

Fig. 6 Tip deflection of L-1, IC $[0.002\ 0\ 0\ 0]^T$

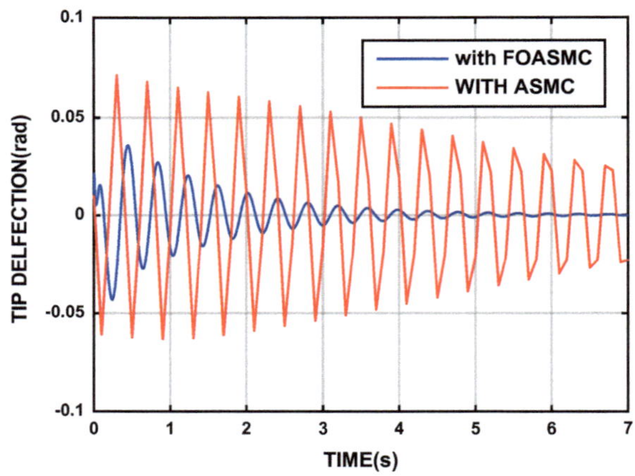

Fig. 7 Tip deflection of L-1, IC $[0.2\ 0.01\ 0\ 0]^T$

7 Conclusions

A fractional-order adaptive sliding mode controller is used in this paper to track the tip position of a two-link flexible manipulator along with suppression of tip deflection. The performances are compared with the integer-order adaptive sliding mode controller. We have designed the FO sliding surfaces using the error dynamics instead of states for better tracking of the desired trajectory. Further, sliding surfaces are made adaptive to the variation of some of the parameters of the system. From the simulation results, it is seen that the FO two-link flexible manipulator with its proposed FO controller produces better tracking along with small tip deflections compared with its integer counterpart. Similar performances are also observed when subjected to

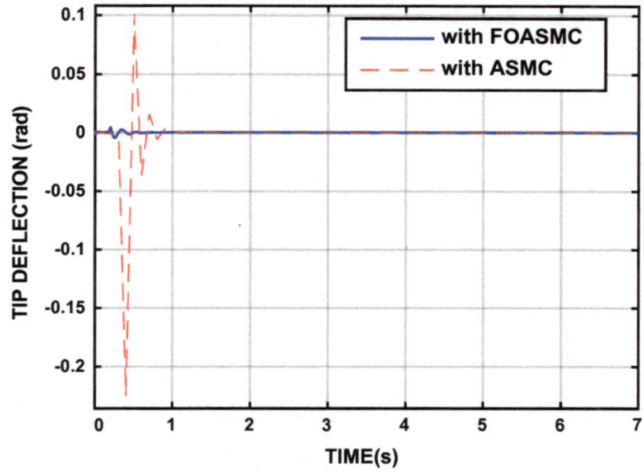

Fig. 8 Tip deflection of L-2, IC $[0.002\ 0\ 0\ 0]^T$

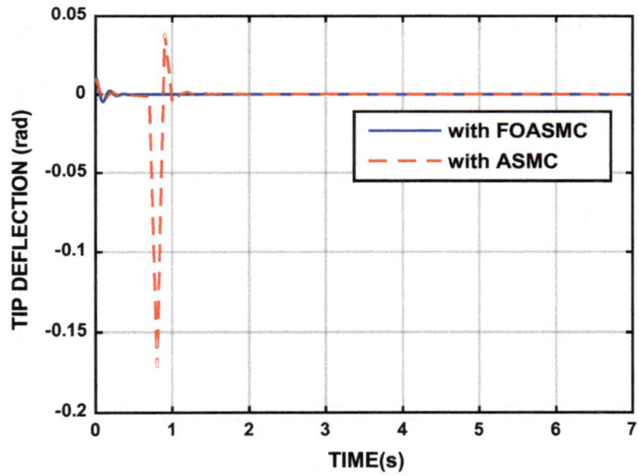

Fig. 9 Tip deflection of L-2, IC $[0.2\ 0.01\ 0\ 0]^T$

different initial conditions. A commensurate-ordered FO model is considered in this paper. However, a noncommensurate-ordered model may have a better representation of the system. So, the same controller can be applied on a noncommensurate-ordered model of the system in future along with its practical implementation.

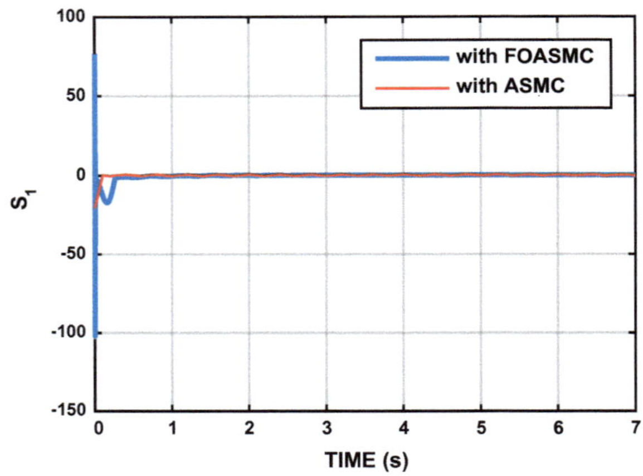

Fig. 10 Sliding surface for L-1, IC $[0.002\ 0\ 0\ 0]^T$

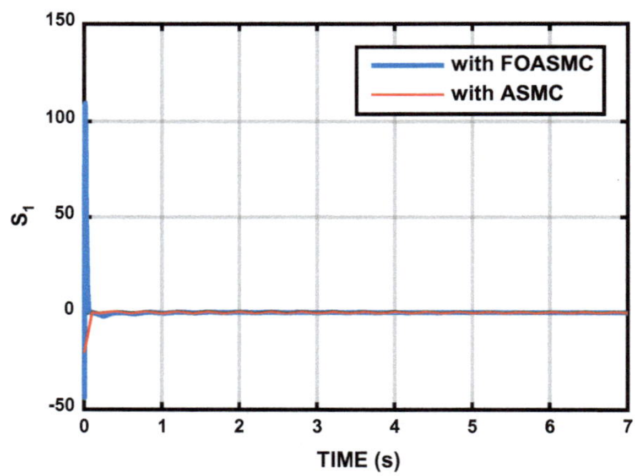

Fig. 11 Sliding surface for L-1, IC $[0.2\ 0.01\ 0\ 0]^T$

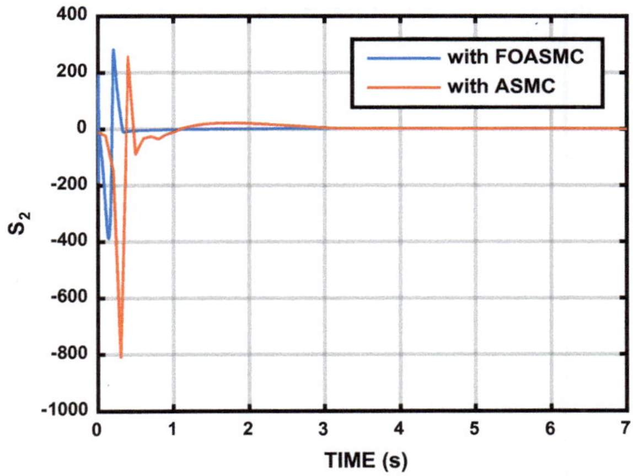

Fig. 12 Sliding surface for L-2, IC $[0.002\ 0\ 0\ 0]^T$

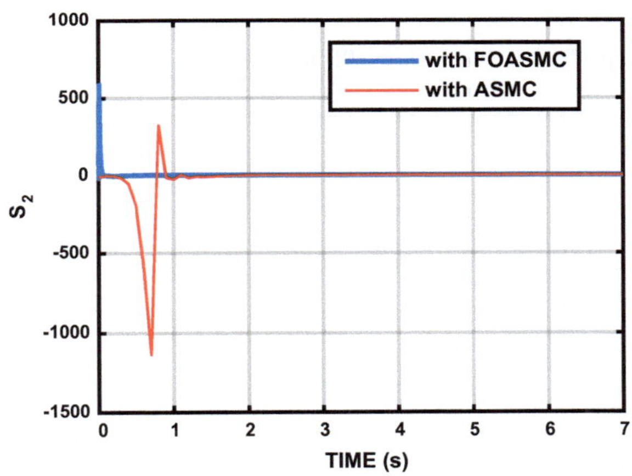

Fig. 13 Sliding surface for L-2, IC $[0.2\ 0.01\ 0\ 0]^T$

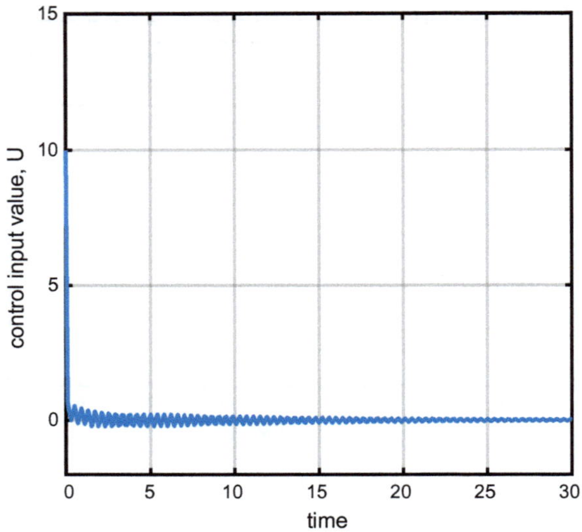

Fig. 14 Control input for L-1, IC $[0.002\ 0\ 0\ 0]^T$ using ASMC

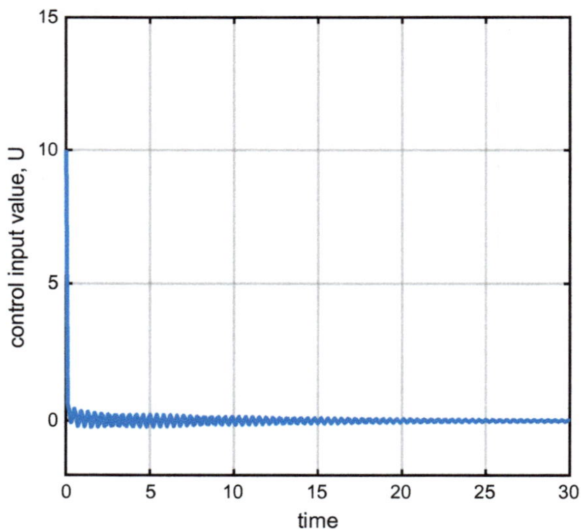

Fig. 15 Control input for L-1, IC $[0.2\ 0.01\ 0\ 0]^T$ using ASMC

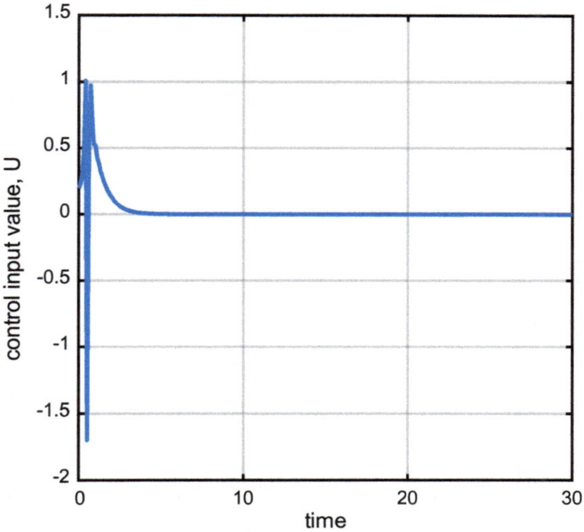

Fig. 16 Control input for L-2, IC $[0.002\ 0\ 0\ 0]^T$ using ASMC

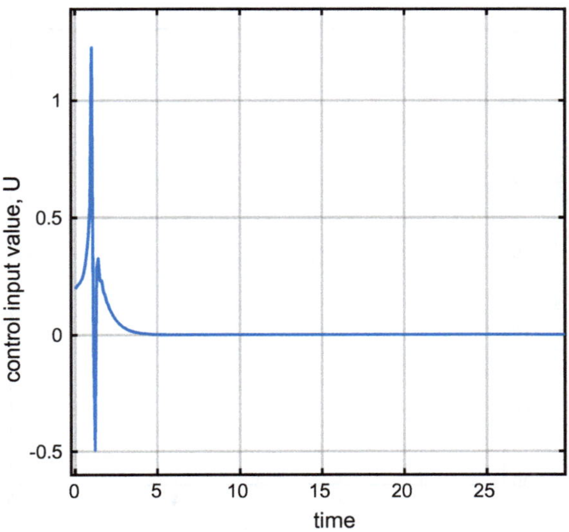

Fig. 17 Control input for L-2, IC $[0.2\ 0.01\ 0\ 0]^T$ using ASMC

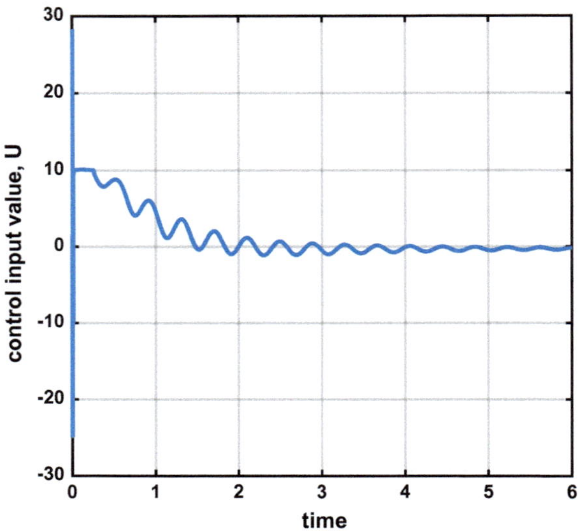

Fig. 18 Control input for L-1, IC $[0.002\ 0\ 0\ 0]^T$ using FOASMC

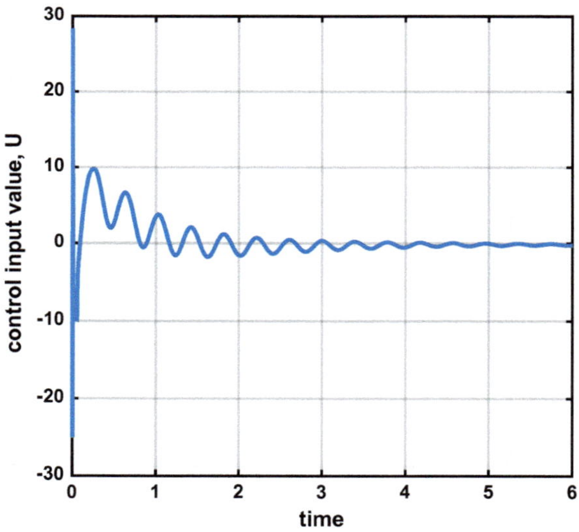

Fig. 19 Control input for L-1, IC $[0.2\ 0.01\ 0\ 0]^T$ using FOASMC

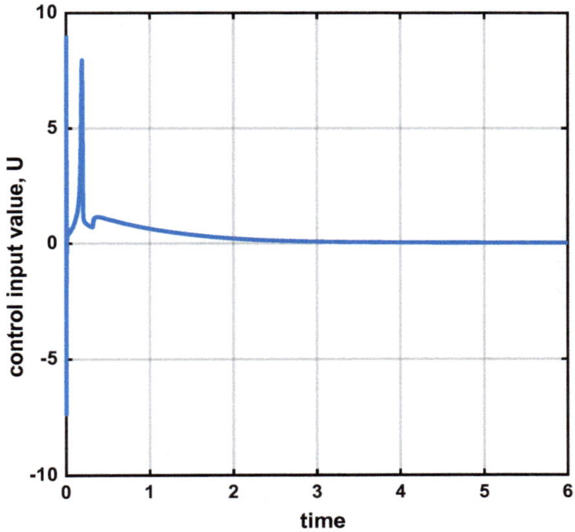

Fig. 20 Control input for L-2, IC $[0.002\ 0\ 0\ 0]^T$ using FOASMC

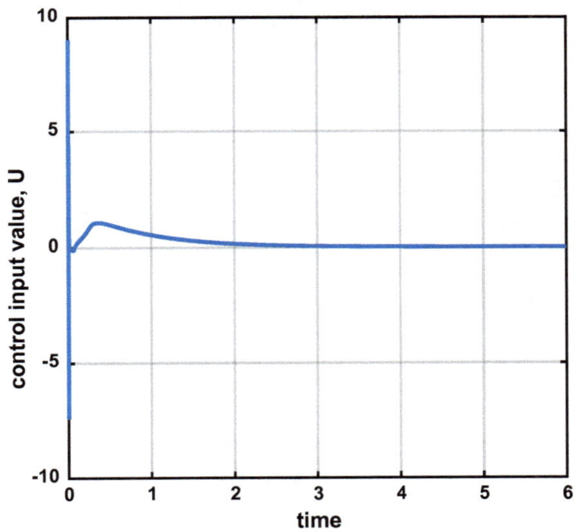

Fig. 21 Control input for L-1, IC $[0.2\ 0.01\ 0\ 0]^T$ using FOASMC

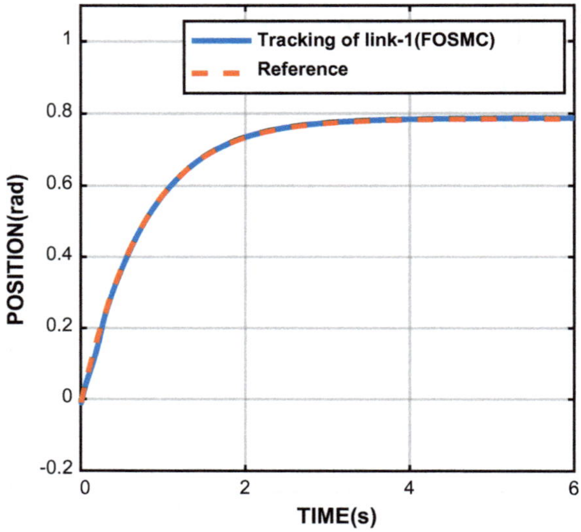

Fig. 22 Tracking of L-1, IC $[0.002\ 0\ 0\ 0]^T$

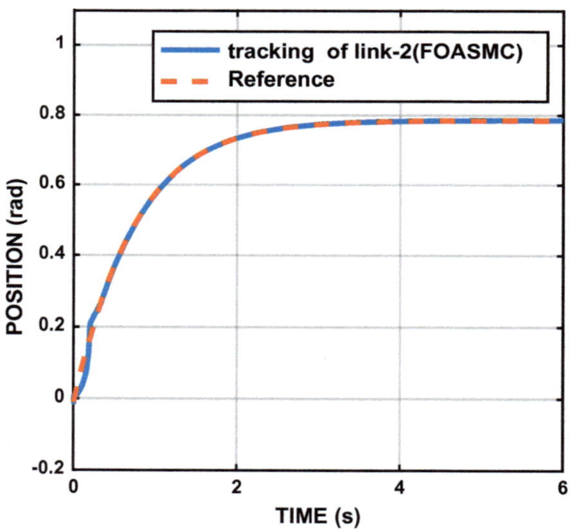

Fig. 23 Tracking of L-2, IC $[0.002\ 0\ 0\ 0]^T$

References

1. Tokhi, M.O., Azad, A.K. (eds.): Flexible Robot Manipulators: Modelling, Simulation and Control, vol. 68. Iet (2008)
2. Kumar, A., Pathak, P.M., Sukavanam, N.: Trajectory control of a two DOF rigid–flexible space robot by a virtual space vehicle. Rob. Autonomous Syst. **61**(5), 473–482 (2013)
3. Tumari, M.M., Ahmad, M.A., Saealal, M.S., Zawawi, M.A., Mohamed, Z., Yusop, N.M.: The direct strain feedback with PID control approach for a flexible manipulator: Experimental results. In 2011 11th International Conference on Control, Automation and Systems (ICCAS), pp. 7–12. IEEE (2011)
4. Zhang, Y.A., Mi, Y.L., Zhu, M., Lu, F.L.: August. Adaptive sliding mode control for two-link flexible manipulators with H/sub/spl infin//tracking performance. In: Machine Learning and Cybernetics, 2005. Proceedings of 2005 International Conference on, vol. 2, pp. 702–707. IEEE (2005)
5. Yang, T.C., Yang, J.C., Kudva, P.: Adaptive control of a single-link flexible manipulator with unknown load. In: IEE Proceedings D (Control Theory and Applications), vol. 138, no. 2, pp. 153–159. IET Digital Library (1991)
6. Li, G., Khajepour, A.: Robust control of a hydraulically driven flexible arm using backstepping technique. J. Sound Vib. **280**(3), 759–775 (2005)
7. Lochan, K., Suklabaidya, S., Roy, B.K.: Sliding mode and adaptive sliding mode control approaches of two link flexible manipulator. In: Proceedings of the 2015 Conference on Advances In Robotics, p. 58. ACM (2015)
8. Mujumdar, A., Tamhane, B. Kurode, S.: Fractional order modeling and control of a flexible manipulator using sliding modes. In: American Control Conference (ACC), 2014, pp. 2011–2016. IEEE (2014)
9. Huh, S.H., Bien, Z.: Robust sliding mode control of a robot manipulator based on variable structure-model reference adaptive control approach. IET Control Theory Appl. **1**(5), 1355–1363 (2007)
10. Rana, D.S.: Mathematical modelling, stability analysis and control of flexible double link robotic manipulator: a simulation approach. Math. Modell. **3**(4) (2013)
11. Mujumdar, A., Kurode, S.: Fractional order sliding mode control for single link flexible manipulator. Multi system conference MSC'13, Hyderabad (2013)

Glucose Regulation in Diabetes Patients Via Insulin Pump: A Feedback Linearisation Approach

Sipon Das, Anirudh Nath, Rajeeb Dey and Saurabh Chaudhury

Abstract The primary objective of the paper is to design a nonlinear control technique for a nonlinear intravenous model of Type 1 diabetes mellitus (T1DM) patient. Input–output feedback linearisation is utilised for deriving the nonlinear control law based on a modified version of Bergman's minimal model augmented with the dynamics of the insulin pump and the meal disturbance. The results depict that the proposed control technique avoids severe hypoglycaemia and postprandial hyperglycaemia in the presence of exogenous meal disturbance as well as parametric uncertainty within a population of 100 virtual T1DM patients (inter-patient variability). The efficacy of the proposed control technique is investigated through variability grid analysis.

Keywords Type 1 diabetes mellitus · Hypoglycaemia · Feedback linearisation Inter-patient variability

1 Introduction

Due to immune-mediated depredation of the pancreatic β-cells resulting in negligible secretion of insulin, blood glucose concentrations cannot be tightly controlled within the safe range (50–180 mg/dl) by the pancreas leading to events of hyperglycaemia (>180 mg/dl) or severe hypoglycaemia (<50 mg/dl) [11]. While hyperglycaemia is associated with long-term health diseases such as heart disease, blindness, kidney failure and nerve damage, hypoglycaemia has the immediate effect that may

S. Das (✉) · A. Nath · R. Dey · S. Chaudhury
National Institute of Technolog, Silchar 788010, Assam, India
e-mail: sipondastit@gmail.com

A. Nath
e-mail: anirudh.nath88@gmail.com

R. Dey
e-mail: rajeeb.iitkgp@gmail.com

S. Chaudhury
e-mail: saurabh@nits.ac.in

D. Deb et al. (eds.), *Innovations in Infrastructure*, Advances in Intelligent Systems and Computing 757, https://doi.org/10.1007/978-981-13-1966-2_5

lead to diabetic coma leading to death [3, 6]. The dependency of T1DM patients on exogenous insulin infusions either through multiple daily insulin injections or insulin infusion pump (IIP) results in hyperglycaemic and hypoglycaemic events. The role of Artificial Pancreas (AP) system is vital in addressing these issues via closed-loop control of the blood glucose concentration that is achieved through sensing the blood glucose by continuous glucose monitoring (CGM) systems and infusing insulin continuously through IIP as determined by the control algorithm [7]. Despite the advances in technology and communications in AP systems, important control challenges exist due to: (i) huge time lags in insulin action, insulin absorption and glucose sensing dynamics, (ii) exogenous disturbances like meals, exercise, stress and sleep and (iii) parametric variability both within a population (inter-patient variability) and within the same T1DM patient (intra-patient variability) [5]. The physiologically based mathematical models of T1DM patients can be broadly categorised into the intravenous models [4] and the subcutaneous models [6, 15] depending upon whether the glucose sensing and insulin infusion route are intravenous (directly into the veins) or subcutaneous (under the skin). In this current work, we have considered popular Bergman's minimal model (BMM) that is not only simplistic in structure but also represents the glucose–insulin regulatory system quite accurately, which have important applications in ICU medications and treatment of diabetes ketoacidosis [20].

Various control algorithm like proportional–integral–derivative (PID), model predictive control (MPC) and fuzzy logic control have been proposed for maintaining the plasma glucose concentration in the safe range in T1DM patients are discussed in [11]. Different variants of sliding mode control (SMC) techniques like back steeping SMC [19], Higher order SMC [14, 16] and super twisting controller [1] were designed for Bergman's minimal model where the issue of inter-patient variability was addressed. It is important to note that the design of the control algorithms based on BMM as reported in [1, 8, 9, 14, 16, 19] did not consider the dynamics of IIP. The novelty in this paper is the consideration of more realistic scenario for AP system by augmenting the pump dynamics [12, 13] to the modified Bergman's minimal model [2], where the parametric variability appearing due to inter-patient variability in T1DM patients is addressed. The design of the nonlinear controller avoids the linearization of the nonlinear dynamics as in [9, 12, 13], thus preserving the nonlinear characteristics of the system. Since control variability grid analysis (CVGA) [18] is a significant tool for the performance assessment of closed-loop control techniques in addressing inter-patient variability, unlike the above-mentioned control algorithms [1, 8, 9, 12–14, 16, 19], here CVGA was performed for 100 virtual T1DM patients.

The paper is structured into four sections. Section 2 deals with the modelling of the glucose–insulin regulatory system of T1DM patients as well as the design of the control law. The simulation results investigating the effectiveness of the proposed scheme is provided in Sect. 3. Section 4 contains the concluding remarks and future scope of the proposed work.

2 Design Methodology

In this section, an integrated mathematical model that is essentially a modification of the BMM is introduced where the insulin pump dynamics, as well as the meal disturbance dynamics, are augmented. The first subsection deals with the mathematical modelling of T1DM patients. The design of the control law is discussed in the succeeding subsection.

2.1 *Mathematical Model for Glucose–Insulin Regulatory System*

In this present work, the BMM [4] is chosen for the model-based controller design. A modified version of the BMM is considered here as in [2]. Further, the modified model is augmented with the insulin pump dynamics as discussed below

$$\dot{G}_p(t) = -I_r(t)G_p(t) - c_1G_p(t) + c_1G_b + D(t) \tag{1}$$

$$\dot{I}_r(t) = -c_2I_r(t) + c_3I_p(t) - c_3I_b, \tag{2}$$

$$\dot{I}_p(t) = -c_4I_p(t) + c_4I_b + \gamma_p[G_p(t) - h_p]^+t, \tag{3}$$

where the BGC is represented by $G_p(t)$, the delayed insulin's effect on BGC by $I_r(t)$ (min^{-1}) and $I_p(t)$ (μU/ml) denotes the plasma insulin concentration. G_b and I_b denote the steady state (basal) value of $G_p(t)$ and $I_p(t)$, respectively. The BMM is essentially a compartmental model composed of the dynamics for glucose homeostasis and insulin kinetics as modelled by Eqs. (1)–(3) representing the plasma glucose compartment, the remote insulin compartment and plasma insulin compartments, respectively. The important physiological parameters of glucose–insulin regulatory system can be expressed in terms of the BMM's parameters [2, 4] directly. The parameter, c_1 (min^{-1}) signifies the insulin-independent glucose utilisation occurring in muscles and liver. Insulin sensitivity is represented by $\frac{c_3}{c_2}$ ($(\mu U/ml)^{-1}\text{min}^{-1}$) and c_4 (min^{-1}) is the rate of degradation of $I_p(t)$. $\gamma_p[G_p(t) - h_p]^+t$ represents the pancreatic actions in maintaining BGC (negligible in T1DM patients). As mentioned in [10], the term $D(t)$ (mg/dl/min) represents the exogenous meal disturbance (the rate at which glucose appears in the plasma glucose compartment) that is modelled by an exponentially decaying function, as follows:

$$\dot{D}(t) = -c_6D(t), \quad c_6 > 0, \tag{4}$$

where the parameter c_6 (min^{-1}) represents the time-to-peak glucose following exogenous glucose disturbance.

In this present work, the above-mentioned modified version of the BMM is augmented with a simple first-order linear state-space model that represents the actuator (insulin pump) dynamics, thus enabling us to design a suitable controller for regulating BGC in T1DM patients for a more realistic scenario. As reported in [12, 13], the pump model is represented as

$$\dot{U}(t) = c_5[-U(t) + U_c(t)] \tag{5}$$

where $\frac{1}{c_5}$ represents the time constant for the pump and 'U' is the actual insulin infusion rate as determined by the command input 'U_c'. All the dynamics of the physiological variables of the glucoregulatory system along with the meal disturbance as well as insulin pump dynamics are expressed in a compact form:

$$\begin{aligned}
\dot{x}_{1_o} &= -x_{1_o}x_{2_o} - c_1x_{1_o} + c_1G_b + x_{5_o}\\
\dot{x}_{2_o} &= -c_2x_{2_o} + c_3x_{3_o} - x_{3_o}I_b\\
\dot{x}_{3_o} &= -c_4x_{3_o} + c_4I_b + x_{4_o}\\
\dot{x}_{4_o} &= -c_5x_{4_o} + c_5U_c\\
\dot{x}_{5_o} &= -c_6x_{5_o}
\end{aligned} \tag{6}$$

where, $x_{1_o}, x_{2_o}, x_{3_o}, x_{4_o}$ and x_{5_o} represents $G_p(t), I_r(t), I_p(t), U(t)$ and $D(t)$, respectively. The state variable appearing in (6) can be expressed as deviation terms from their equilibrium as reported in [2].

$$\begin{aligned}
\dot{x}_{1d} &= -c_1x_{1d} - (x_{1d} + G_b)x_{2d} + x_{5d}\\
\dot{x}_{2d} &= -c_2x_{2d} + c_3x_{3d}\\
\dot{x}_{3d} &= -c_4x_{3d} + x_{4d}\\
\dot{x}_{4d} &= -c_5x_{4d} + c_5U_c\\
\dot{x}_{5d} &= -c_6x_{5d}
\end{aligned} \tag{7}$$

where the deviated states $[x_{1_d}\ x_{2_d}\ x_{3_d}\ x_{4_d}\ x_{5_d}]^T$ are expressed as the difference between the original states $[x_{1_o}\ x_{2_o}\ x_{3_o}\ x_{4_o}\ x_{5_o}]^T$ and their equilibrium $[x_{1_e} = G_b\ x_{2_e} = 0\ x_{3_e} = I_b\ x_{4_e} = 0\ x_{5_e} = 0]^T$ are given as follows:

$$\left[x_{1_o}\ x_{2_o}\ x_{3_o}\ x_{4_o}\ x_{5_o}\right]^T = \left[x_{1_e}\ x_{2_e}\ x_{3_e}\ x_{4_e}\ x_{5_e}\right]^T + \left[x_{1_d}\ x_{2_d}\ x_{3_d}\ x_{4_d}\ x_{5_d}\right]^T, \tag{8}$$

We can rewrite (7) as in the compact form as,

$$\dot{\boldsymbol{x}}_{\boldsymbol{d}}(t) = \boldsymbol{f}(\boldsymbol{x}_{\boldsymbol{d}}, t) + \boldsymbol{g}(\boldsymbol{x}_{\boldsymbol{d}}, t)U_c(t), \tag{9}$$

where, $\boldsymbol{f}(\boldsymbol{x}_{\boldsymbol{d}}, t) = \begin{bmatrix} -c_1x_{1d} - (x_{1d} + G_b)x_{2d} + x_{5d}\\ -c_2x_{2d} + c_3x_{3d}\\ -c_4x_{3d} + x_{4d}\\ -c_5x_{4d}\\ -c_6x_{5d} \end{bmatrix}$, $\boldsymbol{g}(\boldsymbol{x}_{\boldsymbol{d}}, t) = \begin{bmatrix} 0\\ 0\\ 0\\ c_5\\ 0 \end{bmatrix}$, and output

$$y(t) = h(\boldsymbol{x}_{\boldsymbol{d}}, t) = \boldsymbol{C}\boldsymbol{x}_{\boldsymbol{d}}, \quad \boldsymbol{C} = [1\,0\,0\,0\,0] \tag{10}$$

2.2 Feedback Linearisation Control Technique

In this work, a nonlinear control law is proposed based on the input–output feedback linearisation. A conceptual block diagram of the proposed control technique is demonstrated in Fig. 1.

The main idea of this control technique is composed of (i) the choice of appropriate scalar output function and (ii) the linearisation of the input–output mapping of the original nonlinear system by the proposed controller via successive differentiation of the selected output. Here, the output of the system is chosen as $y = h(\boldsymbol{x_d}) = x_{1d}$ (BGC) that can be measured directly via CGM devices.

$$y = h(\boldsymbol{x_d}) = x_{1d} \tag{11}$$

The derivatives of the output function are calculated as

$$y^{(1)} = \mathscr{L}_f h(\boldsymbol{x_d}) = x_{5d} - c_1 x_{1d} - x_{2d}(G_b + x_{1d}), \quad \mathscr{L}_g \mathscr{L}_f h(\boldsymbol{x_d}) = 0 \tag{12}$$

$$y^{(2)} = \mathscr{L}_f^2 h(\boldsymbol{x_d}) = (c_1 + x_{2d})(c_1 x_{1d} - x_{5d} + x_{2d}(G_b + x_{1d})) - c_6 x_{5d} + (G_b + x_{1d})(c_2 x_{2d} - c_3 x_{3d}), \quad \mathscr{L}_g \mathscr{L}_f^{2h}(\boldsymbol{x_d}) = 0 \tag{13}$$

$$y^{(3)} = \mathscr{L}_f^{3h}(\boldsymbol{x_d}) = c_6 x_{5d}(c_1 + c_6 + x_{2d}) - (c_2 x_{2d} - c_3 x_{3d})(c_1 x_{1d} - x_{5d} + (G_b + x_{1d})(c_1 + x_{2d}) + c_2(G_b + x_{1d}) + x_{2d}(G_b + x_{1d})) - c_3(G_b + x_{1d})(x_{4d} - c_4 x_{3d}) - (c_1 x_{1d} - x_{5d} + x_{2d}(G_b + x_{1d}))(c_2 x_{2d} - c_3 x_{3d} + (c_1 + x_{2d})^2) \tag{14}$$

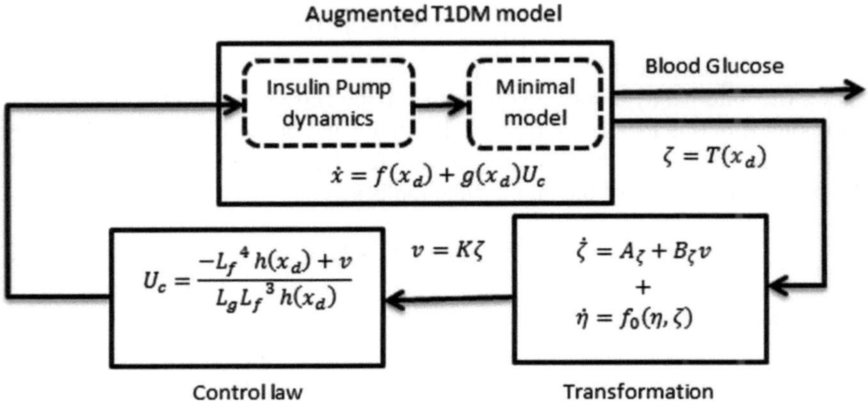

Fig. 1 Feedback linearisation-based nonlinear control

$$\mathscr{L}_g\mathscr{L}_f^{3h}(\boldsymbol{x_d}) = -c_3c_5(G_b + x_{1d}) \neq 0 \tag{15}$$

$$\begin{aligned} y^{(4)} = \mathscr{L}_f^{4h}(\boldsymbol{x_d}) &= ((c_1 + x_{2d})(c_2x_{2d} - c_3x_{3d} + (c_1 + x_{2d})^2) \\ &+ (c_2x_{2d} - c_3x_{3d})(2c_1 + c_2 + 2x_{2d}) \\ &+ c_3(x_{4d} - c_4x_{3d}))(c_1x_{1d} - x_{5d} + x_{2d}(G_b + x_{1d})) \\ &+ (c_2x_{2d} - c_3x_{3d})((G_b + x_{1d})(c_2x_{2d} - c_3x_{3d} + (c_1 + x_{2d})^2) \\ &+ (2G_b + 2x_{1d})(c_2x_{2d} - c_3x_{3d}) - c_6x_{5d} + (c_1x_{1d} \\ &- x_{5d} + x_{2d}(G_b + x_{1d}))(2c_1 + c_2 + 2x_{2d}) \\ &+ c_2(c_1x_{1d} - x_{5d} + (G_b + x_{1d})(c_1 + x_{2d}) + c_2(G_b + x_{1d}) \\ &+ x_{2d}(G_b + x_{1d}))) + (x_{4d} - c_4x_{3d})(c_3(c_1x_{1d} - x_{5d} + x_{2d}(G_b + x_{1d})) \\ &+ c_3(c_1x_{1d} - x_{5d} + (G_b + x_{1d})(c_1 + x_{2d}) + c_2(G_b + x_{1d}) + x_{2d}(G_b + x_{1d})) \\ &+ c_3c_4(G_b + x_{1d})) - c_6x_{5d}(2c_2x_{2d} - 2c_3x_{3d} + c_6(c_1 + c_6 + x_{2d}) \\ &+ (c_1 + x_{2d})^2) + c_3c_5x_{4d}(G_b + x_{1d}) \end{aligned} \tag{16}$$

where $y^{(i)}, i = 1, 2 \ldots, 4$ denote the successive differentiation terms of the output function (10), respectively. Since $\mathscr{L}_g\mathscr{L}_f^{3h}(\boldsymbol{x}_d) \neq 0$ the relative degree of the system (9) is $\rho = 4$. Clearly, since the relative degree ($\rho = 4$) is less than the system's order ($n = 5$), the system (9) can be expressed in the in the transformed coordinate as follows:

$$\dot{\boldsymbol{\zeta}} = \boldsymbol{A}_\zeta\zeta + \boldsymbol{B}_\zeta v \tag{17}$$

$$\dot{\boldsymbol{\eta}} = \boldsymbol{f_0}(\eta, \zeta) \tag{18}$$

$$y = \boldsymbol{C}_\zeta\eta \tag{19}$$

where $\boldsymbol{A}_\zeta = \begin{bmatrix} 0\,1\,0\,0 \\ 0\,0\,1\,0 \\ 0\,0\,0\,1 \\ 0\,0\,0\,0 \end{bmatrix}$, $\boldsymbol{B}_\zeta = \begin{bmatrix} 0 \\ 0 \\ 0 \\ 1 \end{bmatrix}$ and $\boldsymbol{C}_\zeta = \begin{bmatrix} 1\,0\,0\,0 \end{bmatrix}$. In (18), $\dot{\boldsymbol{\eta}} = \boldsymbol{f_0}(\eta, \zeta) = -p_6x_{5_d}$ (i.e. internal dynamics), with respect to the chosen output (10). The equilibrium point $\eta = 0$ is locally asymptotically stable and hence if we can stabilise the transformed system (17) by designing a state-feedback control law $v = -K\zeta$ [17]. The $\rho = 4th$-derivative of the output function is given by

$$y^{(4)} = \mathscr{L}_f^4 h(\boldsymbol{x_d}) + \mathscr{L}_g\mathscr{L}_f^3 h(\boldsymbol{x_d})U_c \tag{20}$$

Let us consider the auxiliary control as

$$y^{(4)} = v \tag{21}$$

Then, the actual controller expression can be computed from (20) using (21) as follows:

$$U_c = \frac{-\mathscr{L}_f^{\rho} h(\boldsymbol{x_d}) + v}{\mathscr{L}_g \mathscr{L}_f^{\rho-1} h(\boldsymbol{x_d})} \tag{22}$$

The auxiliary control input ‘v’ can be designed as

$$v = -k_4 y^{(3)} - k_3 y^{(2)} - k_2 y^{(1)} - k_1 y \tag{23}$$

such that the resulting output dynamics

$$y^{(4)} = -k_4 y^{(3)} - k_3 y^{(2)} - k_2 y^{(1)} - k_1 y \tag{24}$$

is linear and time-invariant with the positive constants k_1, k_2, k_3 and k_4 chosen such that the characteristic polynomial given below is Hurwitz.

$$\lambda^4 + k_4 \lambda^3 + k_3 \lambda^2 + k_2 \lambda + k_1 = 0 \tag{25}$$

The characteristic polynomial (25) is derived from the fact that the output dynamics can be expressed in a matrix form as follows:

$$\dot{\zeta} = A_{cl} \zeta \tag{26}$$

where

$$\zeta = \left[y \; y^{(1)} \; y^{(2)} \; y^{(3)}\right]^T, \quad A_{cl} = \begin{bmatrix} 0 & 1 & 0 & 0 \\ 0 & 0 & 1 & 0 \\ 0 & 0 & 0 & 1 \\ -k_1 & -k_2 & -k_3 & -k_4 \end{bmatrix} \tag{27}$$

The exponential convergence of the output trajectories to the origin is guaranteed by the Routh–Hurwitz criterion. From the Routh–Hurwitz criterion, it is obtained that if the inequality

$$k_4 k_3 > k_2 \tag{28}$$

holds then the matrix A_{cl} is Hurwitz and

$$\lim_{t \to 0} \zeta = 0$$

with exponential rate of convergence.

3 Simulation Results

In this section, the simulation studies are carried out for the performance assessment of the proposed feedback linearisation based nonlinear control technique for regulating plasma glucose concentration in T1DM patients for the augmented minimal model. The T1DM patient is considered to be in the hyperglycaemic state ($x_{1_o} = 230\,\text{mg/dl}$) with no prior exogenous insulin infusions that is reflected by the initial conditions $x_{2_o} = 0$ min^{-1}, $x_{3_o} = 7\ mU/l$ and $x_{4_o} = 0$ mU/l, with high exogenous meal disturbance $x_{5_o} = 10\,\text{mg/dl/min}$. The BMM's parameter values are considered as in Table 1. The controller gains $k_1 = 0.00009$, $k_2 = 0.0058$, $k_3 = 0.15$ and $k_4 = 4.5$ are chosen heuristically for the adjustment in the insulin dosages, u(t) such that the control objectives are satisfied.

It is evident from Fig. 2 that the blood glucose level stays at the hyperglycaemic state in the absence of exogenous insulin infusion, whereas the blood glucose level is brought down below 180 mg/dl within 150 min thereby avoiding postprandial hyperglycaemia and hypoglycaemia. The external insulin infusion by the insulin pump as determined by the control command U_c is depicted in Fig. 3.

Finally, to investigate the controller's robustness to parametric uncertainty representing inter-patient variability, CVGA [18] is carried out. CVGA is essentially the grid representation of the maximum (Y-axis) and minimum (X-axis) blood glucose variations of a virtual T1DM patient during the whole simulation period.

Table 1 Nominal value of the parameters [2, 13]

Parameter	Nominal value	Parameter	Nominal value
c_1	0	c_4	0.2814
c_2	0.0142	c_5	0.5
c_3	1.54×10^{-6}	c_6	0.05

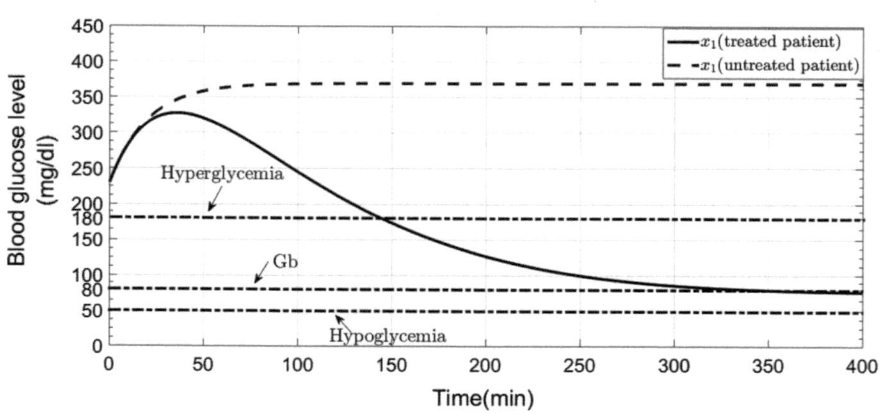

Fig. 2 Plasma glucose concentration of treated and untreated patient

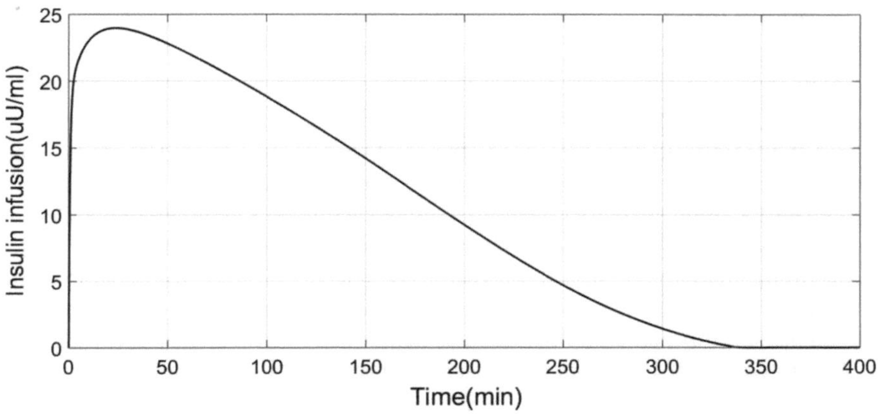

Fig. 3 Exogenous insulin infusion U(t) determined by the insulin pump

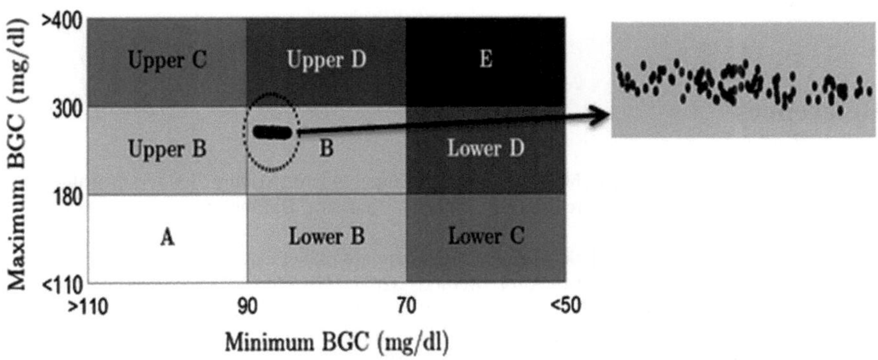

Fig. 4 Control variability grid analysis of 100 closed-loop responses. Each black dot corresponds to 400 min long closed-loop simulation for a single T1DM patient

100 numerical simulations are carried out with random parameters with $\pm 10\%$ variation from the nominal value specified in Table 1 and with the initial conditions $x_{1_o} = 80\,\text{mg/dl}$, $x_{2_o} = 0\,\text{min}^{-1}$, $x_{3_o} = 7\,\text{mU/l}$, $x_{4_o} = 0\text{mU/l/min}$, $x_{5_o} = 0\,\text{mg/dl/min}$ and with an administration of meal disturbance of 10 mg/dl/min at the 100th min. Figure 4 elucidates that all the black dots corresponding to T1DM subjects are confined to the grid B (green zone), thereby ensuring no events of hypoglycaemia during the whole simulation period.

4 Conclusion

A feedback linearisation technique-based nonlinear control law is designed for the plasma glucose regulation problem of an augmented intravenous modified minimal model of T1DM patient by considering the insulin pump dynamics. The closed-loop performance under inter-patient variability ($\pm 10\%$) is investigated. Occurrences of prolonged hyperglycaemia, as well as hypoglycaemia, are completely avoided as validated by the simulation studies. Despite the parametric uncertainty, the controller is able to maintain the plasma glucose of 100 random virtual T1DM patients in the safe range (50–180 mg/dl) as confirmed by control variability grid analysis plot. Although the proposed controller can efficiently handle parametric uncertainty of $\pm 10\%$, to deal with larger parametric uncertainty, robust and adaptive controllers need be designed.

Acknowledgements Authors acknowledge the financial support by **TEQIP-III**, NIT Silchar, 788010, Assam India for this work.

References

1. Ahmad, S., Ahmed, N., Ilyas, M., Khan, W., et al.: Super twisting sliding mode control algorithm for developing artificial pancreas in type 1 diabetes patients. Biomed. Signal Process. Control **38**, 200–211 (2017)
2. Ali, S.F., Padhi, R.: Optimal blood glucose regulation of diabetic patients using single network adaptive critics. Potimal Control Appl. Methods **32**(2), 196–214 (2011)
3. Bequette, B.W., Cameron, F., Buckingham, B.A., Maahs, D.M., Lum, J.: Overnight hypoglycemia and hyperglycemia mitigation for individuals with type 1 diabetes: how risks can be reduced. IEEE Control Syst. **38**(1), 125–134 (2018)
4. Bergmann, R.: Physiologic evaluation of factors controlling glucose tolerance in man. J. Clin. Invest. **68**, 1456–1467 (1981)
5. Bondia, J., Romero-Vivo, S., Ricarte, B., Diez, J.L.: Insulin estimation and prediction: a review of the estimation and prediction of subcutaneous insulin pharmacokinetics in closed-loop glucose control. IEEE Control Syst. **38**(1), 47–66 (2018)
6. Cobelli, C., Dalla, C., Man, G.S.: Diabetes: Models, signals, and control. IEEE Rev. Biomedi. Eng. **2**(3), 54–96 (2009)
7. Cinar, A.: Artificial pancreas systems: an introduction to the special issue. IEEE Control Sys. **38**(1), 26–29 (2018)
8. Cocha, G., Amorena, C., Mazzadi, A., D'Attellis, C.: Geometric adaptive control in type 1 diabetes. In: 12th International Symposium on Medical Information Processing and Analysis, vol. 10160, p. 101600R. International Society for Optics and Photonics (2017)
9. Coman, S., Boldisor, C.: Simulation of an adaptive closed loop system for blood glucose concentration control. Bulletin of the Transilvania University of Brasov. Eng. Sci. Series I **8**(2), 107 (2015)
10. Fischer, U., Schenk, W., Salzsieder, E., Albrecht, G., Abel, P., Freyse, E.J.: Does physiological blood glucose control require an adaptive control strategy? IEEE Trans. Biomed. Eng. **8**, 575–582 (1987)
11. Haidar, A.: The artificial pancreas: how closed-loop control is revolutionizing diabetes. IEEE Control Syst. **36**(5), 28–47 (2016)

12. Hariri, A., Wang, L.Y.: Observer-based state feedback for enhanced insulin control of type i sdiabetic patients. Open Biomed. Eng. J. **5**, 98 (2011)
13. Hariri, A.M.: Identification, state estimation, and adaptive control of type i diabetic patients (2011)
14. Hernandez, A.G.G., Fridman, L., Levant, A., Shtessel, Y., Leder, R., Monsalve, C.R., Andrade, S.I.: High-order sliding-mode control for blood glucose: practical relative degree approach. Control Eng. Practice **21**(5), 747–758 (2013)
15. Hovorka, R., Svačina, Š., Carson, E.R., Williams, C.D., Sänksen, P.H.: A consultation system for insulin therapy. Comput. Methods Programs Biomed. **32**(3), 303–310 (1990). https://doi.org/10.1016/0169-2607(90)90113-N
16. Kaveh, P., Shtessel, Y.B.: Blood glucose regulation using higher-order sliding mode control. Int. J. Rob. Nonlinear Control **18**(4–5), 557–569 (2008)
17. Khalil, H.K., Grizzle, J.: Nonlinear Systems, vol. 3. Prentice hall Upper Saddle River (2002)
18. Magni, L., Raimondo, D.M., Man, C.D., Breton, M., Patek, S., De Nicolao, G., Cobelli, C., Kovatchev, B.P.: Evaluating the efficacy of closed-loop glucose regulation via control-variability grid analysis. J. Diabetes Sci. Technol. **2**(4), 630–635 (2008)
19. Parsa, N.T., Vali, A., Ghasemi, R.: Back stepping sliding mode control of blood glucose for type i diabetes. World Acad. Sci. Eng. Technol. Int. J. Med. Health Biomed. Bioeng. Pharm. Eng. **8**(11), 779–783 (2014)
20. Van Herpe, T., Pluymers, B., Espinoza, M., Van den Berghe, G., De Moor, B.: A minimal model for glycemia control in critically ill patients. In: 28th Annual International Conference of the IEEE Engineering in Medicine and Biology Society, 2006. EMBS'06 , pp. 5432–5435. IEEE (2006)

Delayed State Feedback Controller Design for Inverted Pendulum Using T-S Fuzzy Modeling: An LMI Approach

Rupak Datta, Rajeeb Dey, Baby Bhattacharya and Abanishwar Chakrabarti

Abstract Balancing inverted pendulum along a vertical position with or without cart is a benchmark control problem owing to the fact that out of two equilibrium points, inverted one is open-loop unstable. In this paper, an attempt has been made to stabilize the system with delayed state feedback control strategy using T-S fuzzy modeling in an linear matrix inequality (LMI) framework. Then, by Lyapunov–Krasovskii (L-K) theory, it is proved that the closed-loop system is locally asymptotically stable around its unstable equilibrium point. The result of the control design is validated through closed-loop simulation carried out in MATLAB Simulink.

Keywords Inverted pendulum · Takagi-Sugeno (T-S) fuzzy model · State feedback control · Linear matrix inequality (LMI)

1 Introduction

The inverted pendulum is a benchmark control problem due to the nonlinearity in its dynamics and the existence of unstable equilibrium point [1, 2]. Many immensely complicated system, for instance rockets during takeoff, pulley, automation system, etc., are being motivated by the control of inverted pendulum. Generally, the system constitute of a cart on which a pole (pendulum) is placed and it moves aclinic (linear motion). It occupy two equilibrium points, stable one when pendulum orients

R. Datta (✉) · B. Bhattacharya
Department of Mathematics, NIT Agartala, Agartala, Tripura, India
e-mail: rupak.kls@gmail.com

B. Bhattacharya
e-mail: babybhatt75@gmail.com

R. Dey
Department of Electrical Engineering, NIT Silchar, Silchar, Assam, India
e-mail: rajeeb.iitkgp@gmail.com

A. Chakrabarti
Department of Electrical Engineering, NIT Agartala, Agartala, Tripura, India
e-mail: abanishwar@yahoo.com

D. Deb et al. (eds.), *Innovations in Infrastructure*, Advances in Intelligent Systems and Computing 757, https://doi.org/10.1007/978-981-13-1966-2_6

downwards and unstable one when pendulum is vertically upward. There is always a requirement of an external force apropos to set the pendulum in the upward location [3]. This force may be applied to the cart when cart-pendulum system is considered, while the force is applied to the pendulum hinge if only pendulum system without cart is considered. In the present paper, only pendulum dynamics is considered keeping cart stationary. Thus, in such a scenario, the control input (force) is applied around the hinged point between cart and pendulum.

To control the inverted pendulum system, many usual control skills such as proportional–integral–derivative (PID) controller [4], linear quadratic regulator (LQR) [1], and fuzzy logic controller (FLC) [5] have been used and also they are founded on the conventional control theory, modern control theory, and intelligent control theory have been used to control the inverted pendulum. In [6], a simple conventional PID controller have been designed to stabilize the inverted pendulum system. Applying state-space pole placement technique, Lan and Fei in [7] design a controller for a nonlinear inverted pendulum system. PID, FLC, and state feedback control are compared among themselves and presented in [8]. An intelligent control technique for inverted pendulum system has been proposed in [9].

In all engineering and industrial applications, the presence of complex nonlinear system are ubiquitous. In recent few decades, various nonlinear control methods have been developed to realize near practical situations and (or) responses for the nonlinear system [10]. The demerit of nonlinear control method is that they are not unique for a general nonlinear system or model. Thus, in recent times, fuzzy logic control has emerged as an effective method for obtaining, desired system behavior. Takagi-Segeno (T-S) fuzzy modeling [11] has been widely used to control a nonlinear system in spite of many other available control techniques and it ensures that the stability criteria can be reformation in LMI [12] framework so that it can be proficiently solved by using MATLAB LMI toolbox [13]. Nowadays, for the control of nonlinear system using T-S fuzzy control technique, [14] has drawn much attention over the usual nonlinear control techniques owing to the fact that this modeling approach can represent a complex nonlinear system into local linear models such that Lyapunov-based control strategy for linear systems can be applied directly.

In this paper, our aim is to stabilize the unstable nonlinear inverted pendulum system that has been approximated by using T-S fuzzy modeling approach. The stabilizing law is considered to be two-term control (one is the gain associated with present state while another gain is associated with the delayed states). This control law makes the closed-loop system an infinite dimensional system due to the presence of delayed term in the feedback. Further, the controller design for the considered system is being obtained by fuzzy parallel distributed compensation (PDC) technique using existing stability theories of linear retarded system.

Notations. In this manuscript, the notations are standard. The symmetric term in a matrix is denoted by '$*$' and $sym(A) = A + A^T$.

2 Mathematical model of inverted pendulum

Without any control action, the pendulum in Fig. 1 is unstable in upward (inverted) position and thus it will fall off. To balance the pendulum in a cart-pendulum system, a force is required to apply on the cart to which the pendulum is attached or for pendulum hinged at a fixed point and there are two equilibrium points of this system as discussed above. First, the dynamic modeling of cart-pendulum system is presented and then considering that cart is stationary a simplified model with only rotational variables of the pendulum is obtained. Thus the simplified model has 1 degree-of-freedom (DOF) compared to the original cart-pendulum system which has 2 DOF (linear and rotational motion). Figure 1 shows the original cart-pendulum system. The modeling is carried out using Lagrangian mechanics.

Under equilibrium condition, the law of physics that governs the system shown in Fig. 1 are described below,

$$\widetilde{I}\ddot{\omega} = -\widetilde{I}\left(\widehat{N}\cos\omega + \widehat{P}\sin\omega\right) \tag{1}$$

where $\widetilde{N}$ denote the normal reaction force, $\widetilde{P}$ denote the horizontal force which are acting along the vertical and $\widetilde{I}$ represent moment of inertia. From free body diagram of the pendulum in Fig. 1, normal reaction force $\widetilde{N}$ and horizontal force $\widetilde{P}$ can be calculated by the following equations as:

$$\widetilde{N} = m_p\left(\ell\ddot{\omega}\cos\omega - \ell\dot{\omega}^2\sin\omega + \ddot{x}\right), \quad \widetilde{P} = m_p\left(g + \ell\ddot{\omega}\sin\omega + \ell\dot{\omega}^2\cos\omega\right) \tag{2}$$

Substitute (2) in (1), one can obtain that $(\widetilde{I} + m_p\ell^2)\ddot{\omega} + m_p\ell\ddot{x}\cos\omega = -m_pg\ell\sin\omega$. The Lagrangian is given by $£ = K_e - V_e$, where K_e and V_e represent the sys-

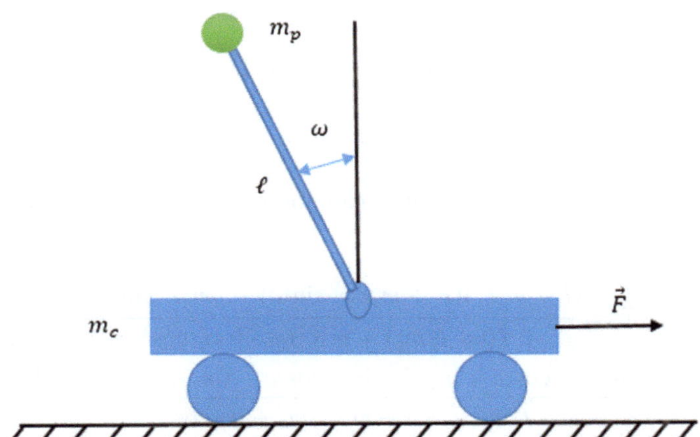

Fig. 1 Inverted pendulum system

tem's kinetic energy and potential energy respectively. The kinetic energy (K_e) of the pendulum in Fig. 1 is represented by $K_e = K_c + K_p$, where $K_c = \frac{1}{2} m_c V_1^2$ and $K_p = \frac{1}{2} m_p V_2^2$ denote the kinetic energy of the cart and the kinetic energy of the pendulum, respectively. Whereas, m_c and $V_1 = \dot{x}$ represent the mass and the velocity of the cart, and m_p,

$V_2 = \left[\left\{ \frac{d}{dt}(x - \ell \sin \omega) \right\}^2 + \left(\frac{d}{dt} \ell \cos \omega \right)^2 \right]^2$ represents the mass and the velocity of the pendulum respectively. Substituting the values of K_c and K_p in K_e, one can obtain that $K_e = \frac{1}{2} \left[m_c \dot{x}^2 + m_p \left\{ (\dot{x} - \ell \dot{\omega} \cos \omega)^2 + (\ell \dot{\omega} \sin \omega)^2 \right\} \right]$.

Now, by the following equation, potential energy, (V_e) can be represented as $V_e = V_{ec} + V_{ep}$, where $V_{ec} = 0$ and $V_{ep} = m_p g \ell \cos \omega$ denote the potential energy of the cart and the potential energy of the pendulum respectively.

Again, substitute the values of K_e and V_e, the Lagrangian can be calculated as

$$£ = \frac{1}{2}(m_c + m_p)\dot{x}^2 - m_p \ell \dot{x} \dot{\omega} \cos \omega + \frac{1}{2} m_p \ell^2 \dot{\omega}^2 - m_p g \ell \cos \omega \tag{3}$$

The Lagrange's equation for generalized coordinate is defined as:

$$\frac{d}{dt}\left(\frac{\partial £}{\partial \dot{x}}\right) - \frac{\partial £}{\partial x} = F, \quad \frac{d}{dt}\left(\frac{\partial £}{\partial \dot{\theta}}\right) - \frac{\partial £}{\partial \theta} = 0 \tag{4}$$

Substitute (3) into (4), one can obtain that

$$\left.\begin{aligned} (m_c + m_p)\,\ddot{x} - m_p \ell \ddot{\omega} \cos \omega + m_p \ell \dot{\omega}^2 \sin \omega = F, \\ \ell \ddot{\omega} - g \sin \omega = \ddot{x} \cos \omega, \end{aligned}\right\} \tag{5}$$

From Fig. 1, force along aclinic way is represented by

$$F = m_c \ddot{x} + \widetilde{N} \tag{6}$$

By solving (4), (5), and (6), one can obtain the dynamical model of the inverted pendulum as:

$$\left.\begin{aligned} \ddot{x} &= \frac{(\widetilde{I} + m_p \ell^2)(F + m_p \ell \dot{\omega}^2 \sin \omega) - g m_p^2 \ell^2 \sin \omega \cos \omega}{\widetilde{I}(m_c + m_p) + m_p \ell^2 (m_c + m_p \cos^2 \omega)}, \\ \ddot{\omega} &= -\frac{m_p \ell [F \cos \omega + m_p \ell \dot{\omega}^2 \sin \omega \cos \omega - (m_c + m_p) g \sin \omega]}{\widetilde{I}(m_c + m_p) + m_p \ell^2 (m_c + m_p \cos^2 \omega)}, \end{aligned}\right\} \tag{7}$$

where the state variables ω, $\dot{\omega}$, and $\dot{x}$ denotes the angle of the pendulum with respect to perpendicular axis, angular velocity of the pendulum, position of the cart and the velocity of the cart, respectively.

In this paper for simplification of the model, it is assumed that the cart is stationary, thus $\dot{x} = 0$ and $\ddot{x} = 0$. Therefore, first part of (7) becomes

$$(\widetilde{I} + m_p\ell^2)(F + m_p\ell\dot{\omega}^2 \sin\omega) - gm_p^2\ell^2 \sin\omega \cos\omega = 0 \tag{8}$$

Invoking stationary assumption of the cart in (8) makes the linear dynamics zero. Thus, the following state variables are considered $x_1 = \omega$ (angle of the pendulum from the vertical position) and $x_2 = \dot{\omega}$ (angular velocity of the pendulum) to obtain the state-space model of the second of (7). Further, control input is the force applied at the hinged point between the cart and the pendulum, i.e., $u = F$. The system in (7) can now be represented by new state variables $x(t) = \left[x_1^T(t) \;\; x_2^T(t)\right]^T$ and the input variable $u = F$ as:

$$\dot{x}(t) = \begin{bmatrix} x_2 \\ -\dfrac{m_p\ell[\frac{1}{2}m_p\ell x_2^2 \sin 2x_1 - (m_c + m_p)g \sin x_1]}{\widetilde{I}(m_c + m_p) + m_p\ell^2(m_c + m_p \cos^2 x_1)} \end{bmatrix} + \begin{bmatrix} 0 \\ -\dfrac{m_p\ell \cos x_1}{\widetilde{I}(m_c + m_p) + m_p\ell^2(m_c + m_p \cos^2 x_1)} \end{bmatrix} u(t). \tag{9}$$

3 T-S Fuzzy Model for Nonlinear Systems

Consider a nonlinear system as:

$$\dot{y}(t) = h\left(t, y(t), \eta(t)\right), \qquad t \geq 0, \tag{10}$$

where $y(t) \in \mathbb{R}^n$ $(n = 2)$ denotes the states, $\eta(t) \in \mathbb{R}^m$ $(m = 1)$ denote the control input, 'h' denote the nonlinear continuous function. If one chooses the control input $\eta(t) = Ky(t) + K_\tau y(t - \tau)$ and if it is applied to the system in (10), then the closed-loop form of (10) becomes a retarded or time-delay system as below

$$\dot{y}(t) = h\left(t, y(t), y(t - \tau)\right), \qquad t \geq 0, \tag{11}$$

$$y(t) = \lambda(t), \qquad -\tau \leq t \leq 0$$

where $y(t) = \lambda(t)$ is the vector valued initial condition on $[-\tau, 0]$.

The system in (10) can be linearized by the T-S fuzzy model which is represented in the following fuzzy IF-THEN rules:

Rule l: IF $d_1(t)$ is M_{l1} and ...and $d_q(t)$ is M_{lq} THEN

$$\dot{y}(t) = A_l x(t) + B_l \eta(t), \quad t \geq 0, \tag{12}$$

where $d_1(t), d_2(t), \ldots, d_q(t)$ denote the premise variables, M_{lk}, $l = 1, 2, 3, \ldots, n$, $k = 1, 2, 3, \ldots, q$ denote the fuzzy membership functions, n denote the number of fuzzy IF-THEN rules and q is the number of premise variables. A_l, B_l are the system matrices with appropriate dimensions.

If $d_k(t) = d_k^0$ given, where d_k^0 are singletons, then the truth values of $\dot{y}(t)$ for each lth fuzzy rule is calculated as:

$$z_l(d(t)) = \left(M_{l1}(d_1(t)) \wedge \ldots \wedge M_{lq}(d_q(t))\right), l = 1, 2, \ldots, n, \tag{13}$$

where $d(t) = \left[d_1(t)\ d_2(t) \ldots d_q(t)\right]^T$, $M_{l1}(d_1(t)), \ldots, M_{lq}(d_q(t))$ is the membership grade of $d(t)$ in M_{lk} and $\wedge$ denote the "min" operation.

Next, using centroid method for defuzzification, the overall output of (12) is calculated as:

$$\dot{y}(t) = \frac{\sum_{l=1}^{n} z_l(d(t)) \{A_l y(t) + B_l \eta(t)\}}{\sum_{l=1}^{n} z_l(d(t))} = \sum_{l=1}^{n} w_l(d(t)) \{A_l y(t) + B_l \eta(t)\}, \tag{14}$$

where $w_l(d(t)) = \dfrac{z_l(d(t))}{\sum_{l=1}^{n} z_l(d(t))}$, $\forall\ t$ and $l = 1, 2, \ldots, n$, is represent the fuzzy weighting function and by definition $w_l(d(t)) \geq 0$ and $\sum_{l=1}^{n} w_l(d(t)) = 1$.

To stabilize the system in (14), a state feedback fuzzy controller is designed via parallel distributed compensation (PDC). The controller gains are by solving the LMI conditions derived for the system in the next section. T-S fuzzy PDC controller rules are given as,

Rule j: IF $d_1(t)$ is M_{k1} and ...and $d_q(t)$is M_{kq} THEN

$$\eta(t) = K_k y(t) + K_{\tau k} y(t - \tau), \quad t \geq 0,\ k = 1, 2, \ldots, q \tag{15}$$

After defuzzification, the output of the control rule (15) can be obtained as:

$$\eta(t) = \sum_{k=1}^{q} w_k(d(t)) \left[K_k y(t) + K_{\tau k} y(t - \tau)\right] \tag{16}$$

where K_k, $K_{\tau k}$, $(k = 1, 2, \ldots, q)$ are controller gains to be determined and 'τ' is the time delay in the state.

Substitute (16) into (14), the closed-loop form of (14) as:

$$\dot{y}(t) = \sum_{l=1}^{n} \sum_{k=1}^{q} w_l(d(t)) w_k(d(t)) \{(A_l + B_l K_k) y(t) + B_l K_{\tau k} y(t - \tau)\}, \quad t \geq 0 \tag{17}$$

Furthermore, to avoid the notational complexity w_l is used instead of $w_l(d(t))$. Moreover, the following integral inequality lemma is used to obtain our results in the successive section.

Lemma 1 ([15], Wirtinger inequality)*: Given a symmetric matrix $S > 0$, scalars $\gamma,\ \delta > 0$ and the vector function $y : [\gamma, \delta] \to \mathbb{R}^n$ such that the following inequality holds:*

$$\int_{\gamma}^{\delta} \dot{y}^T(s)S\dot{y}(s)ds \geq \frac{1}{\delta-\gamma}\chi_1^T(t)S\chi_1(t) + \frac{3}{\delta-\gamma}\chi_2^T(t)S\chi_2(t) \tag{18}$$

where $\chi_1 = \left[y(\delta) - y(\gamma)\right]$ *and* $\chi_2 = \left[y(\delta) + y(\gamma) - \frac{2}{\delta-\gamma}\int_{\gamma}^{\delta} y(s)ds\right]$.

4 Delayed State Feedback Stabilization

A new state feedback stabilization criteria for the system in (17) is derived in this section using Lemma 1.

Theorem 1 *Given $\tau,\ \rho > 0$, the closed-loop T-S fuzzy system (17) is asymptotically stable with feedback gains $K_k = N_k Y^{-1}$, $K_{\tau k} = N_{\tau j} Y^{-1}$, $(k = 1, 2, \ldots, q)$ if there exist symmetric matrices $0 < \widehat{P} \in \mathbb{R}^{2n\times 2n}$, $0 < \widehat{Q} \in \mathbb{R}^{n\times n}$, $0 < \widehat{R} \in \mathbb{R}^{n\times n}$, any matrices $Y \in \mathbb{R}^{2n\times 2n}$ the following LMIs are satisfies:*

$$\widehat{\Omega}_{ll} - \Gamma^T\overline{R}\Gamma < 0, \ \ for\ \ l = 1, 2, \ldots, n \tag{19}$$

and

$$\widehat{\Omega}_{lk} + \widehat{\Omega}_{kl} - 2\Gamma^T\overline{R}\Gamma, \ \ for\ \ 1 \leq l < k \leq n \tag{20}$$

with

$$\begin{aligned}\widehat{\Omega}_{lk} = sym\left\{\Pi_{11}^T\widehat{P}\Pi_{22}\right\} + e_1^T\widehat{Q}e_1 - e_2^T\widehat{Q}e_2 + \tau^2 e_4^T\widehat{R}e_4 \\ + 2\rho\left\{e_1^T\left(A_l + B_l N_k\right)e_1 + e_1^T B_l N_k e_2 - e_1^T Y e_4\right. \\ \left. + e_4^T\left(A_l + B_l N_k\right)e_1 + e_4^T B_l N_{\tau k} e_2 - e_4^T Y e_4\right\},\end{aligned}$$

$\Pi_{11} = \left[e_1^T\ e_3^T\right]$, $\Pi_{22} = \left[e_4^T\ (e_1^T - e_2^T)\right]^T$, $\overline{R} = diag\left\{\widehat{R},\ 3\widehat{R}\right\}$, $\Gamma = [\zeta_1^T\ \zeta_2^T]^T$, $\zeta_1 = [I\ \ -I\ \ 0\ \ 0]$, $\zeta_2 = [I\ \ I\ \ -2I\ \ 0]$, *and* $e_s = \left[e_{n\times(s-1)n}\ I_{n\times n}\ 0_{n\times(4-s)n}\right]^T$, $s = 1, 2, \ldots, 4$ *represent the block matrices.*

Proof Choose a L-K functional as:

$$V(y_t) = V_1(y_t) + V_2(y_t) \tag{21}$$

where

$$V_1(y_t) = \varpi^T(t)P\varpi(t),$$
$$V_2(y_t) = \int_{t-\tau}^{t} y^T(s)Wy(s)ds + \tau\int_{-\tau}^{0}\int_{t+\lambda}^{t} \dot{y}^T(s)S\dot{y}(s)dsd\lambda$$

with $\varpi(t) = \begin{bmatrix} y^T(t) & \int_{t-\tau}^{t} y^T(s)ds \end{bmatrix}^T$ and $P = \begin{bmatrix} P_{11} & P_{12} \\ * & P_{22} \end{bmatrix}$.

Now, differentiating (21) with respect to time along with (17), one can obtain

$$\dot{V}(y_t) = \dot{V}_1(y_t) + \dot{V}_2(y_t) \tag{22}$$

where

$$\dot{V}_1(y_t) = 2\varpi^T(t)P\dot{\varpi}(t) \tag{23}$$

Next, the following terminology of vectors and matrices are introduce for simplicity, $\psi(t) = \begin{bmatrix} y^T(t) \; y^T(t-\tau) \; \frac{1}{\tau}\int_{t-\tau}^{t} y^T(s)ds \; \dot{y}^T(t) \end{bmatrix}^T$, $e_s = \begin{bmatrix} e_{n\times(s-1)n} \; I_{n\times n} \; 0_{n\times(4-s)n} \end{bmatrix}^T$, $s = 1, 2, \ldots, 4$.

Further with the help of the vectors $\psi(t)$ and e_s, (23) can be rewritten as

$$\dot{V}_1(y_t) \le \psi^T(t)\left[sym\left\{\Pi_{11}^T P\Pi_{22}\right\}\right]\psi(t) \tag{24}$$

Similarly, we proceed as above $\dot{V}_2(x_t)$ can be obtained as:

$$\dot{V}_2(y_t) = \psi^T(t)\left\{e_1^T We_1 - e_2^T We_2 + \tau^2 e_4^T Se_4\right\}\psi(t) - \tau\int_{t-\tau}^{t} \dot{y}^T(s)S\dot{y}(s)ds \tag{25}$$

Now, we apply Lemma 1 to estimate the integral term in the right-hand side (RHS) of (25), yields that

$$-\tau\int_{t-\tau}^{t} \dot{y}^T(s)S\dot{y}(s)ds \le -\psi^T(t)\begin{bmatrix}\zeta_1 \\ \zeta_2\end{bmatrix}^T\begin{bmatrix} S & 0 \\ * & 3S\end{bmatrix}\begin{bmatrix}\zeta_1 \\ \zeta_2\end{bmatrix}\psi(t) = -\psi^T(t)\Gamma^T\widetilde{R}\Gamma\psi(t) \tag{26}$$

where, $\zeta_1 = [e_1 - e_2]$, $\zeta_2 = [e_1 + e_2 - 2e_3]$, $\Gamma = [\zeta_1^T \;\; \zeta_2^T]^T$ and $\widetilde{R} = diag\{S, \; 3S\}$.

Moreover, introducing some free matrices M_l, $l = 1, 2$, (17) can be written as

$$2\sum_{l=1}^{n}\sum_{k=1}^{q}\psi^T(t)\left[\left(e_1^T M_1^T + e_4^T M_2^T\right)\left\{(A_l + B_l K_k)e_1 + B_l K_{\tau k}e_2 - e_4\right\}\right]\psi(t) = 0 \tag{27}$$

Finally, adding (24), (25), (26) and (27), yields

$$\dot{V}(y_t) \le \sum_{l=1}^{n} w_l^2 \psi^T(t)\left[\Omega_{ll} - \Gamma^T \widetilde{R} \Gamma\right]\psi(t) + \sum_{l=1}^{n}\sum_{l<k}^{q} w_l w_k \psi^T(t)\left[\Omega_{lk} + \Omega_{kl} - 2\Gamma^T \widetilde{R}\Gamma\right]\psi(t) \quad (28)$$

where $\Omega_{lk} = sym\left\{\Pi_{11}^T P \Pi_{22}\right\} + e_1^T W e_1 - e_2^T W e_2 + \tau^2 e_4^T S e_4$ $+ 2\left\{\left(e_1^T M_1^T + e_4^T M_2^T\right)\left((A_l + B_l K_k)e_1 + B_l K_{\tau k} e_2 - e_4\right)\right\}$.

Therefore, the system (17) is stable if the inequality

$$\dot{V}(y_t) \le \psi^T(t)\left[\sum_{l=1}^{n} w_l^2\left\{\Omega_{ll} - \Gamma^T \widetilde{R}\Gamma\right\} + \sum_{l=1}^{n}\sum_{l<k}^{q} w_l w_k \left\{\Omega_{lk} + \Omega_{kl} - 2\Gamma^T \widetilde{R}\Gamma\right\}\right]\psi(t) < 0 \quad (29)$$

is satisfied. That is,

$$\Omega_{ll} - \Gamma^T \widetilde{R}\Gamma < 0, \ \ for \ \ l = 1, 2, \ldots, n \quad (30)$$

and

$$\Omega_{lk} + \Omega_{kl} - 2\Gamma^T \widetilde{R}\Gamma < 0, \ \ for \ \ 1 \le l < k \le n \quad (31)$$

where Ω_{lk} is defined in above.

From (30) and (31), one can see that the element $\left\{\tau^2 e_4^T S e_4 - (M_2 + M_2^T)\right\}$ is present in block (4,4). The term $-(M_2 + M_2^T)$ should be negative definite because S is symmetric positive definitive and the scalar $\tau \neq 0$. Therefore, it signify that M_2 is non-singular. Let $M_2^{-1} = X$ and $M_1 = \rho M_2$, where ρ is a scalar parameter. Further, pre and post multiplication of both (30) and (31) performed $diag\{X, X, X, X\}$ and its transpose respectively and denote the variables $\widehat{P}_{11} = XP_{11}X^T$, $\widehat{P}_{12} = XP_{12}X^T$, $\widehat{P}_{22} = XP_{22}X^T$, $\widehat{Q} = XWX^T$, $\widehat{R} = XSX^T$, $N_k = K_k X^{-1}$ and $N_{\tau k} = K_{\tau k} X^{-1}$, we can easily obtain the LMIs in (19) and (20) respectively. This complete the proof.

5 Simulation Result

We suppose that the pendulum moving in upright plane and the force applied at the hinged point between the cart and pendulum are the control input. The pendulum should come back to its equilibrium point (steep position) after the force being applied within hypothecated time and in upright direction it should not move more than the hypothecated angle. After approximating, the highly nonlinear system (9) by T-S fuzzy modeling, we simulate the local linearized model (32) and (33) with known parameters are mass of the pendulum $m_p = 0.2\,\text{kg}$, mass of the cart $m_c = 2\,\text{kg}$, length of the pendulum $\ell = 0.5\,\text{m}$, moment of inertia $\widetilde{I} = 0.09\,\text{kg}\,\text{m}^2$ and acceleration due to gravity $g = 9.81\,\text{m/s}^2$.

Suppose that the pendulum is on the cart system and it will work within the region $[-\pi, \ \pi]$ and then extend the result to $y_1(t) \in [-\pi, \ \pi]$ except for a small slice close to $\pm\frac{\pi}{2}$. Since for $y_1 = \pm\frac{\pi}{2}$, the system is unstable. Therefore, for balancing the system in the region $\frac{\pi}{2}(\mid y_1 \mid < \pi)$, we linearized the system (9) by using the T-S model as:

$$\textbf{Rule-1: IF } y_1(t) \textbf{ is about `0' THEN } \dot{y}(t) = A_1 y(t) + B_1 \eta(t) \tag{32}$$

$$\textbf{Rule-2: IF } y_1(t) \textbf{ is about } \pm\frac{\pi}{2}(\mid y_1 \mid < \frac{\pi}{2}) \textbf{ THEN } \dot{y}(t) = A_2 y(t) + B_2 \eta(t) \tag{33}$$

where $A_1 = \begin{bmatrix} 0 & 1 \\ \dfrac{(m_c + m_p)g}{(\widetilde{I} + m_p\ell^2)(m_c + m_p)} & 0 \end{bmatrix}$, $B_1 = \begin{bmatrix} 0 \\ \dfrac{-m_p\ell}{(\widetilde{I} + m_p\ell^2)(m_c + m_p)} \end{bmatrix}$,

$$A_2 = \begin{bmatrix} 0 & 1 \\ \dfrac{2g(m_c + m_p)}{\pi\left\{\widetilde{I}(m_c + m_p\ell) + m_p\ell^2(m_c + m_p\beta^2)\right\}} & 0 \end{bmatrix},$$

$$B_2 = \begin{bmatrix} 0 \\ \dfrac{-m_p\ell\beta}{\widetilde{I}(m_c + m_p\ell) + m_p\ell^2(m_c + m_p\beta^2)} \end{bmatrix}, \ \beta = \cos(88^\circ), \ y(t) = [y_1^T \ y_2^T].$$

The membership function for this fuzzy model are given in Fig. 2a.

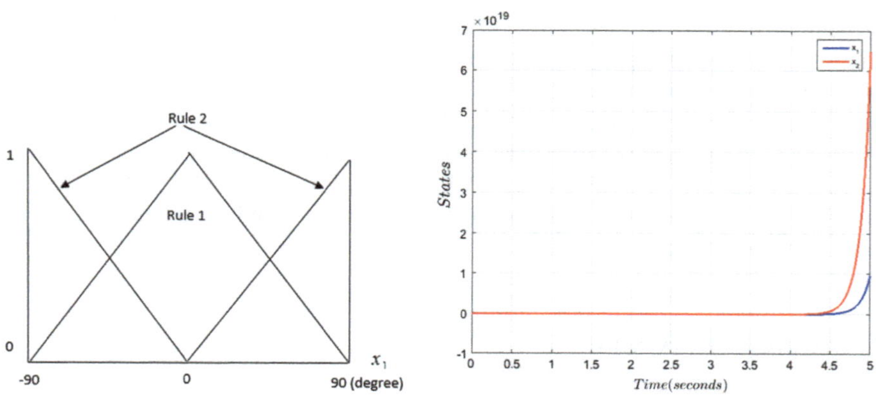

(a) Fuzzy Membership function for state y_1.

(b) Responses of the state without controller.

Fig. 2 Fuzzy membership function and state responses without controller

Taking $\rho = 1$ and the different values of τ, state feedback gain matrix are calculated using Theorem 1 and which are presented in Table 1. Simulation of the system in (32) and (33) is performed with parameters are given in above and choose $y(t) \in [-88°, 88°]$ as initial states. Figure 2b shows the state responses of inverted

Table 1 Calculated feedback gain matrices for different time delay

Time delay	Feedback gain matrices
$\tau = 4$	$K_1 = [-333.1743 \quad -19.3002]$, $K_2 = [-282.8861 \quad -16.3929]$ $K_{\tau 1} = [-0.0610 \quad -0.0044]$, $K_{\tau 2} = [-0.2089 \quad -0.03858]$
$\tau = 5$	$K_1 = [-330.5092 \quad -19.2049]$, $K_2 = [-281.8243 \quad -16.4909]$ $K_{\tau 1} = [-0.0368 \quad -0.0125]$, $K_{\tau 2} = [-0.1566 \quad -0.0267]$
$\tau = 7.55$	$K_1 = [-3291417 \quad -19.3894]$, $K_2 = [-279.9060 \quad -16.5314]$ $K_{\tau 1} = [-0.0886 \quad -0.0203]$, $K_{\tau 2} = [-0.1409 \quad -0.0372]$

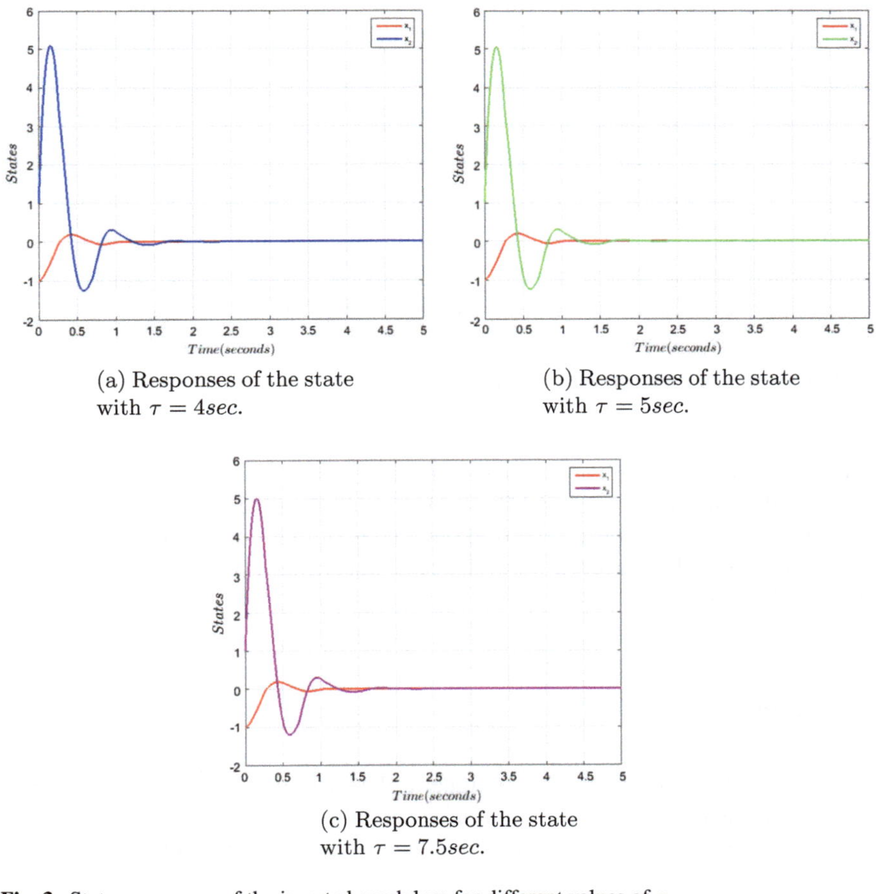

(a) Responses of the state with $\tau = 4sec.$

(b) Responses of the state with $\tau = 5sec.$

(c) Responses of the state with $\tau = 7.5sec.$

Fig. 3 States responses of the inverted pendulum for different values of τ

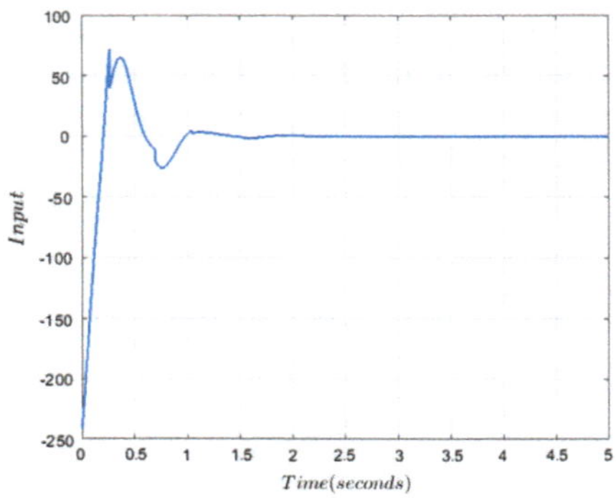

(a) Responses of the input for $\tau = 4sec$.

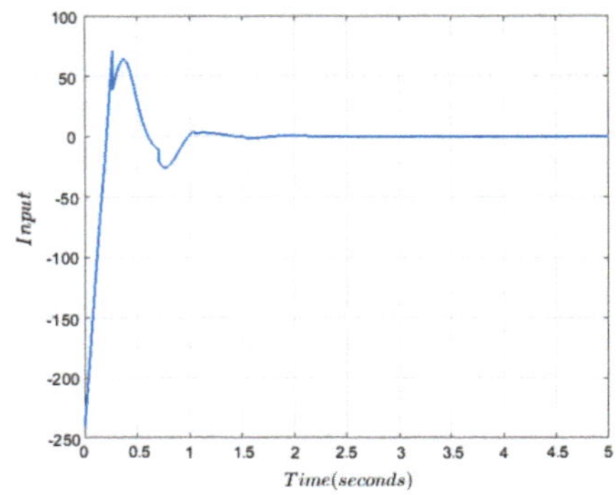

(b) Responses of the input for $\tau = 5sec$.

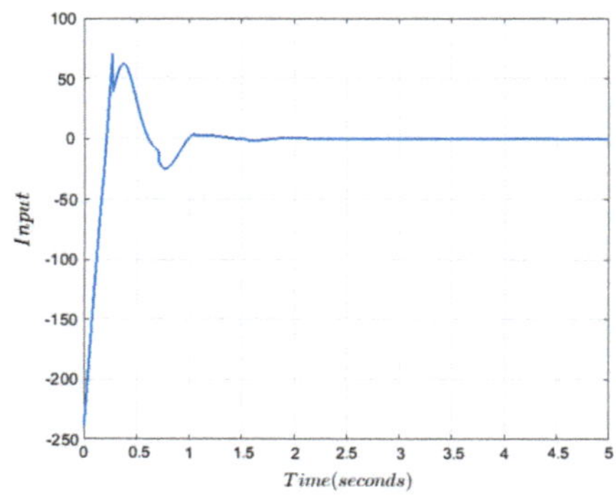

(c) Responses of the input for $\tau = 7.5sec$.

Fig. 4 Responses of the input with different values of τ

pendulum without controller, whereas Fig. 3a–c shows the state responses of the system with PDC controller for different values of the time delay τ. Figure 4a–c represents the input responses for different values of the time delay τ. Simulation results reveals that the proposed controller can balance the pendulum for initial states $y(t) \in [-88^\circ, 88^\circ]$.

6 Conclusion

By using the T-S fuzzy control technique, a delayed state feedback controller is designed in this paper for highly nonlinear pendulum system. First, mathematical model of inverted pendulum is derived and then linearizing this nonlinear model by

T-S fuzzy modeling technique. Finally, simulation result shows that the proposed delayed state feedback controller balance the inverted pendulum system along a vertical position at the equilibrium points.

References

1. Prasad, L.B., Tyagi, B., Gupta, H.O.: Optimal control of nonlinear inverted pendulum dynamical system with disturbance input using PID controller & LQR. In: IEEE International Conference on Control System, Computing and Engineering (ICCSCE), pp. 540–545 (2011)
2. Cannon, R.H.: Dynamics of Physical Systems. Courier Corporation (2003)
3. Ogata, K., Yang, Y.: Modern Control Engineering, vol. 4. Prentice Hall India (2002)
4. Faizan, F., Farid, F., Rehan, M., Mughal, S., Qadri, M.T.: Implementation of discrete PID on Inverted pendulum. In: 2nd International Conference on Education Technology and Computer (ICETC), vol. 1, pp. V1–48 (2010)
5. Liu, Y., Chen, Z., Xue, D., Xu, X.: Real-time controlling of inverted pendulum by fuzzy logic. In: IEEE International Conference on Automation and Logistics, ICAL'09, pp. 1180–1183 (2009)
6. Huang, C.H., Wang, W.J., Chiu, C.H.: Design and implementation of fuzzy control on a two-wheel inverted pendulum. IEEE Trans. Ind. Electron. **58**(7), 2988–3001 (2011)
7. Lan, Y., Fei, M.: Design of state-feedback controller by pole placement for a coupled set of inverted pendulums. In: 10th International Conference on Electronic Measurement & Instruments (ICEMI), vol. 3, pp. 69–73 (2011)
8. Huang, C.E., Li, D.H., Su, Y.: Simulation and robustness studies on an inverted pendulum. In: 30th Chinese Control Conference (CCC), pp. 615–619 (2011)
9. Prasad, L.B., Gupta, H.O., Tyagi, B.: Intelligent control of nonlinear inverted pendulum dynamical system with disturbance input using fuzzy logic systems. In: International Conference on Recent Advancements in Electrical, Electronics and Control Engineering (ICONRAEeCE), pp. 136–141 (2011)
10. Khalil, H.K.: Noninear Systems. Prentice-Hall, New Jersey (1999)
11. Tagaki, T., Sugeno, M.: Fuzzy identification of systems and its application to modelling and control. IEEE Trans. Syst. Man Cybern. **15**(1), 116–132 (1985)
12. Boyd, S., El Ghaoui, L., Feron, E., Balakrishnan, V.: Linear Matrix Inequalities in System and Control Theory, vol. 15. Siam (1994)
13. Gahinet, P., Nemirovskii, A., Laub, A.J., Chilali, M.: The LMI control toolbox. In: Proceedings of the 33rd IEEE Conference on Decision and Control, vol. 3, pp. 2038–2041 (1994)
14. Tanaka, K., Wang, H.O.: Fuzzy Control Systems Design and Analysis: A Linear Matrix Inequality Approach. Wiley, New York (2004)
15. Seuret, A., Gouaisbaut, F.: Wirtinger-based integral inequality: application to time-delay systems. Automatica **49**(9), 2860–2866 (2013)

Failure Reconfiguration of Pumps in Two Reservoirs Connected to Overhead Tank

Ravi Patel, Anil Gojiya and Dipankar Deb

Abstract This paper makes two main contributions. The first one is a modeling of groundwater forced into overhead tanks with single and two groundwater reservoirs. The second contribution is to formulate failure reconfiguration using DC gain technique for the stable linear MISO or MIMO system. This work deals with the failure reconfiguration of pumps in two natural water reservoirs connected to single overhead tank system. Pump failure happens when debris material available in water arrives at impeller of the pump and is stuck at a certain speed and leads to reduced output performance. Proposed failure reconfiguration scheme provides redundancy in output, and the effect of failure is compensated by the remaining pump. The simulation has been done in MATLAB/Simulink.

Keywords DC gain · Controller reconfiguration · Underground water reservoirs
Water distribution

1 Introduction

All physical systems are susceptible to the occurrence of faults and disturbances. Most of the systems and their operations require autonomy against faults and disturbances [1]. The controller needs to have an ability to recover faults and disturbances to maintain normal operation as stated by [2]. When a fault happens in the system, the control structure or law needs to respond accordingly to compensate for the possibility of the reduction in the output performance due to that fault. Controller reconfiguration techniques are used to accommodate system fault and disturbance

R. Patel (✉) · A. Gojiya · D. Deb
Institute of Infrastructure Technology Research and Management,
Ahmedabad 380026, Gujarat, India
e-mail: ravi.patel.16me@iitram.ac.in

A. Gojiya
e-mail: anil.gojiya.16me@iitram.ac.in

D. Deb
e-mail: dipankardeb@iitram.ac.in

D. Deb et al. (eds.), *Innovations in Infrastructure*, Advances in Intelligent Systems and Computing 757, https://doi.org/10.1007/978-981-13-1966-2_7

recovery in such a way that the system behaves as normal operating condition and the output is smoothly delivered.

Controller reconfiguration problem can be divided into two different levels: (1) Plant level reconfiguration, (2) Supervision level reconfiguration [3]. There are several approaches to controller reconfiguration. Adaptive methods are used to handle small faults in actuators and sensors by changing adaptive controller parameters and no changes are required in controller structure in such cases [4, 5]. Fault tolerance control schemes are developed to handle the faults [6]. Another approach is proposed to extend the control loop by a reconfiguration block [7]. Controller reconfiguration is done to replace the nominal controller with the other controller which is adjusted to the failure part to maintain desired output. Block diagram of controller reconfiguration is shown in Fig. 1.

A model-based controller reconfiguration technique is described using virtual actuator when the actuator fault happens [8, 9]. The virtual actuator is a linear dynamical system designed online during the system operation based on the plant model. Disturbance decoupling methods for controller reconfiguration is presented by [10]. In that case, adaptive control failure reconfiguration is used.

This paper proposes a novel failure reconfiguration technique for a stable linear system using DC gain of the system using two pumps, at most one of which may encounter a fault at any given time. Assume that both pumps of the system can individually navigate the system output to the desired steady-state value. If one of the pumps is stuck at a certain position, then the shortfall in output is compensated by the remaining pump. The system output remains same in steady state.

This paper is organized as follows. Section 2 presents the problem statement and control objective of the paper. Section 3 provides the modeling of the overhead water tank with single and two natural water reservoirs. Section 4 introduces a novel failure reconfiguration technique using DC gain for stable linear time-invariant systems. Section 5 presents the simulation results. The conclusion and future works are available in the last section.

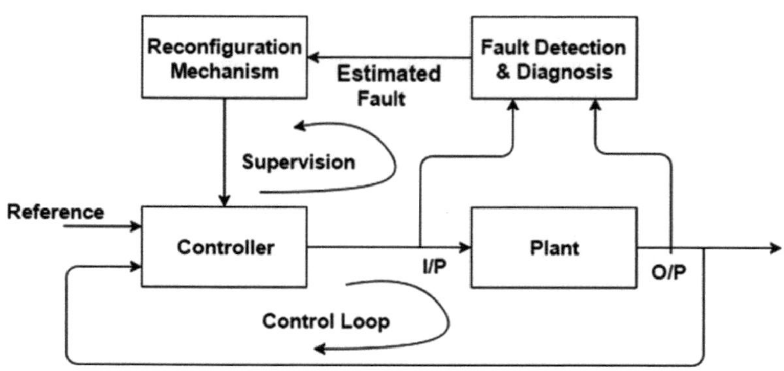

Fig. 1 Block diagram of controller reconfiguration

2 Problem Formulation

The aim of an effective water distribution network is to deliver water to a consumer with sufficient pressure, quality, and quantity. Good water distribution system has a continuous supply of water from the source to consumers. Natural reservoirs of water connected to a distribution system mainly depends on the level of the source, geographical area, and other local conditions. There are several methods of source water forced into the distribution system as explained in [11]: (i) Gravity System, (ii) Pumping System, and (iii) Combination of Gravity and Pumping System.

The gravity system is suitable when the natural water reservoir is at a sufficiently high altitude above the required water distribution area. The water flows in the main distribution line due to gravitational force and such a system is suitable when the source of supply has sufficient height. This is a reliable and economical process of water distribution system as detailed in [12]. In the pumping system, water is directly pumped from the natural water reservoir at ground level as shown in Fig. 2. It is also called pumping without a storage tank. Disadvantages of this system is that high lift pumps are required and if the pump fails, then the entire distribution network is hampered, and so it is not generally preferred [11].

2.1 Combination of Gravity and Pumping System

This system is more suitable for an efficient water distribution network and is the most commonly used system. The water of the natural reservoir is pumped and stored in overhead tank situated at sufficient height as shown in Fig. 3. Then, water is supplied to consumers by the concept of the gravity system. During low demand periods, water is stored in an overhead tank and is supplied during high demand period. It is an economical, efficient and reliable system. A detailed description of this arrangement is provided in [13].

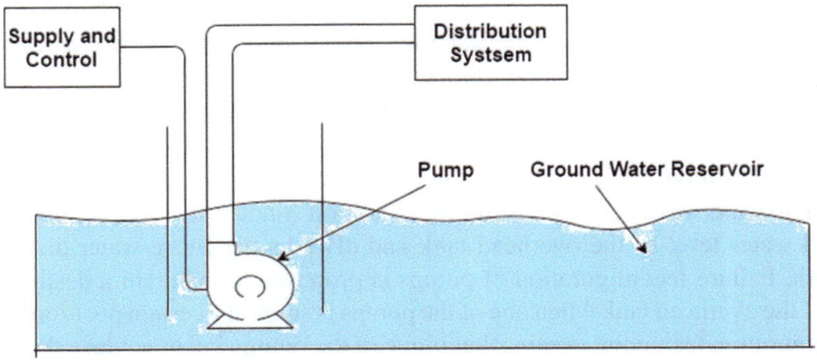

Fig. 2 Pumping system

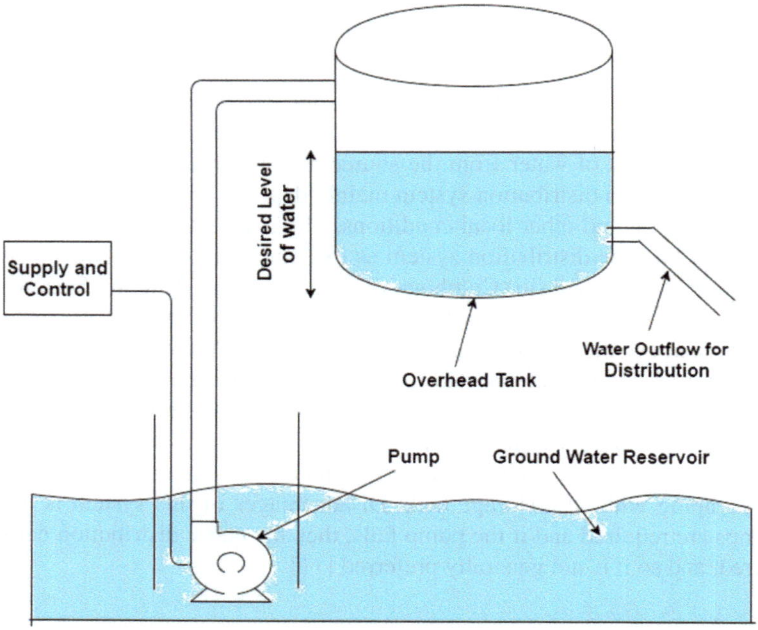

Fig. 3 Gravity and pumping system

2.2 *Objective and Problem Statement*

Water demand is more for large metropolitan areas. Therefore, multiple groundwater reservoirs are used to meet the water demand. For the sake of illustration of our proposed failure compensation control methodology, we consider that there are two natural groundwater reservoirs and single overhead tank for the water distribution system as shown in Fig. 4. Reservoirs have individual pumps and are operated simultaneously so as to get the water forced into the overhead tank.

Our objective is to maintain the water level of the overhead tank through simultaneous operation of both the pumps. If one of the pumps get stuck at a certain position due to any reason such as some debris material in the pump impeller, there would be an impact on the impeller speed and pump flow rate would be reduced due to the stuck condition. As a result, less amount of water would be pumped from that particular water reservoir. Pump failure condition hinders the achievement of the desired water level of the overhead tank and disturbs the entire water distribution network. Failure reconfiguration of pumps is proposed to maintain a desired water level of the overhead tank when one of the pumps is stuck at a certain position. Failure reconfiguration technique ensures that if one of the pumps fail to achieve the desired operation, then the other pump tries to compensate that failure to some extent, so as to achieve the same desired output for the entire system.

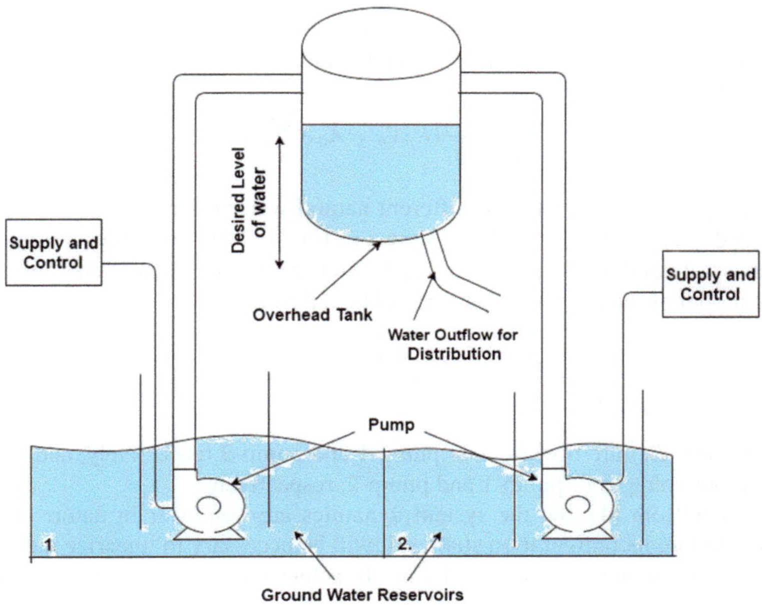

Fig. 4 Overhead tank with two reservoir water supply

3 Modeling of Overhead Water Tank With Natural Reservoirs

Consider a system that has a single natural water reservoir with a pump and overhead tank. The input is supplied by the pump which lifts water to the overhead tank. Stored water in an overhead tank is passed through the orifice to the distribution network. The mathematical model of the system is determined by considering mass balance principle and related volumetric flows in the tank as described in [14]. A dynamical equation of the rate of change of water volume in an overhead tank is

$$\frac{\mathrm{d}V}{\mathrm{d}t} = A_t \frac{\mathrm{d}L}{\mathrm{d}t} = F_{\text{in}} - F_{\text{out}}, \tag{1}$$

where V is the volume of the overhead tank, A_t and L are the area and level of the water in the overhead tank, respectively, F_{in} and F_{out} are the groundwater inflow rate from the pump and outflow of the water from overhead tank, respectively, and can be defined as

$$F_{\text{in}} = K_p V_p \ , \ \ F_{\text{out}} = A_0 V_0, \tag{2}$$

where K_p, V_p, A_0 and V_0 are pump flow rate constant in cm^3/s, pump input voltage, area of orifice in cm^2, and water outflow velocity in cm/s^2.

Using Bernoulli's equation, water outflow velocity can be expressed as $V_0 = \sqrt{2gL}$, where g is the gravitational constant in cm/s^2. Substituting outflow velocity, (2) in the (1), we get

$$\frac{\mathrm{d}L}{\mathrm{d}t} = \frac{1}{A_t}(K_p V_p - A_0\sqrt{2gL}). \tag{3}$$

The proposed system has two different natural water reservoirs with individual pumps and single overhead tanks as shown in Fig. 4. The two pumps are operated simultaneously and try to achieve desired water level of overhead tanks. A mathematical model of the proposed system is obtained from (3)

$$\frac{\mathrm{d}L}{\mathrm{d}t} = \frac{1}{A_t}\left(\frac{K_{p1}V_{p1} + K_{p2}V_{p2}}{2} - A_0\sqrt{2gL}\right), \tag{4}$$

where K_{p1} and K_{p2} are flow rates of pump-1 and pump-2 respectively, and V_{p1} and V_{p2} are input voltages of pump-1 and pump-2, respectively.

It is clear from (4) that the system dynamics are nonlinear in nature. For the proposed failure reconfiguration studies, it will be necessary to linearize the system at the equilibrium point L_0, V_{p10} and V_{p20}. To achieve the desired operating level of overhead tank, (L_0), $\frac{\mathrm{d}_L}{\mathrm{d}_t}$ in (4) is equated to zero, and we get

$$K_{p1}V_{p10} + K_{p2}V_{p20} = 2A_0\sqrt{2gL_0}. \tag{5}$$

Linearization of (4) is done using (5) and Taylor's series approximation is as follows:

$$\frac{\mathrm{d}L}{\mathrm{d}t} = \frac{1}{2A_t}(u_1 + u_2) - \frac{A_0\sqrt{2g}L}{2A_t\sqrt{L_0}}, \tag{6}$$

where $K_{p1}V_{p2} = u_1$ and $K_{p2}V_{p2} = u_2$ are considered as the two inputs.

Transfer functions of the system are derived from (6) is as follows:

$$\frac{L(s)}{u_1(s)} = \frac{L(s)}{u_2(s)} = \frac{K_{\mathrm{dc}}}{1 + s\tau}, \tag{7}$$

where DC gain, K_{dc}, of the transfer functions and the time constant τ are defined as

$$K_{\mathrm{dc}} = \frac{\sqrt{L_0}}{\sqrt{2}A_0\sqrt{g}}, \quad \tau = \frac{\sqrt{2}A_t\sqrt{L_0}}{A_0\sqrt{g}}. \tag{8}$$

In the next section, we present a failure reconfiguration method to recover the desired output in the presence of a fault in one of the inputs, by appropriately providing a suitable control allocation from the other input which is in normal condition.

4 Failure Reconfiguration

Failure reconfiguration technique is used to maintain the system output at the desired level even under failure condition of inputs to the system. A failure reconfiguration scheme is proposed for the stable linear time-invariant system using DC gain approach. When a stable control system is excited with constant input, the response obtained is a certain level at steady state.

DC Gain is defined as the ratio of the steady-state output of a stable system to steady-state input of the system. Steady-state output of any system using final value theorem is

$$Y(\infty) = \lim_{s\to 0} sG(S)U, \tag{9}$$

where $Y(\infty)$, $G(S)$ and U are steady-state output, open loop transfer function and input of the system respectively as described by [15]. From the final value theorem, DC gain (g) is defined as

$$g = \frac{Y(\infty)}{sU} = \lim_{s\to 0} G(S). \tag{10}$$

Let us consider a linear time-invariant system in state space form whose dynamics are given by

$$\begin{aligned} \dot{x} &= Ax + B_1u_1 + B_2u_2, \\ y &= Cx + D_1u_1 + D_2u_2. \end{aligned} \tag{11}$$

where $x \in \mathscr{R}^n$, $u_1, u_2 \in \mathscr{R}$. The input u_1 is generated using actuator-1 and u_2 is generated by another actuator-2.

Assume that the system is stable and controllable. If the system is not stable, a preliminary feedback mechanism can be applied to stabilize the system and then the reconfiguration controller using DC gain can be applied to the stabilized system. Let us assume that the desired steady-state value is in the linear active region of both actuators, i.e., both the actuator can individually navigate the system output to the desired steady-state value. However, for the long life of the proposed system, it is expected to use two pumps, instead of only one so that a single pump is not always operating at an undesirably high speed to provide the net output. That is, we have actuator redundancy. Steady-state output of the system is given by

$$\begin{aligned} y_{ss} &= -(CA^{-1}B_1 + D_1)u_{1ss} \pm \left(-(CA^{-1}B_2 + D_2)u_{2ss}\right), \\ &= g_1u_{1ss} \pm g_2u_{2ss} \end{aligned} \tag{12}$$

where $g_1 = -CA^{-1}B_1 + D_1$, $g_1 = -CA^{-1}B_2 + D_2$ are the DC gains and $\pm$ sign depends on the specific system. For the proposed system, this particular sign is positive. The system (11) is compared to the proposed dynamics presented in (6), and we get

$$A = \frac{-A_0\sqrt{2g}}{2A_t\sqrt{L_0}} \ , \ B_1, B_2 = \frac{1}{2A_t} \ , \ C = 1 \ , \ D_1, D_2 = 0. \tag{13}$$

The steady-state actuator outputs u_{1ss} and u_{2ss} are determined based on the user priority. In some cases, only one pump is used under normal condition or the redundant actuator may deliver a fixed input based on other system requirements such as system safety. In practice, any of these actuators may get stuck at a particular position and then the output needs to be controlled with the remaining active actuator. The objective is to compute the output of the redundant pump such that the system output remains same as before.

Theorem 1 *Consider the system dynamics given by (11) with the desired steady-state position given by (12). Let us assume that the actuator-1 is at an arbitrary position, $\bar{u}_1$, then the steady-state output of the actuator-2 is given by*

$$u_{2r} = \frac{1}{g_2}(g_1 u_{1ss} + g_2 u_{2ss} - g_1 \bar{u}_1). \tag{14}$$

Proof Without loss of generality, let us assume that actuator-1 is stuck at position $\bar{u}_1$ and the actual position of the valve is available for measurement. The steady-state output of the system due to the stuck valve is given by

$$y_{ss}^f = g_1 \bar{u}_1 + g_2 u_{2ss}. \tag{15}$$

The control task is to compute the output of the redundant actuator, u_{2r} such that the steady-state output of the system after recovery, referred by y_{ss}^r, remains same as the normal condition y_{ss}, that is $y_{ss}^r = y_{ss}$, such that

$$y_{ss}^r = g_1 \bar{u}_1 + g_2 u_{2r}. \tag{16}$$

Substituting (12) in (16) and since $y_{ss}^r = y_{ss}$, we get

$$g_1 u_{1ss} + g_2 u_{2ss} = g_1 \bar{u}_1 + g_2 u_{2r}. \tag{17}$$

Rearranging (17), we get

$$u_{2r} = \frac{1}{g_2}(g_1 u_{1ss} + g_2 u_{2ss} - g_1 \bar{u}_1). \tag{18}$$

5 Simulation Results

Failure reconfiguration scheme is applied to the combined water distribution system of two natural water reservoirs in the overhead distribution tank system. Both pumps are operated simultaneously to achieve desired level of the overhead tank for efficient water distribution system. Modeling of the system has been done in Sect. 3. All the parameters and constants value are given in Table 1.

DC gain is obtained from the transfer functions expressed in (7). We have considered nonidentical pump case for the simulation.

In this case, assume that the pumps are nonidentical, and so the pump flow rate constants and input voltages are different. Linearization of the model is about an operating point with $L_0 = 15\,\text{cm}$ and the pump voltage $V_{p1} = 4.628$ V and $V_{p1} = 6.066$ V. Substituting the values of Table 1 in (8), the transfer functions of the system are as defined in (7) where

$$K_{p1} \neq K_{p2}\ ,\ \ V_{p1} \neq V_{p2}.$$

Choose a operating input voltage of pump-1 (V_{p1}) to be 4.62 V. From (5), the operating input voltage of pump-2 (V_{p2}) is calculated to be 6.066 V. Consider that pump-1 is stuck at 20 s. Therefore, the pump flow rate constant is reduced from 3.3 to $2\,\text{cm}^3/s$ Pump-1 input (u_1) when it is in normal condition (blue) and in stuck condition (red) at 25 s, are shown in Fig. 5. The steady-state output of the system in normal operation (y_{ss}) and with stuck condition (y_{ss}^f) are shown in Fig. 6. Due to stuck condition of pump-1, the level of overhead tank decreases to around 10.5 cm which is not desirable [11].

The redundant input of the pump-2 (u_{2r}) and the steady-state output of the system after recovery (y_{ss}^r) are shown in Figs. 7 and 8 respectively. The input voltage of pump-2 increases from 6.06 to 10 V at 20 s and is able to maintain the desired level of overhead tank at 15 cm.

Table 1 Parameters and constants

Name	Notation	Value	Unit
Pump-1 flow rate constant	K_{p1}	3.3	cm^3/s
Pump flow rate constant (identical)	K_p	3.3	cm^3/s
Pump-2 flow rate constant	K_{p2}	2.5	cm^3/s
Area of orifice	A_0	0.1781	cm^2
Gravitational constant	g	981	cm/s^2
Area of overhead tank	A_t	15.51	cm^2
Desired level of water	L_0	15	cm
Dc gain of pump	K_{dc}	0.4909	–

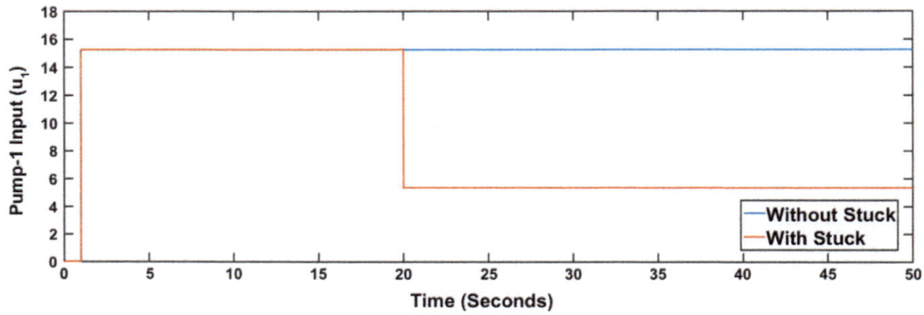

Fig. 5 Pump-1 input at different conditions

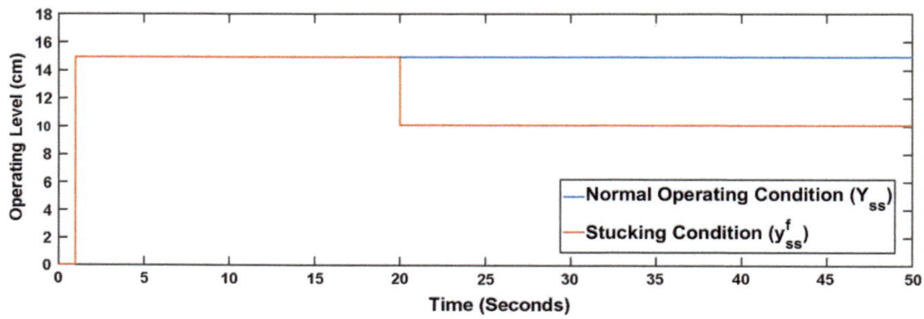

Fig. 6 Operating level (pre and post failure)

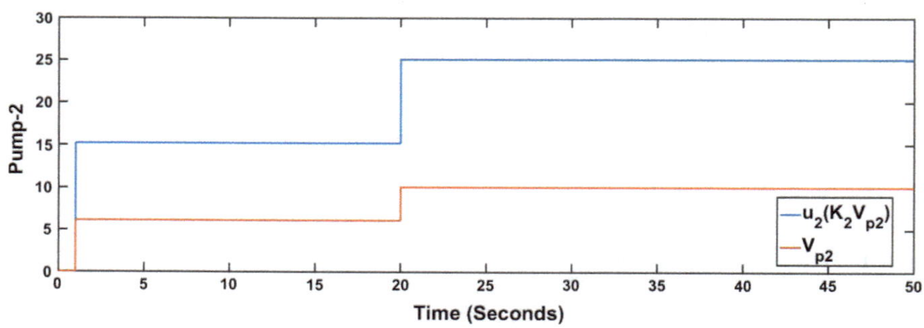

Fig. 7 Pump-2 input after reconfiguration

6 Conclusion

The effective water distribution network is very important for any city or village. The water of natural reservoirs is pumped into the overhead tank for an effective water distribution system. In this work, we have derived a mathematical model of two tank reservoirs connected with an overhead tank. If one of the pumps in a reservoir is stuck at a certain position, the entire distribution network is disturbed.

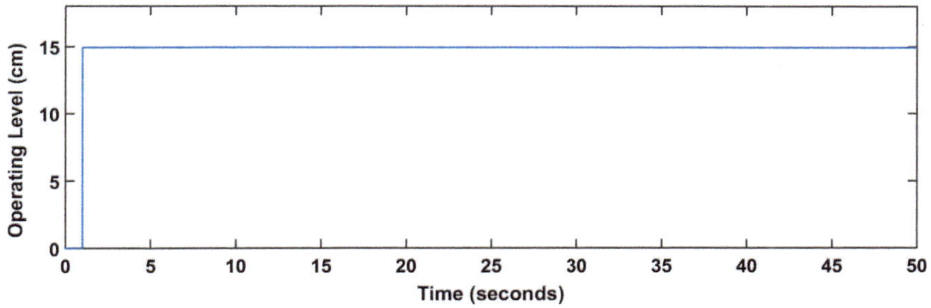

Fig. 8 Operating level after reconfiguration

Therefore, a novel failure reconfiguration technique for linear systems is proposed using DC gain approach, and this reconfiguration method is applied to the water distribution system. Under the stuck condition of any pump, reduction in output performance is compensated through the remaining pump by suitably changing the input voltage. Performance of the proposed controller reconfiguration technique is validated through MATLAB/Simulink simulation. This work can be extended for failure reconfiguration of n number of pumps present in the n number of reservoirs (1 each per reservoir), wherein $m < n$ number of pumps fail. The remaining $n - m$ pumps are reconfigured to achieve the same system output as the case when all the n pumps function normally.

References

1. Alcaraz-Mejia, M., Lopez-Mellado, E., Ramirez-Trevino, A.: Redundancy based controller reconfiguration for fault recovery of manufacturing systems. In: 2007 IEEE International Conference on Automation Science and Engineering (2007)
2. Burns, A., Wellings, A.: Real-Time Systems and Programming Languages. Addison-Wesley, Harlow (2009)
3. Lunze, J., Rowe-Serrano, D., Steffen, T.: Control reconfiguration demonstrated at a two-degrees-of-freedom helicopter model. Eur. Control Conf. **2003**, 2254–60 (2003)
4. Ioannou, P., Sun, J.: Robust Adaptive Control. Dover Publications, New York (2013)
5. Gao, Z., Antsaklis, P.: Stability of the pseudo-inverse method for reconfigurable control systems. Int. J. Control **53**, 717–729 (1991)
6. Blanke, M.: Diagnosis and fault-tolerant control. Berlin [u.a.]: Springer (2010)
7. Lunze, J.: Control Reconfiguration. Institute of Automation and Process Control (2002)
8. Richter, J., Lunze, J.: Reconfigurable control of Hammerstein systems after actuator faults. IFAC Proc. **41**(2), 3210–3215 (2008)
9. Richter, J., Lunze, J., Schlage, T.: Control reconfiguration of a thermofluid process by means of a virtual actuator. IET Control Theory Appl. **1**(6), 1606–1620 (2007)
10. Lunze, J., Steffen, T.: Control reconfiguration after actuator failures using disturbance decoupling methods. IEEE Trans. Autom. Control. **51**(10), 1590–1601 (2006)
11. Mala-Jetmarova, H., Barton, A., Bagirov, A.: A history of water distribution systems and their optimization. Water Sci. & Technol. Water Supply. **15**(2), 224–235 (2015)

12. Swamee, P., Sharma, A.: Gravity flow water distribution system design. J. Water Supply: Res. Technol. **49**(4), 169–179 (2000)
13. Tiwari, K., Jatale, A., Khandelwal, S.: Assessment of public water distribution system of Indore City. India. Int. J. Eng. Innov. Technol. (IJEIT) **2**(8), 169–179 (2013)
14. Shingare, P., Joshi, M.: Model identification of coupled two tanks. IFAC Proceedings (2007)
15. Patranabis, D.: Principle of Process Control. Tata McGraw-Hill Education 45 (1981)

Simplified Takagi-Sugeno Fuzzy Regulator Design for Stabilizing Control of Electromagnetic Levitation System

Ravi V. Gandhi and Dipak M. Adhyaru

Abstract This paper presents a simplified yet an effective design of the Takagi-Sugeno (T-S) fuzzy regulator based on pole-placement approach. In this research, the designed regulator is implemented for a class of nonlinear and unstable Electromagnetic Levitation System (EMLS). The primary objective of the designed regulator is to maintain the position of the levitated object for a set of operating points without changing the regulator gains. Investigation reveals that the proposed regulator offers robust performance under the large variation of the operating conditions. The simulation results for the EMLS are presented to validate the performance effectiveness of the considered regulator design compared to the regulator designed by well-known LMI approach under the diversified operating conditions with or without enormous variations of the vertical payload disturbance.

Keywords Takagi-Sugeno fuzzy regulator · Electromagnetic levitation system
Pole-placement · Stability analysis

1 Introduction

Fuzzy regulators have gained more attention for the linear as well as nonlinear control applications based on the Sugeno-type Fuzzy Logic Controller (FLC) [1]. Most of the real-time control applications have nonlinear behavior which can be approximated by the fuzzy blending of local subsystems so-called Takagi-Sugeno fuzzy model as discussed in [2]. Once T-S fuzzy model is obtained, the desired regulator for the overall closed-loop system can be obtained using the Parallel Distributed Compensation (PDC) approach [3]. Analysis of the stability for the PDC-based T-S fuzzy regula-

R. V. Gandhi (✉) · D. M. Adhyaru
Institute of Technology, Nirma University, S.G Highway,
Ahmedabad 382481, Gujarat, India
e-mail: ravi.gandhi09@gmail.com

D. M. Adhyaru
e-mail: dipak.adhyaru@nirmauni.ac.in

D. Deb et al. (eds.), *Innovations in Infrastructure*, Advances in Intelligent Systems and Computing 757, https://doi.org/10.1007/978-981-13-1966-2_8

tors and fuzzy observers for the conventional, optimal, or robust control problems can be solved based on the well-known LMI approach with the Quadratic Lyapunov Function (QLF) [4–6]. Many researchers have introduced different methodologies to confirm the stability of the T-S model-based control system without solving the set of LMIs, i.e., polynomial approach [7, 8], super stability approach [9], and eigenvalue approach with guaranteed stability of the T-S regulators [10, 11].

Frictionless tremendous features and variety of real-time applications like magnetic suspension, magnetic bearing, MAGLEV trains, rocket launcher...etc. [12], based on Magnetic Levitated (MAGLEV) technology attracts researchers to perform more and more research in this area. Dynamics and control aspects of such a nonlinear and unstable system are discussed in [13]. Various modeling aspects for the linear and nonlinear EMLS to represent the behavior of the different approximations of electromagnet parameters have been detailed in [14–16]. Investigation of various literatures shows great revolution of the FLC-based intelligent control in the area of EMLS since last two decades.

FLCs may be well suited for the stabilization of EMLS compared to the conventional PID controllers or the lead-lag compensators as discussed in [17, 18]. Many researchers proposed PID like fuzzy controller design to combine the advantages of conventional and intelligent controllers [18–20]. Neuro-fuzzy controllers [21] and advanced type-2 fuzzy controllers [22] have been designed for the precise control of EMLS for the fix operating conditions. As discussed in [13], the track irregularities or payload variation may disturb the operating conditions of the MAGLEV system. To solve this issue, T-S fuzzy regulator using the LMI approach for a set of operating conditions was designed with decay rate (β) of 1 and 0.8 [23].

In this research, the T-S fuzzy regulator has been designed using the simplified pole-placement approach for a set of the operating points. Key features of the proposed approach are as follows: (*i*) Regulator gains for each rules (i.e., $i=1, 2, 3, 4$) can be simply obtained based on pole-placement approach which reduces the reliance on LMIs, (*ii*) No need to recalculate the regulator gains again and again for the different cases as discussed, (*iii*) Offers performance robustness under variation of the payload disturbance of 0–40%, and (*iv*) System stability in the large operating region is guaranteed.

2 Electromagnetic Levitation System (EMLS)

EMLS is class of a nonlinear and unstable system by nature [15, 16]. This nature demands the effective regulator design to maintain the air gap (y) between the electromagnet and the steel ball around the desired operating position (y_0). Under the linear relationship between the coil voltage and coil current, the current controlled structure of EMLS is shown in Fig. 1 [23]. The parameters used for the simulation of EMLS are mentioned in Table 1.

As shown in the above structure, the ball position ($y(t)$) is controlled by varying the coil current ($i(t)$) by the proposed regulator. Let F_d be the major vertical disturbance

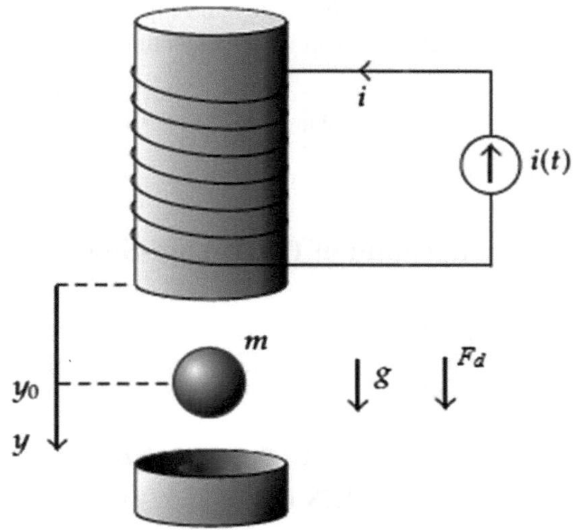

Fig. 1 Electromagnetic levitation system [23]

Table 1 Simulation parameters of EMLS [23]

Parameters	Value	Units
m	0.05	kg
g	9.8	m/s^2
k_s	0.001	Ns/m
λ	0.46	H
μ	2	m^{-1}

due to variation of the payload whose effect needs to be minimized by the proposed regulator based on feedback signals, i.e., air gap position, $y(t)$, and air gap velocity, $\dot{y}(t)$.

Using the first principle of physical system modeling, force balance equation on the spherical ball with the vertical disturbance (F_d) can be expressed by (1) [19, 23]

$$\frac{d^2 y}{dt^2} = g - C_1 \frac{dy}{dt} - C_0 \left(\frac{i}{1 + \mu\, y} \right)^2 + \frac{F_d}{m} \tag{1}$$

Here, $x \in R^n$ $u \in R^m$ and $Y \in R^q$, are the state vector, input and output vectors, respectively, for the considered EMLS. Let define the variables for the state-space modeling as follows $x_1 = y, x_2 = \dot{y}, u = i, Y = x_1$. The nonlinear state-space model using the defined variables for the EMLS can be represented by (2)

$$\dot{x} = \begin{bmatrix} x_2 \\ g - C_0 \left(\frac{u}{1+\mu\, x_1} \right)^2 - C_1 x_2 + \frac{F_d}{m} \end{bmatrix} \tag{2}$$

Operating points for states and input are (x_{10}, x_{20}) and u_0, respectively, for the EMLS model (2). It can be obtained under steady-state conditions as follows [23]:

$$x_{10} = y_0, x_{20} = 0, u_0 = (1 + \mu\, y_0)\sqrt{\frac{g}{C_0}} \tag{3}$$

3 Background of the Takagi-Sugeno (T-S) Fuzzy Regulator

T-S fuzzy regulator design begins with the development of the T-S fuzzy model for the given nonlinear system [2]. The *i*th rule of T-S fuzzy system can be written as follows [2, 5]:

$$\begin{aligned} &\textbf{IF } Z_1 \text{ is } M_{1\mathrm{i}} \textbf{ AND} \ldots \textbf{AND } Z_\mathrm{p} \text{ is } \mathrm{M}_{\mathrm{pi},} \\ &\textbf{THEN } \dot{x}(t) = A_i x(t) + B_i u(t), \quad i = 1, 2, \ldots, r. \end{aligned} \tag{4}$$

Here, M_{ji} is fuzzy set; Z_1, ..., Z_p are the nonlinear terms well known as premise variables; $A_i \in R^{nxn}$, $B_i \in R^{nxm}$ and $C_i \in R^{qxn}$ are, respectively, the state, input, and output matrices of *i*th linear subsystem with No. of rules (r). The overall state model for the given system (4) can be expressed as follows [2]:

$$\dot{x}(t) = \sum_{i=1}^{r} h_i(Z)[A_i x(t) + B_i u(t)], \quad i = 1, 2, \ldots, r \tag{5}$$

where

$$h_i(Z) = \frac{w_i(Z)}{\sum_{i=1}^{r} w_i(Z)}, \sum_{i=1}^{r} h_i(Z) = 1, h_i(Z) \geq 0$$

$$w_i(Z) = \prod_{j=1}^{r} M_{ji}(Z_j), \sum_{i=1}^{r} w_i(Z) > 0, \quad w_i(Z) \geq 0$$

Design the PDC-based state feedback regulator for the *i*th rule of T-S fuzzy regulator can be formed as follows [3]:

$$\begin{aligned} &\textbf{IF } Z_1 \text{ is } M_{1\mathrm{i}} \textbf{ AND} \ldots \textbf{AND } Z_\mathrm{p} \text{ is } M_{\mathrm{pi},} \\ &\textbf{THEN } u(t) = -K_i x(t), \quad i = 1, 2, \ldots, r. \end{aligned} \tag{6}$$

The overall T-S fuzzy regulator based on PDC approach can be expressed by (7) [3]

$$u(t) = -\sum_{i=1}^{r} h_i(Z) K_i x(t), \quad i = 1, 2, \ldots, r \tag{7}$$

Combining and simplifying (5) and (7), the overall T-S fuzzy model-based regulator system can be expressed as follows [2, 3]:

$$\dot{x}(t) = \sum_{i=1}^{r} \sum_{j=1}^{r} h_i(Z) h_j(Z) \left[A_i - B_i K_j \right] x(t) \tag{8}$$

The above dynamics of (8) can be expanded as follows with $G_{ij} = A_i - B_i K_j$ [5],

$$\begin{aligned} \dot{x}(t) &= \sum_{i=1}^{r} h_i(Z) h_i(Z) G_{ii} x(t) \\ &\quad + 2 \sum_{i=1}^{r} \sum_{i<j} h_i(Z) h_j(Z) \left[\frac{G_{ij} + G_{ji}}{2} \right] x(t) \end{aligned} \tag{9}$$

Lemma 1 *[5] The equilibrium of the continuous T-S fuzzy system described by (9) is globally asymptotically stable if there exists a common positive definite matrix P such that*

$$\begin{aligned} &G_{ii}^T P + P G_{ii} < 0, \\ &\left[\tfrac{G_{ij}+G_{ji}}{2} \right]^T P + P \left[\tfrac{G_{ij}+G_{ji}}{2} \right] \leq 0, \quad i < j, s.t. h_i \sqcap h_j \neq \phi \end{aligned} \tag{10}$$

4 Simplified Takagi-Sugeno Fuzzy Regulator Design

From (10), it is clear that the operating point for the EMLS is not at the origin such that $f(0, 0) \neq 0$. Defining new variables based on the following state transformation [23 are:

$$\bar{x}_1 = x_1 - x_{10}, \bar{x}_2 = x_2 - x_{20}, \bar{u} = u^2 - u_0^2 \tag{11}$$

Simplifying (1)–(3) would result in the new transformed nonlinear state model as (12) with Membership Functions (MFs) for Z_1 and Z_2 are shown in Fig. 2 [23]

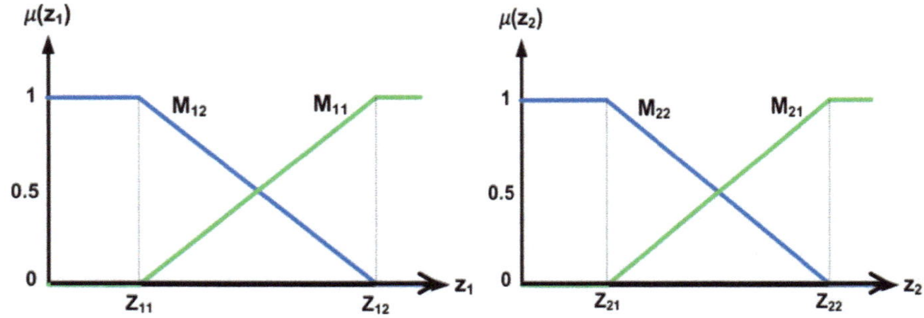

Fig. 2 Membership Function (MF) for premise variables: $\mathbf{Z}_1 \in [\mathbf{Z}_{11}, \mathbf{Z}_{12}]$ and $\mathbf{Z}_2 \in [\mathbf{Z}_{21}, \mathbf{Z}_{22}]$

$$\dot{\bar{x}} = \begin{bmatrix} 0 & 1 \\ Z_1(\bar{x}_1, x_{10}) & -C_1 \end{bmatrix} \bar{x} + \begin{bmatrix} 0 \\ Z_2(\bar{x}_1, x_{10}) \end{bmatrix} \bar{u} \tag{12}$$

Various state and input matrices for different subsystems using the model rules based on (4) are described below

$$A_1 = A_2 = \begin{bmatrix} 0 & 1 \\ Z_{11} & -C_1 \end{bmatrix}, B_1 = B_3 = \begin{bmatrix} 0 \\ Z_{21} \end{bmatrix},$$

$$A_3 = A_4 = \begin{bmatrix} 0 & 1 \\ Z_{12} & -C_1 \end{bmatrix}, B_2 = B_4 = \begin{bmatrix} 0 \\ Z_{22} \end{bmatrix}$$

The set of control rules for the T-S fuzzy regulator design based on (7) are as follows:

IF Z_1 is M_{11} **AND** Z_2 is M_{21} **THEN** $u = -K_1 x$
IF Z_1 is M_{11} **AND** Z_2 is M_{22} **THEN** $u = -K_2 x$
IF Z_1 is M_{12} **AND** Z_2 is M_{21} **THEN** $u = -K_3 x$
IF Z_1 is M_{12} **AND** Z_2 is M_{22} **THEN** $u = -K_4 x$

By expanding the (8) with No. of control rules as $r = 2$, the overall closed-loop dynamics can be expressed by (13) in the simple expanded form:

$$\begin{aligned} \dot{x}(t) = \{ & h_1^2(A_1 - B_1 K_1) + h_2^2(A_2 - B_2 K_2) + h_1 h_2 (A_1 - B_1 K_2) \\ & + h_1 h_2 (A_2 - B_2 K_1)\} x(t) \end{aligned} \tag{13}$$

Remark 1 It can be observed from (13) that if the regulator gains (i.e., K_1, K_2, K_3, K_4) are chosen such that all the subsystems (i.e., $G_{ii} = A_i - B_i K_i$ and

$G_{ij} = A_i - B_i K_j$) gets stable, then the overall closed-loop system presented by (8) can be asymptotically stable in large based on Lemma 1.

Remark 2 Using above Remark 1, the T-S regulator gain can be easily obtained using the pole-placement method. In case of LMIs, No. of LMIs would rapidly increase or change with an increase in r as well as change in the ranges of premise variables.

Remark 3 Using the concept of Lemma 1, if the eigenvalues for all the closed-loop subsystems (i.e., G_{ii}) are placed at a same location with sufficient left half of s-plane such that all the coupled subsystems (i.e., G_{ij}) gets stabilized, then the overall closed-loop system presented by (9) can be stabilized in large.

The recommended location of the common eigenvalue $(\sigma_1\sigma_2 \ldots \sigma_n)$ for all the closed-loop subsystems to obtain pole-placement-based regulator gains is mentioned below

$$\lambda_C = \alpha * [\sigma_1\sigma_2 \ldots \sigma_n] \tag{14}$$

where

λ_C Recommended location of eigenvalues of all the subsystems,
α Eigen multiplier to update the λ_C as per the requirement, $\alpha > 0$.

5 Simulation Results

The nominated T-S fuzzy regulator is employed to stabilize the nonlinear and unstable EMLS with initial state condition of $[0.04\ 1]^{\mathrm{T}}$ for two cases under the set of operating points (y_0), (*i*) without payload (F_{d}), (*ii*) with payload (F_{d}) variation of 0–40%. The obtained results for both the cases are compared with the results obtained by the regulator designed by LMI approach with decay rate (β) of 1 and 0.8, respectively. The initial step is to place the eigenvalues of all the subsystems with sufficient left half of s-plane. After a long iteration, selecting the values of $\alpha = 5$, $\sigma_1 = -5$, and $\sigma_2 = -10$ would result into the desired regulator gains (K_i). The closed-loop eigenvalues based on K_i, for all the subsystems are shown in Table 2.

In case (*i*) without payload, the set of operating points are considered as, y_0: {0.04, 0.1, 0.05, and 0.08} with the equal interval of 1 s. The proposed regulator smoothly tracks the change of operating conditions (y_0) without disturbing the transient behavior compare to former regulators as shown in Fig. 3. It is observed that for the similar conditions, LMI-based regulators with $\beta = 1$ and 0.8 having the high overshoot and low-speed problem during the first change of operating point. However, after the first change in the y_0, all the regulators work almost similar. Even the response of air gap speed (x_2) for the proposed regulator is quite good compared to the LMI-based regulators.

In case (*ii*) with payload variation in range of 0–40%, the set of operating points are considered as, y_0: {0.04, 0.1, 0.05, and 0.08} with the equal interval of 0.5 s.

Table 2 Eigenvalues of closed-loop subsystems with $r = 4$ based on the proposed approach

Subsystems (G_{ii})	Eigenvalues (λ_{ii})	Subsystems $Q_{ij} = \left(\frac{G_{ij}+G_{ji}}{2}\right), i < j$	Eigenvalues (λ_{ij})
G_{11}	−25, −50	Q_{12}	−22.83, −62.74
G_{22}	−25, −50	Q_{13}	−25.00, −50.00
G_{33}	−25, −50	Q_{14}	−22.91, −62.66
G_{44}	−25, −50	Q_{23}	−22.71, −62.86
–	–	Q_{24}	−25.00, −50.00
–	–	Q_{34}	−22.78, −62.79

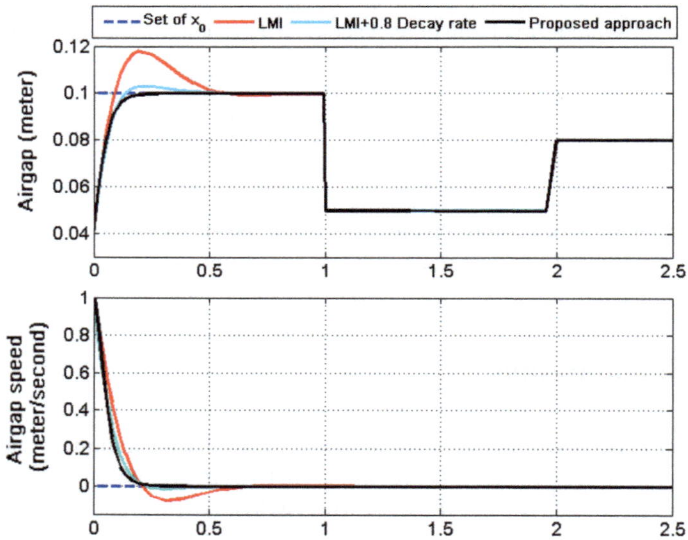

Fig. 3 Response of EMLS for set of operating points without variation in payload

Payload disturbance pattern is set as {10, 25, 40, and 15%} with the identical interval of 0.5 s which is quite fast and high leveled. The proposed regulator efficiently tracks the change of operating conditions (y_0) even under the presence of the vertical payload without any problem compared to the former regulators as shown in Fig. 4. It is observed that for the similar conditions, LMI-based regulator with $\beta = 1$ almost fails to regulate the EMLS with very large overshoot and permanent offset while the regulator with $\beta = 0.8$ having the moderate overshoot as well as the steady-state problems.

The pole-placement-based simple design of T-S fuzzy regulator provides the robust performance to the EMLS under most diverse conditions like change in the operating conditions and large quick variation of the disturbance. For the proposed approach, the regulator gains have not been changed for the different cases as dis-

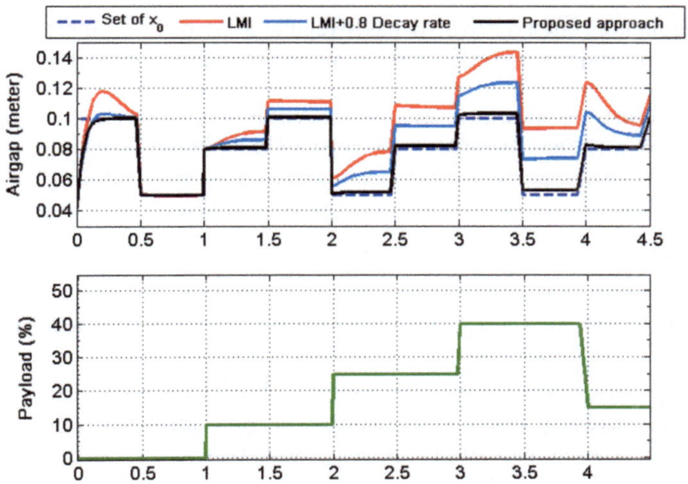

Fig. 4 Response of air gap for set of operating points with 0–40% variation in payload

cussed earlier which need to change in case of LMI-based regulator design for $\beta = 1$ and $\beta = 0.8$. This change of the regulator gains would add the computation burden to run the large set of LMIs for different cases.

There is the existence of positive definite matrix P as mentioned below such that all the subsystems get asymptotically stabilized in the left half of the s-plane as shown in Table 2. Hence, the designed T-S fuzzy regulator would stabilize the considered EMLS for a set of operating conditions.

$$P = \begin{bmatrix} 0.367 & -5.1042 \\ -5.1042 & 85.6754 \end{bmatrix}, \quad P > 0$$

6 Conclusion

The proposed research work discusses the pole-placement-based simplified design of T-S fuzzy regulator for a control application like EMLS without solving the LMIs. It is one of the multipurpose regulator designs which may be suitable for the stabilizing control as well as disturbance rejection control for nonlinear and unstable systems. It has been observed during the simulation that the designed regulator works satisfactorily to achieve multiple control objectives without changing the regulator gains for different requirements like change in the operating points, payload variation…etc. The proposed regulator gives superior transient performance as compared to LMI-based regulator under fixed as well as change in operating conditions. Also, the performance robustness against payload variation of 0–40% is achieved by the proposed regulator compared to LMI-based regulator. Also, the asymptotical stability in the large operating region based on the proposed control approach is confirmed.

Acknowledgements The proposed research work is a part of the full-time Ph.D. of Ravi V. Gandhi under the Visvesvaraya Ph.D. Scheme which is governed by the M.H.R.D., India.

References

1. Takagi, T., Sugeno, M.: Fuzzy Identification of systems and its applications to modeling and control. IEEE Trans. Syst. Man Cybern. **15**, 116–132 (1985)
2. Tanaka, K., Wang, H.O.: Fuzzy Control Systems Design and Analysis: A Linear Matrix Inequality Approach. Wiley, pp. 1–49 (2001)
3. Wang, H.O., Tanaka, K., Griffin, M.F.: Parallel distributed compensation of nonlinear systems by Takagi-Sugeno Fuzzy Model. In: Proceedings of 4th IEEE International Conference Fuzzy Systems, Yokohama, Japan, pp. 531–538 (1995)
4. Niemann, D., Li J., Wang, H.O.: Parallel distributed compensation for Takagi-Sugeno Fuzzy Models: new stability conditions and dynamic feedback designs. In: 14th Proceedings of International Federation of Automatic Control (IFAC) World Congress, Beijing, pp. 5374–5398 (1999)
5. Tanaka, K., Wang, H.O.: Fuzzy regulators and fuzzy observers: a linear matrix inequality approach. In: Proceedings of the 36th Conference on Decision & Control, San Diego, California, USA, pp. 1315–1320 (1997)
6. Tanaka, K., Taniguchi, T., Wang, H.O.: Robust and optimal fuzzy control: a linear matrix inequality approach. In: International Federation of Automatic Control (IFAC). World Congress, Beijing, pp. 213–218 (1999)
7. Chung, H., Chan, S.: Stability analysis of fuzzy control systems using polynomial methods. Int. J. Innov. Comput. Inf. Control ICIC Int., **5**, 1–12 (2009)
8. Chung, H., Chan, S.: An alternate approach to the analysis of a T-S Fuzzy Model. Int. J. Innov. Comput. Inf. Control ICIC Int. **4**, 1–13 (2008)
9. Talagaev, Y.: An approach to analysis and stabilization of Takagi-Sugeno fuzzy control systems via super stability conditions. IFAC-Papers Online **48**, 426–433 (2015)
10. Feng, G., Cao, S., Ress, N.: Design of fuzzy control systems with guaranteed stability. Fuzzy Sets Syst. **85**, 1–10 (1997)
11. Zhu, K., Gonska, H.: Eigenvalue constraints for the stability of T-S fuzzy models. In: American Control Conference (ACC), Seattle, Washington, USA, pp. 87–88 (2008)
12. Yaghoubi, H.: Practical applications of Magnetic Levitation Technology. Iran Maglev Technology (IMT), Iran, pp. 1–56 (2012)
13. Goodall, R.: Dynamics and control requirements for EMS Maglev suspensions. In: Proceedings of 18th International Conference on Magnetically Levitated Systems and Linear Drives, Shanghai, China, pp. 926–934 (2004)
14. Gandhi, R.V., Adhyaru, D.M.: Modeling of voltage controlled and current controlled electromagnetic levitation system based on novel approximation of coil inductance. In: International Conference Control, Automation and Robotics (ICCAR), New Zealand, pp. 212–217 (2018)
15. Gandhi, R.V., Adhyaru, D.M.: Feedback linearization based optimal controller design for electromagnetic levitation system. In: IEEE, International Conference Control Instrumentation Communication and Computational Technologies, India, pp. 36–41 (2016)
16. Gandhi, R.V., Adhyaru, D.M.: Novel approximation based dynamical modelling and nonlinear control of electromagnetic levitation system. Int. J. Comput. Syst. Eng. Indersci. (In Press, https://doi.org/10.1504/ijcsyse.2018.10008868)

17. Unni, A.C., Junghare, A.S., Mohan, V., Ongsakul, W.: PID, fuzzy and LQR controllers for magnetic levitation system. In: International Conference on Cogeneration, Small Power Plants and District Energy (ICUE 2016), BITEC, Thailand, pp. 1–5 (2016)
18. Benomair, M., Tokhi, M.O.: Control of single axis magnetic levitation system using fuzzy logic control. In: Science and Information Conference, London, UK, pp. 14–18 (2015)
19. Gandhi, R.V., Adhyaru, D.M.: Pre-fuzzy-PID controller for effective control of electromagnetic levitation system. In: 4th Indian Control Conference (ICC), India, pp. 113–118 (2018)
20. Gandhi, R.V., Adhyaru, D.M.: Hybrid intelligent controller design for an unstable electromagnetic levitation system: a fuzzy interpolative controller approach. Int. J. Autom. Control Indersci. (article in press)
21. Wu, S., Wu, C.: Affined-type neural-fuzzy gap control of electromagnetic suspension systems. In: IEEE Conference, China, pp. 173–178 (2006)
22. Umar, F., Jason, G., Mohamed, E., Ghulam, A., Muhammad, U.: An interval type-2 fuzzy regulator for magnetic levitation system. In: Proceeding of the IEEE 28th Canadian Conference on Electrical and Computer Engineering, Canada, pp. 424–431 (2015)
23. Santim, M.P., Teixeira, M., Souza, W., Cardim, R., Assuncao, E.: Design of Takagi-Sugeno fuzzy regulator for a set of operation points. Math. Prob. Eng. **2012**, 1–17 (2012)

AGC of Thermal-Split Shaft Gas Turbine System Integrating IPFC and Ultra-Capacitor

Arindita Saha, Lalit Chandra Saikia, Rumi Rajbongshi, Debdeep Saha and Washima Tasnin

Abstract The present work deals with an unequal two-area multi-source thermal-split shaft gas turbine (SSGT)-based automatic generation control (AGC) system. Each of the thermal units is associated with both reheat turbine and generation rate constraint (GRC). The whale optimization algorithm (WOA) is used for simultaneous optimization of controller gains and other parameters. Through investigations it is observed that combination of integer order proportional–integral–derivative with filter (PIDN) controller and fractional-order integral-derivative (FOID) controller (PIDN-FOID) provides upperhand over classical controllers with respect to lesser peak deviations, settling time, and smaller values of cost function. Different analyses also show that involvement of interline power flow controller (IPFC) and ultra-capacitor (UC) one at a time or both together improves the system dynamics with much-reduced oscillations, lesser peak deviations, and settling time. Sensitivity analysis at higher value of step load perturbation (SLP) is performed for checking the robustness of controller.

Keywords Automatic generation control · Generation rate constraint
Interline power flow controller · PIDN-FOID controller · Ultra-capacitor · Whale optimization algorithm

A. Saha (✉) · L. C. Saikia · R. Rajbongshi · D. Saha · W. Tasnin
Department of Electrical Engineering,
National Institute of Technology Silchar, Silchar, Assam, India
e-mail: sahaarindita.91@gmail.com

L. C. Saikia
e-mail: lcsaikia@yahoo.com

R. Rajbongshi
e-mail: rumi.nits90@gmail.com

D. Saha
e-mail: saha_debdeep_RS@yahoo.com

W. Tasnin
e-mail: washima.nits@gmail.com

D. Deb et al. (eds.), *Innovations in Infrastructure*, Advances in Intelligent Systems and Computing 757, https://doi.org/10.1007/978-981-13-1966-2_9

1 Introduction

The basic aim of an interconnected system is to facilitate with electric energy at nominal value of both frequency and terminal voltages. This is dependent upon the balance between the amount of power being generated and demanded. If a situation arises where the amount of power generated is less compared to amount of power being demanded, then there is reduction in speed as well as frequency of the generators and vice versa [1] and this is reflected as deviations in frequency. Here, lies the concept of automatic generation control (AGC) whose main objective is to maintain steady-state error as zero. Nanda et al. [2] have undergone analysis for a two-area hydrothermal system under both continuous and discrete mode in presence of nonlinearity like generation rate constraint (GRC). Nanda et al. [3] have studied an unequal three-area system having only thermal units as generating units in presence of GRC. Gozde et al. [4] have studied gain scheduling of a two-area thermal system but in presence of another nonlinearity called governor dead band. Bhatt et al. [5] have studied a two-area system considering mixed generating units like thermal, hydro, and diesel. There are mainly two types of gas turbine (GT) and they are single shaft GT and split shaft GT (SSGT). In single shaft GT, both generator and turbine are on same shaft, but for SSGT, generator is connected to turbine via gearbox. Shankar et al. [6] have initiated the use of SSGT in the field of AGC for a two-area hydrothermal-SSGT system. The authors in [6] have undergone studies for a simple system without considering any type of nonlinearities like GRC. Thus, there lies a scope of studying SSGT with conventional thermal units but in the presence of GRC.

Various flexible AC transmission devices (FACTS) play a major role in maintaining power flow in an interconnected system. Interline power flow controller (IPFC) in general form is a combination of several dc-to-ac converters which provides series compensation for different lines. Rahman et al. [7] have utilized IPFC for the study of AGC under conventional environment. Similarly, Chidambaram et al. [8] have considered IPFC for the study of AGC under deregulated environment. Thus, the study of AGC under conventional environment integrating IPFC for the considered system needs to be studied.

The energy storage device ultra-capacitor (UC) is also popularly known as supercapacitor is based on the conception of double-layer effect. It has high specific energy, specific power energy, high charge, and discharge efficiency as well as longer life. Das et al. [9] have used UC for an isolated system of load frequency control, whereas Saha et al. [10] have incorporated UC in their study of AGC but under deregulated environment. Thus, there lies an option to use UC under conventional environment of AGC.

Most of the literatures in the field of AGC report about the use of classical controllers, namely integral (I), proportional-integral (PI), and proportional–integral–derivative with filter (PIDN). In recent days, Saha et al. [11] have suggested the combination of IO controller PIDN and fractional-order controller ID (PIDN-

FOID) under deregulated environment of AGC. Therefore, the use of PIDN-FOID in conventional AGC system requires to be considered.

Many literatures in the field of AGC marks the use of some optimization techniques like bacterial foraging-based optimization technique [3], craziness-based particle swarm optimization [4], quasi-oppositional harmony search algorithm [6], and many more. Most of the techniques suffer from disadvantages like getting entrapped in neighboring optimal and have sluggish convergence. Mirjalili et al. [12] have put forward whale optimization algorithm (WOA) that supersedes the drawbacks as it carries out exploration as well as exploitation separately using two varied set of equations. The use of WOA in conventional system with SSGT, IPFC, and UC are still unseen. According to the above review of previous literatures, the following are the key objectives of the present article:

(a) Development of an unequal two-area AGC system with thermal units in area-1 and thermal-SSGT in area-2 in presence of GRC.
(b) Simultaneous optimization of I, PI, PIDN, and PIDN-FOID controllers with the help of WOA and evaluation of dynamic responses in order to find the best controller amongst them.
(c) Study the effect of IPFC, UC one at a time and both together on system dynamics using the best-obtained controller from (b).
(d) Sensitivity analysis is performed to check the robustness of controller gains and parameters for higher value of step load perturbation (SLP).

2 System Under Consideration

An unequal two-area AGC system with multi sources are considered for the study. The two areas are in the capacity ratio of 1:3. In these two-area systems, the area-1 comprises of thermal units and area-2 comprises of both thermal and SSGT units. The area participation factors (apfs) considered for each generating units for each areas are $apf_{11} = apf_{12} = 0.5$ (for area-1) and $apf_{21} = 0.4$, $apf_{22} = 0.6$ (for area-2). The nominal system is integrated with IPFC, UC one after another as well as both together to reduce oscillations between areas. The complete transfer function of two-area systems incorporating IPFC and UC are shown in Fig. 1. The nominal values various parameters of thermal and SSGT units are obtained from [2, 6], respectively. Similarly, the nominal values for IPFC and UC are obtained from [7, 10], respectively. An attempt is made to make system realistic with proper consideration of GRC of 3%/min for thermal units. A step load perturbation (SLP) of 1% is taken as disturbance in area-1. The controller gains and parameters are optimized making use of WOA considering integral squared error as cost function as given by (1).

$$J_{\mathrm{ISE}} = \int_{0}^{T} \left(\Delta f_1^2 + \Delta f_2^2 + \Delta \mathrm{P}_{\mathrm{tie1-2}}^2\right) \mathrm{d}t \tag{1}$$

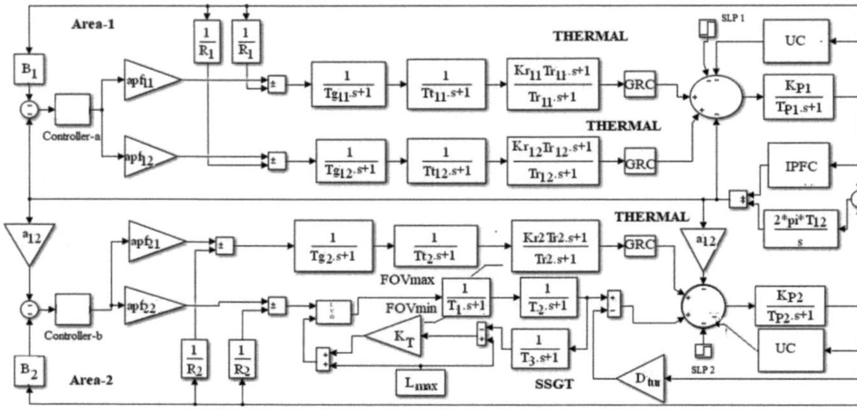

Fig. 1 Transfer function model of complete system with both IPFC and UC

3 PIDN-FOID Controller

The secondary controller used for the considered system is a combination integer-order (IO) and fractional-order (FO) controller named as PIDN-FOID controller. In PIDN-FOID controller, the IO controller is PIDN and FO controller is $I^{\lambda}D^{\mu}$. The detailed block diagram of the controller is given in [11]. The transfer function blocks of PIDN and FOID controllers are given by (2) and (3), respectively.

$$C_i(s) = \frac{(K_{Pi} + K_{Di}N_i)s^2 + (K_{Ii} + N_i)s + K_{Ii}N_i}{s\,(s + N_i)} \tag{2}$$

$$F_i(s) = K_{ii}/s^{\lambda i} + K_{di}s^{\mu_i} \tag{3}$$

In (3), λ and μ as fractional order of integrator and derivative part, respectively, of $I^{\lambda}D^{\mu}$ controller. Therefore,

$$U_i(s) = (C_i(s)F_i(s)) * R_i(s) \tag{4}$$

$$U_i(s) = \left\{ \begin{array}{l} \left(\frac{(K_{Pi}+K_{Di}N_i)s^2+(K_{Ii}+N_i)s+K_{Ii}N_i}{s\,(s+N_i)} \right) \\ * \left(K_{ii}/s^{\lambda i} + K_{di}s^{\mu_i} \right) \end{array} \right\} * ACE_i \tag{5}$$

$R_i(s)$ and $U_i(s)$ are input and output from PIDN-FOID controller, respectively. K_{Pi}, K_{Di}, K_{Ii}, and N_i are tuning parameters of the PIDN controller and K_{ii}, λ_i, K_{di}, and μ_i are the tuning parameters of $I^{\lambda}D^{\mu}$ controller for *i*th areas. The gains and parameters are optimized with the help of WOA by keeping minimum value of J_{ISE} given by (1) subject to (6). The lower and upper limits of each parameter are superscripted in (6) and these are set by experience and different set of trials.

$$K_{Pi}^{\min} \leq K_{Pi} \leq K_{Pi}^{\max}, K_{Ii}^{\min} \leq K_{Ii} \leq K_{Ii}^{\max}, K_{Di}^{\min} \leq K_{Di} \leq K_{Di}^{\max}, N_i^{\min} \leq N_i \leq N_i^{\max},$$
$$K_{ii}^{\min} \leq K_{ii} \leq K_{ii}^{\max}, K_{di}^{\min} \leq K_{di} \leq K_{di}^{\max}, \lambda_i^{\min} \leq \lambda_i \leq \lambda_i^{\max}, \mu_i^{\min} \leq \mu_i \leq \mu_i^{\max} \quad (6)$$

The specified range of frequency $[\omega_l, \omega_h]$ considered for analyses is [0.01, 50]. The basic details of FO controller and PIDN-FOID are given in [11].

4 Whale Optimization Algorithm (WOA)

The WOA technique was proposed by Mirjalili et al. [12] based on the attacking behavior of humpback whales. This algorithm maintains a balance between both exploration and exploitation with the help of two different set of equations. Therefore, it has an advantage of not getting trapped in local optima. The detailed explanation of WOA is given in [12]. The input variables of WOA are number of search agents, maximum iterations, and $\vec{a}$ which reduces from 2 to 0 during iterations. And, the output variables are the controller gains and parameters.

5 Results and Analysis

5.1 *Performance Comparison of I, PI, PIDN, and PIDN-FOID Controllers with 1% SLP in Area-1*

The unequal two-area system with thermal units in area-1 and thermal-SSGT units in area-2 is considered for the study. The few secondary controllers I, PI, PIDN, and PIDN-FOID are considered one at a time and their gains and other parameters are optimized with the help of WOA considering ISE as cost function. The optimum values of each controller gains and parameters and corresponding J_{ISE} values are provided in Table 1. Using these optimum values, dynamic responses are plotted and shown in Fig. 2 (only two). Critical analysis of Fig. 2 infers that PIDN-FOID provides better performance than other classical controllers with regard to reduced peak deviations and settling time. It is also observed from Table 1 column 3 that J_{ISE} value corresponding to PIDN-FOID controller is least compared to other controllers. Thus, it can be said that the performance of PIDN-FOID is better.

5.2 *System Dynamics Using PIDN-FOID Controller with IPFC, UC*

The system considered for the study in Sect. 5.1 is now integrated with IPFC, UC and both IPFC and UC one at a time. PIDN-FOID is considered as the secondary

Table 1 Optimum values of controller gains and parameters and corresponding J_{ISE} value

Controllers	Optimum gains/other parameters	J_{ISE}
I	$K_{I1}^{*}=0.5276$ $K_{I2}^{*}=0.2356$	0.0131
PI	$K_{PI}^{*}=0.0001$ $K_{P2}^{*}=0.0174$ $K_{I1}^{*}=0.4712$ $K_{I2}^{*}=0.0824$	0.0084
PIDN	$K_{P1}^{*}=0.0001$ $K_{P2}^{*}=0.0001$ $K_{I1}^{*}=0.9524$ $K_{I2}^{*}=0.4168$ $K_{D1}^{*}=0.2996$ $K_{D2}^{*}=0.9751$ $N_{1}^{*}=96.00$ $N_{2}^{*}=95.00$	0.0058
PIDN-FOID	$K_{P1}^{*}=0.0036$ $K_{P2}^{*}=0.0458$ $K_{I1}^{*}=0.6214$ $K_{I2}^{*}=0.6254$ $K_{D1}^{*}=0.8638$ $K_{D2}^{*}=0.9565$ $N_{1}^{*}=86.36$ $N_{2}^{*}=89.36$ $K_{i1}^{*}=0.5214$ $K_{i2}^{*}=0.5890$ $\lambda_{1}^{*}=0.0012$ $\lambda_{2}^{*}=0.0015$ $K_{d1}^{*}=0.7324$ $K_{d2}^{*}=0.8115$ $\mu_{1}^{*}=0.2135$ $\mu_{2}^{*}=0.6560$	0.0049

controller. For each case, the gains and parameters of controller are optimized using WOA and optimum values are tabulated in Table 2. Corresponding to these optimum values, the dynamic responses are obtained and compared in Fig. 3 with that obtained without IPFC and UC. It is observed that system dynamics improves in presence of IPFC, UC or both IPFC and UC. But, critical observation reveals that the most improvement in system dynamics is observed in presence of both IPFC and UC in terms much less peak deviations, settling time, and magnitude of oscillations.

5.3 *Sensitivity Analysis with Higher Value of SLP*

The robustness of PIDN-FOID controller gains and other parameters are checked for higher value of SLP like 5% in area-1. The controller gains and other parameters are optimized using WOA (optimum values are not shown). The dynamic responses of the system at changed 5% SLP condition are compared for optimum values corresponding to change 5% SLP condition and nominal condition in Fig. 4. It is noticed that responses are almost the same. Thus, further resetting is not required for higher value of SLP.

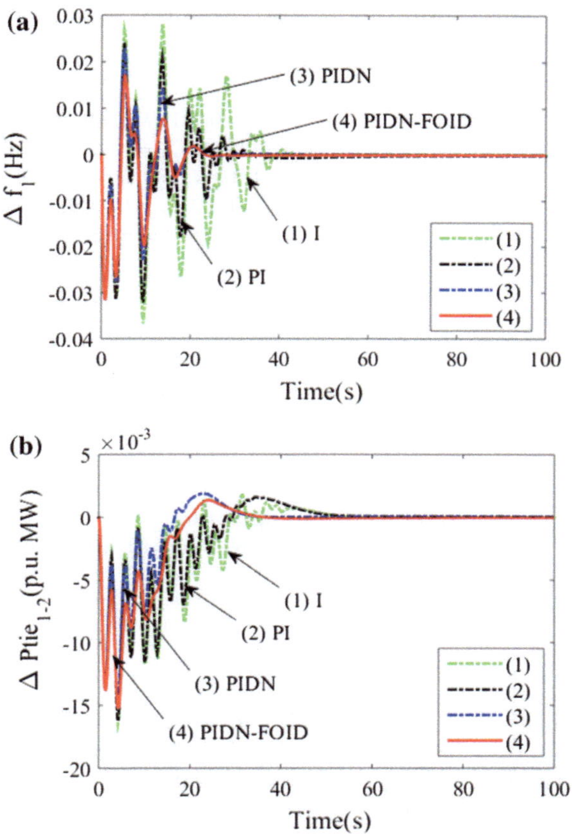

Fig. 2 Dynamic response comparison of I, PI, PIDN, and PIDN-FOID controllers for **a** Δf_1 versus time, **b** ΔPtie_{1-2} versus time

6 Conclusion

An analysis of two-area multi-source thermal-SSGT system integrating IPFC and UC are done in this article. The controller gains and other parameters are optimized using WOA technique. It is observed that PIDN-FOID controller performance is much better than other classical controllers like I, PI, and PIDN. The system is integrated with IPFC, UC and both IPFC and UC one at a time using PIDN-FOID controller. Analysis says that system dynamics shows improvement with much-reduced peak deviations, settling time, and intensity of oscillations in each of the cases. But, the best dynamic responses are obtained when the system is integrated with both IPFC and UC. It is observed that the controller gains and parameters are robust to higher value of SLP and requires no further resetting. Thus, it is recommended to use both IPFC and UC for such combination of the system.

Table 2 Optimum values of PIDN-FOID controller gains and parameter with IPFC, UC and both IPFC and UC

Conditions	Optimum gains/other parameters
With IPFC only	K^*_{P1} =0.9213 K^*_{P2} =0.7728 K^*_{I1}=0.9521 K^*_{I2} =0.1144 K^*_{D1} =0.4427 K^*_{D2}=0.5421 N^*_1 =42.00 N^*_2 =46.00 K^*_{i1} =0.9369 K^*_{i2} =0.9996 λ^*_1=0.0073 λ^*_2 =0.0009 K^*_{d1}=0.9965 K^*_{d2}=0.9990 μ^*_1=0.2135 μ^*_2=0.0093
With UC only	K^*_{P1} =0.0001 K^*_{P2}=0.6750 K^*_{I1}=0.9789 K^*_{I2} =0.0002 K^*_{D1} =0.0002 K^*_{D2} =0. 9453 N^*_1 =69.00 N^*_2 =79.00 K^*_{i1}=0.9658 K^*_{i2} =0.9447 λ^*_1 =0.0002 λ^*_2 =0.9416 K^*_{d1} =0.9784 K^*_{d2} =0.9521 μ^*_1 =0.0002 μ^*_2=0.0093
With both IPFC and UC	K^*_{P1}=0.6744 K^*_{P2} =0.4629 K^*_{I1}=0.6058 K^*_{I2}=0.3315 K^*_{D1}=0.3271 K^*_{D2}=0.4107 N^*_1 =61.65 N^*_2 =52.71 K^*_{i1} =0.2772 K^*_{i2}=0.4852 λ^*_1=0.0016 λ^*_2= 0.0017 K^*_{d1} =0.5974 K^*_{d2} =0.6907 μ^*_1 =0.0062 μ^*_2 = 0.0024

Appendix

f =60 Hz; loading=50%; K_{pi} =120 Hz/p.u. MW; T_{pi} =20 s; T_{12} =0.086 p.u. MW/rad; H_i =5 s; $D_i = 8.33 \times 10^{-3}$ p.u. MW/Hz; $B_i = \beta_i$ =0.425 p.u. MW/Hz; R_i =2.4 Hz/p.u. MW; T_{gi} =0.08 s, T_{ti} =0.3 s, K_{ri} =0.5, T_{ri} =10 s; $T_1 = T_2$ =1.5 s, T_3 =3 s, $L_{\max}$ =1, K_T =1, $\mathrm{FOV}_{\min}$ =−0.02, $\mathrm{FOV}_{\max}$ =1, D_{tur} =0 p.u.; K_{uci} =−0.7, T_{uci} =0.9; K_{IPFC} =1, T_{IPFC} =0.01 s.

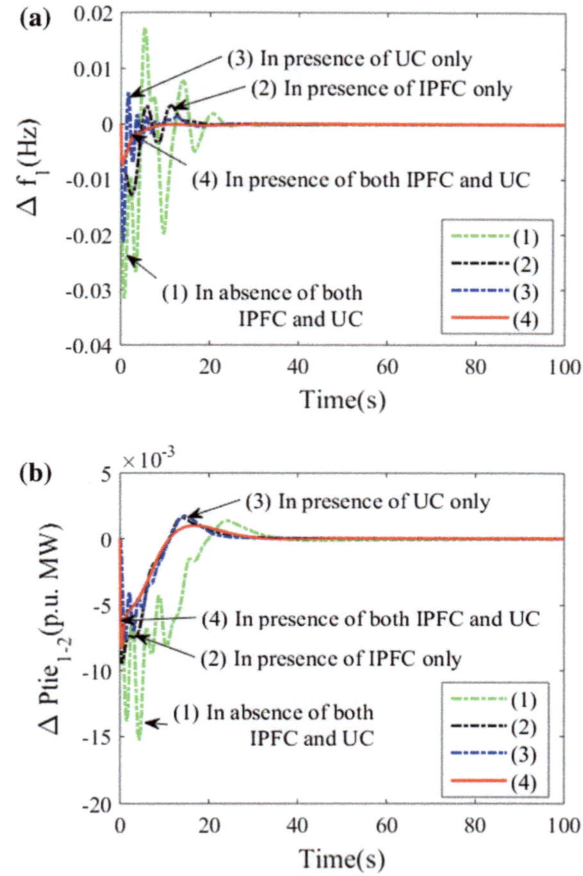

Fig. 3 Dynamic response comparison of responses with and without IPFC, UC and both IPFC and UC using PIDN-FOID controller **a** Δf_1 versus time, **b** $\Delta P_{\text{tie}1-2}$ versus time

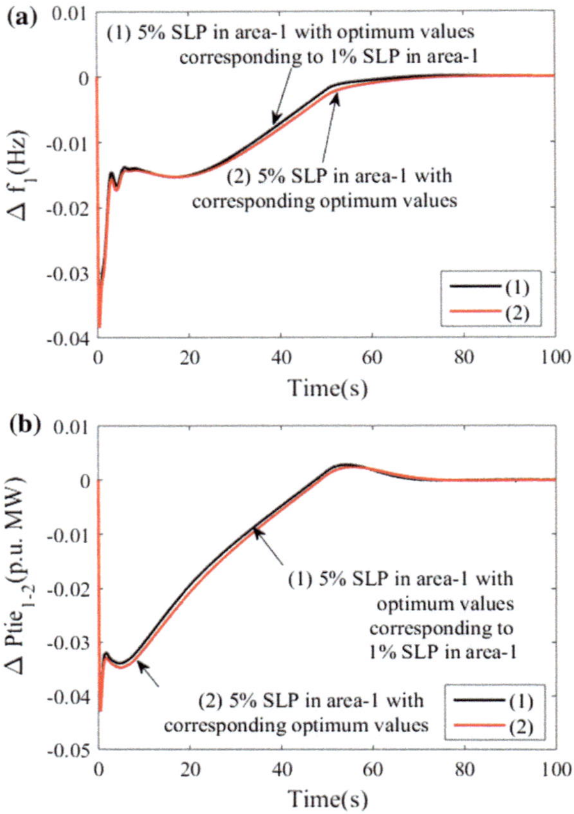

Fig. 4 Dynamic response comparison of responses with higher value of SLP **a** Δf_1 versus time, **b** ΔPtie_{1-2} versus time

References

1. Elgerd, O.I.: Electric Energy Systems Theory, pp. 299–362. McGraw-Hill, New York (1982)
2. Nanda, J., Mangla, A., Suri, S.: Some new findings on automatic generation control of an interconnected hydrothermal system with conventional controllers. IEEE Trans. Energy Convers. **21**(1), 187–194 (2006)
3. Nanda, J., Mishra, S., Saikia, L.C.: Maiden application of bacterial foraging based optimization technique in multiarea automatic generation control. IEEE Trans. Power Syst. **24**(2), 602–609 (2009)
4. Gozde, H., Cengiz, T.M.: Automatic generation control application with craziness based particle swarm optimization in a thermal power system. Int. J. Electr. Power Energy Syst. **33**(1), 8–16 (2011)
5. Bhatt, P., Roy, R., Ghoshal, S.P.: GA/particle swarm intelligence based optimization of two specific varieties of controller devices applied to two-area multi-units automatic generation control. Int. J. Electr. Power Energy Syst. **32**(4), 299–310 (2010)
6. Shankar, G., Mukherjee, V.: Quasi oppositional harmony search algorithm based controller tuning for load frequency control of multi-source multi-area power system. Int. J. Electr. Power Energy Syst. **75**, 289–302 (2016)
7. Rahman, A., Saikia, L.C., Sinha, N.: Maiden application of hybrid pattern search-biogeography based optimisation technique in automatic generation control of a multi-area system incorporating interline power flow controller. IET Gener. Transm. Distrib. **10**(7), 1654–1662 (2016)
8. Chidambaram, I.A., Paramasivam, B.: Optimized load-frequency simulation in restructured power system with Redox flow batteries and interline power flow controller. Int. J. Electr. Power Energy Syst. **50**, 9–24 (2013)
9. Das, D.Ch., Roy, A.K., Sinha, N.: GA based frequency controller for solar thermal–diesel–wind hybrid energy generation/energy storage system. Int. J. Electr. Power Energy Syst. **43**(1), 262–279 (2012)
10. Saha, A., Saikia, L.C.: Utilisation of ultra-capacitor in load frequency control under restructured STPP-thermal power systems using WOA optimised PIDN-FOPD controller. IET Gener. Transm. Distrib. **11**(13), 3318–3331 (2017)
11. Saha, A., Saikia, L.C.: Combined application of redox flow battery and DC link in restructured AGC system in the presence of WTS and DSTS in distributed generation unit. IET Gener. Transm. Distrib. https://doi.org/10.1049/iet-gtd.2017.1203
12. Mirjalili, S., Lewis, A.: The whale optimization algorithm. Adv. Eng. Softw. **95**, 51–67 (2016)

An Efficient Unbalanced Load Flow for Distribution Networks

Tanmoy Malakar and Ujjwal Ghatak

Abstract This paper presents a new unbalanced three-phase load flow solution technique for electric power distribution systems (DS). A new matrix formulation is utilized to develop the proposed Distribution Load Flow (DLF) approach. The proposed algorithm is composed of two forward sweep steps; unlike conventional backward forward sweep approach. Here, the traditional backward sweep step is replaced by a novel forward sweep step. The bus voltages are determined in the first forward sweep and branch currents are calculated in the second forward sweep. In this paper, the DLF formulation is presented for balanced and unbalanced radial networks. Simulation results prove the accuracy of the proposed method with excellent convergence characteristics for distribution networks with varied complexity and size. Results obtained with the proposed method are compared with other methods, and it is found that the proposed approach has made the algorithm faster than the other contemporary methods.

Keywords Unbalanced load flow · Distribution systems · Computations

1 Introduction

The Power Flow (PF) study is the most basic and important analysis tool for power system networks. It is an essential component for solving many power system problems and because of this, PF computation cannot be avoided. PF studies are mainly executed to determine some unknown quantities associated with power systems such as load bus voltages and their angles, bus injections, line flows, losses, etc. The digital computer solution approach of PF problem was first reported in 1950s [1, 2]

T. Malakar (✉) · U. Ghatak
Electrical Engineering Department,
National Institute of Technology Silchar, Silchar, Assam, India
e-mail: m_tanmoy1@rediffmail.com; malakar.tanmoy@yahoo.com

U. Ghatak
e-mail: contactujjwalranaghatak@gmail.com

D. Deb et al. (eds.), *Innovations in Infrastructure*, Advances in Intelligent Systems and Computing 757, https://doi.org/10.1007/978-981-13-1966-2_10

and since then numerous techniques have evolved. Admittance matrix type formulation approaches [3–6] were successfully implemented in solving PF studies of complex power transmission systems. These are well-established methods and have been trusted by the power system practitioners over the last four decades. They show excellent convergence characteristics for PF problems associated with power transmission/sub-transmission networks. However, they fail grossly when dealing with Distribution Systems (DSs) [7] because of the following reasons:

- Most of the DS are radial and weakly linked.
- High R/X ratio.
- The DS are mostly unbalanced.
- Transposition in DS is rare practice.

Different techniques were proposed in the literature for solving PF problems for both balanced and unbalanced DS. Among these, formulations based on modification of existing methods [8–11] were the early attempts for solving DLF problems. Off late, the KCL-KVL-based forward–backward sweep approaches [12–28] have evolved as most conventional DLF techniques. Some research [29–32] shows DLF formulation using bi-quadratic equations.

The KCL-KVL-based Backward–Forward (B/F) sweep approach was possibly attempted first by Shirmohammadi in their works in [12, 13] for DLF solution. Further improvement of this method is reported in [14] to make it adaptive for modeling slow dynamics on distribution system. A new DLF approach utilizing simple algebraic expressions of receiving end voltages is reported in [15]. A direct approach of PF solution for unbalanced three-phase distribution system is proposed in [16]. The topological characteristics of distribution system are utilized to develop the bus injection to branch current matrix (BIBC) and branch current to bus voltage matrix (BCBV) and in the later stage, these two matrices are multiplied to obtain PF solution. In [17], a new DLF approach based on branch impedance matrix (Z_{BR}) is proposed. Another novel DLF approach is proposed in [18] by exploiting the radial structure of DS. The approach utilizes radial configuration matrix to carry out the backward/forward iterative steps. A loop analysis-based PF method is presented in [19]. A new approach to analyze radial and weakly meshed distribution system supplying nonlinear loads is reported in [20]. The process is named as backward sweep, because of the unique equations involved for RDS and meshed networks.

An efficient matrix transformation technique is reported in [21] to make backward forward sweep faster. Most recently, another promising DLF algorithm is proposed in [23], where simultaneous calculations for all laterals of distribution networks can be avoided. The advantage of using such an approach reduces the size of the problem and shaves off the computation burden. However, such an approach is applicable only for Radial Distribution Systems (RDSs).

A sweep based three-phase PF method is presented in [24] to analyze distribution network with meshes/loops and transformers around the mesh structure. A generalized single equation load flow technique is proposed in [25] which are based on impedance matrix formulation. Another efficient approach to solve multi-phase distribution system load flow problem is presented in [26]. An extension of popular backward/forward sweep PF algorithm is discussed in [27] where a new loop-based analysis technique is used to improve the convergence.

It is revealed from the literature review that backward–forward sweep technique can be implemented either directly [12–15] or using matrix formulation [16–27] approach. In this paper, an efficient load flow algorithm is proposed using a unique matrix formulation approach. The proposed approach is based on two matrices, namely FFP matrix and SFP matrix. The former is needed for calculating the bus voltages and the latter helps to find the branch currents in the developed approach. The unique property of the proposed PF model is that it replaces the conventional backward sweep step by a simple forward sweep step and this uniqueness has made the algorithm superior than other contemporary methods. The test results show excellent convergence characteristics for distribution networks with varied complexity and size.

2 Mathematical Formulation

2.1 Modeling of Distribution Lines

With the help of this model, the relation between nodal voltages and line currents for a three-phase line can be obtained. Figure 1 shows a section of unbalanced three-phase line comprising self and mutual impedances of a DS. The voltages at bus j of the line section is shown in [17] and represented in Eq. (1).

$$\begin{bmatrix} V_a \\ V_b \\ V_c \end{bmatrix} = \begin{bmatrix} V_A \\ V_B \\ V_C \end{bmatrix} - \begin{bmatrix} Z_{aa\_n} & Z_{ab\_n} & Z_{ac\_n} \\ Z_{ba\_n} & Z_{bb\_n} & Z_{bc\_n} \\ Z_{ca\_n} & Z_{cb\_n} & Z_{cc\_n} \end{bmatrix} \cdot \begin{bmatrix} I_{Aa} \\ I_{Bb} \\ I_{Cc} \end{bmatrix} \tag{1}$$

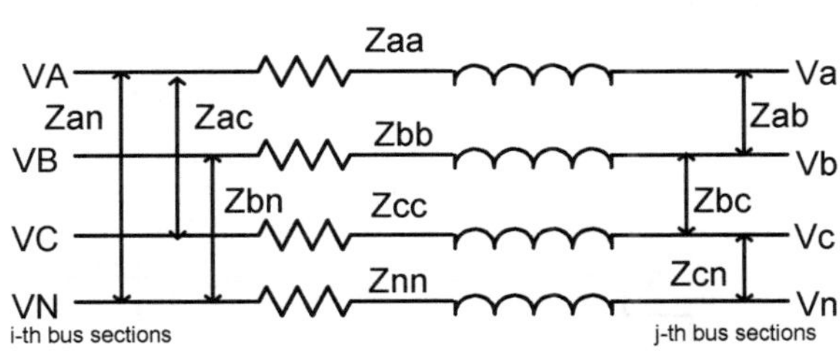

Fig. 1 A three-phase line section

2.2 The Proposed PF Solution Method

Like most of the conventional distribution load flow (DLF) methods, the proposed approach is based on the calculation of branch currents and bus voltages by KCL and KVL, respectively. The developed formulation is based on double forward sweep steps. Here, the conventional backward sweep step for calculating branch currents is replaced by a novel forward sweep step. In the first forward sweep, the bus voltages are determined using KVL by assuming all branch currents are initially set as zero and calculation proceeds in forward path, i.e., from start node to end node. After obtaining the bus voltages from first forward sweep step, the branch currents are calculated using KCL in the second forward sweep step. The calculation begins with the first branch and continues up to the last branch for a given distribution network. These two forward sweep steps are formulated using two matrices, namely First Forward Path (FFP) matrix and Second Forward Path (SFP) matrix, respectively. The bus voltage calculation is done using the former and the branch currents are determined using the second matrix. In this section, derivation of these two matrices will be discussed in detail with the help of a sample distribution system as shown in Fig. 2.

A. Formation of FFP matrix and calculation of $[V_B]$

From Fig. 2, if voltage of ith bus is $V_{Bi,}$ branch current of kth branch is I_{Bk}, and impedance of that branch is Z_k, then, the nodal voltage expressions of the network can be written in the following way:

$$V_{B2} = V_{B1} - I_{B1}Z_1 \quad (2)$$

$$V_{B3} = V_{B2} - I_{B2}Z_2 \quad (3)$$

$$V_{B4} = V_{B3} - I_{B3}Z_3 \quad (4)$$

$$V_{B5} = V_{B4} - I_{B4}Z_4 \quad (5)$$

$$V_{B6} = V_{B3} - I_{B5}Z_5 \quad (6)$$

Substituting V_{B2} from (2) into (3)

$$V_{B3} = V_{B1} - I_{B1}Z_1 - I_{B2}Z_2 \quad (7)$$

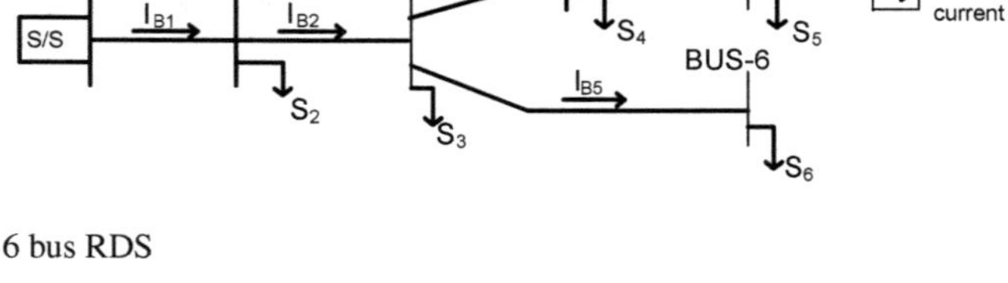

Fig. 2 A 6 bus RDS

Similarly, all other bus voltages can be expressed using substation bus voltage (V_{B1})

$$V_{B4} = V_{B1} - I_{B1}Z_1 - I_{B2}Z_2 - I_{B3}Z_3 \tag{8}$$

$$V_{B5} = V_{B1} - I_{B1}Z_1 - I_{B2}Z_2 - I_{B3}Z_3 - I_{B4}Z_4 \tag{9}$$

$$V_{B6} = V_{B1} - I_{B1}Z_1 - I_{B2}Z_2 - I_{B5}Z_5 \tag{10}$$

The above equations are written in matrix form.

$$\begin{bmatrix} V_{B2} \\ V_{B3} \\ V_{B4} \\ V_{B5} \\ V_{B6} \end{bmatrix} = \begin{bmatrix} V_{B1} \\ V_{B1} \\ V_{B1} \\ V_{B1} \\ V_{B1} \end{bmatrix} - \begin{bmatrix} Z_1 & 0 & 0 & 0 & 0 \\ Z_1 & Z_2 & 0 & 0 & 0 \\ Z_1 & Z_2 & Z_3 & 0 & 0 \\ Z_1 & Z_2 & Z_3 & Z_4 & 0 \\ Z_1 & Z_2 & 0 & 0 & Z_5 \end{bmatrix} \begin{bmatrix} I_{B1} \\ I_{B2} \\ I_{B3} \\ I_{B4} \\ I_{B5} \end{bmatrix} \tag{11}$$

In compact form

$$[V_B] = [V_{B1}] - [FFP].[I_B] \tag{12}$$

Here, $[V_B]$ is the vector of bus voltages except the supply bus voltage and $[V_{B1}]$ is the supply bus voltage vector, [FFP] is FFP matrix, and $[I_B]$ is the vector of branch currents.

B. Formation of SFP matrix and calculation of $[I_B]$

In Fig. 2, if S_i is the load connected to ith bus, then, the branch current of first branch (I_{B1}) can be calculated as

$$I_{B1} = \sum_{i=2}^{n} \left(\frac{S_i}{V_{Bi}} \right)^* \tag{13}$$

where n is total No. of buses present in the distribution network. Equation (13) can be expressed as

$$I_{B1} = \left(\begin{bmatrix} S_2 & S_3 & S_4 & S_5 & S_6 \end{bmatrix} . \begin{bmatrix} 1/V_{B2} \\ 1/V_{B3} \\ 1/V_{B4} \\ 1/V_{B5} \\ 1/V_{B6} \end{bmatrix} \right)^* \tag{14}$$

From Fig. 2, the branch currents can be calculated by applying KCL

$$I_{BI} = I_{B1} - 0 \tag{15}$$

$$I_{B2} = I_{B1} - \left(\frac{S_2}{V_{B2}}\right)^* \tag{16}$$

$$I_{B3} = I_{B1} - \left(\frac{S_2}{V_{B2}} + \frac{S_6}{V_{B6}} + \frac{S_3}{V_{B3}}\right)^* \tag{17}$$

$$I_{B4} = I_{B1} - \left(\frac{S_2}{V_{B2}} + \frac{S_3}{V_{B3}} + \frac{S_4}{V_{B4}} + \frac{S_6}{V_{B6}}\right)^* \tag{18}$$

$$I_{B5} = I_{B1} - \left(\frac{S_2}{V_{B2}} + \frac{S_3}{V_{B3}} + \frac{S_4}{V_{B4}} + \frac{S_5}{V_{B5}}\right)^* \tag{19}$$

In matrix form

$$\begin{bmatrix} I_{B1} \\ I_{B2} \\ I_{B3} \\ I_{B4} \\ I_{B5} \end{bmatrix} = \begin{bmatrix} I_{B1} \\ I_{B1} \\ I_{B1} \\ I_{B1} \\ I_{B1} \end{bmatrix} - \left(\begin{bmatrix} 0 & 0 & 0 & 0 & 0 \\ S_2 & 0 & 0 & 0 & 0 \\ S_2 & S_3 & 0 & 0 & S_6 \\ S_2 & S_3 & S_4 & 0 & S_6 \\ S_2 & S_3 & S_4 & S_5 & 0 \end{bmatrix} \begin{bmatrix} 1/V_{B2} \\ 1/V_{B3} \\ 1/V_{B4} \\ 1/V_{B5} \\ 1/V_{B6} \end{bmatrix} \right)^* \tag{20}$$

In compact form

$$[I_B] = [I_{B1}] - \left([\text{SFP}].[V_B] \bullet\ {}^\wedge(-1)\right)^* \tag{21}$$

$$[I_{B1}] - [I_B] = \left([\text{SFP}].[V_B] \bullet\ {}^\wedge(-1)\right)^* \tag{22}$$

$$[\Delta I_B] = \left([\text{SFP}].[V_B] \bullet\ {}^\wedge(-1)\right)^* \tag{23}$$

Here, $[\Delta I_B]$ is the deviation of branch currents from I_{B1} and [SFP] is the SFP matrix containing the complex loadings of distribution buses. Dots are used to indicate the element-wise inversion operation in the bus voltage vector $[V_B]$.

Using Eqs. (12), (14), it can be rewritten as

$$I_{B1} = \left(\begin{bmatrix} S_2 & S_3 & S_4 & S_5 & S_6 \end{bmatrix}.([V_{B1}] - [\text{FFP}].[I_B]) \bullet^\wedge (-1)\right)^* \tag{24}$$

In compact form

$$I_{B1} = \left([S_i].[V_B] \bullet^\wedge (-1)\right)^* \tag{25}$$

Similarly, the Eq. (23) can be rewritten as

$$[\Delta I_B] = \left([\text{SFP}].([V_{B1}] - [\text{FFP}].[I_B]) \bullet^\wedge (-1)\right)^* \tag{26}$$

Therefore, the branch currents can be calculated using the following expression:

$$[I_B] = [I_{B1}] - [\Delta I_B] \tag{27}$$

C. Computation procedure of the proposed PF solution method

Step 1: Set all branch currents to zero initially.
Step 2: Calculate the current at branch 1 (I_{B1}) and vector of branch current deviations $[\Delta I_B]$ using Eqs. (25) and (26), respectively.
Step 3: Calculate the vector of branch currents $[I_B]$ using Eq. (27).
Step 4: Check the convergence using the following expression:

$$\max\left([I_B]^k - [I_B]^{k-1}\right)\langle\text{tolerance}$$

where k is the No. of iterations. If the condition is satisfied, go to Step 5, else repeat from Step 2.
Step 5: Calculate the final bus voltages using Eq. (12).
Step 6: Stop.

D. Formulation for three-phase networks

In this section, we discuss the formulation for three-phase networks using the developed approach. Let, a branch, B_i connected between two buses is a three-phase line, then the branch current I_{Bi} is a (3×1) matrix as shown in Eq. (28).

$$I_{Bi} \Rightarrow \begin{bmatrix} I_{Bi\_a} \\ I_{Bi\_b} \\ I_{Bi\_c} \end{bmatrix} \tag{28}$$

Each element of the SFP matrix shown in Eq. (20) is replaced by a (3×3) matrix as shown below.

$$S_i \Rightarrow \begin{bmatrix} S_{i\_a} & 0 & 0 \\ 0 & S_{i\_b} & 0 \\ 0 & 0 & S_{i\_c} \end{bmatrix} \tag{29}$$

The diagonal entries represent the load demand of phases a, b, and c, respectively, for load bus i. The FFP matrix is required to be modified accordingly. Here, each element in FFP matrix is a 3×3 impedance matrix as mentioned in Eq. (1). The supply bus voltage vector is also written in terms of its phase components as shown in Eq. (30).

$$[V_{B1}] = \begin{bmatrix} V_{B1\_a} \\ V_{B1\_b} \\ V_{B1\_c} \end{bmatrix} \tag{30}$$

Equation (25) used for calculating branch current I_{B1} has to be modified accordingly and I_{B1} is now a (3×1) matrix. The three-phase load at bus i is represented by a (3×3) matrix and the mathematical expression for I_{B1} vector is shown in Eq. (31).

$$\begin{bmatrix} I_{B1\_a} \\ I_{B1\_b} \\ I_{B1\_c} \end{bmatrix} = \left(\begin{bmatrix} S_{2\_a} & 0 & 0 & S_{3\_a} & \cdot\cdot \\ 0 & S_{2\_b} & 0 & 0 & \cdot\cdot \\ 0 & 0 & S_{2\_c} & 0 & \cdot\cdot \end{bmatrix} \left[\begin{bmatrix} V_{B1\_a} \\ V_{B1\_b} \\ V_{B1\_c} \\ V_{B1\_a} \\ V_{B1\_b} \\ V_{B1\_c} \\ \cdot \\ \cdot \end{bmatrix} - [\mathrm{FFP}] \begin{bmatrix} I_{B1\_a} \\ I_{B1\_b} \\ I_{B1\_c} \\ I_{B2\_a} \\ I_{B2\_b} \\ I_{B2\_c} \\ \cdot \\ \cdot \end{bmatrix} \right] \bullet \hat{}(-1) \right)^{*} \tag{31}$$

Here, small dots (.) are used to indicate other elements of same sequence.

3 Result Analysis

The proposed algorithm is coded using version 13.2 of MATLAB and implemented in Intel Core i3, 2.4 GHz processor, with 2 GB RAM digital computer. The developed approach is investigated on standard balanced and unbalanced RDS cases. This section highlights the performance of the approach in terms of accuracy and convergence rate and also the comparison is drawn with other existing methods.

3.1 Three-Phase Balanced RDS

In this section, the experimentation is made on two standard balanced RDS networks as Case 1 and Case 2. In Case 1, a 28 bus RDS [33] and in Case 2, 69 bus RDS [34] is considered. The voltage solutions obtained using the proposed LF approach for both the cases are found to be same as reported by other methods. The highest value of voltage deviation is obtained as 0.0675 and 0.0973 v for 28 bus and 69 bus RDS, respectively, when compared with original Backward–Forward (B/F) approach. The performance comparisons with other existing techniques are shown in Table 1. For comparison, the original B/F approach [12] along with other methods mentioned in Table 1 are also programmed and implemented for both the test cases. Results show that the proposed approach converges faster with less iteration compared to other methods. The computing time savings with respect to the original B/F approach are also mentioned in Table 1.

Table 1 Performance comparison for balanced RDS

Methods	Case 1: 28 bus RDS			Case 2: 69 bus RDS		
	Iter. No.	CPU time (s)	CPU time saving (%)	Iter. No.	CPU time (s)	CPU time saving (%)
B/F [12]	4	0.051	–	4	0.094	–
Direct approach [16]	4	0.034	33	4	0.066	29.7
IRLF [21]	3	0.022	56	3	0.035	63.54
Method in [22]	4	0.025	50.9	4	0.046	51.0
Proposed method	3	0.017	66.6	3	0.028	70.28

3.2 Three-Phase Unbalanced RDS

As majority of the DS in country like India operates under unbalanced conditions, therefore it is important to verify the effectiveness of the proposed approach in solving the PF problems for such networks. In view of this, the proficiency of the proposed approach is further verified in this section by solving LF problems for two different unbalanced (19 bus and 25 bus) RDS networks. The total real and reactive power load of 19 bus unbalanced RDS is 365 KW and 177 KVAR, respectively. On the other hand, the 25 bus unbalanced RDS supplies 3.24 MW and 2.39 MVAR loads. The systems data are taken from [35]. The results obtained with the proposed method are compared with other methods in Table 2. Like earlier, other methods including original B/F method were coded in MATLAB and implemented for both test systems for comparison. The savings in computing time with respect to B/F approach are mentioned in Table 2. Moreover, the LF solution accuracy of the proposed method compared to the original B/F approach is presented in Table 3 in respect to the maximum voltage deviations.

4 Conclusions

This paper presents a new and efficient unbalanced PF solution method to solve DLF problems. The developed algorithm is based on double forward sweep steps instead of conventional backward–forward sweep approach. The mathematical model of the proposed DLF technique for balanced and unbalanced networks are discussed in this paper and tested on various standard test cases. The algorithm is capable to produce accurate results in lesser computation time. The reason being; first, the replacement of conventional backward sweep step with relatively simpler forward sweep steps and second, the omission of conventional branch numbering phase. It is observed that

Table 2 **a** Performance comparison for 19 bus unbalanced RDS. **b** Performance comparison for 25 bus unbalanced RDS

Methods	**a** 19 bus unbalanced RDS		
	Iter. No.	CPU time (s)	CPU time saving (%)
B/F [12]	4	0.10	–
Direct approach [16]	4	0.060	40
IRLF [21]	3	0.051	49
Method in [22]	3	0.053	47
Proposed method	3	0.034	66
Methods	**b** 25 bus unbalanced RDS		
	Iter. No.	CPU time (s)	CPU time saving (%)
B/F [12]	4	0.120	–
Direct approach [16]	4	0.080	33.33
IRLF [21]	4	0.062	48.33
Method in [22]	4	0.066	45
Proposed method	3	0.040	66.66

Table 3 Maximum voltage deviations w.r.t. B/F approach

Test system	Maximum voltage magnitude (volt.) deviation			Maximum voltage angle (degree) deviation		
	Phase A	Phase B	Phase C	Phase A	Phase B	Phase C
19 bus unbalanced RDS	0.0035	0.005	0.005	5.63×10^{-8}	2.07×10^{-7}	2.65×10^{-8}
25 bus unbalanced RDS	0.0043	0.0062	0.006	6.14×10^{-8}	3.08×10^{-7}	3.10×10^{-8}

the algorithm gives accurate results for both balanced and unbalanced RDS cases. From these, it can be concluded that the proposed DLF method is efficient, faster, and accurate in solving DLF problems of power systems.

References

1. Dunstan, L.E.: Digital load flow studies. Trans. Amer. Inst. Elect. Eng. **73**, 825–831 (1954)
2. Ward, J.B., Hale, H.W.: Digital computer solution of power flow problems. Trans. Amer. Inst. Elect. Eng. **75**, 398–404 (1956)
3. Stevenson, W.D.: Elements of Power System Analysis. McGraw-Hill (1982)
4. Tinney, W.G., Hart, C.E.: Power flow solutions by Newton's method. IEEE Trans. Power Apparatus Syst. **PAS-86**, 1449–1457 (1967)
5. Stott, B.: Decoupled newton load flow. IEEE Trans. Power App. Syst. **PAS-91**, 1955–1959 (1972)

6. Stott, B., Alsac, O.: Fast decoupled load flow. IEEE Trans. Power App. Syst. **PAS-93**, 859–869 (1974)
7. Ulas, E., Hocaoglu, M.H.: Distribution systems forward/backward sweep-based power flow algorithms: a review and comparison study. EPCS. **37**, 91–110 (2008)
8. Teng, J.H.: A modified Gauss-Seidel algorithm of three-phase power flow analysis in distribution networks. Elect. Power Energy Syst. **24**(2002), 97–102 (2002)
9. Zhang, F., Cheng, C.S.: A modified Newton method for radial distribution system power flow analysis. IEEE Trans. Power Syst. **12**, 389–397 (1997)
10. Costa, V.M.D., Martins, N., Pereira, J.L.: Development in the Newton Raphson power flow formulation based on current injections. IEEE Trans. Power Syst. **14**, 1320–1326 (1999)
11. Lin, W.-M., Teng, J.-H.: Three-phase distribution network fast-decoupled power flow solutions. Elect. Power Energy Syst. **22**, 375–380 (2000)
12. Shirmohammadi, D., Hong, H.W., Semlyen, A., Luo, G.X.: A compensation-based power flow method for weakly meshed distribution and transmission networks. IEEE Trans. Power Syst. **3**, 753–762 (1988)
13. Cheng, C.S., Shirmohammadi, D.: A three-phase power flow method for real time distribution system analysis. IEEE Trans. Power Syst. **10**, 671–679 (1995)
14. Zhu, Y., Tomsovic, K.: Adaptive power flow method for distribution systems with dispersed generation. IEEE Trans. Power Deliv. **17**, 822–827 (2002)
15. Ghosh, S., Das, D.: Method for load-flow solution of radial distribution networks. IEE Proc. Gene. Trans. Distrib. **146**, 641–648 (1996)
16. Teng, J.H.: A direct approach for distribution system load flow solutions. IEEE Trans. Power Deliv. **18**, 882–887 (2003)
17. Chen, T.-H., Yang, N.-C.: Three-phase power-flow by direct Z_{BR} method for unbalanced radial distribution systems. IET Gener. Transm. Distrib. **3**, 903–910 (2009)
18. Alhajri, M.F., El-Hawary, M.E.: Exploiting the radial distribution structure in developing a fast and flexible radial power flow for unbalanced three-phase networks. IEEE Trans. Power Deliv. **25**, 378–389 (2010)
19. Wu, W.C., Zhang, B.M.: A three-phase power flow algorithm for distribution system power flow based on loop-analysis method. Int. J. Electr. Power Energy Syst. **30**, 8–15 (2008)
20. Augugliaro, et al.: A backward sweep method for power flow solution in distribution networks. Int. J. Electr. Power Energy Syst. **32**, 271–280 (2010)
21. Singh, S., Ghose, T.: Improved radial load flow method. Int. J. Electr. Power Energy Syst. **44**, 721–727 (2013)
22. Aravindhababu, P., Ganapathy, S., Nayar, K.R.: A novel technique for the analysis of radial distribution systems. Int J. Elect. Power Energy Syst. **23**, 167–171 (2001)
23. Eltantawy, A.B., Salama, M.A.: A novel zooming algorithm for distribution load flow analysis for smart grid. IEEE Trans. Smart Grid. **5** (2014)
24. Murat, D., et al.: A robust multiphase power flow for general distribution networks. IEEE Trans. Power Syst. **25** (2010)
25. Sergio, S., da Silva, L.C.P., Romero, R.: Generalised single-equation load flow method for unbalanced distribution systems.: IET Gen. Trans. Distrib. **5**, 347–355 (2011)
26. Ilhan, K., et al.: Multiphase load-flow solution for large-scale distribution systems using MANA.: IEEE Trans. Power Deliv. **29** (2014)
27. Ju, Y, et al.: Loop-analysis-based continuation power flow algorithm for distribution networks. IET Gener. Transm. Distrib. **8**, 1284–1292 (2014)
28. Das, B.: Uncertainty modeling of wind turbine generating system in power flow analysis of radial distribution network. EPSR. **111**, 141–147 (2014)
29. Ulas, E., Hocaoglu, M.H.: A new power flow method for radial distribution systems including voltage dependent load models.: EPSR. **76**, 106–114 (2005)
30. Satyanarayana, S., et al.: An efficient load flow solution for radial distribution network including voltage dependent load models. Electr. Power Compon. Syst. **35**(5), 539–551 (2007)
31. Nagaraju, K., et al.: A novel load flow method for radial distribution systems for realistic loads. Electr. Power Compon. Syst. **39**(2), 128–141 (2011)

32. Li, H., et al.: A load flow method for weakly meshed distribution networks using powers as flow variables. Int. J. Elect. Power Energy Syst. **58**, 291–299 (2014)
33. Das, D., Nagi, H.S., Kothari, D.P.: Novel method for solving radial distribution networks. IEE Proc-Gener. Transm. Distrib. **141**(4), 291–298 (1994)
34. Savier, J.S., Das, D.: Impact of network reconfiguration on loss allocation of radial distribution systems.: IEEE Trans. Power Deliv. **22**(4), 2473–2480 (2007)
35. Subrahmanyam, J.B.V., Radhakrishna, C.: A simple method for feeder reconfiguration of balanced and unbalanced distribution systems for loss minimization. Electr. Power Compon. Syst. **38**(1), 72–84 (2009)

Line Stability Indices and Contingency Screening by Sensitivity Factors Based Static Voltage Stability Study

Rinkesh A. Jain and Darshan B. Rathod

Abstract Due to ever increasing load demand, currently, power system has become stressful in context with reactive power management and voltage control which may in turn lead to voltage instability problems in electrical power system. If the voltage stability is not evaluated and the problems occurred are not attended timely, sequential outages of components of power system may occur and this may lead to voltage collapse or blackout. Therefore, detect voltage collapse point, voltage stability is needed. Several voltage stability indices have been suggested in the literature to assess static and dynamic voltage stability of power system based on power flow through transmission line. In the present work, three voltage stability indices, i.e., Line Stability Index (L_{mn}), Fast Voltage Stability Index (FVSI), and Voltage Collapse Proximity Index (VCPI) have been used to detect voltage stability status. Contingency screening has been carried out using Linear Sensitivity Factors (LSFs) based on Z-bus. Voltage stability indices and contingency screening results have been obtained on IEEE 14-bus test system.

Keywords Linear Sensitivity Factors (LSFs) · Voltage stability indices · Z-bus method

Acronyms

VCPI	Voltage collapse proximity index
L_{mn}	Line stability index
LODF	Line outage distribution factor
θ	Impedance angle

R. A. Jain (✉) · D. B. Rathod
PG Student, Electrical Engineering, Shantilal Shah Engineering College, Bhavnagar, Gujarat, India
e-mail: rinkeshjain72@yahoo.com

D. B. Rathod
e-mail: darshanbr2012@gmail.com

D. Deb et al. (eds.), *Innovations in Infrastructure*, Advances in Intelligent Systems and Computing 757, https://doi.org/10.1007/978-981-13-1966-2_11

varphi Power factor angle

1 Introduction

Voltage stability is the ability of a power system to maintain steady voltages at all buses in the system after being subjected to a disturbance from a given initial operating condition [1]. Nowadays, voltage instability is a challenging problem in a power system. Insufficient of supply and unnecessary absorption of reactive power cause voltage instability in a power system. Voltage instability may result in the voltage collapse if necessary actions are not taken immediately to restore the system voltage within limits.

Voltage stability study has been classified into static and dynamic analysis. For static voltage stability study, algebraic equations are solved. Therefore, static voltage stability analysis is computationally less complex than dynamic analysis [2]. In this paper, static voltage stability analysis has been carried out to identify voltage stability status.

Static voltage stability analysis is to be done with various methods such as PV and QV curves [3], reduced Jacobian matrix based modal analysis [4] but these methods are time consuming for interconnected system network. Nowadays, numbers of voltage stability indices such as VCPI, L_{mn}, FVSI, Line Stability Factor (LQP), and New Voltage Stability Index (NVSI) have been used to assess system voltage stability status [5]. In the present paper, simulation and result analysis of L_{mn}, FVSI, and VCPI are carried out for static voltage stability study. To identify the distance from the particular current operating point to the point of voltage collapse, these all indices are used.

Contingency means unpredictable event/outage and it may be caused by line outage or change in generation in the system which could lead the voltage instability. Contingency analysis can be done using different methods such as AC power flow, LSFs [6], line stability indices [7], various artificial intelligence techniques, etc. In the present paper, contingency analysis has been carried out using LSFs based on Z-bus method. From the contingency analysis, power system operator can know about the effect on power system when an outage of any particular line or generator occurs.

2 Indices Formulation

2.1 Line Stability Index (L_{mn})

Moghavemmi [8] has expressed L_{mn} by analyzing power flow in a single transmission line. The L_{mn} index is detected the distance from current operating point to the point of voltage collapse. The L_{mn} value is 0 in no load condition and 1 in collapse condition. L_{mn} is given by Eq. (1)

$$L_{mn} = \frac{4xQ_r}{[V_s \sin(\theta - \delta)]^2} \tag{1}$$

where θ is impedance angle, Q_r is the receiving bus reactive power flow in pu, x is line reactance in pu, δ is angular difference between sending end and receiving end bus voltage, and V_s is sending end bus voltage in pu.

2.2 *Fast Voltage Stability Index (FVSI)*

Musirin [9] has used the same concept of power flow through a single transmission line and the derived FVSI can be given by Eq. (2)

$$\text{FVSI} = \frac{4|Z|^2 Q_r}{|V_s|^2 x} \tag{2}$$

where Z is the line impedance in pu.

FVSI can be calculated based on the above equation which depends on the reactive power flow through transmission line. The line has index value is close to 1 indicates that line goes into instability condition and may cause voltage collapse.

FVSI can also use to identify weak bus based on a maximum allowable load on the bus in a system. The weakest bus in the system is considered as a bus which has minimum the value of maximum allowable reactive load in the whole system.

2.3 *Voltage Collapse Proximity Index*

Moghavvemi [8] has expressed VCPI index for the investigate voltage stability of each line based on the same concept power transfer through the line. VCPI is given by Eq. (3)

$$\text{VCPI(P)} = \frac{P_r}{P_{r(\max)}} \tag{3}$$

where P_r is the active power flow at the receiving end bus in pu and $P_{r(\max)}$ is the maximum active power transferred through the line in pu is given by Eq. (4).

$$P_{r(\max)} = \frac{V_s^2}{Z} \cdot \frac{\cos\varphi}{4\cos^2\left(\frac{\theta-\varphi}{2}\right)} \tag{4}$$

where $\varphi = \tan^{-1}\frac{Q_r}{P_r}$ = power factor angle.

With the increasing power transfer through a line, VCPI index value is increased when the VCPI reaches to 1, and the system voltage collapses. VCPI is adequate for indicating voltage collapse in the line.

3 Linear Sensitivity Factors (LSFs)

There are thousands of possible outages on a daily basis. It is difficult to solve the outage with less time in a power system. With the help of LSFs, it is possible to get the quick and fast solution of possible overloads [6]. These factors give approximate change in line flow with the change in generation or outage of any line. These factors are basically two types:

- Line Outage Distribution Factor (LODF)
- Generation Shift Factor (GSF)

In this paper, only LODF is used for contingency screening.

3.1 LODF Using Z-Bus Element

When the transmission circuit is lost, the LODF is used to verify the overload of the line [10]. It is given by Eq. (5),

$$L_{ij,mn} = -\frac{Z_a}{Z_c}\left[\frac{(Z_{im} - Z_{in}) - (Z_{jm} - Z_{jn})}{Z_{th,mn} - Z_a}\right] \quad (5)$$

where mn = outage line, ij = ne whose post-outage power flow is to be checked, Z_a = impedance of outage line, Z_c = impedance of line under consideration, $Z_{im}, Z_{in}, Z_{jm}, Z_{jn}$ = off-diagonal elements of Z-bus.

$$Z_{th,mn} = Z_{mm} + Z_{nn} - 2 * Z_{mn} \quad (6)$$

where Z_{mm}, Z_{nn} = diagonal elements of Z-bus.

Post-outage power flow in line *i–j* due to outage of line *m–n* is given by Eq. (7),

$$P'_{ij} = P_{ij} + L_{ij,mn}P_{mn} \quad (7)$$

where I_{ij} = pre-outage power flow in line *i–j*, I_{mn} = pre-outage power flow in line *m–n*.

4 Test Result and Discussion

In order to demonstrate the effectiveness of voltage stability indices and LSFs numerical analysis have been made for the IEEE 14-bus test system.

4.1 Base Case Loading

Base loading means that load at all buses to prespecified load of IEEE 14 bus test system [11]. Lines with smaller voltage stability indices have much more voltage stability margin whereas larger voltage stability indices indicate that lines are heavily stressed condition and further addition of load line goes into voltage collapse condition.

Different voltage stability indices have been calculated for IEEE 14 bus system are presented in Table 1. It is observed from this Table 1 that line 7–8 has the highest index value compared to another line because compensator is connected to bus 8. In base case loading, no one line becomes stressful. Therefore, more reactive power can supply in base the case.

4.2 Global Load Increase

Global load increase, it means that equal percentage of active, reactive or apparent load are increased on all buses. Loading means increased active or reactive load on the load bus of IEEE 14 bus test system from its base loading.

From Table 2, it is observed that L_{mn} and FVSI have a similar result for all types of loading for 20 and 40% loading. If the load is increased, the indices value is increased in all three cases but L_{mn} and FVSI gives system status when reactive load increases.

Table 1 Voltage stability indices for IEEE 14—bus test system with base case loading

Line	L_{mn}	FVSI	VCPI
1–2	0.0688	0.0645	0.0451
1–5	0.0203	0.0184	0.0275
2–4	0.0227	0.0211	0.02
3–4	0.0365	0.0377	0.0252
4–7	0.0901	0.0899	0.045
5–6	0.0767	0.0759	0.038
6–13	0.0388	0.0381	0.0252
7–8	0.111	0.111	0.0557
9–14	0.0431	0.0424	0.0281
13–14	0.0228	0.0224	0.0148

Table 2 Indices result of line 13–14 for global loading

	20% loading			40% loading		
Types of loading	L_{mn}	FVSI	VCPI	L_{mn}	FVSI	VCPI
Reactive	0.0373	0.0367	0.0821	0.052	0.0512	0.0924
Active	0.0267	0.0261	0.0864	0.0314	0.0307	0.1011
Apparent	0.0449	0.044	0.0984	0.061	0.0597	0.1211

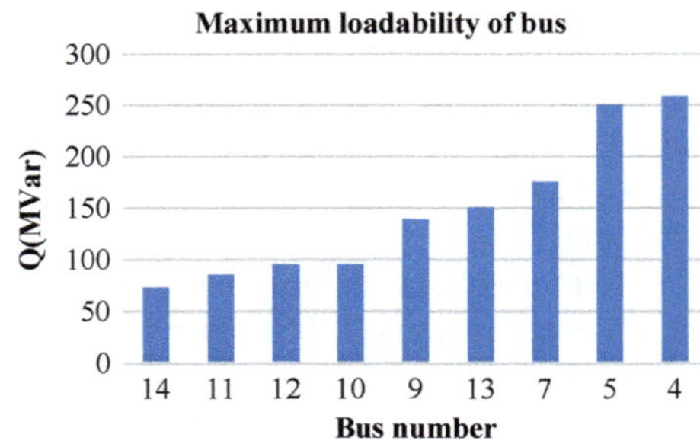

Fig. 1 Maximum loadability of load buses in IEEE 14 bus system

VCPI is more suitable for active and apparent loading condition. In all these buses, the voltage is more than 0.9 pu in all cases so that the system voltage remains in stable condition. Table 2 shows that in power system, till 40% load increase, no lines become critical because the voltage stability index for 40% loading is less than one. So for global loading, system maintain stability for up to 40% or more loading.

Voltage stability indices can also determine the weakest bus in the system and it is based on the maximum loadability of a bus. Arrange maximum loadability of the bus in ascending order and lowest maximum loadability of bus is considered as the weakest bus.

From Fig. 1, it is shown that bus 14 has the less reactive loading. So, bus 14 is considered as the weakest bus in the system and this bus is also far away from the generator so less active power reach to bus 14. Bus 4 is the largest loadability compared to all load buses. So, this bus is the strongest bus in the system and bus 4 is near to generator so it contains large active power from the generator.

Table 3 shows the voltage stability indices for maximum loading condition of load bus and it shows that line 9–10 and 10–11 are having largest indices value for loading on bus 10. So, these lines become critical for loading on bus 10 (Fig. 2).

Figure 3 shows that line 6–13 is the critical line for loading on bus 13 and when this load is 240 MVAR, this line achieves FVSI value near to unity.

The chart presented in Fig. 4 shows the value of VCPI in each variation of reactive load for bus 13. Line 13–14 needs some higher value of reactive loading to attain unity value of VCPI index.

Table 3 Line stability indices for IEEE 14 test system

Load (MVAR)	Line	L_{mn}	FVSI	VCPI
Q10 = 94.8	9–10	0.5269	0.5301	0.5217
	10–11	0.5411	0.5346	0.5095
Q11 = 85	6–11	0.8089	0.8244	0.7647
	10–11	0.5431	0.606	0.5801
Q14 = 72.8	9–14	0.833	0.8837	0.7971
	13–14	0.7924	0.8193	0.7439
Q4 = 250	3–4	0.4648	0.4925	0.4445
	2–4	0.4215	0.4017	0.4941
Q13 = 150	6–13	0.5866	0.6186	0.5391
	12–13	0.4389	0.4914	0.3035
Q7 = 175	7–8	1.0139	1.0139	0
	4–7	0.3784	0.3773	0.4022
Q12 = 95	12–13	1.0579	0.8722	0.2544
	6–12	0.6019	0.6413	0.1337
Q5 = 258	5–6	0.5992	0.59	0.726
	2–5	0.4528	0.4406	0.3654

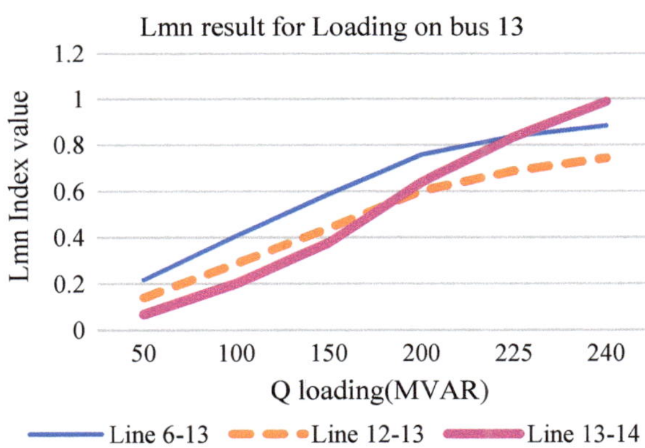

Fig. 2 L_{mn} versus reactive load variation for bus 13

Table 4 shows that lines which have L_{mn} value is near to unity for maximum reactive loading and these all lines are considered as the most critical line for individual bus loading.

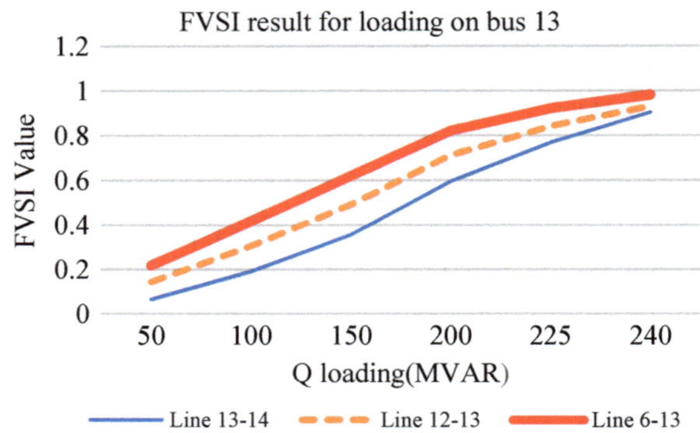

Fig. 3 FVSI versus reactive load variation for bus 13

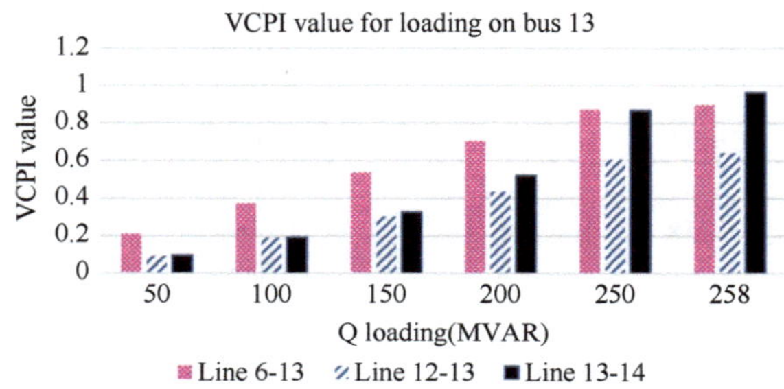

Fig. 4 VCPI versus reactive load variation for bus 13

Table 4 Critical line for individual bus loading based on L_{mn}

Bus	Maximum reactive load	Critical line
14	120	13–14
13	240	13–14
12	90	12–13
10	148	10–11
7	175	7–8
11	195	6–11
5	340	5–6
4	440	4–7

5 Contingency Screening

Steps to follow for performing contingency screening by using LSFs based on Z-bus method.

Step 1. Obtain Z-bus for IEEE 14 bus test system using MATLAB coded program.
Step 2. Calculate pre-outage power flow for all lines.
Step 3. Find out LSFs from Z-bus element for particular line outage.
Step 4. Calculate post-outage power flow using LSFs.
Step 5. Compare post-outage power flow with MW limit of line.

Table 5 Contingency screening of IEEE 14 bus system using LSFs

Contingency (outage)	Most severe line	Rank	Power in pu
2–1	5–1	1	2.1885
5–1	2–1	2	2.1885
6–5	2–1	3	1.5017
11–6	2–1	4	1.4803
13–6	2–1	5	1.4796
14–13	2–1	6	1.4794
12–6	2–1	7	1.4782
13–12	2–1	8	1.4780
8–7	2–1	9	1.4779
11–10	2–1	10	1.4767
10–9	2–1	11	1.4758
14–9	2–1	12	1.4751
9–4	2–1	13	1.4745
7–4	2–1	14	1.4697
9–7	2–1	15	1.4696
4–3	2–1	16	1.4277
4–2	5–4	17	1.3278
3–2	2–1	18	1.3327
5–2	2–1	19	1.3303
5–4	2–1	20	1.2977

Step 6. Extract lines which exceed its MW limit and choose one line which is maximum MW overloading margin from MW limit. Consider this line is the most stressed line for particular line outage.
Step 7. Repeat steps 1–6 for all line outages.
Step 8. Arrange the line with MW margin in descending order.
Step 9. Top order outage is considered as the most critical line outage.

Table 5 shows that in most of the line outage case, line 2–1 becomes most critical line because this line is directly connected to a bus where large generator connected. So, that for every outage, most of active power supply by generator is flowing through line 2–1 so this line is the most stressful line.

6 Conclusion

This paper presents a comparative study and analysis of the different voltage stability indices for static voltage stability assessment. Voltage collapse occurs when system is heavily loaded, the voltage magnitude decreases. These indices can predict voltage stability of power system under all operating condition. These indices can be used to identify critical lines and weakest bus in the system. Based on simulation result, it can be concluded that to identify voltage stability status, VCPI gives accurate result when active power change in the system while L_{mn} and FVSI gives accurate for reactive power change in the system. From the contingency screening, it can be concluded that line 1–2 is a most severe line and which is overloaded in all the line outages and line 1–2 is most severe outage in the system.

References

1. Hill, J.P., David, P.Kundur: Definition and classification of power system stability. IEEE Trans. Power Syst. **19**(2), 1387–1401 (2004)
2. Reis, C., Barbosa, F.P.M., Member, S.: Line indices for voltage stability assessment. In: IEEE Bucharest Power Tech Conference, pp. 1–6 (2009)
3. Chiang, H., Jumeau, R.J.: Toward a practical performance index for predicting voltage collapse in electric power systems. IEEE Trans. Power Syst. **10**(2), 584–592 (1995)
4. Ayasun, S., Nwankpa, C.O., Kwatny, H.G.: Voltage stability toolbox for power system education and research. **49**(4), 432–442 (2006)
5. Yari, S.: Assessment of line stability indices in detection of voltage stability status. pp. 7–11 (2017)
6. Wood, B.F.W., Allen, J.: Power Generation, Operation, and Control. Wiley
7. Tiwari, R., Niazi, K.R., Gupta, V.: Electrical power and energy systems line collapse proximity index for prediction of voltage collapse in power systems. Elsevier Electr. Power Energy Syst. **41**(1), 105–111. Oct, 2012
8. Moghavvemi, F.M.O.M.: Technique for contingency monitoring and voltage collapse prediction. In: IEEE Proceedings—Generation Transmission and Distribution, vol. 145, no. 6 (1998)
9. Musirin S.K.N.A., Zin, A.M., Khairuddin, A.: A comparison of voltage stability indices 1. IEEE 8th International Conference on Power and Energy Engineering Conferences, pp. 30–34. March, 2014
10. Grainger, W.D.S.J.J.: Power System Analysis. McGraw-Hill (1994)
11. https://www2.ee.washington.edu/research/pstca/pf14/pg_tca14bus.htm

Comparative Analysis on Security-Constrained Optimal Power Flow Using Linear Sensitivity Factors-Based Contingency Screening

Darshan B. Rathod and Rinkesh A. Jain

Abstract Security-Constrained Optimal Power Flow (SCOPF) is a significant tool used for analysis of power system operation and planning. This paper presents a solution of SCOPF considering critical contingencies simulated on IEEE 30-bus system. The main objective of the presented work is to minimize the total generation cost. An interior point algorithm has been used to find out a feasible and optimal solution with minimum computational time for secured power system operation. Contingency screening for SCOPF formulation has been accomplished with the help of Linear Sensitivity Factors (LSFs) obtained from the Z-bus algorithm. Comparative analysis has been carried out for the results obtained with those of other techniques published in the literature for same test cases.

Keywords Contingency screening · Interior point algorithm · LSFs · Optimal power flow (OPF) · SCOPF

1 Introduction

SCOPF plays a major role in the power system economic operation and security study. Minimization of generation cost without breaching security constraints is a challenging task for power system analyst in a large interconnected power system having many threats under normal and stressed conditions. SCOPF is an extension of OPF, wherein OPF problem is augmented to consider contingency cases such as line outage, generation outage, or important outage of any important components of power system to attain an operating point which is secure and optimal [1]. Thus SCOPF problem is highly constrained, nonlinear, and non-convex problem of the power system [2]. With increasing the size of network and more contingencies to

D. B. Rathod (✉) · R. A. Jain
Shantilal Shah Engineering College, Bhavnagar, Gujarat, India
e-mail: darshanbr2012@gmail.com

R. A. Jain
e-mail: rinkeshjain72@yahoo.com

D. Deb et al. (eds.), *Innovations in Infrastructure*, Advances in Intelligent Systems and Computing 757, https://doi.org/10.1007/978-981-13-1966-2_12

be considered, SCOPF problem becomes more complex and computational time increases. The overall SCOPF problem has been based on two types of security analysis; steady state and dynamic security [3]. In steady-state security-based SCOPF, the system operators monitor the power flow with stimulating contingencies to analyze any case of overload and voltage violations.

Network designers and system operators typically rely on the use of conventional methods, including Newton's method [4], quadratic programming [4], interior point method [5], etc., to solve OPF/SCOPF problem. Recently, many meta-heuristic algorithms are proposed to solve power system problems. They have a good performance in finding the global optimum solution but require more computational time as well as are not adaptive to online interface of the dynamical behavior of power system. That is why still conventional OPF algorithms are widely used due to computational efficiency and strong theoretical background [6].

This paper is organized as follows: Sect. 2 presents a summary of SCOPF problem formulation. Contingency screening is carried out in Sect. 3. Section 4 shows results of SCOPF for IEEE 30-bus test system.

2 SCOPF Problem Formulation

SCOPF problem deals with adjusting controlling parameters of the system to acquire optimal solution while considering security. But one concern that needs to be taken care of is while acquiring optimal solution of SCOPF the security of power system is not violated [1]. That means, if there is any outage occurs during operation, the system can somehow manage the condition and stay intact even after outage.

For ensuring the security of the system, contingency constraints for critical contingencies should be identified and SCOPF problem is to be formulated. Some possible controllable parameters are:

- Generator MW outputs.
- Generator voltages.
- Transformer tap ratio.
- Switched capacitor settings.

SCOPF can be implemented by first formulating OPF. For OPF problem, the objective function is generation cost to be minimized can be given by Eq. (1).

$$f = \sum_{i} \mathrm{K}_i(\mathrm{P_{Gi}}) \tag{1}$$

where K_i is the cost function for generating $\mathrm{P_{Gi}}$ power at the respective bus. A quadratic generation cost function is considered and given by Eq. (2).

$$f(\mathrm{P_{Gi}}) = \mathrm{a}_i\mathrm{P}_{\mathrm{Gi}}^2 + \mathrm{b}_i\mathrm{P_{Gi}} + \mathrm{c_i} \tag{2}$$

The solution must satisfy constraints at the pre-contingency state and constraints at the post-contingency state.

Equality Constraints at Pre-Contingency

The active and reactive power balance equations in the pre-contingency case are given by Eqs. (3) and (4).

$$\sum_{i=1}^{\mathrm{NG}} \mathrm{P}_{\mathrm{Gi}}^{0} - \sum_{j=1}^{\mathrm{NLoad}} (\mathrm{P}_{\mathrm{Load}j}) - \mathrm{P}_{\mathrm{Loss}}^{0} = 0 \tag{3}$$

$$\sum_{i=1}^{\mathrm{NG}} \mathrm{Q}_{\mathrm{Gi}}^{0} - \sum_{j=1}^{\mathrm{NLoad}} (\mathrm{Q}_{\mathrm{Load}j}) - \mathrm{Q}_{\mathrm{Loss}}^{0} = 0 \tag{4}$$

Equality Constraints at Post-Contingencies

The active and reactive power balance equations in the post-contingency case are given by Eqs. (5) and (6).

$$\sum_{i=1}^{\mathrm{NG}} \mathrm{P}_{\mathrm{Gi}}^{\mathrm{k}} - \sum_{j=1}^{\mathrm{NLoad}} (\mathrm{P}_{\mathrm{Load}j}) - \mathrm{P}_{\mathrm{Loss}}^{\mathrm{k}} = 0 \tag{5}$$

for $k = 1,2, \ldots \mathrm{K}$

$$\sum_{\mathrm{i}=1}^{\mathrm{NG}} \mathrm{Q}_{\mathrm{Gi}}^{\mathrm{k}} - \sum_{\mathrm{j}=1}^{\mathrm{NLoad}} (\mathrm{Q}_{\mathrm{Load}j}) - \mathrm{Q}_{\mathrm{Loss}}^{\mathrm{k}} = 0 \tag{6}$$

for $\mathrm{k} = 1,2, \ldots \mathrm{K}$

Inequality Constraints

Inequality constraints such as limit on voltage at generator bus, limit on real power generation and transformer tap ratio limit must be taken into consideration. Limits on control variables are given by:

Limit on voltage at generator bus is given by Eq. (7).

$$\mathrm{V}_i^{\min} \le \mathrm{V}_i \le \mathrm{V}_i^{\max}, \text{ for } i = 1, \ldots, \mathrm{N} \tag{7}$$

Limit on real power generation is given by Eq. (8).

$$\mathrm{P}_{\mathrm{Gi}}^{\min} \le \mathrm{P}_{\mathrm{Gi}} \le \mathrm{P}_{\mathrm{Gi}}^{\max}, \text{ for } i = 1, \ldots, \mathrm{NG} \tag{8}$$

Transformer tap ratio limit is given by Eq. (9).

$$\mathrm{a}_i^{\min} \le \mathrm{a}_i \le \mathrm{a}_i^{\max}, \text{ for } i = 1, \ldots, \mathrm{NT} \tag{9}$$

Functional constraints

Functional constraints such as limit on voltage at PQ bus, reactive power generation limit, line overflow limit, etc., are also considered for SCOPF problem.

Limit on voltage at PQ bus is given by Eq. (10).

$$V_{PQi}^{min} \leq V_{PQi} \leq V_{PQi}^{max} \text{ , for } i = 1, \ldots, N \tag{10}$$

Line overflow limit is given by Eq. (11).

$$S_i^{min} \leq S_i \leq S_i^{max} \text{ , for } i = 1, \ldots, N \tag{11}$$

Reactive power generation limit is given by Eq. (12).

$$Q_i^{min} \leq Q_i \leq Q_i^{max} \text{ , for } i = 1, \ldots, NG \tag{12}$$

3 Contingency Screening

The contingency screening analysis is carried out using LSFs based on Z-bus [7]. LSFs approximate the change of branch power flow depending on the shift of generation or outage of any other branch. These factors are of two types

- Line Outage Distribution Factor (LODF) [8].
- Generation Shift Factor (GSF) [8].

3.1 LODF Using Z-Bus

For simulating outage of branch, it is convenient to calculate LODF. This factor illustrates the change of power flow in the line ij when the outage of line mn occurred.

LODF is represented in Eq. (13).

$$L_{ij,mn} = \frac{\Delta P_{ij}^{mn}}{P_{mn}^{0}} \tag{13}$$

LODF using Z-bus method is given by Eq. (14).

$$L_{ij,mn} = -\frac{Z_a}{Z_c}\left[\frac{(Z_{im} - Z_{in}) - \left(Z_{jm} - Z_{jn}\right)}{Z_{th,mn} - Z_a}\right] \tag{14}$$

where lower index *mn* shows outage line and *ij* shows line whose post-outage power flow is to be checked, Z_a = impedance of outage line, Z_c = impedance of line under consideration, $Z_{im}, Z_{in}, Z_{jm}, Z_{jn}$ are off-diagonal elements of Z-bus.

Post-outage power flow in line *ij* due to outage of line *mn* is given by Eq. (15).

$$P'_{ij} = P_{ij} + L_{ij,mn} P^0_{mn} \tag{15}$$

3.2 GSF Using Z-Bus

For simulating outage of generator, it is convenient to calculate GSF. This factor describes the change of power flow in the line ij rely on the change of generation in the bus m. GSF represented in Eq. (16).

$$K^m_{ij} = \frac{\Delta P_{ij}}{\Delta P_m} \tag{16}$$

This factor gives approximation in-line flow due to change in generation on particular bus. GSF using Z-bus method is given by Eq. (17).

$$K_{ij,m} = \frac{Z_{im} - Z_{jm}}{Z_c} \tag{17}$$

where *m* shows bus whose generation is to be changed and *ij* shows line whose post-outage power flow is to be checked, Z_c = impedance of line under consideration, Z_{im}, Z_{jm} are off-diagonal elements of Z-bus.

Post-outage line flow in line *ij* due to change in generation at bus *m* is calculated by Eq. (18).

$$P'_{ij} = P^0_{ij} + K_{ij,m}\, \Delta P_m \tag{18}$$

4 Results and Discussions

Simulation of IEEE 30-bus standard test system is carried out for OPF and SCOPF as per loading conditions considered in [9]. Formulation of Z-bus is done using step-by-step method and from Z-bus LSFs are calculated. MATPOWER 6.0 toolbox [10] has been used to obtain OPF and SCOPF solution. The objective function considered here is minimization of the total generation cost. The line data and bus generation data for IEEE 30-bus system is taken from [9]. Two different cases are simulated and the results have been analyzed.

1. OPF and SCOPF case considering seven critical outages/contingencies.
2. OPF and SCOPF case considering a generator outage.

Table 1 Details for case 1

Control parameters		Line outages
Generator bus voltage	VG1, VG2, VG5, VG8, VG11, VG13	1, 2, 4, 5, 7, 33, 35
Generator active power injection	PG2, PG5, PG8, PG11, PG13	

Table 2 Details for case 2

Control parameters		Generator outage
Generator bus voltage	VG1, VG2, VG5, VG8, VG11, VG13	13
Generator active power injection	PG2, PG5, PG8, PG11, PG13	

4.1 Case 1

Here, OPF-base case and SCOPF problem are simulated for IEEE 30-bus system using MATPOWER. For SCOPF case, total of seven critical line outages obtained from LSFs are chosen as shown in Table 1. For secure and reliable operation of power system without violation of any security constraint, a generation rescheduling needs to be done. The test results are shown in Table 3 and that reveals due to rescheduling the cost of generation gets increased.

4.2 Case 2

Here, outage of generator at bus 13 is simulated using MATPOWER and details of case 2 is given in Table 2. By observing results of generator outage case, generation cost is increased compared to line outage case.

The control parameter's value for the optimal solution obtained in both the cases has been shown in Table 3. Rated line data for both cases are considered to be their maximum line flow limits.

Total generation cost of OPF and SCOPF for both the cases are represented in Fig. 1. As seen from Fig. 1, generation cost for SCOPF case is little more than OPF case.

Active power generation at different buses is shown in the Fig. 2. This graph gives comparative results of both OPF and SCOPF case for case 1 for active power generation. Comparative study of test case 1 with gradient method is given in Table 4.

A comparative study of other algorithms available in the literature for OPF and SCOPF has been given in Tables 5 and 6. Comparison shows that IPM gives better performance compared to other methods found in literature.

Table 3 Control parameters for both cases

Optimal control variable	Case 1		Case 2	
	OPF	SCOPF	OPF	SCOPF
PG2	48.87	58.45	50.24	62.49
PG5	21.52	24.12	21.95	25.07
PG8	22.23	35.99	25.52	41.32
PG11	12.27	16.52	13.34	18.7
PG13	11.36	14.23	0	0
VG1	1.06	1.058	1.06	1.06
VG2	1.042	1.044	1.043	1.044
VG5	1.015	1.009	1.015	1.012
VG8	1.02	1.025	1.018	1.023
VG11	1.082	1.062	1.06	1.06
VG13	1.071	1.093	1.041	1.044
Losses (MW)	9.456	7.263	9.879	7.426
Fuel cost ($/h)	802.2	812.1	805.98	818.86

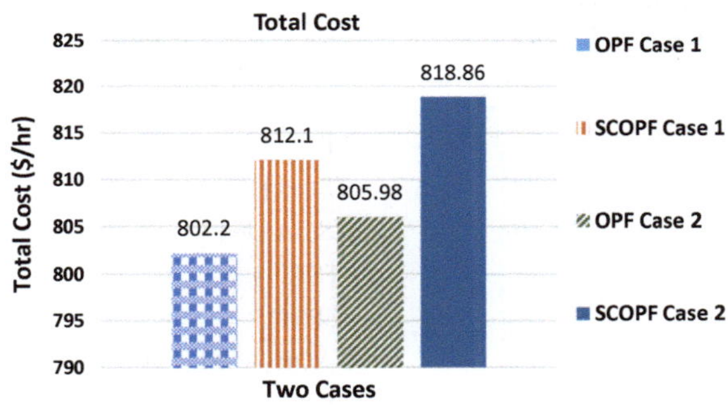

Fig. 1 Representation of total cost for test cases

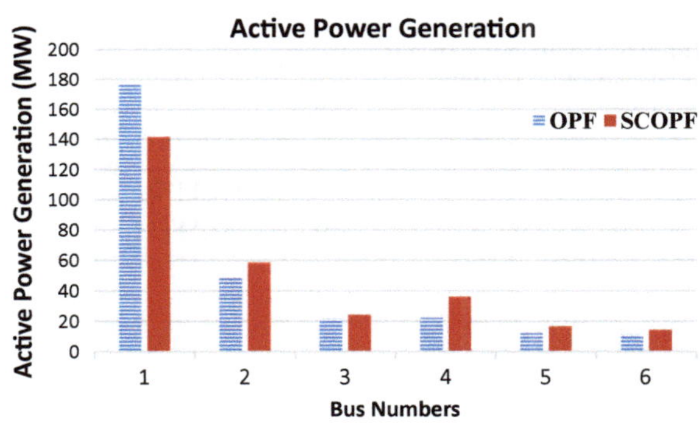

Fig. 2 Representation of active power generation for case 1

Table 4 Comparative study of test case 1

Optimal Solution	OPF		SCOPF	
	IPM	Gradient [1]	IPM	Gradient [1]
PG2	48.87	48.84	58.45	57.56
PG5	21.52	21.51	24.12	24.56
PG8	22.23	22.15	35.99	35
PG11	12.27	12.14	16.52	17.93
PG13	11.36	12	14.23	16.91
VG1	1.06	1.05	1.06	1.05
VG2	1.042	1.038	1.044	1.033
VG5	1.015	1.011	1.009	1.005
VG8	1.02	1.019	1.025	1.023
VG11	1.06	1.09	1.06	1.09
VG13	1.06	1.09	1.06	1.08
Losses (MW)	9.456	9.48	7.263	7.11
Fuel cost ($/h)	802.2	802.4	812.1	813.74

Table 5 Comparative results of test case 1—OPF

Technique	Fuel cost ($/h)
IPM	802.2
Gradient [1]	802.4
EP [11]	802.907
TS [11]	802.502
TS/SA [11]	802.788
ITS [11]	804.556
IEP [11]	802.465
SADE_ALM [12]	802.404
pSADE_ALM [12]	802.405

Table 6 Comparative results of test case 1—SCOPF

Technique	Fuel cost ($/h)
IPM	812.1
Gradient [1]	813.74
SADE_ALM [12]	834.54
pSADE_ALM [12]	826.97

5 Conclusion

This paper presents, a method for the solution of security-constrained OPF problem using IP algorithm. The results of the different cases of the IEEE 30-bus test system show the potential of the suggested method for the SCOPF problem. LSFs based contingency screening technique using Z-bus is very fast and fairly accurate for further SCOPF implementation. By observing all the results, it can be concluded that the cost of generation in SCOPF is found to be little more than that of OPF in order to make the system secure. This extra cost is a difference between the security and economy operation. Interior point method's convergence time is less. Thus, this method satisfies the basic necessity of having a very fast static assessment of power system security.

References

1. Alsac, O., Stott, B.: Optimal load flow with steady-state security. IEEE Trans. Power Appar. Syst. **PAS-93**(3), 745–751 (1974)
2. Monticelli, A., Pereira, M.V.F., Granville, S.: Security-constrained optimal power flow with post-contingency corrective rescheduling. IEEE Trans. Power Syst. **2**(1), 175–180 (1987)
3. Momoh, J.A., El-Hawary, M.E., Adapa, R.: A review of selected optimal power flow literature to 1993 part I: nonlinear and quadratic programming approaches. IEEE Trans. Power Syst. **14**(1), 96–103 (1999)
4. Monoh, J.A., Ei-Hawary, M.E., Adapa, R.: A review of selected optimal power flow literature to 1993 part II: Newton, linear programming and interior point methods. IEEE Trans. Power Syst. **14**(1), 105–111 (1999)
5. Capitanescu, F., Glavic, M., Ernst, D., Wehenkel, L.: Interior-point based algorithms for the solution of optimal power flow problems. Electr. Power Syst. Res. **77**(5–6), 508–517 (2007)
6. Gunda, J., Djokic, S., Langella, R., Testa, A.: Comparison of conventional and meta-methods for security-constrained OPF analysis. In: 2015 AEIT International Annual Conference (AEIT), pp. 58–65 (2015)
7. J. J. Grainer and W. Stevenson, *Power System Analysis*. McGraw-Hill, 1994
8. Wood, A.J., Wollenberg, B.F.: Power Generation, Operation, and Control. Wiley, New York (1984)
9. Christie, R: Power Systems Test Case Archive. Washington University, 2013. www.ee.Washington.edu/research/pstca
10. Zimmerman, R.D., Murillo-Sánchez, C.E., Thomas, R.J.: MATPOWER's extensible optimal power flow architecture. In: 2009 IEEE Power and Energy Society General Meeting, PES'09, no. 2, pp. 1–7 (2009)
11. Ongsakul, W., Tantimaporn, T.: Optimal power flow by improved evolutionary programming. Electr. Power Compon. Syst. **34**(1), 79–95 (2006)
12. Thitithamrongchai, C., Eua-Arporn, B.: Security-constrained optimal power flow: a parallel self-adaptive differential evolution approach. Electr. Power Compon. Syst. **36**(3), 280–298 (2008)

Probabilistic Evaluation of Seismically Isolated Building Using Quintuple Friction Pendulum Isolator

Ankit Sodha, Sandip A. Vasanwala and Devesh Soni

Abstract Multiple-level performance of building provided with seismic isolation by Quintuple Friction Pendulum Isolator (QTFP) is investigated multiple-level seismic hazards (design basis earthquake (DBE), service level earthquake (SLE), and maximum considered earthquake (MCE)). Five independent pendulum mechanisms along with six concave surfaces are included in this isolator. The paper describes nonlinear mathematical model and seismic response of isolated with Quintuple friction pendulum bearing. To control the demand of super-structure in different hazard levels, QTFP represents a novel invention of multi-phase adaptive friction pendulum isolation system. Total six design configurations of Quintuple Friction Pendulum isolator results from two different displacement capacities and three types of effective period and effective damping are considered. Probabilistic study of engineering demand parameters like isolator displacement, base shear for the isolated building, and top floor absolute acceleration under different levels of seismic hazards are to be carried out. It is found that, Quintuple Friction Pendulum isolator is very effective in all different level of seismic hazards under all seismic demand parameters.

Keywords Seismic isolation · Quintuple friction pendulum system · Probabilistic evaluation · Multi-hazard level earthquake

1 Introduction

Research works in the last one decade is focused on the utilization of frictional type of base isolation. Its effectiveness for a wide range of frequency input is the reason for

A. Sodha (✉) · S. A. Vasanwala
Applied Mechanics Department, Sardar Vallabhbhai National Institute of Technology Surat Ichchhanath, Surat, Gujarat 395007, India
e-mail: aankitsodha@gmail.com

D. Soni
Department of Civil Engineering, Sardar Vallabhbhai Patel Institute of Technology, Vasad, Anand District, Gujarat 388306, India

D. Deb et al. (eds.), *Innovations in Infrastructure*, Advances in Intelligent Systems and Computing 757, https://doi.org/10.1007/978-981-13-1966-2_13

its popularity in the domain of base isolation. Resilient-friction base isolator system [1] and friction pendulum system [2] are the force restoring devices, performing based on the concept of sliding friction. A variety of restoring force devices such as Recently, a concept of multiple friction pendulum systems has been introduced with a view to reduce heat due to friction, to increase displacement capacity and to exhibit adaptive behavior compare to friction pendulum system [3]. With reference to above advantages, Double Concave Friction Pendulum (DCFP) bearing with two concave surfaces [4] and Triple Friction Pendulum (TFP) bearings with four sliding surfaces have been developed [5–7]. To optimize the performance in terms of coefficients of friction for different sliding surfaces and radius of curvature, the DCFP and TFP offers the designer extra design parameters. Further, the isolation system with multiple surfaces having elliptic sliding geometry has also been studied to avoid tuning of isolation frequency with excitation frequency [8].

Very recently developed Quintuple Friction Pendulum (QTFP) bearing [9] is an extension of the FP isolator consisting of four spherical sliding surfaces. It offers a more complex multi-stage behavior than the FP isolator, which may be implemented to control the response of the isolated structure when advanced performance objectives are considered or when the isolator displacement demand needs to be within acceptable limits during very strong seismic event.

2 Adaptive Behavior Quintuple Friction Pendulum Bearing

Quintuple Friction Pendulum (QTFP) isolator is an extended technology of Triple Friction Pendulum (TFP) isolator having nine stage sliding regimes operation with five effective pendula. Quintuple Friction Pendulum system shows highly adaptive behavior, due to its distinct hysteretic properties for different stages of displacement. As shown in Fig. 1, d_i is the displacement capacity for surface i, R_i is the radius of curvature of surface i, and μ_i is the coefficient of friction at the sliding interface. The motion due to internal construction of these bearings with different combinations of sliding surfaces, resulting in changes of stiffness and damping [4].

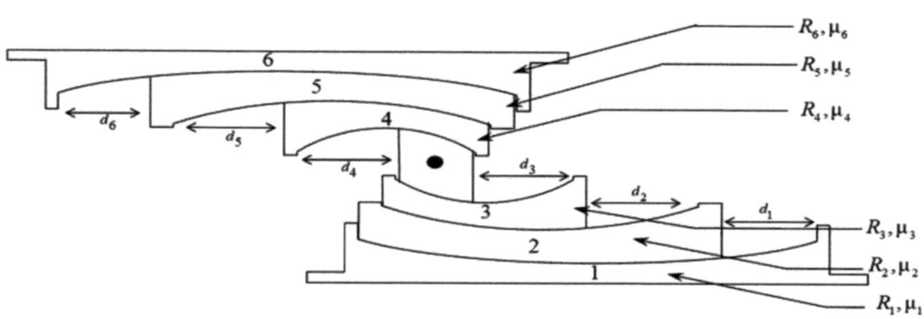

Fig. 1 Schematic diagram of QTFP

2.1 Different Stages of Sliding for QTFP

A generic model comprises the sloshing, impulsive, and rigid masses in terms of liquid mass and QTFP bearing is considered to explore the seismic performance of Isolated Building using QTFP. Different levels of excitation at different stages of sliding have been illustrated (refer Fig. 1):

Stage I: Motion starts on surfaces 3 and 4. Motion occurs on surfaces 3 and 4.
Stage II: Motion stops on 4 and starts on 5. Motion occurs on surfaces 3 and 5.
Stage III: Motion stops on 3 and starts on 2. Motion occurs on surfaces 2 and 5.
Stage IV: Motion stops on 5 and starts on 6. Motion occurs on surfaces 2 and 6.
Stage V: Motion stops on 2 and starts on 1. Motion occurs on surfaces 1 and 6.
Stage VI: Motion reaches end on 6 and stops. Motion starts on 5. Motion occurs on surfaces 1 and 5.
Stage VII: Motion reaches end on 1 and stops. Motion starts on 2. Motion occurs on surfaces 2 and 5.
Stage VIII: Motion reaches end on 5 and stops. Motion starts on 4. Motion occurs on surfaces 2 and 4.
Stage IX: Motion reaches end on 2 and stops. Motion starts on 3. Motion occurs on surfaces 3 and 4.

3 Governing Equation of Motion and Its Solution

For an N-story superstructure total dynamic Degree-of-freedom (DOF) considering one DOF at each floor is $N+5$, owing to five friction pendulum elements connected in series in the QTFP. Therefore, the governing equations of motion in matrix structure are expressed as

$$[M]\{\ddot{u}\} + [K]\{\dot{u}\} + [C]\{u\} = -[M]\{r\}(\ddot{u}_b + \ddot{u}_g) \tag{1}$$

where $[M]$, $[K]$, and $[C]$ are N sized square matrices for mass, stiffness, and damping of the superstructure, respectively. $\{u\} = \{u1, u2, \ldots uN\}$ is the vector of relative displacement of the superstructure; $u1, u2, \ldots uN$ is the floor lateral displacement relative to the base; $\ddot{u}g$ is the ground acceleration of earthquake; and here, derivative with respect to time is indicated by over dots; $\{r\} = \{1, 1,\ldots\}$ is the influence coefficient vector; $\ddot{u}b$ is the base mass acceleration relative to ground.

To solve the above equations, the incremental form of Newmark's step-by-step method is adopted. It was assumed that the linear variation of acceleration with small time interval was 1×10^{-6} s. Further, for each time step of incremental hysteretic displacement components in hysteretic model are solved by fourth-order Runge–Kutta method.

4 Earthquake Ground Motions

The present study is focused on the investigation of multi-stage performance of seismically isolated Building over different hazard levels, using Quintuple Friction Pendulum Isolator (QTFP). To achieve this objective, a suite of time history developed for Boston, Seattle, and Los Angeles which represents a range of seismic hazard levels from seismic Zone 2 to Zone 4 has been selected. They are comprised of three probabilities of occurrence: service level earthquake (50% in 50 years), design basis earthquake (10% in 50 years), and maximum considered earthquake (2% in 50 years) and (Somerville et al. [10]). It provides an effective mean of study for the multi-stage performance of QTFP with various frequency content and intensities. Table 1 shows the suite of 60-time histories for Los Angeles used in this study.

Table 1 Multiple hazard level ground motions [10]

SLE		DBE		MCE	
Record Label	Scale Factor	Record Label	Scale Factor	Record Label	Scale Factor
LA41	0.590	LA01	0.461	LA21	1.283
LA42	0.333	LA02	0.676	LA22	0.921
LA43	0.143	LA03	0.393	LA23	0.418
LA44	0.112	LA04	0.488	LA24	0.473
LA45	0.144	LA05	0.302	LA25	0.868
LA46	0.159	LA06	0.234	LA26	0.944
LA47	0.337	LA07	0.421	LA27	0.927
LA48	0.308	LA08	0.426	LA28	1.330
LA49	0.318	LA09	0.520	LA29	0.809
LA50	0.546	LA10	0.360	LA30	0.992
LA51	0.781	LA11	0.665	LA31	1.297
LA52	0.632	LA12	0.970	LA32	1.297
LA53	0.694	LA13	0.678	LA33	0.782
LA54	0.791	LA14	0.657	LA34	0.681
LA55	0.518	LA15	0.533	LA35	0.992
LA56	0.379	LA16	0.580	LA36	0.101
LA57	0.253	LA17	0.569	LA37	0.712
LA58	0.231	LA18	0.817	LA38	0.776
LA59	0.769	LA19	1.019	LA39	0.500
LA60	0.478	LA20	0.987	LA40	0.657

5 Numerical Study and Results

To study adaptive behavior of QTFP, the example building was subjected to three levels of ground excitations, LA 42 (SLE), LA 1 (DBE), and LA 27 (MCE). The hysteresis behavior for each level of earthquake is shown in Figs. 2, 3, and 4 for the QTFP1-6 having time period of 4–6 s and damping of 10 to 20% as shown in Table 2.

The isolator deforms into sliding regimes I and II, with rigid linear behavior of force deformation. As radius of curvature of inner surfaces is relatively smaller, the isolator stiffens during SLE. The isolator deforms into sliding regime III and IV with rigid bi-linear behavior during DBE. The isolator deforms into sliding regimes V–IX.

For SLE, The inner sliding surfaces 3 and 4 with lower value of coefficient of friction get activated, as the lateral force acting on isolator is comparatively low (Fig. 3).

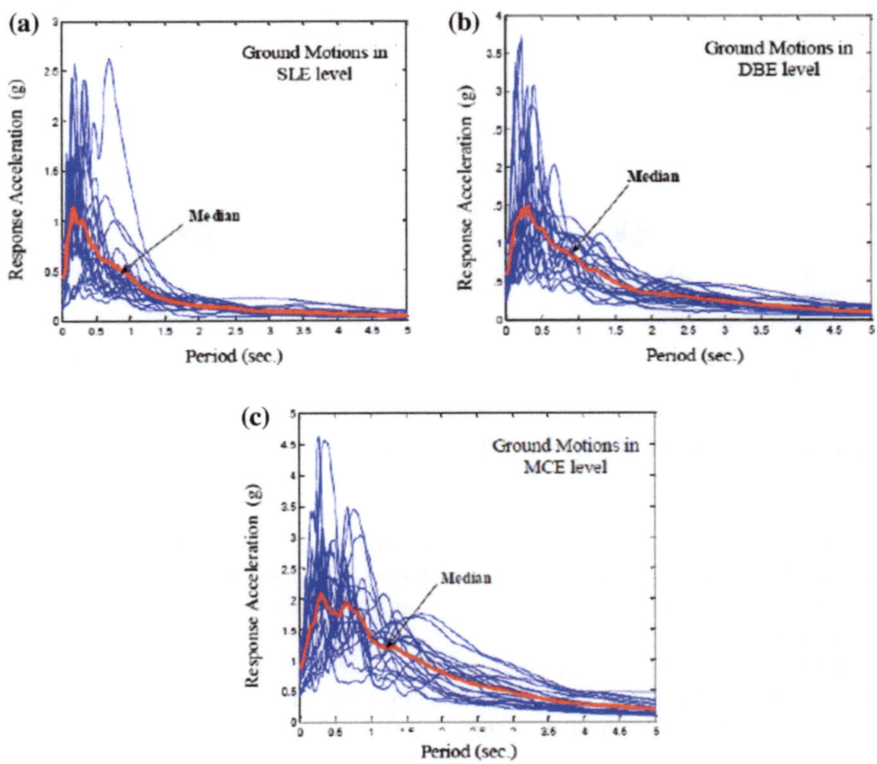

Fig. 2 Response acceleration spectra. **a** SLE, **b** DBE, and **c** MCE

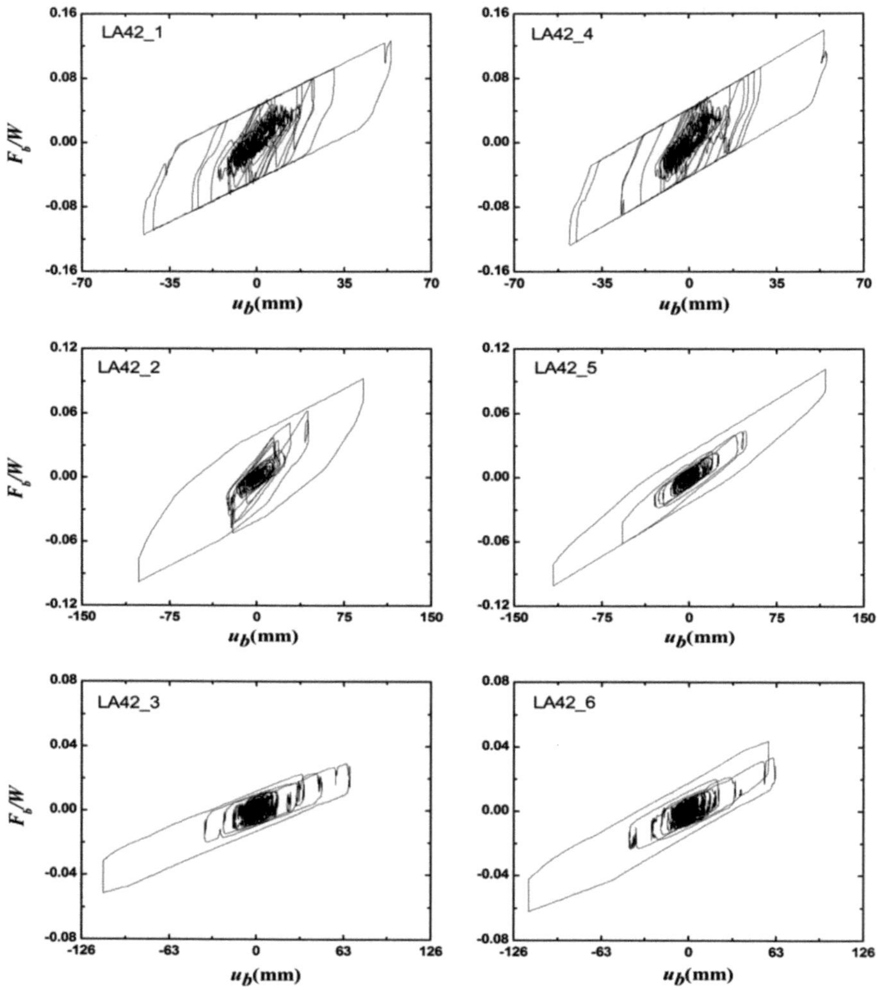

Fig. 3 Hysteretic behavior of QTFP1 to QTFP6 under LA 42 (SLE)

For DBE earthquake, due to increased lateral force after motion stops on surfaces 3, it starts on surfaces 2 further increases in lateral force the motion occurs on surfaces 2 and 5 then again motion stops on 5 and starts on 6 then motion occurs on surfaces 2 and 6 (Fig. 4).

The lateral force is so large to start motion on surface 1. Surface 1 has the largest friction during MCE event. Increasing in lateral force further, the motion occurs on surfaces 1 and 6; and the motion on surface 2 stops (Fig. 5).

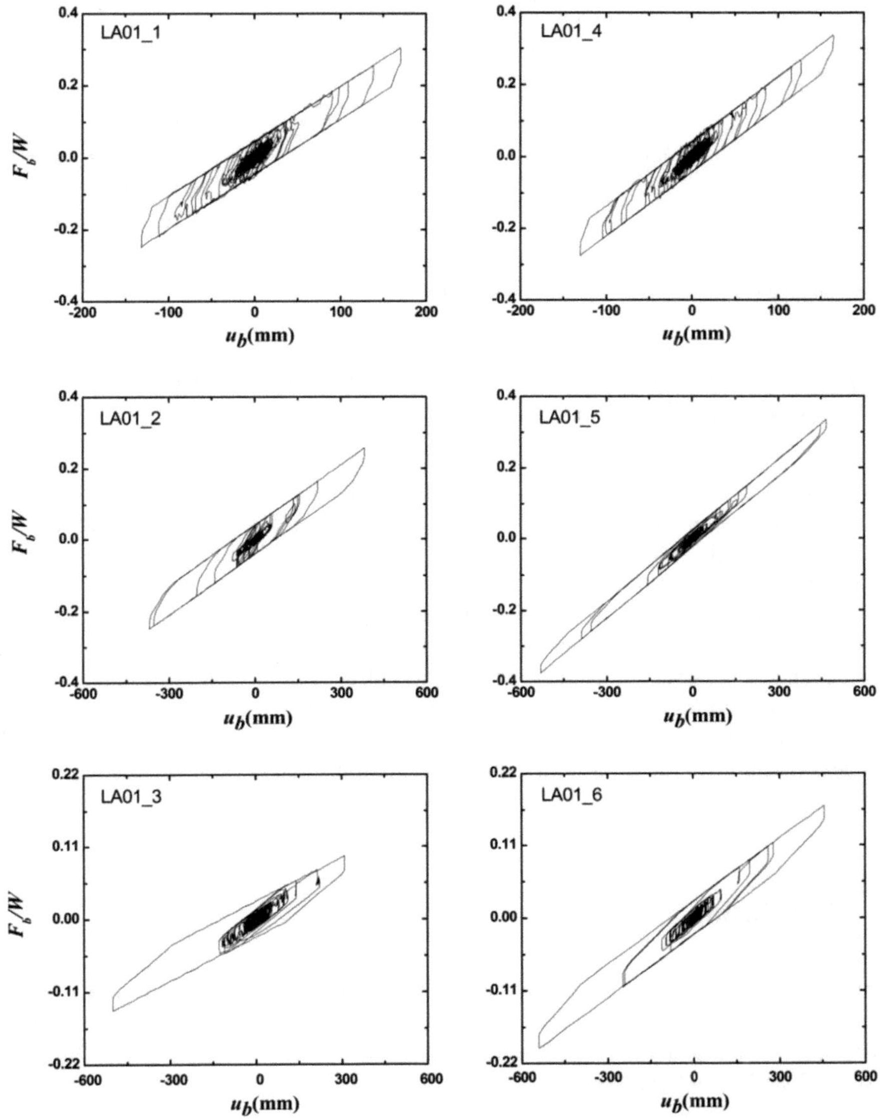

Fig. 4 Hysteretic behavior of QTFP1 to QTFP6 under LA 1 (DBE)

Table 2 Properties of the six QTFP design configurations

Q	ξ	T	R_1	R_2	$R_{3=4}$	R_5	R_6	μ_1	μ_2	$\mu_{3,4}$	μ_5	μ_6
1	10	4	2.8	0.8	0.1	0.7	2	0.09	0.05	0.01	0.04	0.06
2	15	5	4.5	2	0.25	1.5	4	0.09	0.06	0.01	0.04	0.07
3	20	6	7.5	3.5	0.3	4	6	0.1	0.05	0.01	0.04	0.09
4	10	4	2.5	0.8	0.1	0.7	2	0.07	0.05	0.01	0.04	0.05
5	15	5	4.5	2.7	0.25	2	3.5	0.07	0.04	0.01	0.05	0.06
6	20	6	6.5	3.5	0.3	3	5	0.07	0.05	0.01	0.04	0.06

ξ = damping (%), Q = QTFP, T = time period (s)

5.1 Performance of QTFP Under Annual Probability of Exceedance

The responses are categorized for three different annual probabilities of earthquake occurrences, named MCE (2% in 50 year), DBE (10% in 50 year), and SLE (50% in 50 year). Seismic responses of Configurations QTFP1 to QTFP6 are plotted as per suits of time histories of annual probability of exceedance.

As shown in Fig. 6, three responses termed as peak acceleration, normalized base shear, and peak isolator displacement have been plotted against annual probability of exceedance. It is observed from the graph that peak accelerations and normalized base shear are maximum for minimum values of damping, time period and displacement capacity. In contrast with these observations, peak isolator displacement is witnessed to be minimum for the minimum values of damping, time period, and displacement capacity.

6 Conclusion

Presented in this paper is the behavior of QTFP system using nonlinear time history analysis with multi-hazard level excitation and directivity focusing earthquakes. The MDOF system is analyzed for 60 multi-hazard earthquake records consist of 20 SLE, 20 DBE, and 20 MCE records;

1. The QTFP bearing stiffens at low input with service level earthquake (SLE), softens with increasing input of design basis earthquake (DBE), It gets stiffens at higher levels of input with maximum considered earthquake (MCE) again. Thus, it shows highly adaptive behavior, despite being a passive system.
2. With the increase in displacement capacity, isolator displacement increases and base shear decreases for all three events.
3. The QTFP can be very useful in controlling the response parameters of a seismically isolated building due to its six spherical sliding surfaces, five effective

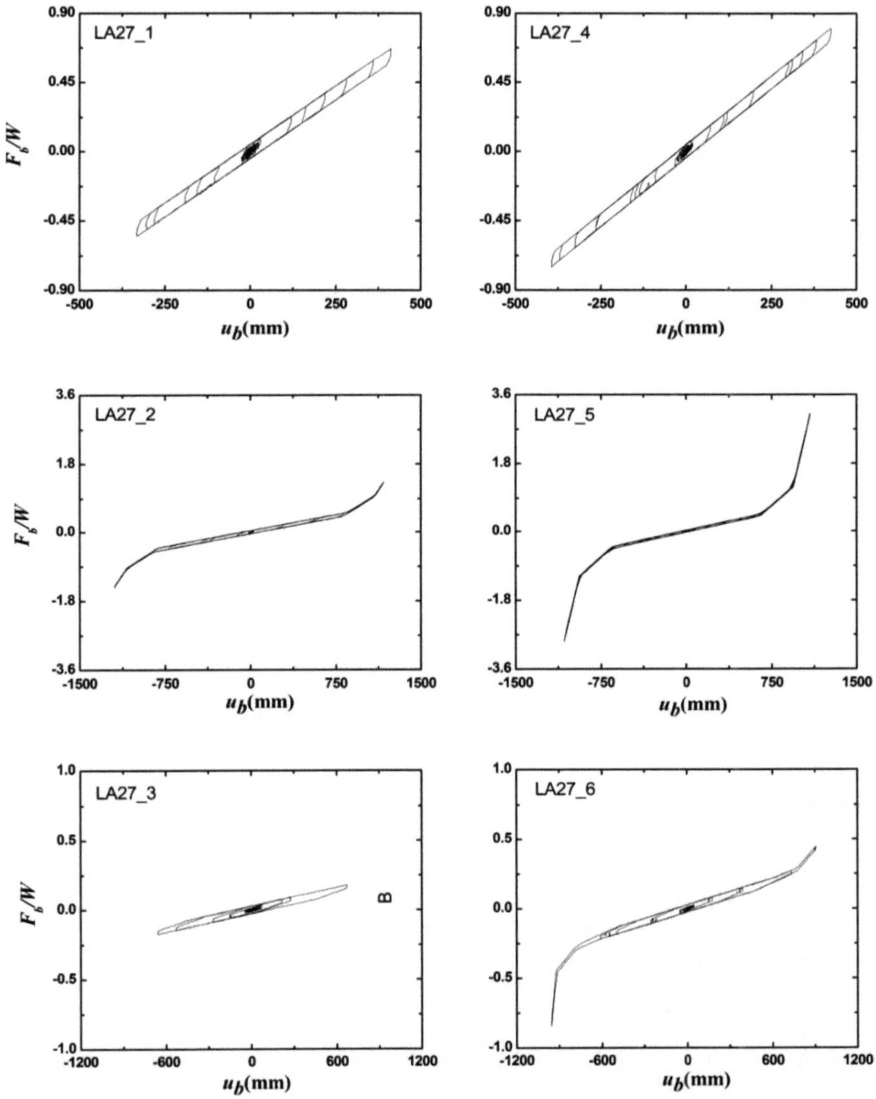

Fig. 5 Hysteretic behavior of QTFP1 to QTFP6 under LA 27 (MCE)

pendula, which can provide an engineer greater flexibility for the selection various design parameters to optimize the isolator performance.

4. Maximum peak accelerations and normalized base shear with minimum peak isolator displacement are observed for minimum values of damping, time period and displacement capacity.

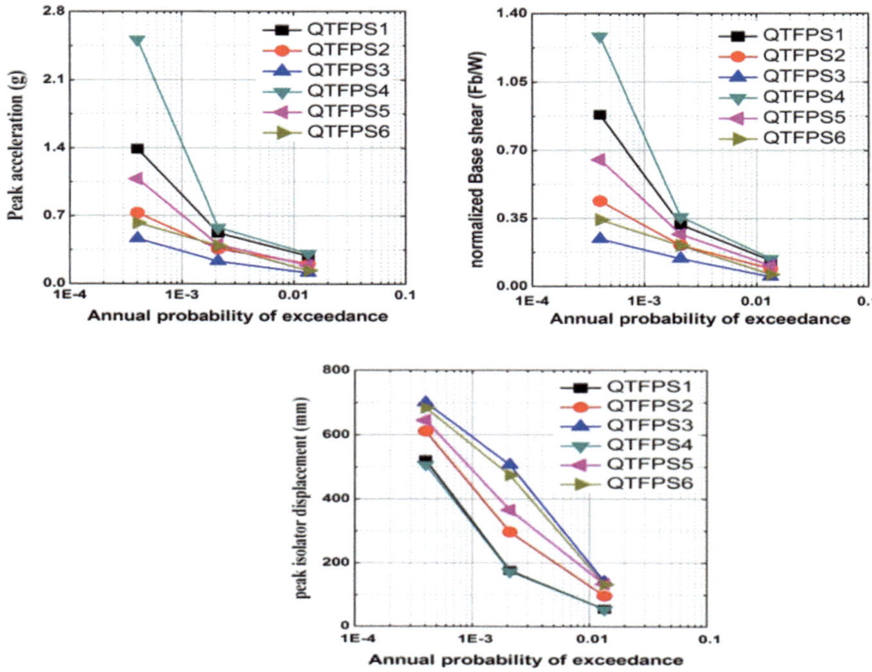

Fig. 6 Comparison of seismic responses for annual probability of exceedance of QTFP1 to QTFP6

References

1. Mostaghel, N., Khodaverdian, M.: Dynamics of resilient-friction base isolator (R-FBI). Earthq. Eng. Struct. Dyn. **15**(3), 379–390 (1987)
2. Zayas, V.A., Low, S.S., Mahin, S.A.: A simple pendulum technique for achieving seismic isolation. Earthq. Spectra **6**(2), 317–333 (1990)
3. Tsai, C.S., Lin, Y.C., Su, H.C.: Characterization and modeling of multiple friction pendulum isolation system with numerous sliding interfaces. Earthq. Eng. Struct. Dyn. **39**(13), 1463–1491 (2010)
4. Fenz, D.M., Constantinou, M.C.: Behaviour of the double concave friction pendulum bearing. Earthq. Eng. Struct. Dynam. **35**(11), 1403–1424 (2006)
5. Fenz, D.M., Constantinou, M.C.: Modeling triple friction pendulum bearings for response-history analysis. Earthq. Spectra **24**(4), 1011–1028 (2008)
6. Becker, T.C., Mahin, S.A.: Approximating peak responses in seismically isolated buildings using generalized modal analysis. Earthq. Eng. Struct. Dyn. **42**(12), 1807–1825 (2013)
7. Dhankot, M.A., Soni, D.P.: Behaviour of triple friction pendulum isolator under forward directivity and fling step effect. KSCE J. Civil Eng. **21**(3), 872–881 (2017)
8. Soni, D.P., Mistry, B.B., Jangid, R.S., Panchal, V.R.: Seismic response of the double variable frequency pendulum isolator. Struct. Control Health Monit. **18**(4), 450–470 (2011)

9. Lee, D., Constantinou, M.C.: Quintuple friction pendulum isolator-behavior, modeling and validation. Technical Report MCEER-15-0007, Multidisciplinary Center for Earthquake Engineering Research, University at Buffalo, Buffalo, NY (2015)
10. Somerville, P., Anderson, D., Sun, J., Punyamurthula, S., Smith, N.: Generation of ground motion time histories for performance-based seismic engineering. In: Proceedings 6th National Earthquake Engineering Conference. Seattle, Washington (1998)

Seismic Vulnerability Assessment of Mid-rise Reinforced Concrete Building in Ahmedabad

Ronak Motiani, Dharmil Joshi, Sandip A. Vasanwala, Kavan Bhatt and Jaimin Korat

Abstract The city of Ahmedabad is one of the most urbanized and crowded cities in the western part of India. The vulnerability associated with this city has increased due to haphazard development, owing to rapid urbanization and industrial development. This article focuses on evaluating the performance of an existing mid-rise RC frame building in a densely populated area of the city. The selected stock building is precisely modeled in SAP2000 for carrying out its pushover analysis from which the capacity spectrum is obtained as per the recommendations of ATC-40. Finally, the Fragility curve is produced as per the guidelines of HAZUS and the damageability function of the building is calculated. From the results of fragility curve and damage probability function, it is deduced that the building will suffer extensive damage during a seismic event.

Keywords Fragility · Seismic vulnerability · Reinforced concrete · Capacity spectrum · Extensive damage

1 Introduction

India is the second most populated country in world with a total population of about 1.3 billion. The city of Ahmedabad, situated on the banks of river Sabarmati is the seventh largest metropolitan city of India and is one of the most densely populated

R. Motiani (✉) · D. Joshi · K. Bhatt · J. Korat
Pandit Deendayal Petroleum University, Gandhinagar, Gujarat 382421, India
e-mail: rnkmotiani@gmail.com

D. Joshi
e-mail: joshidharmil@gmail.com

K. Bhatt
e-mail: bhattkavan.bhatt@gmail.com

J. Korat
e-mail: jaiminkorat@gmail.com

S. A. Vasanwala
S.V National Institute of Technology, Surat, Gujarat 395007, India
e-mail: vsandip18@gmail.com

D. Deb et al. (eds.), *Innovations in Infrastructure*, Advances in Intelligent Systems and Computing 757, https://doi.org/10.1007/978-981-13-1966-2_14

cities of India. Due to rapid growth in population, the city has encountered haphazard development, owing to which risk during seismic events has increased. Ahmedabad faced the terror of earthquake destruction during Bhuj earthquake, 2001 (Mw = 7.7) which caused a property damage of $5.5 billion and around 2500 people lost their lives. It is difficult to predict the occurrence of an earthquake and safeguard losses so the best that can be done is to assess the probability of loss and capitalize on the preventive measures.

In recent times with the development of civilization, concern for safety has increased. To ensure safety, lots of research has been carried out from nonlinear analysis to multi-criteria decision making technique like AHP to fuzzy AHP. Tesfamaraim and Saatcioglu [1] assessed risk of RC building using fuzzy AHP by calculating weights and contribution of various factors related with risk. Singh et al. [2], explained role of unreinforced masonry wall (URM) in failure mode of building. Many codes like ATC-40 [3] and FEMA 356 [4] has explained number of methodologies for the assessment of risk and vulnerability of structures.

The basic aim of these methodologies is to predict the estimation of losses during an earthquake and provide a suitable methodology for facing these challenges. A number of methods have been proposed for the computation of vulnerability by developing different vulnerability functions based on the available data sets and suitable methods. These methods have been adopted in numerous studies like for computation of seismic vulnerability and risk assessment of large and small target areas using GIS and other tools. Nath et al. [5] computed the seismic vulnerability in the city of Kolkata using AHP and obtained cumulative damage probabilities in terms of none, slight, moderate, extensive, and complete as in Barbat et al. [6] for the model building types in the city.

Kamatchi et al. [7] showed vulnerability of buildings in Ahmedabad using site-specific earthquake and proposed suitable mitigation measures. Trivedi [8] carried out number of hazard studies in Ahmedabad and came with PGA and liquefaction map of the city. In Ahmedabad, the probability of recurrence of earthquake is low but the city is under high risk due to poor seismic design of structures. Very limited studies have been done till date in Ahmedabad regarding its risk and vulnerability to a seismic event despite of its possibility to suffer high risk. This study aims to get vulnerability of building in Ahmedabad constructed after the Bhuj earthquake.

2 Vulnerability

The seismic vulnerability of a building is its propensity to suffer a certain level of damage when subjected to a defined intensity of seismic event [9]. Buildings have different degrees of vulnerability to specific earthquake ground motion based on their geometrical or qualitative characteristics (such as height of the building, plan irregularities, elevation configurations, vertical setbacks and soft storey), and structural characteristics (such as material of construction, mass, stiffness of members, quality

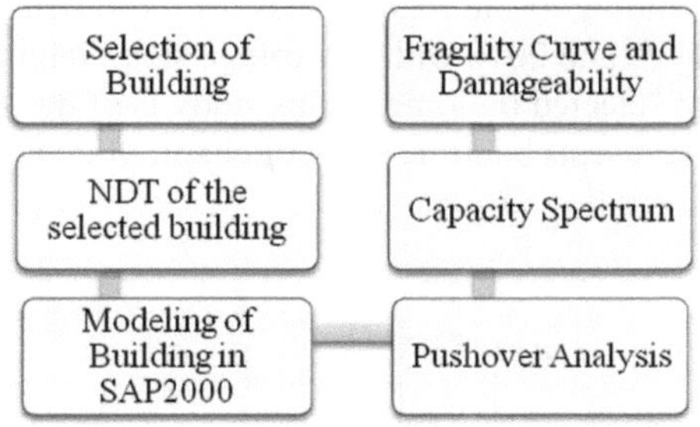

Fig. 1 Vulnerability assessment method

of construction, strength, intrinsic ductility, state of stress, seismic displacements, nonlinear behavior parameters, etc.).

Various methods existing for the vulnerability assessment are: (a) Expert opinion. (b) Empirical methods. (c) Analytical methods. (d) Hybrid methods, which are also explained in brief in technical document on seismic vulnerability assessment methods for buildings by IIT [10]. Vulnerability of a building can be calculated by expert opinion method based on ATC-13 [11]. The empirical methods were prevalent before the development of advanced computers and involve complex analysis. After the advancement of computers, the application of analytical methods increased with nonlinear analysis of structures and various other complex mathematical formulations. The Hybrid method focusing on the combination of two or more than two methods is described above for predicting the vulnerability with higher accuracy. Mai et al. [12] showed an alternate way to find vulnerability using fragility curve which was developed using Monte Carlo simulation and kernel density estimation. Fan [13] explained the need of progressive as a follow up to seismic vulnerability analysis for important structures like bridge. Xu and Hu [14] studied vulnerability of towers of cable-stayed bridge and gave an idea about vulnerability of it. Whereas, few such studies has been carried for Ahmedabad.

The seismic vulnerability in vigorous construction defying adequate seismic codes is continuously increasing in Ahmedabad leading to the continual increase in the seismic risk of the city. The vulnerability function of a building, in this case, is computed from its capacity to resist any seismic action as shown in Fig. 1. In this study, the displacement-based capacity curve is used to find the capacity of the building to resist seismic action.

3 Building Specification

Structural framework taken into consideration for assessing vulnerability in mid-rise reinforced concrete buildings which is the most prevalent building typology in the city of Ahmedabad. One of the most emblematic and representative regions of Ahmedabad city is that of Nikol, where a survey was conducted to find out the stock building of the region. Most of these structures were constructed in between years 1990 and 2010 with an average story height of four storeys. Figure 2 shows the plan of selected building in this study is of the same region built in the year 2007 which represents building stock of Nikol. M-25 Grade of concrete and Fe-415 grade of steel

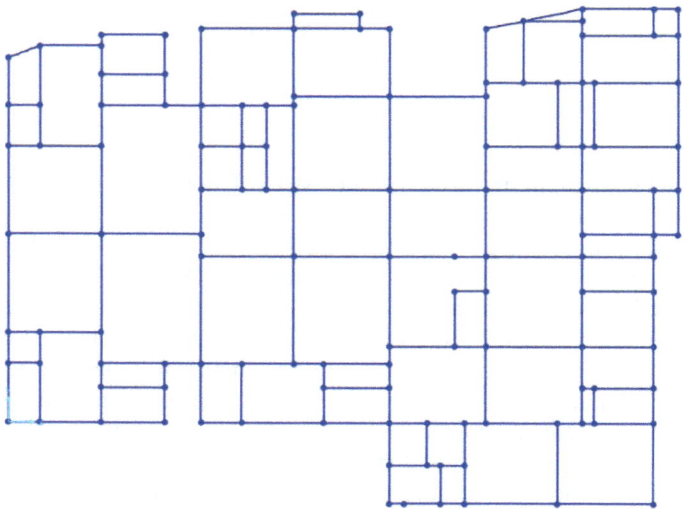

Fig. 2 Plan of the building

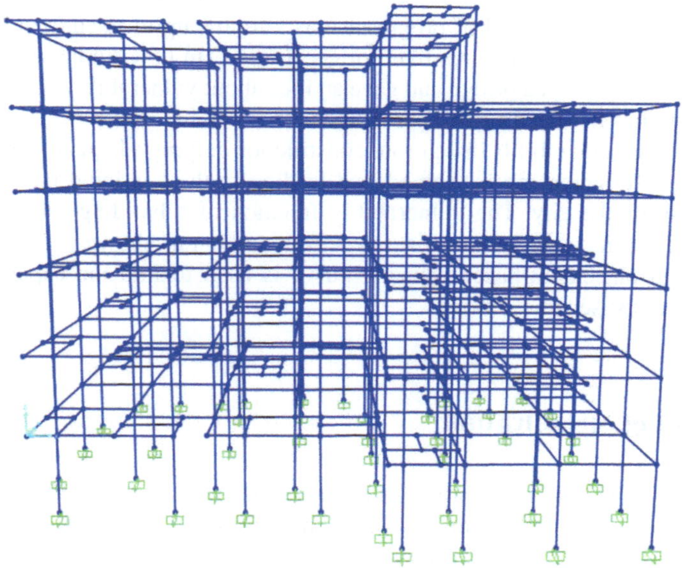

Fig. 3 SAP model

Table 1 Load cases for analysis

Load case	Load combination
1	$D.L + I.L$
2	$1.5\,(D.L + I.L)$
3	$1.2\,(D.L + I.L \pm E.Q)$
4	$1.5\,(D.L \pm E.Q)$
5	$0.9\,(D.L \pm 1.5E.Q)$

is used for construction. The building is mixed usage with part of its ground floor being used for commercial space.One half of the building is G+4 building which is used as a commercial space while other half is G+5. The design details of the building were modeled in SAP2000 as shown in Fig. 3. The detailed design drawing showed the excess overhang of beams at multiple locations and excess cantilever with lack of proper reinforcement and detailing in the design for the cantilever and overhang.

The entire loading system of the building was assigned loads as per IS-875(Part 1) [15] and the load combinations as per IS-1893 [16] as shown below in Table 1. Time period of the building as per IS-1893 $T = 0.09h\sqrt{d}$ is obtained as 0.324, whereas the time period obtained from the modal analysis carried out in SAP2000 is 0.425.

Here, the DL consists of dead loads of wall, slab, floor finish and roof treatment. Whereas the imposed load consists of live load and roof live load.

4 Non Destructive Testing

In order to get an idea of current scenario of building vulnerability it is necessary to study existing strength of concrete. Nondestructive testing which gives degraded strength was performed using rebound hammer and ultrasonic pulse velocity instrument. After conducting the tests on the beams and columns as per the specifications of IS-13311: Part-1 [17] and IS-1311: Part-2 [18], the degraded strength of the building was calculated as per the guidebook on nondestructive testing of concrete structures, TCS No. [19] from the Eq. (1).

$$S = -24.1 + 1.24R + 0.058V^4 \tag{1}$$

where R = Rebound Number and V = Velocity observed in UPV. The degraded strength calculated by this technique comes out to be 19.84 N/mm^2. Design strength of concrete used for construction of the building was M-25, whereas from the results of NDT the observed degraded strength of the concrete is 19.84 N/mm^2, thus M-20 is taken for the modeling in SAP2000.

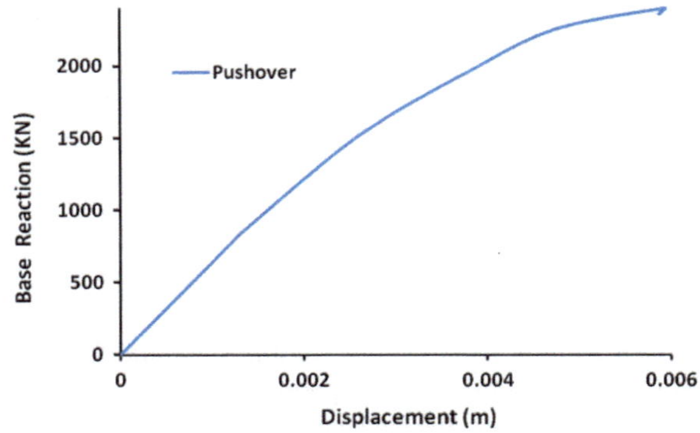

Fig. 4 Pushover curve

5 Push Over analysis

Nonlinear static or pushover analysis in which the structure is subjected to monotonically increasing horizontal load was performed on the stock building. Lateral loads were continually increasing until the structure reaches its ultimate condition during the analysis performed in SAP2000. The pushover analysis in SAP2000 is based on the FEMA 356 and ATC-40 for obtaining the lateral load displacement curve known as Pushover Curve. It can be performed as displacement controlled or force controlled. The method of displacement control is used when the building is subjected to a seismic event and it suffers some damage or when specified drifts are checked. In this case, the amount of loads is not known. The second approach is used when the structure is predicted to withstand the loading. For analysis of the building in this article, the first method is used.

Figure 4 shows the pushover curve of displacement versus base shear for the building under investigation as per FEMA 356. Moment rotation values are assigned to the member section based on the reinforcement details and the section of the members at the location of the hinges. The values are generated based on the material model of concrete by using equilibrium and compatibility equations. The column sections are assigned with P-M2-M3 hinges moment rotation characteristics. The beam sections are assigned with the flexural M2 hinges [20].

6 Capacity Spectrum

The capacity spectrum method of ATC-40 is an efficient alternative to nonlinear time history analysis and it is used in HAZUS [21] methodology for earthquake loss estimation. The performance point of the building under investigation is obtained from the intersection of the acceleration displacement response spectrum (ADRS) with the capacity curve. The yield spectral acceleration, yield spectral displacement, ultimate spectral acceleration, and ultimate spectral displacement can be obtained from the capacity curve.

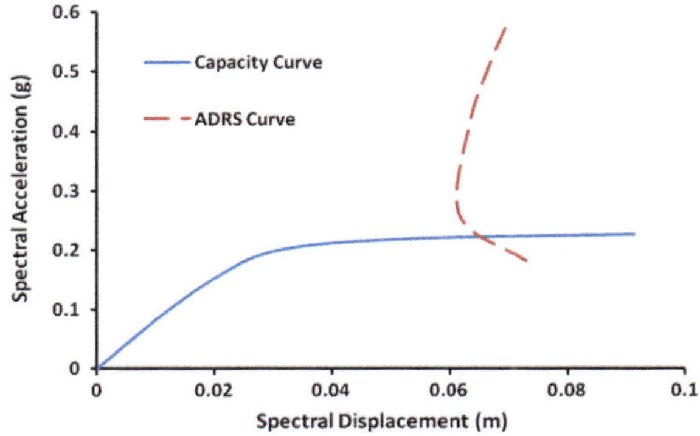

Fig. 5 Capacity spectrum

Yield capacity of a structure is the point where the building begins to yield and starts to deform in a nonelastic manner. As the building deforms more and more on application of further gravity and lateral load, the building reaches a point of ultimate capacity beyond which it cannot further yield and collapses. The spectral displacement at these points is yield displacement and ultimate displacement

Figure 5 shows the capacity spectrum as per ATC-40 of the selected building obtained from SAP2000.

7 Fragility Curve

The fragility curve is an essential step as per HAZUS methodology for estimation of Seismic vulnerability and risk by nonlinear analysis. Fragility curve of a building is the lognormal distribution of probability of that building for reaching or exceeding the damage states with respect to the Potential Earth Science Hazards (PESH) parameters of spectral acceleration, peak ground acceleration (PGA), or spectral displacement. The damage states are divided into four categories—Slight, Moderate, Extensive, and Complete having same characteristics [6]. The PESH parameter used for structural and nonstructural damage in this article is that of spectral displacement. The spectral displacement for the corresponding damage states is assigned as per the values of yield and ultimate displacement obtained from the capacity curve. Table 2 shows the formulation of spectral displacement for the respective damage states. Where $D_y = Yield\ displacement\ D_u = Ultimate\ Displacement$. This damage states and its corresponding spectral displacement is used to calculate the probability

Table 2 Spectral displacement for different damage states

Slight Damage	Moderate damage	Extensive damage	Complete damage
$0.7D_y$	D_y	$D_u + 0.25(D_u - D_y)$	D_u

of exceedance as per the HAZUS methodology using Eq. (2)

$$\mathbf{P}\left[\frac{\mathbf{ds}}{\mathbf{Sd}}\right] = \varphi\left[\frac{\mathbf{1}}{\beta\mathbf{ds}}\mathbf{In}\frac{\mathbf{Sd}}{\mathbf{Sd, ds}}\right] \tag{2}$$

where $\mathbf{P}\left[\frac{\mathbf{ds}}{\mathbf{Sd}}\right]$ = probability of being in or exceeding a damage state
$\frac{Sd}{Sd,ds}$ = median value of spectral displacement at which the building reaches the threshold of the damage state, ds
βds = standard deviation of natural logarithm of spectral displacement of damage state, ds
φ = Standard normal cumulative distribution function.

Figure 6 shows the fragility curve developed from the above formula for the four damage states. The area under each curve shows the probability of damage for the corresponding damage state.The probability of damage under each corresponding state is calculated by forming a damage probability matrix as shown in Table 3 from the performance point obtained from capacity curve, yield spectral displacement and ultimate spectral displacement. From this fragility curve and the damage probability matrix, it is evident that the building selected has higher probability of suffering from an extensive damage during a seismic event.

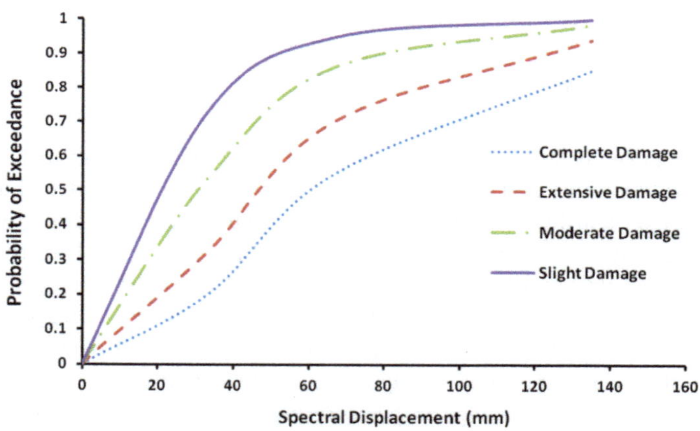

Fig. 6 Fragility curve

Table 3 Damage probability matrix

Damage state	Percentage (%)
No damage	5.37
Slight damage	8.42
Moderate damage	16.02
Extensive damage	36.83
Complete damage	33.36

8 Results and Discussion

Mid-rise RC frame buildings, the prevalent stock buildings of Ahmedabad is surveyed with detail in a representative region of Nikol, Ahmedabad. Most of the buildings are found to be designed only for gravity loads and lack specific detailing to sustain the earthquake loads. The selected stock building, modeled in SAP2000 for pushover analysis shows early failure under moderate loading conditions. Although the building is designed as per strong column, weak beam theory the plastic hinge formation occurs at critical location starting from the ground floor in very early stages. The building lacks proper design methodology which results into its increase in probability of suffering heavy damage.

Damage state probability from the fragility curves gives the probability of 36% for the building to suffer extensive damage and 33% probability to suffer complete collapse. The reason for this lack in strength and resistance to loading is imposed on the extensive overhang of beams, irregular floor plan, vertical setback, and extensive cantilever portion for which no adequate provision is given in the structural design.

9 Conclusion

This article discusses the vulnerability of a selected stock building of Ahmedabad city by generating pushover curve, capacity curve and fragility curves. The degree of damage probability is expressed in terms of probability of reaching the damage state. The study concludes that:

- The concrete existing strength has degraded considerably over a period of short time span to its design life.
- In spite of having the design as per weak beam, strong column theory, the building will suffer severe damage during a seismic activity.
- This kind of buildings in Ahmedabad city will suffer major economic and life losses unless effective mitigation and retrofitting measures are not taken.

This study focuses on analysis of mid-rise RC frame building which is most prevalent in the city and comments on its vulnerability. However, in order to work out effective mitigation technique, it is necessary to prepare risk map of entire city. An effective method that consumes less time needs to be developed.

References

1. Tesfamaraim, S., Saatcioglu, M.: Seismic risk assessment of RC buildings using fuzzy synthetic evaluation. J. Earthq. Eng. **12**(7), 1157–1184 (2008)
2. Singh, Y., Lang, D.H., Prasad, J., Deoliya, R.: An analytical study on the seismic vulnerability of masonry buildings in India. J. Earthq. Eng. **17**(3), 399–422 (2013)

3. ATC: Evaluation and Retrofit of Concrete Buildings, vol. 1. ATC-40 Report, Applied Technology Council, Redwood City, CA (1996)
4. FEMA: Commentary for the Seismic Rehabilitation of Buildings. FEMA-356, Federal Emergency Management Agency, Washington, DC (2000)
5. Nath, S., Adhikari, M., Devaraj, N., Maiti, S.: Seismic vulnerability and risk assessment of Kolkata City, India. Nat. Hazards Earth Syst. Sci. Discuss. **2**, 3015–3063 (2014)
6. Barbat, A.H., Carreño, M.L., Pujades, L.G., Lantada, N., Cardona, O.D., Marulanda, M.C.: Seismic vulnerability and risk evaluation methods for urban areas. a review with application to a pilot area. Struct. Infrastruct. Eng. **6**(1–2), 17–38 (2010)
7. Kamatchi, P., Raju, K.R., Ravisankar, K., Iyer, N.R.: Design of visco-elastic dampers for RC frame for site-specific earthquake. J. Inst. Eng. (India): Ser. A **97**(4), 359–366 (2016)
8. Trivedi, S.: Soil amplification studies for ahmedabad region. In: International Conference on Current Trends in Technology, pp. 1–6. Nirma Uuiversity, Ahmedabad (2011)
9. Lang, K., Bachmann, H.: On the seismic vulnerability of existing unreinforced masonry buildings. J. Earthq. Eng. **7**(03), 407–426 (2003)
10. National Disaster Management Authority Government of India: Seismic Vulnerability Assessment of Building Types in India. National Disaster Management Authority Government of India (2013)
11. Applied Technology Council: Earthquake damage evaluation data for California. Applied Technology Council, Redwood City (1985)
12. Mai, C., Konakli, K., Sudret, B.: Seismic fragility curves for structures using non-parametric representations. Front. Struct. Civil Eng. **11**(2), 169–186 (2017)
13. Fan, L.: Life cycle and performance based seismic design of major bridges in china. Front. Archit. Civil Eng. China **1**(3), 261–266 (2007)
14. Xu, Y., Hu, S.: Seismic design of high-rise towers for cable-stayed bridges under strong earthquakes. Front. Archit. Civil Eng. China **5**(4), 451–457 (2011)
15. Indian Standard: Code of Practice for Design Loads (Other than Earthquake) for Buildings and Structures (1987)
16. Indian Standard: Criteria for Earthquake Resistant Design of Structures: Part I, 5th revision. Bureau of Indian Standard, New Delhi (2002)
17. Indian Standard: Non-destructive Testing of Concrete, Part 1: Ultrasonic Pulse Velocity. Bureau of Indian Standard, New Delhi (1992)
18. Indian Standard: Non-destructive Testing of Concrete, Part 2: Rebound Hammer. Bureau of Indian Standard, New Delhi (1992)
19. TCS No.: Guidebook on Non-destructive Testing of Concrete Structures (2002)
20. Naik, P., Annigeri, S.: Performance evaluation of 9 storey RC building located in North Goa. Proc. Eng. **173**, 1841–1846 (2017)
21. FEMA: Earthquake loss estimation methodology Hazus-MH 2.1. Federal Emergency Management Agency, Washington, DC (2013)

Evaluation of Lateral Stability of the Diagrid Tall Structure Under Different Earthquake Forces

Swaral R. Naik and S. N. Desai

Abstract Diagrid structural mechanism, being aesthetically appealing, structurally efficient and material saving, has gained a lot of attention and acceptance among the architects and structural designers engaged in modern era infrastructures in densely packed urban areas. Conventional way of providing seismic stability to the structure is either by bracing systems or shear wall system. Diagrid system combines the responsibilities of the vertical columns and the bracings as one single structural member—a diagonal member only. This has not only shown increase in the lateral stability of the structure, but it is also proven for reduction in material usage up to 20%. Diagrid is more acceptable nowadays mainly due to the column-free spaces and freedom in interior planning that it can offer to the inhabitants. This study focuses on behaviour of a Diagrid structure subjected to different seismic forces. An existing Diagrid tall structure (Hearst tower, NY, USA) and a similar dimensioned regular conventional structure are compared under various existing earthquake shakings. The time history dynamic analysis is done using FE software SAP2000. The Diagrid structure showed higher lateral seismic stability than the conventional system.

Keywords Diagrid · Tall structure · Seismic analysis · SAP2000

1 Introduction

In urban areas, tall structures have become the only option for providing houses as well as the corporate offices using lesser land. Many structural systems have evolved since about five decades to make the tall structures more stable and efficient

S. R. Naik (✉) · S. N. Desai
Applied Mechanics Department, SVNIT, Surat, India
e-mail: naikswaral@gmail.com

S. N. Desai
e-mail: snd@amd.svnit.ac.in

S. R. Naik
Applied Mechanics Department, GEC, Bharuch, India

D. Deb et al. (eds.), *Innovations in Infrastructure*, Advances in Intelligent Systems and Computing 757, https://doi.org/10.1007/978-981-13-1966-2_15

[1]. Framed shear truss, Bracing systems, framed tube system, truss tube system, bundled tube systems, outrigger system, shear wall system, etc. [2] have been in the choices of the structural designers and architects with respect to their ability to resist large lateral forces (mainly seismic and wind) [3, 4].

There is number of tall structures which are constructed with the above mechanisms and they are still standing, satisfying their purpose and serving the habitants. Diagrid system, a newcomer to the above list, has started showing its capabilities and strength. A conventional structural system will have a framed beam-column structure, mainly resisting gravity loads, assisted by an auxiliary lateral load resisting system [5]. But a Diagrid will have a single system which will be resisting all the gravity loads as well as the lateral loads (Fig. 1). The diagonal columns will contribute to take the gravity load and the lateral load, by simply transforming them into axial forces. Unlikely to beam-column framing, a Diagrid has a triangulated framing. This triangulation is called a Diagrid module. A single module may consist of more than one storey, depending on the Diagrid angle, θ. Many researchers have worked to evaluate an optimum angle of Diagrid (between 60 and 70°), mainly considering wind force [6]. Diagrid node is the main junction which distributes both the gravity as well as the lateral loads to the diagonal members (see Fig. 2). In conventional moment frame system, external forces are transferred through beam-column junction, creating bending moments in vertical columns, whereas in Diagrid system all the forces are transferred through a node, generating axial forces in the diagonal columns [7]. Thus, the diagonal columns (either of RCC or Steel) being very strong in resisting axial forces, they are more efficiently utilized—a major reason behind the ancient structures lasting since centuries (least bending moments in main structural members, pure axial forces only). Till now, around 20 buildings are already constructed using Diagrid systems. Most of these structures are in the regions of heavy wind gust and are optimized with respect to resist wind forces [8]. However, the response of a tall structure under earthquake forces is a critical parameter for design, as a quake does not offer much time for the panicked occupants to escape. Thus, evaluation of response of the tall Diagrid structure under earthquake forces is an essential study. Unfortunately, there is still a large research gap in this area of seismic evaluation of Diagrid structures.

This research mainly includes dynamic time history analysis of Diagrid tall structure under various earthquake motions using FE software SAP2000. The building under the study should not be imaginary or non-practical (as many researchers have mistaken before) here we have taken as reference, Hearst Magazine Tower situated at NY, USA. Actual data of this tall structure is procured and modelled accordingly in the software. Later, a same dimensioned simple moment frame tall structure is also modelled with the same geometry, loads and member sizes data. These two models are analyzed for different known earthquake time histories and the responses are compared.

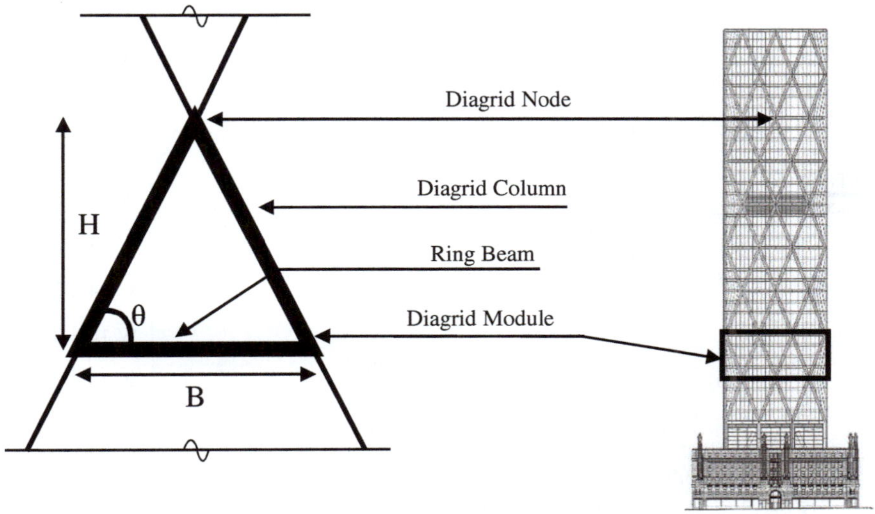

Fig. 1 A typical diagrid building elevation and its components [9]

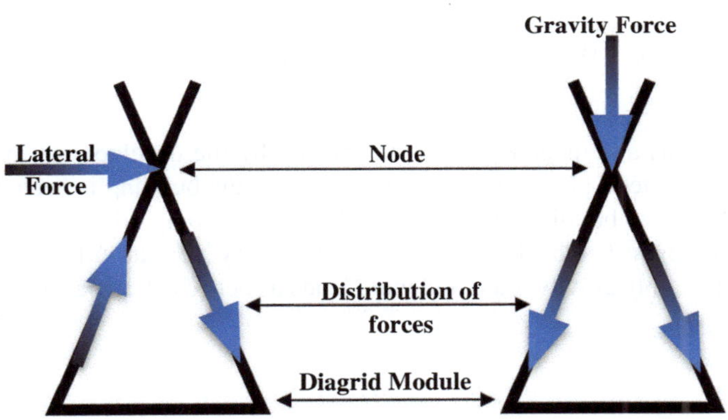

Fig. 2 Distribution of external forces in a typical diagrid module

2 Details of Hearst Tower, NY, USA

A symmetric rectangular plan shape and comparatively low height are the main reasons behind finalizing Hearst Tower as the reference Diagrid building for the analysis. All the geometrical and structural data are procured [10]. Neglecting the interior architectural features and ancillary minor details, a main structural framing is modelled in the software (Fig. 3).

The building consists of the first floor of about 80 ft. height, which is kept as an open space for architectural requirement. This floor is constructed using mega

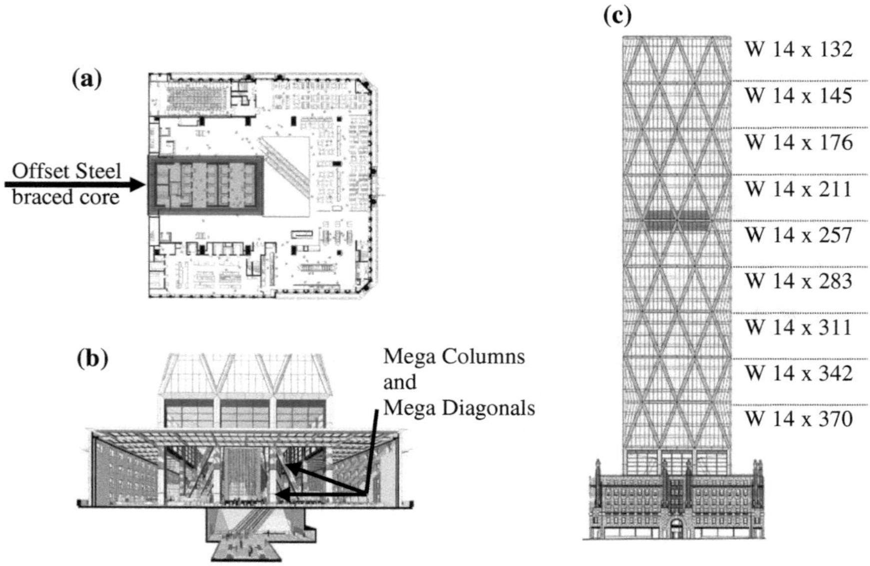

Fig. 3 **a** Typical plan with offset steel braced core. **b** First floor with columns and mega diagonals, omitted for modelling. **c** Diagrid member section details [9]

vertical columns and mega diagonals (as named by the developer of the building Norman and Fosters) [11]. This first 80 ft height of the building is neglected in the analysis. Thus, the building modelled in SAP2000 has only 36 floors, having only Diagrid through the height. The building is designed by Norman & Fosters company, considering mainly gravity loads and wind loads as per ASCE codes for Manhattan, NY, USA.

3 Details of the Seismic Evaluation

The finite element model constructed in SAP2000 software has all the major features of the actual Hearst tower. Also, for comparative study, a same dimensioned simple moment frame building is also constructed in SAP2000.

3.1 Earthquake Data Inputs and Details

The earthquake time histories here considered are, as per Table 1. Behaviour of a structure under various earthquakes is analyzed and compared. The earthquake time history is the actual acceleration versus time data that are captured at the earthquake

Table 1 Details of Hearst Tower, NY, USA [12]

Parameter	Actual hearst building
Story nos.	36 (Diagrid) + 10 = 46
Plan dimension	48 m × 37 m
Total height	148.86 m (Diagrid) + 34.14 m (80 ft)
Diagrid module	Triangle with (b × h) = 12.12 m × 16.54 m, α = 70°
Load per floor area	Average (LL) 5 KN/m^2
Diagrid member sizes	Varying from top to bottom, W 14 × 132 to W 14 × 370, AISC sections
Peripheral ring beam size	W 30 × 90
Floor beam size	W 21 × 55 and W 21 × 48
Slab/Deck details	*Vulcraft* 3VLI Deck (6″ + additional 2″ for fire), 12′ span, Max^{m} load 110 PSF (5.86 KN/m^2)
Material used	Structural steel A992 Grade ASTM
Central core	Offset steel braced core

station. This data has thousands of values at a regular time interval (say 0.02 or 0.005 or 0.01 s etc) which results into a trace as shown in Fig. 4. Peak Ground Acceleration (PGA) and its occurrence time, are the value that matters the most in seismic analysis. The structure will also show a similar kind of graph, of displacement versus time for the respective time history. In the graph, the peak deflection of the structure may or may not be occurring at the same time of the PGA occurrence. Though the structure is going to experience shaking, the main concern is the maximum deflection while shaking. Structure designers consider this value of deflection for the seismic design.

Moreover, whether a structure is stiffer or flexible depends on its value of natural frequency (first modal frequency). Structure with lower natural frequency is flexible than the structure with a higher natural frequency. This parameter is also taken into consideration for comparing the structures.

3.2 Details of SAP2000 Models

The SAP2000 Diagrid model is made with the same exact data available of the original Hearst Tower, NY, USA. This model is analyzed using above-mentioned earthquake time histories. The Diagrid model is also compared with the conventional moment frame building. The SAP2000 models are as shown in Fig. 5.

The Diagrid model consists of only the Diagrid part of the actual Hearst tower. The Diagrid model also consists of an offset braced steel core (see Fig. 3b). All the diagonal members, ring beams, floor beams and slab details are as per the actual Hearst tower data (see Table 2).

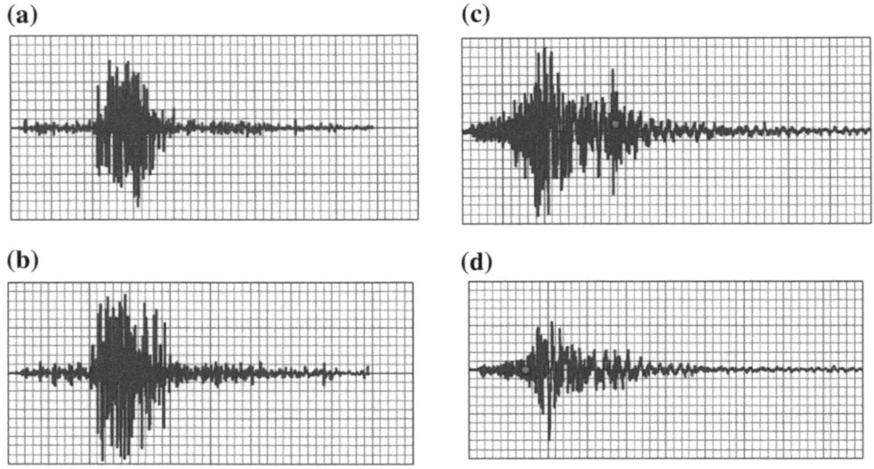

Fig. 4 Earthquake time history samples, acceleration versus time data. **a** Bhuj earthquake, Component 0, **b** Bhuj Earthquake, Component 90, **c** Yermo earthquake, Component 0, **d** Yermo earthquake, Component 90

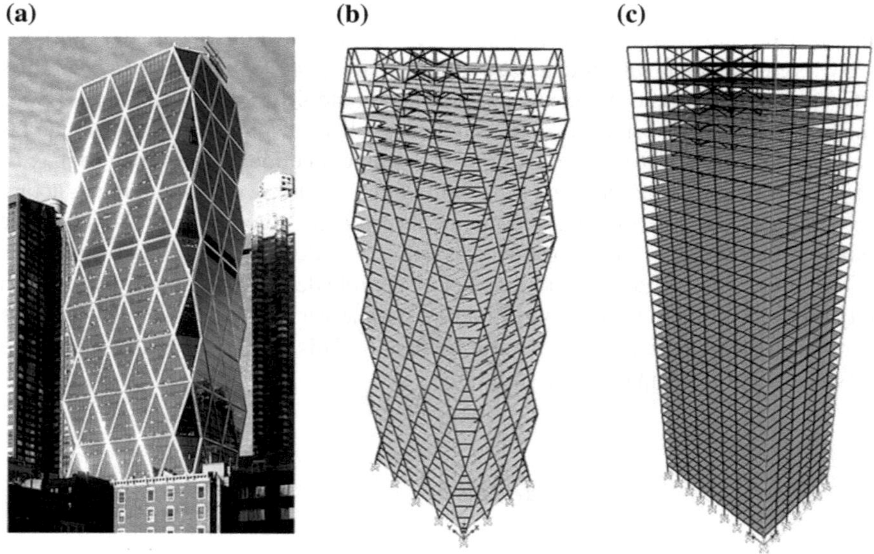

Fig. 5 3D view of the **a** Hearst tower, **b** diagrid model, **c** simple conventional model

The Simple model is made as same as the Diagrid structure. All the details are kept same except the orientation of the peripheral diagonals. The simple model consists of vertical peripheral columns.

Table 2 Details of earthquake time histories considered

Earthquake	Duration (s)	PGA (cm/s/s)	Time (s)	Component (°)
Petrolia, CA, USA	60	649.442	3.28	90
Newhall, CA, USA	60	578.19	4.32	0
Altadena, CA, USA	40	438.9	2.84	0
El Centro, CA, USA	39.1	428.09	5.94	90
Chamoli, UK, India	24.34	352.83	4.6	90
Yermo, CA, USA	80	240.016	16.32	90
Bhuj, GJ, India	133.5	103.82	46.94	0

The dead weight and the weight due to live loads on both the models come out to be almost the same. This is a major criterion for comparing lateral stiffness of a structure. Here, weight (mass) being equal, the only difference in two models is their lateral stiffness. Thus, this study mainly focuses on achieving higher lateral strength keeping the same material usage. The preliminary design check of these models is done using CSI ETABS software.

4 Results and Discussion of the Dynamic Time History Analysis

The first major comparing parameter is the natural frequency of the structures. After the complete modal analysis, the first mode for both the structure is in the X direction, i.e. the direction having lesser plan dimension. This satisfies the conceptual knowledge regarding the shaking modes and modal frequencies. The first modal frequency (natural frequency) of the Diagrid model is 0.214 Hz, whereas the same modal frequency for the simple building is 0.1 Hz (Fig. 6).

Further, both the models are analyzed using different time histories. It is to be understood that the maximum top displacement of the structure may or may not be occurring at the same time instant of the PGA. The top displacement also depends on the duration of the earthquake, intensity and occurrence of PGA. But the graph of top displacement versus time will be of a similar shape of the respected time history. The following figures (Fig. 7) shows some of the top displacement versus time results for both the models.

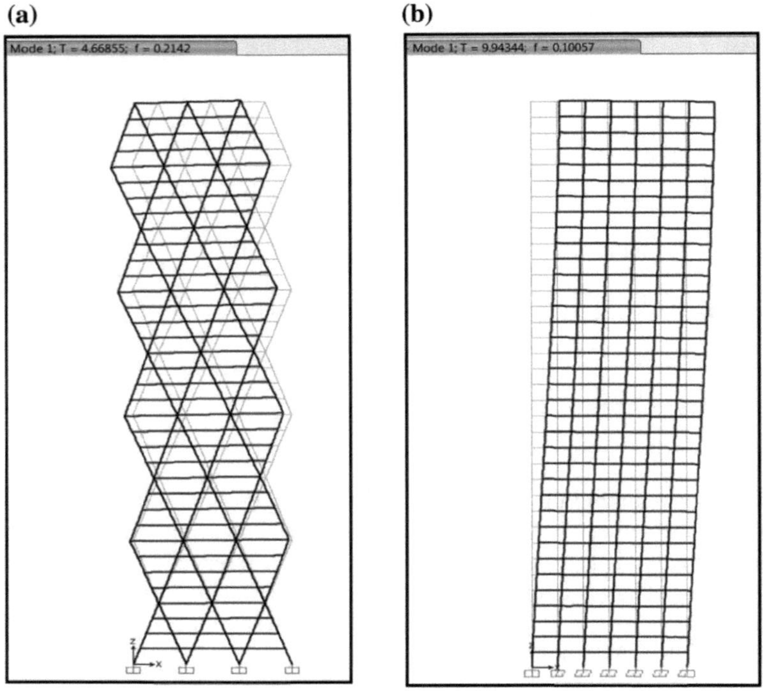

Fig. 6 First modal deflection of **a** diagrid model, **b** simple model

4.1 Top Displacement Results

Top displacement values for Diagrid model for all the earthquakes are much lesser than the values for a simple moment frame model. This is because of the orientation of the load carrying member being inclined in Diagrid model. The diagonal orientation of columns is converting all the gravity as well as lateral forces into axial forces. The same earthquake (say Bhuj EQ) is deflecting both the models very differently.

In Diagrid, the model shows higher lateral stiffness which also reveals from its more of a shear mode deflection pattern, whereas a simple model having lower lateral stiffness exhibits a bending mode deflection pattern (see Fig. 8).

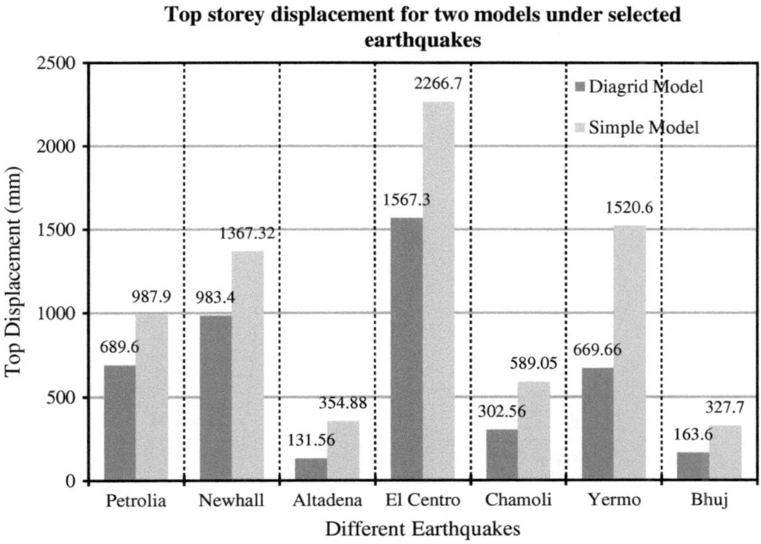

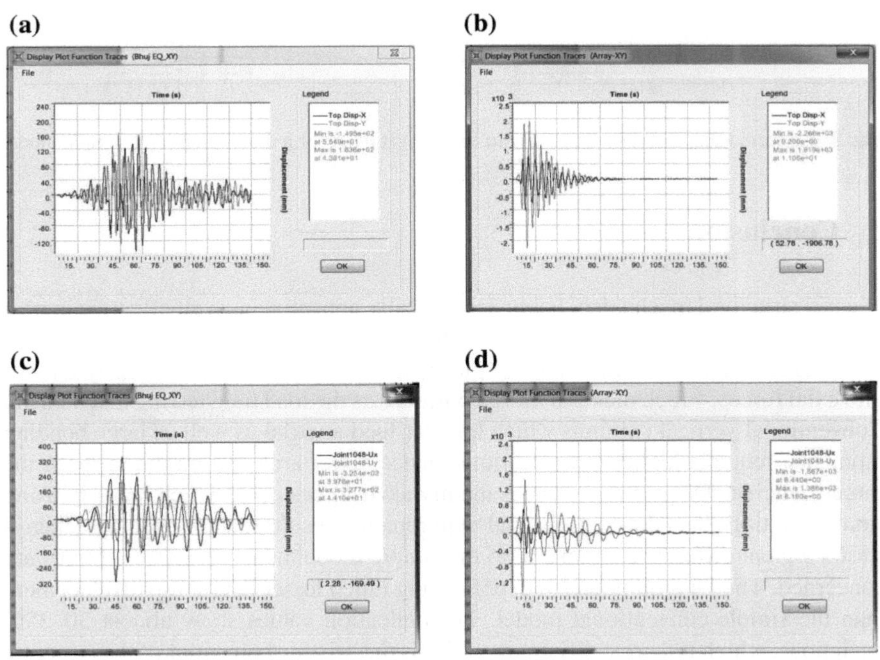

Fig. 7 Some of the top displacement versus time results of **a** Diagrid model Bhuj EQ, **b** Diagrid model El Centro EQ, **c** simple model Bhuj EQ, **d** simple model El Centro EQ

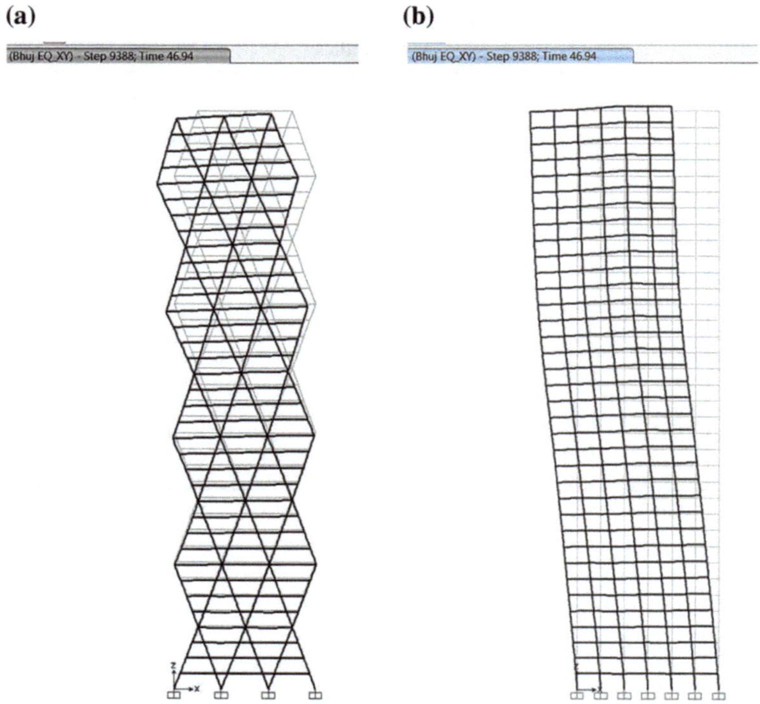

Fig. 8 Deflected shape of the models due to Bhuj EQ at PGA for **a** diagrid model, **b** simple model

5 Conclusion

Diagrid structural mechanism being aesthetically appealing as well effective in providing higher strength, should be adopted for high rise infrastructures. The peripheral diagonal column, because of their unique orientation, transfers all the imposed loads into axial forces, which makes it more optimized as the load transferring mechanism. Conventional vertical columns which transfer load as axial as well as large bending moments, requires higher cross sections, hence larger area and inertia to resist the forces, whereas Diagrid experiences mainly axial forces only, which helps in using smaller sections than the conventional requirement. Results have shown that Diagrid shows a great difference compared to the simple model as far as seismic forces are concerned. The Diagrid model is experiencing much lesser lateral top displacement than the simple conventional model. The deflection values show almost 30–35% deviations, which is a great difference as far as the seismic behaviour of the structure is a concern. Comparatively, lower PGA earthquakes with longer duration can lead to larger top displacements than the higher PGA with short duration. Looking to the top displacement versus time graph (see Fig. 6), it is evident that after the occurrence of PGA, the Diagrid model comes to rest position faster than the simple model, which makes Diagrid much damped than the simple building. This result can lead to optimize the Diagrid mechanism to be used for even much higher tall infrastructures.

References

1. Rahimian, A.: Stability of diagrid structures. Int. J. High-Rise Build. **5**(4) (2016) CTBUH
2. Boake, T.M.: Diagrid Structures-Systems, Connections, Details. Birkhauser, Germany (2014)
3. Al-Kodmany, K., Ali, M.: An overview of structural and aesthetic developments in tall buildings using exterior bracing and diagrid systems. Int. J. High-Rise Build. **5**(4) (2016) CTBUH
4. Korsavi, S., Maqhareh, M.R.: The evolutionary process of diagrid structure towards architectural, structural and sustainability concepts: reviewing case studies. J. Archit. Eng. Tech (2014)
5. Soo, K.J., Sik, K.Y., Hee, L.S.: Structural schematic design of a tall building in asan using the diagrid system. In: Conference Proceedings. CTBUH (2008)
6. Boake, T.M.: The Emergence of the diagrid—It's all about the node. Int. J. High-Rise Build. **5**(4) (2016) CTBUH
7. Zhao, F., Zhang, C.: Diagonal arrangements of diagrid tube structures for preliminary design. Struct. Design Tall Spec. Build. (2014)
8. Moon, K.S.: Structural Design and construction of complex-shaped tall buildings. IACSIT Int. J. Eng. Technol. (2015)
9. https://www.fosterandpartners.com/projects/hearst-headquarters/#drawings
10. Lucas, J.M.: Report on The Hearst Tower, Mechanical, Electrical and Structural details
11. Rahimian, A., Eilon, Y.: Hearst headquarters: innovation and heritage in harmony. In: Conference Proceedings. CTBUH (2008)
12. Mele, E., Toreno, M., Brandonisio, G., De Luca, A: Diagrid structures for tall buildings: case studies and design considerations. Struct. Des. Tall Spec. Build. (2012)

Analysis of the Behavior of High-Rise Structures with Viscoelastic Dampers Installed at Various Locations

Snehal N. Raut, Rohan Majumder, Aman Jain and Vinay Mehta

Abstract It is essential to analyze the behavior of each and every tall structure in the perspective of structural engineering when subjected to severe ground motion or earthquake. Vibratory forces are produced due to the result of such earthquakes at the base of the structure. Oscillations are created in buildings as a result of these vibrations. Such oscillations may cause severe damage to the structure. From the ground level, the vibrations get transmitted up to the top of the building and since the structural mass which creates lateral forces on the frame, the moment resisting capacity gets diminished for that of building components such as columns, beams, etc. This paper gives an idea of the different research works executed on a multi-storey building frame by taking various parameters into account. It explains the results of a study on the seismic behavior of a tall structure (G+15) installed with a damper. From a previously conducted experiment, it has been expressed that the stiffness and the strength of the structure enhances considerably with the use of such dampers. The current work aims to analyze the behavior of a multi-storeyed building for the most efficient location of viscoelastic damper in the structural system. A standard finite element software is used to carry out the analysis. Various parameters like lateral storey drift, base shear, time period, modal shapes, etc., are being analyzed and evaluated for the damper being placed at different locations.

Keywords Ground motion · Dampers · Base shear · Modal shapes

S. N. Raut
Department of Civil Engineering, Ballarpur Institute of Technology, Chandrapur 442901, Maharashtra, India
e-mail: snehalr1212@gmail.com

R. Majumder (✉) · A. Jain · V. Mehta
Department of Civil Engineering, Sir Padampat Singhania University, Udaipur 313601, Rajasthan, India
e-mail: rohan.majumder@spsu.ac.in; majumder1989@gmail.com

A. Jain
e-mail: amanjain@spsu.ac.in

V. Mehta
e-mail: vinay.mehta@spsu.ac.in

D. Deb et al. (eds.), *Innovations in Infrastructure*, Advances in Intelligent Systems and Computing 757, https://doi.org/10.1007/978-981-13-1966-2_16

1 Introduction

In the last few years, quite a large number of devastating earthquakes have occurred at various places all over the world. It resulted in the collapse of buildings and structures which ultimately leads to the loss of human life. Such events explain the hazardous effects of strong seismic vibrations and that the structures like RC residential buildings, dams, bridges, flyovers, etc., are required to be designed with utmost care to save them from the deadly effects of such earthquakes. Seismic response control is now an integral part of structural design approach and is widely acceptable all over the world. It is frequently applied in the field of structural engineering. Recently, researchers have paid strong attention to the development of several structural response control techniques. Passive, active, and semi-active are the three most commonly used techniques of structural control. Special attention is also being given to the improvement of wind and seismic responses of buildings and bridges. Civil engineers are putting in great efforts to develop the structural control concept into a usable technology and to incorporate such control devices within the structures.

Most of the structural systems which are designed for gravity load may not be capable enough to withstand the lateral load. Even if the structure is designed to withstand the lateral load, it will substantially increase the cost of the structure with the increase in the number of storeys. As the seismic load is proportional to the mass of the structure, such structures are made to be light and flexible with low natural damping. Newly constructed high-rise buildings are incorporated with artificial dampers for vibration control by means of energy dissipation. A good example of the passive damping system is the Tuned Mass Damper (TMD). It utilizes a secondary mass attached to the main structure by means of a spring and dashpot to mitigate the dynamic response of the structure.

Diclelia and Mehta 2007 have carried out a parametric study of steel chevron-braced frame system equipped with and without viscoelastic damper when subjected to seismic load [1]. Viscoelastic damper, (Soong and Spencer 2002) consists of viscoelastic layers bonded with steel plates with viscous materials such as copolymers or glassy substances [2]. The energy is dissipated in the form of shear deformation when mounted on a structure and they are highly dependent on ambient temperature and frequency excitation. Choi and Kim proposed the new method of installing viscoelastic cables into building for seismic control of structures. Chen et al. have analytically studied the seismic performance of Wenchuan hospital structure using viscoelastic damper [3]. Viscous dampers work based on fluid flow through orifices. Viscous damper (Feng Qian et al. 2012) consists of viscous wall, piston with a number of small orifices, cover filled with a silicon, or some liquid materials like oil, through which the fluid pass from one side of the piston to the other [4]. Stefano et al. have manufactured the viscoelastic damper and it is implemented in a three-storey building structure for seismic control of structure with additional viscous damper [5].

The main aim of this work is to generate fundamental research information on the seismic performance of building structural systems possessing passive damping devices installed within the frame. In this work, computer models of building with infill wall and damper systems are created for investigation. The effect of important parameters such as damping properties, locations, and configurations of the dampers are studied and the research findings are utilized to propose more effective damping system for seismic mitigation. This work investigates the seismic response of building structures with bare frame, infill wall, and embedded dampers. The main response parameters are tip deflection of the structure. Viscoelastic damper is being considered in the study and is installed at different locations. Response spectra method of dynamic analysis is used.

2 Modeling of the Proposed Building Frame with Viscoelastic Dampers

A general G+15 RC building bare frame with infill wall as diagonal strut is considered. Viscoelastic dampers are used for seismic vibration mitigation. The dampers are placed at different locations. Viscoelastic (VE) damper is a passive type of energy dissipation device.

G+15 RC frame building has been modeled as 3D space frame model with six degree of freedom at each node using SAP 2000 software for stimulation of behavior under gravity and seismic loading. The support condition is considered as fully fixed. The Response Spectrum Method of Analysis is utilized as per IS: 1893-2002 [6]. Typical building plan is shown in Fig. 1.

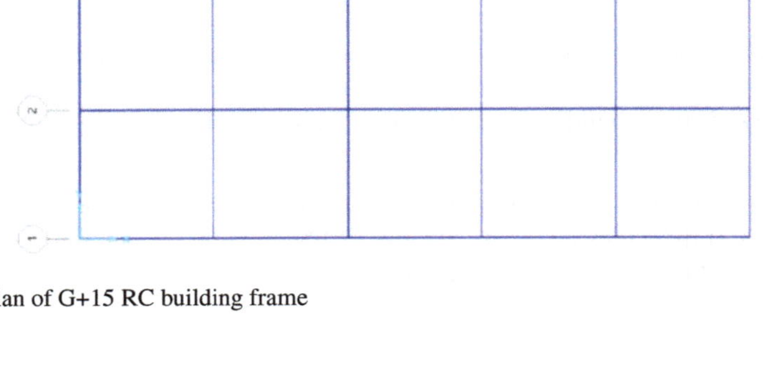

Fig. 1 Plan of G+15 RC building frame

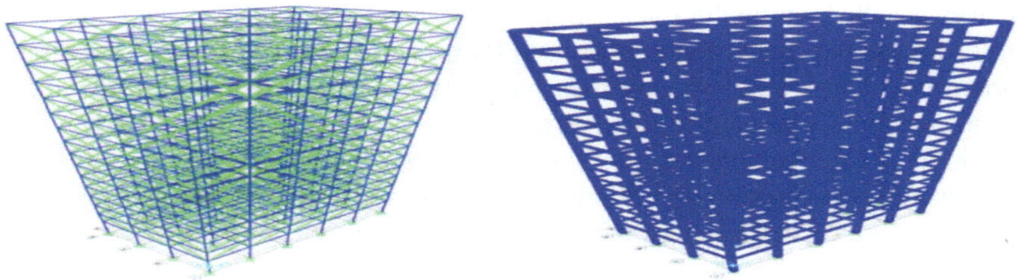

Fig. 2 3D view of G+15 storey building with bare frame and with infill wall

Building/Site Properties considered
Details of building: G+15 storeyed building, plan dimension: 25 m × 15 m, 5 m span in each direction, outer wall thickness: 230 mm, inner wall thickness: 230 mm, floor height: 3 m, parking floor height: 3 m.
Seismic Properties
Seismic zone: IV, zone factor: 0.24, importance factor: 1.0, Response reduction factor R: 5, soil type: medium.
Material Properties
Material grades of M35 and Fe500 were used for the design.
Loading on Structure
Dead load: self-weight of structure, weight of 230 mm wall: 13.8 kN/m^2, Live load: For G+15: 2.5 kN/m^2, roof: 1.5 kN/m^2, wind load: not considered, seismic load: Seismic Zone: IV.
Preliminary Sizes of members
Column: 600 mm × 400 mm, beam: 400 mm × 250 mm, slab thickness: 125 mm.

The models created are as follows:

Model 1: G+15 storey bare frame building.
Model 2: G+15 storey building with infill wall.
Model 3: G+15 storey bare frame building with viscoelastic damper—first position.
Model 4: G+15 storey bare frame building with viscoelastic damper—second positions.
Model 5: G+15 storey bare frame building with viscoelastic damper—third positions.
Model 6: G+15 storey building with infill wall with viscoelastic damper—first position.
Model 7: G+15 storey building with infill wall with viscoelastic damper—second position.
Model 8: G+15 storey building with infill wall with viscoelastic damper—third position.

The prime factor in determining the efficiency of any damper is its properties. The stiffness of the damper K is 56,794,365.92 kN/m while the damping coefficient C is 18,459,853.62 (Figs. 2, 3, 4, 5, and 6).

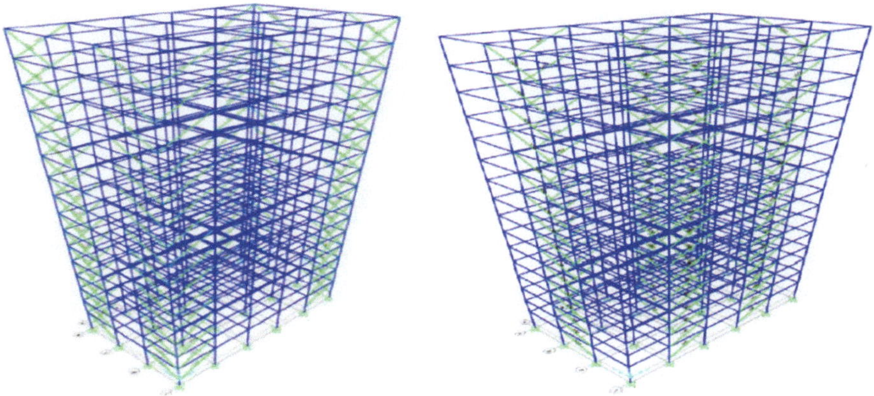

Fig. 3 3D view of G+15 storey bare frame building with viscoelastic damper at 1st and 2nd position

Fig. 4 3D view of G+15 storey bare frame building with viscoelastic damper at 3rd position

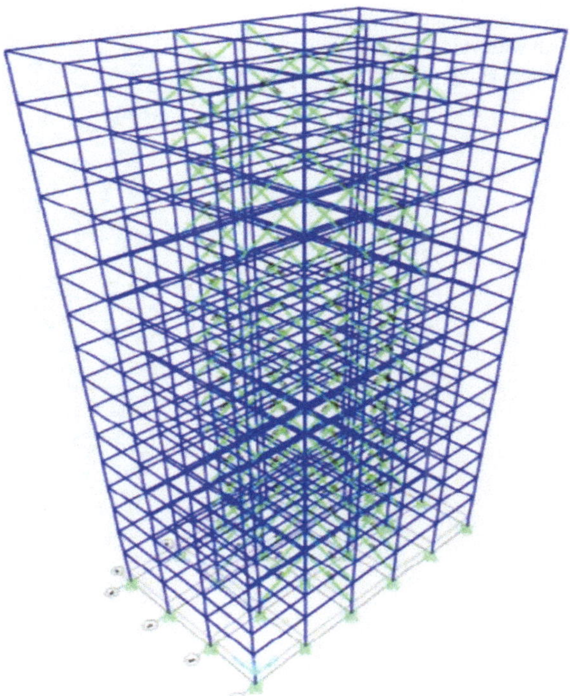

2.1 *Modal Periods and Frequencies*

Modal analysis is the study of dynamic properties of a system in the frequency domain (Figs. 7 and 8).

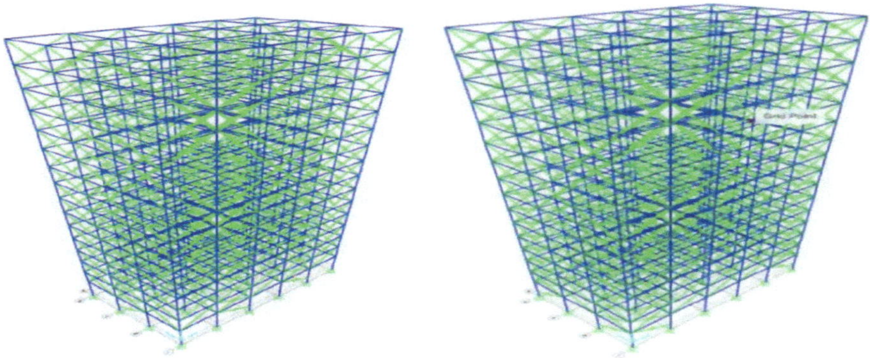

Fig. 5 3D view of G+15 storey building with infill wall with viscoelastic damper at 1st and 2nd position

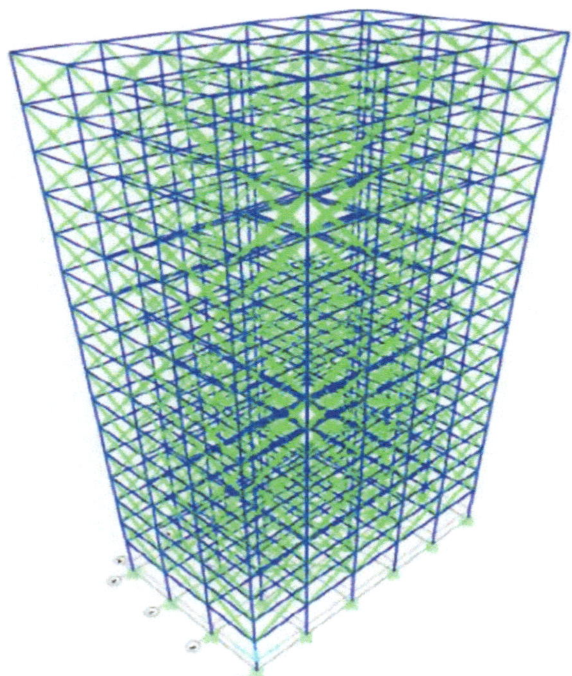

Fig. 6 3D view of G+15 storey building with infill wall with viscoelastic damper at 3rd position

2.2 Base Shear

Base shear is the total design lateral force at the base of a structure (Figs. 9 and 10).

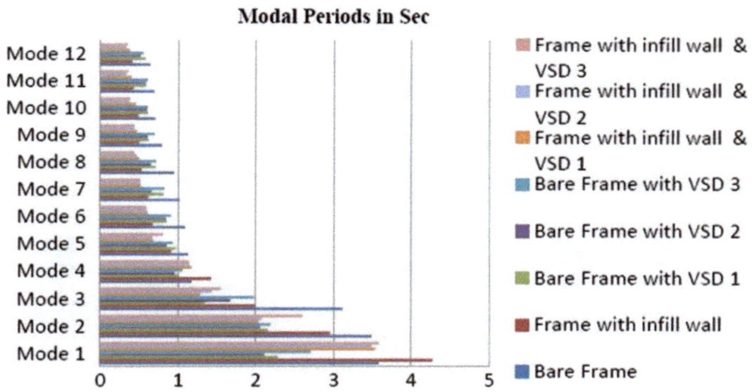

Fig. 7 Comparison of modal periods

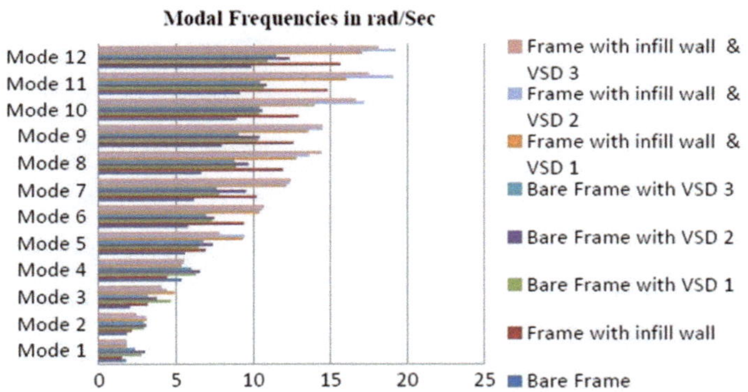

Fig. 8 Comparison of modal frequencies in rad/s

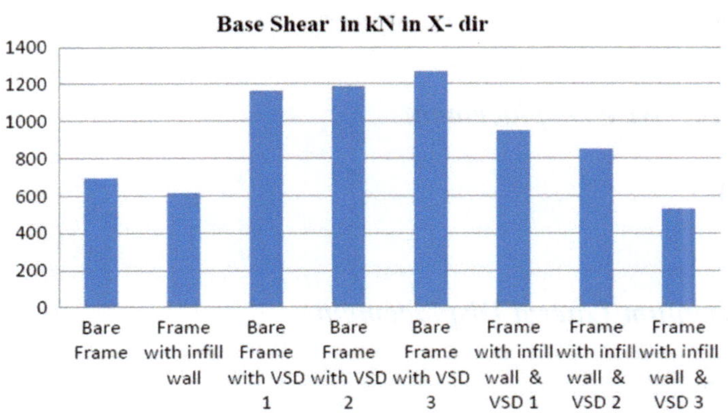

Fig. 9 Base shear in kN in *X*-direction

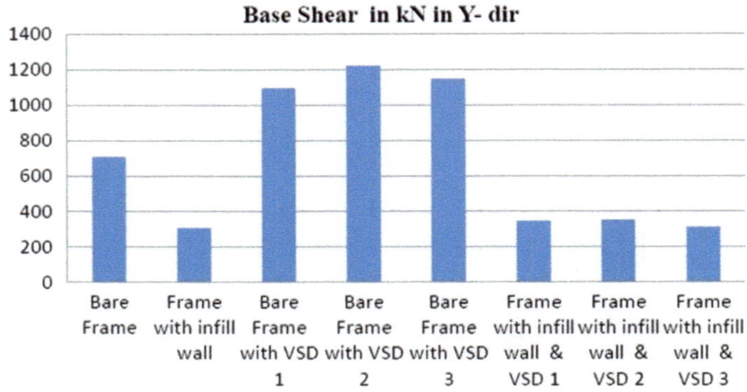

Fig. 10 Base shear in kN in *Y*-direction

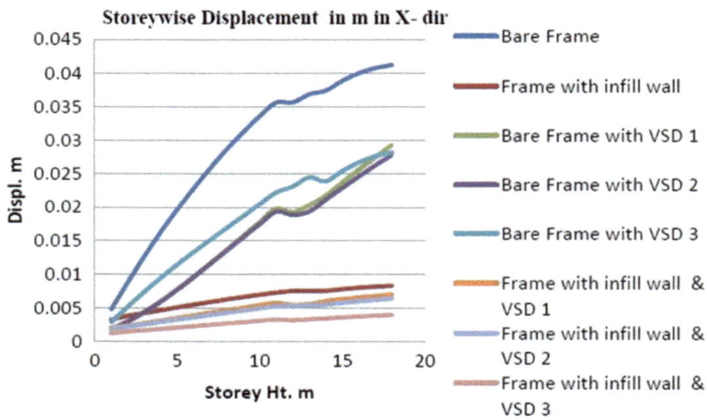

Fig. 11 Storey-wise displacement in m in *X*-direction

2.3 *Storey-Wise Displacement*

See (Fig. 11).

2.4 *Maximum Lateral Displacement*

See (Fig. 12).

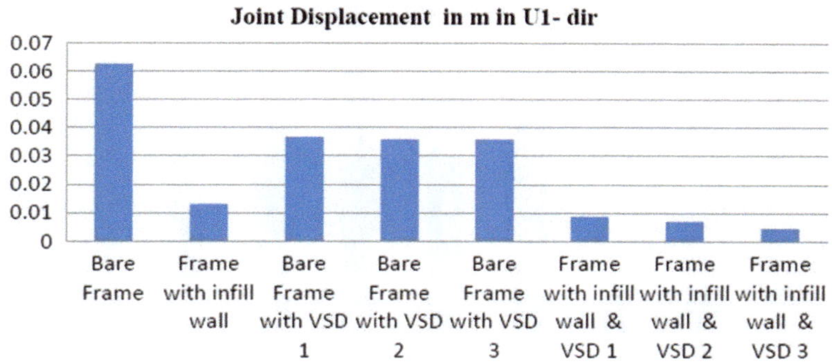

Fig. 12 Joint displacement in m in U-1 direction

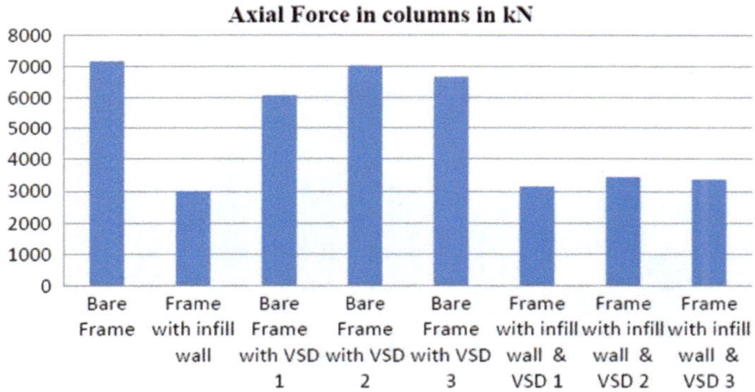

Fig. 13 Axial force in columns in kN

2.5 Forces in Column

Columns are mainly subjected to axial load (Figs. 13, 14, and 15).

2.6 Forces in Beams

See (Figs. 16 and 17).

3 Discussions and Conclusions

Modeling and analysis of bare frame, frame with infill wall, bare frame with viscoelastic damper—position 1, bare frame with viscoelastic damper—position 2, bare

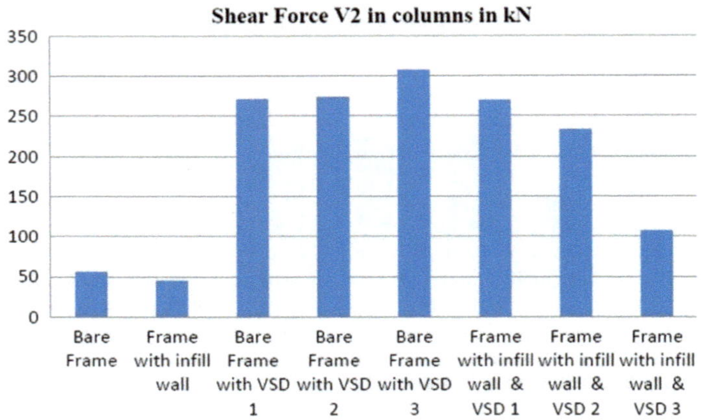

Fig. 14 Shear force $V2$ in columns in kN

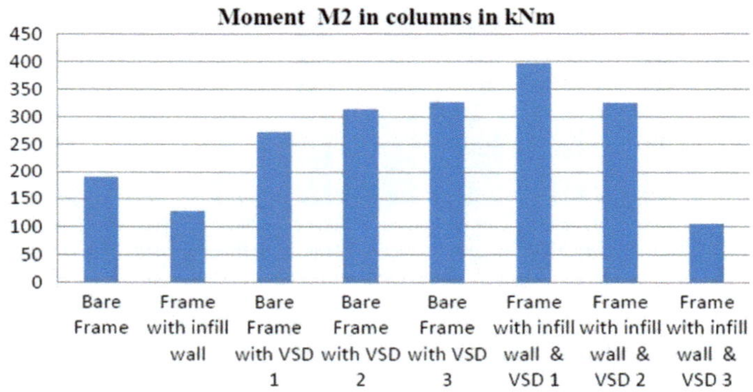

Fig. 15 Moment $M2$ in columns in kN-m

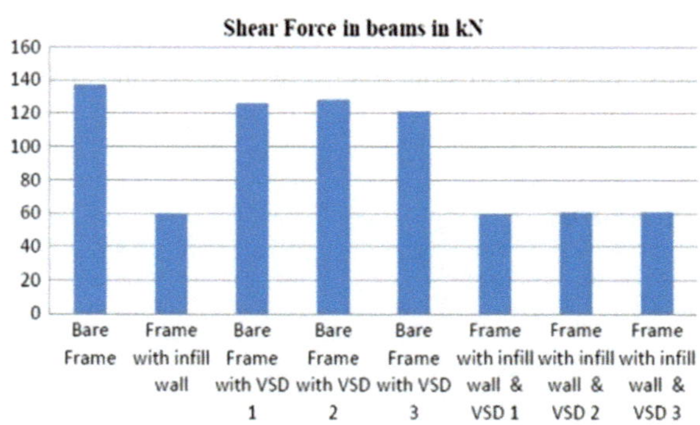

Fig. 16 Shear force in beams in kN

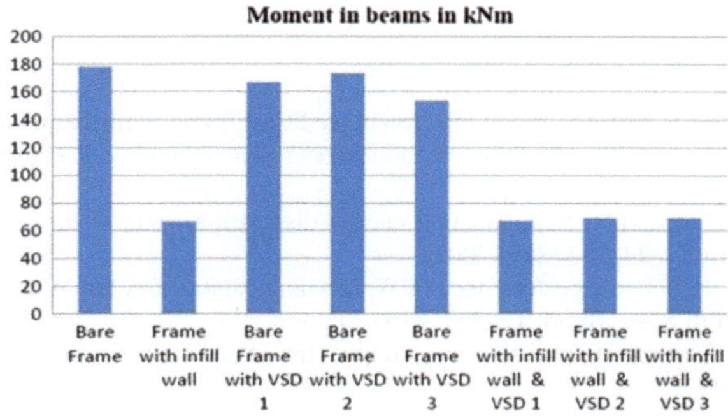

Fig. 17 Moment in beams in kN-m

frame with viscoelastic damper—position 3, frame with infill wall and viscoelastic damper—position 1, frame with infill wall and viscoelastic damper—position 2, frame with infill wall and viscoelastic damper—position 3 is carried in SAP 2000. Modal period is maximum for frame with infill wall model. Modal frequency is maximum for bare frame with viscoelastic damper—position 2 model. Bar diagram shows base shear is high for bare frame with viscoelastic damper models and less for frame with infill wall and viscoelastic model. Storey-wise displacement is maximum for bare frame model. Bare frame with viscoelastic damper reduces storey-wise displacement up to 30–40%, whereas frame with infill wall and viscoelastic damper reduces storey-wise displacement by more than 60%. Hence, response of the structure is reduced by combination of infill wall and dampers. Maximum lateral displacement is maximum for bare frame model. Bare frame with viscoelastic damper reduces the lateral displacement up to 30–40%, whereas frame with infill wall and viscoelastic damper reduces the lateral displacement by more than 80%. Hence, the response of the structure is reduced by combination of infill wall and dampers. Axial force in columns is maximum for the bare frame model. Shear force and moments in columns are maximum for bare frame with viscoelastic damper model. Also, shear force and moments in beams are maximum for bare frame with viscoelastic damper. The results of this study show how the structural response can be significantly reduced without increasing the structural stiffness by using viscoelastic damper. It has been found that the viscoelastic damper can be successfully used to control the vibration of the structure. Displacement is controlled with damper in the inner position in the structure. Therefore, the damper should be placed in the inner position for best control of the first mode. It is observed that the displacement is controlled by a substantial amount by using infill wall and dampers combination, whereas displacement to a considerable amount by using the dampers alone. The analytical study on the effect of viscoelastic damper in high-rise structures has been done. The parameters like base shear, storey displacement, and frequency have been compared.

References

1. Diclelia, M.: Seismic performance of chevron braced steel frames with and without viscous fluid dampers as a function of ground motion and damper characteristics. J. Constr. Steel Res. **63**, 1102–1115 (2007)
2. Soong, T.T., Spencer, B.F.: Supplemental energy dissipation: state-of-the-art and state of-the practice. Eng. Struct. **24**, 243–259 (2002)
3. Chen, X.W., et al.: Seismic performance of Wenchuan hospital structure with viscous damper'. Struct. Design Tall Spec. Build. **19**, 397–419 (2010)
4. Qian, F. et al.: Testing of fluid viscous damper. In: 15th World Conference on Earthquake Engineering (2012)
5. Stefano, S., et al.: A five-step procedure for the dimensioning of viscous dampers to be inserted in building structures. J. Earthq. Eng. **14**, 417–447 (2010)
6. IS: 1893: (Part 1) Criteria for Earthquake Resistant Design of Structures—general provisions and buildings. Bureau of Indian Standards, New Delhi (2002)

Assessment of Municipal Solid Waste as Backfill Material in Reinforced Earth Wall

Kinjal H. Gajjar, Alpa J. Shah and Manish V. Shah

Abstract The main objective of research work is to convert the existing Pirana landfill (Ahmedabad, Gujarat) municipal solid waste (MSW) into a suitable engineering backfill material for reinforced earth wall through segregation and infusement of synthetic reinforcement at suitable positions. For this purpose of evaluating the suitability of MSW as a backfill material, load settlement and load displacement tests were performed on sand, MSW, sand sandwiched (20 mm) between MSW and mixture of 1 (sand):3 (MSW) using a large- sized model test setup. Results were compared with sand being a standard backfill material. A further objective is to evaluate the effect of degree of saturation (Sr) with the optimum backfill composition found by varying the degree of saturation as 40, 50, and 60%. The results depicted a mixture of 1 (sand):3 (MSW) to be satisfactory. With the increase in the percentage of Sr, greater settlement and displacement of fill material and reinforced earth wall, respectively, were observed at lesser loads. The study reveals that mixture of 1 (sand):3 (MSW) can be recommended as suitable fill material under dry conditions.

Keywords Municipal solid waste · Reinforced earth wall · Backfill material

1 Introduction

Mechanically stabilized reinforced earth wall (MSE) is a reinforced soil mass that will be sufficiently stable by itself without the need of any means of external support. As the soil in an MSE wall is stabilized internally by the insertion of reinforcement, the wall is termed as "internally stabilized wall". During 17th and 18th century,

K. H. Gajjar (✉) · A. J. Shah · M. V. Shah
Applied Mechanics Department, L.D. College of Engineering, Ahmedabad, India
e-mail: gajjarkinjal1@hotmail.com

A. J. Shah
e-mail: alpa.jogin@gmail.com

M. V. Shah
e-mail: mvs2212@yahoo.co.in

D. Deb et al. (eds.), *Innovations in Infrastructure*, Advances in Intelligent Systems and Computing 757, https://doi.org/10.1007/978-981-13-1966-2_17

wooden sticks, planks, pegs, bamboos, wire mesh etc were used for reinforcing mud dikes, to control erosion and landslides. Modern state-of-art of soil reinforcement for retaining structure construction was coined by French engineer cum architect Henri Vidal in the 1960s. His research led to the Invention and development of Reinforced Earth, a system in which steel strip reinforcement is used. The rapid compliance of soil reinforcement can be accredited to a number of factors, including esthetics, reliability, low cost, simple construction technique, and the ability to adapt to different site conditions. However, these benefits have often been limited by the availability of good quality granular material to overcome pore water pressure [1]. One possible solution for reinforcing soils is the usage of permeable geo-synthetics that behaves as reinforcement as well as lateral drains [2]. Based on the concept described and explained in both theoretical and experimental measurements by Yang and Singh [3] and Schlosser and Long [4], the reinforcement produces an apparent cohesion on account of its tensile capacity. This concept may be most suitable for geogrid reinforcement on account of their strain compatibility with soils, extensibility as well as the coverage ratio of the geogrid. Moreover, only the mobilized pull force in the reinforcement is responsible for the apparent cohesion. In present, there has been a focus on using sustainable materials for backfill in MSE wall. Considering the focus towards reuse of waste, the reckoning of the potential of municipal solid waste (MSW) as backfill material assumes importance due to the ever-rising demand of inadequate land for its disposal [5]. Owing to the nonplastic nature and high value of angle of internal friction, municipal solid waste may prove to be a better option to be used as a backfill material in RE wall. Hence, the present study inspects the MSW as a potential backfill composition.

2 Materials

Sand used in the research as a fill material in RE wall model was procured from Mahudi, Gujarat and municipal solid waste was procured from Pirana landfill site, Ahmedabad. The classification of soil was found as poorly graded sand (SP). Geogrid used as reinforcement was procured from local manufacturing company having tensile strength of 60 kN/m in machine direction and 20 kN/m in cross machine direction. It was manufactured with high tenacity knitted polyester fiber with a treated coating and high molecular weight. Tables 1 and 2 show the index and engineering properties of different backfill materials used in this study. Tables 3 and 4 show the chemical properties of MSW which were found from proximate analysis and Energy dispersive X-ray spectroscopy (EDXS).

Table 1 Index properties of different backfill materials

Test	IS code	Symbol	Value		
			Sand	MSW	1 (Sand):3 (MSW)
Particle size distribution parameters	IS: 2720-4:1985	C_u	2.90	3.428	3.216
		C_c	1.117	1.166	1.210
Specific gravity	IS: 2720-3:1980	G	2.66	2.25	2.33
Dry density	IS: 2720-14:1983	γ_{max} (kN/m^3)	18.70	13.94	15.41
	IS: 2720-14:1983	γ_{min} (kN/m^3)	15.20	10.98	12.28

Table 2 Direct box shear test results (IS 2720-13:1986)

Parameter	Value					
	Sand	MSW	1 (Sand):3 (MSW)	1 (Sand):3 (MSW) with 40% Sr	1 (Sand):3 (MSW) with 50% Sr	1 (Sand):3 (MSW) with 60% Sr
c (kN/m^2)	0	3.471	2.77	2.20	2.00	1.70
ϕ	32.29°	34.76°	36.54°	34.49°	33.97°	33.38°

Table 3 Proximate analysis of municipal solid waste

Parameter	Value (%)
Moisture	0.80
Volatile matter	11.60
Ash	10.55
Fixed carbon	14.83

Table 4 EDAX test results of municipal solid waste

Element	Value (% weight)
Oxygen (O)	60.56
Magnesium (Mg)	1.7
Aluminum (Al)	5.32
Silicon (Si)	19.54
Potassium (K)	1.45
Calcium (Ca)	5.80
Iron (Fe)	5.73

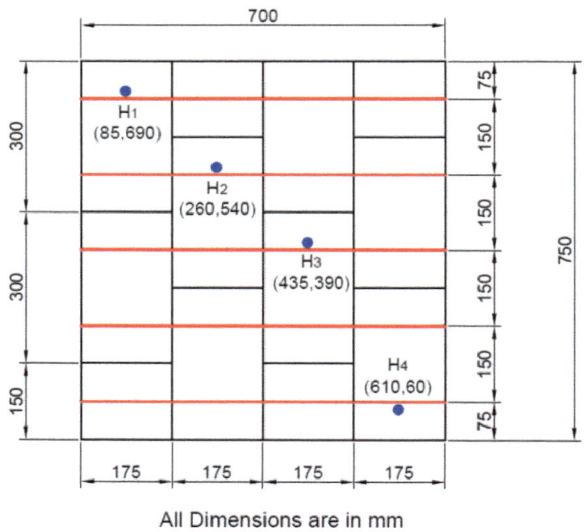

Fig. 1 Schematic diagram of facing panel

3 Test Setup

Model tests were performed in a steel tank of size 700 mm × 990 mm × 750 mm. Sides of the tank were furnished with perpex sheet in order to observe the failure surface. A wooden facing panel of 30 mm thickness was fixed to the open face of the tank model. The panel was made up of 8 number of blocks having a size of 300 mm × 175 mm and 4 number of blocks having size of 150 mm × 175 mm. According to BS 8006:2010 [6], five layers of geogrid having length of 525 mm and a spacing of 150 mm were fixed to the facing panel by means of clamps. To carry out the tests on reinforced earth wall model, the vertical downward load was applied by means of a mechanical jack. To measure the displacement, totally six dial gauges were set up in RE Wall model, out of which 2 dial gauges were set up in order to measure vertical settlement of backfill material and four dial gauges, viz., H_1, H_2, H_3, H_4 were set up in a way to measure the horizontal displacement of RE wall. Figure 1 shows a schematic diagram of the facing panel and shows the position of the horizontal displacement dial gauges.

4 Test Procedure

The tank was filled in layers by compacting the backfill material to achieve the relative density of 80% as per IRC SP 102–2014 [7]. Geogrids were used as reinforcing

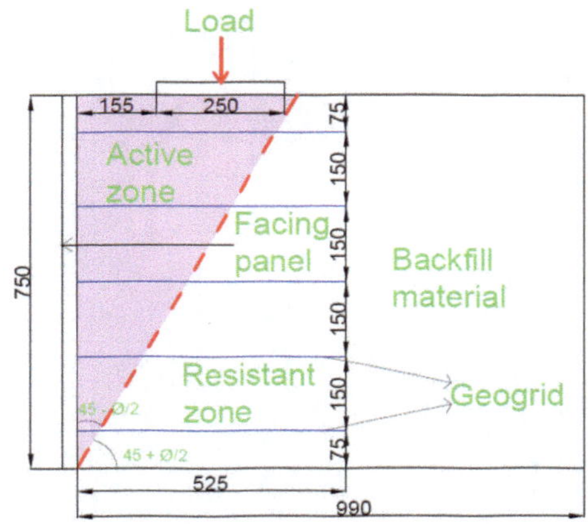

Fig. 2 Active and resistant zone in RE wall

element. In sandwich technique, geogrid was embedded in 20 mm thick layer of sand, while the rest of the backfill used was MSW. A plate of size 250 × 250 mm having 25 mm thickness was placed on RE wall model having firm contact with backfill material, in order to have uniform transfer of load. The loads were applied on the active zone which is inclined at 45° + $\phi/2$ from horizontal as shown in Fig. 2. From theoretical calculations, the RE wall can resist maximum load of 19.60 kN approximately. Hence, for the experimental purpose, the load is applied in increments of about 1/10 of this load, i.e., 1.96 kN. When there is no substantial increase in settlement or the rate of settlement reduces to 0.02 mm per minute as given in IS 1888–1982 [8] and the deformation rate becomes nearly constant, the dial gauge readings were noted. The readings were noted until the wall displaces to a greater extent or when the proving ring rebounds. Load versus displacement graph and load versus settlement graphs were then plotted to know the load displacement characteristics of RE wall and settlement characteristics of backfill material Fig. 3 represents the reinforced earth wall test model.

5 Results, Analysis, and Discussion

5.1 Effect of Different Backfill Materials on Displacement of RE Wall

From Figs. 4, 5, 6, and 7, it can be seen that the displacement in second and third panel is much more as compared to first and last panel due to the reason that the first and last panel receives the confinement from the top and bottom portion of fabricated tank

Fig. 3 Reinforced earth wall test model

while second and third panel receive confinement only from panels above and below it. Sand and mixture of 1 (Sand):3 (MSW) possessing minimum and maximum angle of shearing resistance exhibits maximum and minimum displacement, respectively, at all dial gauge locations. The overall average lateral displacement of RE Wall having mixture of 1 (sand):3 (MSW) as a backfill material is 0.83 mm, which is half the average lateral displacement observed with sand sandwiched between MSW as a backfill, whereas it is 1.45 and 3.22 times lesser as compared to MSW and Sand as a backfill material, respectively.

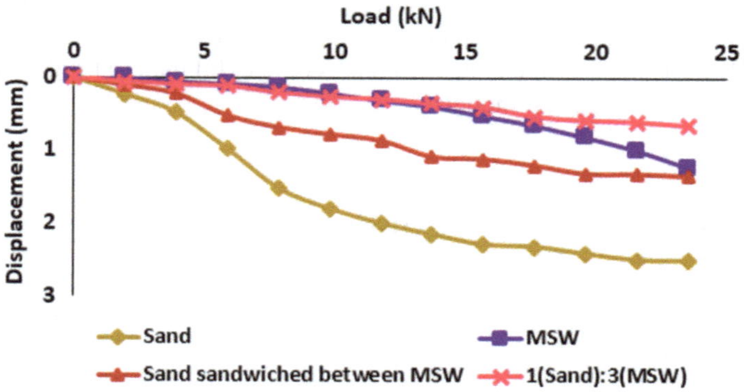

Fig. 4 Load versus lateral displacement characteristics for different backfill materials at H_1

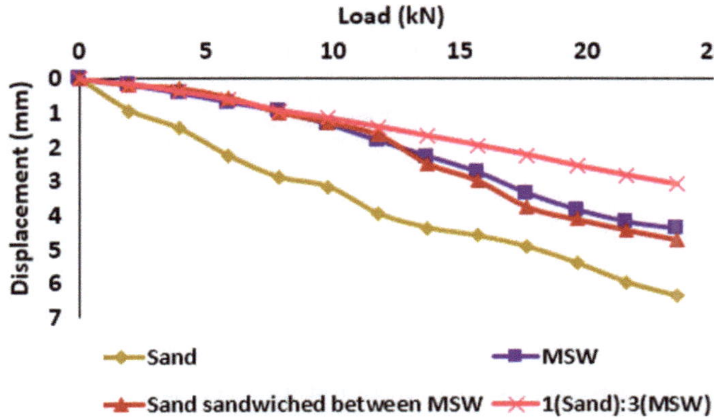

Fig. 5 Load versus lateral displacement characteristics for different backfill materials at H_2

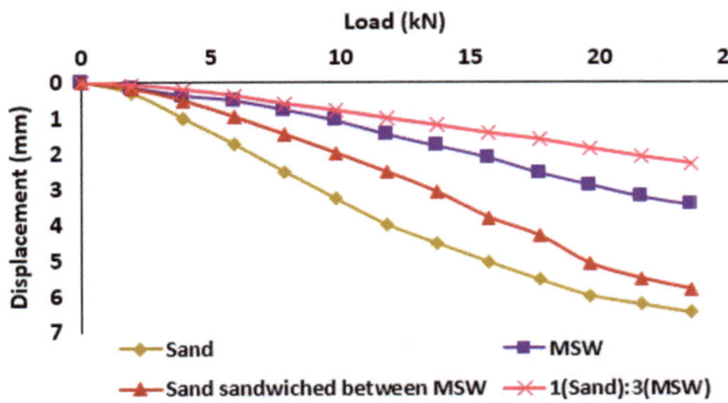

Fig. 6 Load versus lateral displacement characteristics for different backfill materials at H_3

5.2 *Effect of Degree of Saturation of Mixture 1 (Sand):3 (MSW) as a Backfill Material on Displacement Characteristics of RE Wall*

From Figs. 8, 9, 10 and 11, it can be seen that for mixture of 1 (Sand):3 (MSW) with 50% and 60% degree of saturation (Sr), the backfill material undergoes a settlement of nearly 50 mm just under the load of 13.73 and 7.85 kN, respectively, which is considered to be the ultimate failure criteria for the soil as per IS 1888:1982 [8]. Thus, the comparison of lateral displacement at a particular load could not be made. At 7.85 kN load, when the degree of saturation of 1 (Sand):3 (MSW) mixture is increased from 0 to 40%, an increment of 13% is observed in the average lateral displacement of RE Wall, whereas this increment rises to 2.34 and 3.15 times when the Sr is kept at 50 and 60%, respectively. From Figs. 8, 9, 10 and 11, it is clear that

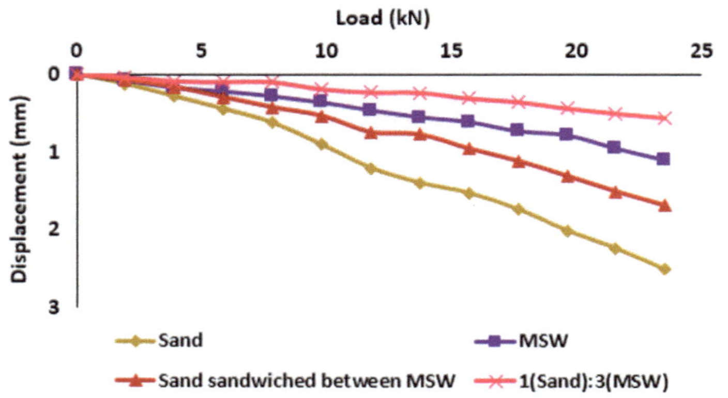

Fig. 7 Load versus lateral displacement characteristics for different backfill materials at H_4

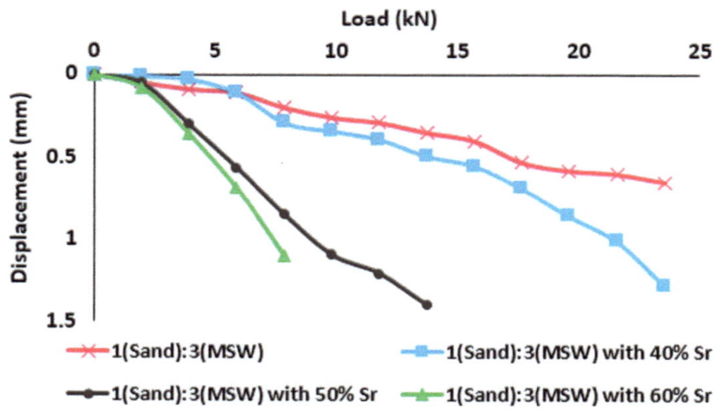

Fig. 8 Load versus lateral displacement characteristics for different backfill materials at H_1

as the wetness of the backfill material increases, the lateral displacement of wall also increases. This proves the suitability of dry backfill material in reinforced earth wall.

5.3 Settlement Characteristic of Different Backfill Materials

Figure 12 presents the settlement characteristics of different backfill material in the modeled reinforced earth wall. The maximum and minimum settlement of 17.425 and 9.50 mm is observed in MSW and mixture of 1 (Sand):3 (MSW) as a backfill material, respectively, under a load of 23.53 kN. Settlement of Sand and mixture of 1 (Sand):3 (MSW) as a backfill material is nearly equal and is approximately 83 and 73% less as compared to MSW and Sand sandwiched between MSW as a backfill.

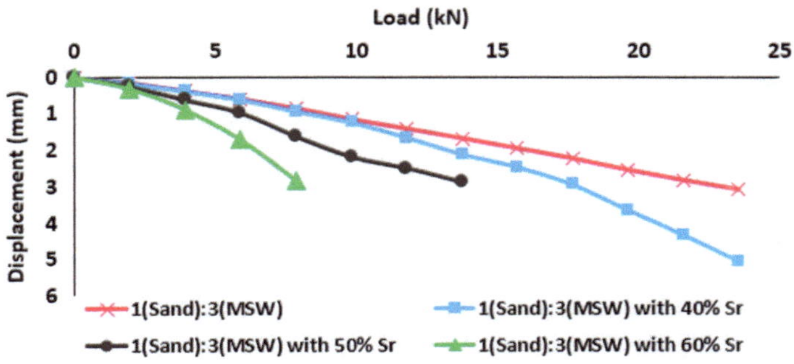

Fig. 9 Load versus lateral displacement characteristics for different backfill materials at H_2

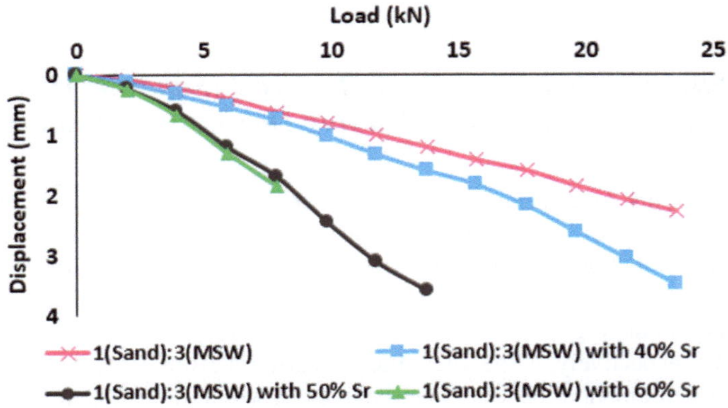

Fig. 10 Load versus lateral displacement characteristics for different backfill materials at H_3

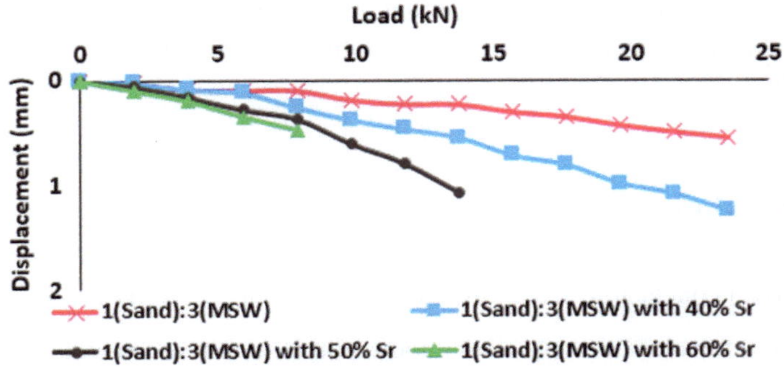

Fig. 11 Load versus lateral displacement characteristics for different backfill materials at H_4

The reason behind this is that sand and MSW both are having the coarser particles but during compaction process, some fractions of particles of MSW may get converted into finer particles, resulting in a higher settlement of the MSW as a backfill.

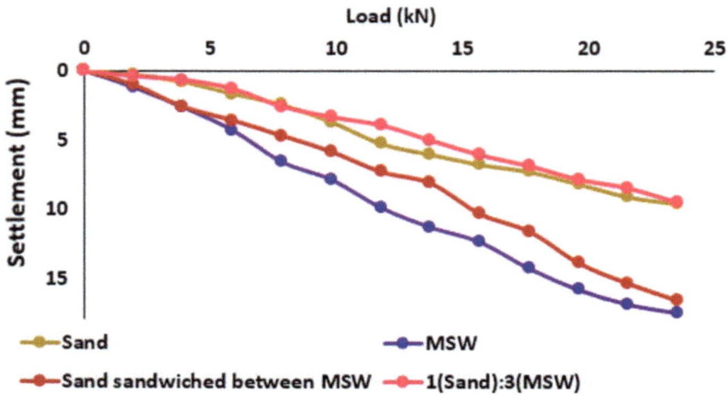

Fig. 12 Load versus settlement characteristics for different backfill materials

5.4 Settlement Characteristic of Mixture 1 (Sand):3 (MSW) as Backfill Material with Varying Degree of Saturation

Figure 13 presents the effect of degree of saturation on settlement characteristics of 1 (Sand):3 (MSW) mixture as a backfill material. From the figure, it can be seen that for a mixture of 1 (Sand):3 (MSW) in dry state and with 40% Sr, the settlement at 23.53 kN load is 9.50 and 16.34 mm, respectively. Whereas for 50 and 60% Sr, the backfill material undergoes a settlement of nearly 50 mm just under the load of 13.73 and 7.85 kN, respectively. As per the settlement criteria given in the IS 1888:1982 [7] after 50 mm settlement of backfill material, the test was terminated. At 7.85 kN load, when the degree of saturation of 1 (Sand):3 (MSW) mixture is increased from 0 to 40%, an increment of 32% is observed in the settlement of backfill material, whereas this rises to 10.5 and 21 times when the Sr is kept at 50 and 60%, respectively. Hence, it is clear that as we increase the degree of saturation, the settlement also increases significantly.

6 Conclusion

The important conclusions from this study are summarized below,

- Angle of internal friction plays an important role in reinforced earth wall. As the angle of internal friction increases, facia displacement decreases.
- The overall average displacement of RE wall having 1 (Sand):3 (MSW) as a backfill material is very less as compared to Sand, MSW, and the sand sandwiched between MSW.

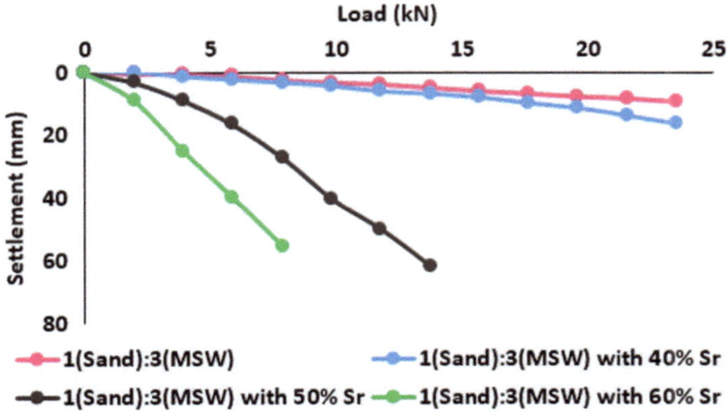

Fig. 13 Load versus settlement characteristics for different backfill materials

- With the increases in degree of saturation of mixture 1 (Sand):3 (MSW) as a backfill material, displacement of RE wall and settlement of backfill material increases, significantly.

The above large model study reveals that MSW can be suitably used as fill material through segregation. In this study, old MSW was used which already has undergone periodical physical and chemical changes, and leachate with 22–25% of organic matter with intermediate degree of saturation shows consistent adoptable results. The present study shows the suitability of MSW of Pirana as an engineered backfill material for RE wall.

References

1. Elias, V., Christopher, B.B.: Mechanically Stabilized Earth Walls and Reinforced Soil Slopes, Design and Construction Guidelines. Federal Highway Administration, Washington DC, Sa-96-071 (1996)
2. Zornberg, J.G., Mitchell, J.K.: Reinforced soil structures with poorly drained backfills. Part I: Reinforcement Interact. Functions Geosynthetics Int. **1**(2), 103–148 (1994)
3. Yang, Z., Singh, A.: Strength and deformation characteristics of reinforced sand. Ph.D. thesis, University of California, Los Angeles (1974)
4. Schlosser, F., Long, N.-T.: Recent results in French research on reinforced earth., J. Constr. Div. ASCE **100**(3), 223–237 (1972)
5. Shafie P.: Construction of pavements and highways using municipal solid waste materials to protect the environment, J. Fundam Appl. Sci. **8**, 1774–1783 (2016)
6. BS 8006:1995.: Strengthened/Reinforced Soils and Other Fills (1995)
7. IRC: SP: 102-2014.: Guidelines for Design and Construction of Reinforced Soil Walls (2014)
8. IS 1888:1982.: Method of Load Test on Soils (1982)

Feasibility Analysis of Decentralized Wastewater Treatment Systems

Shrutaswinee Hazarika and Devanshu Pandit

Abstract This paper aims to identify key factors that affect the performance and sustainability of decentralized wastewater treatment plants. In view of this, a feasibility analysis of three decentralized treatment plants of Ahmedabad, Gujarat based on various technologies is carried out. A list of quantitative and qualitative parameters such as efficiency of the technology, land use, ease of operation, requirement of technical expertise, effluent quality, capital investment, operation and management and user perspective are considered. The technologies assessed are fluidized aerobic bio-reactor (FAB), activated sludge process (ASP) and decentralized wastewater treatment system (DEWATS). The results show that there is no technology that can be used universally but to ensure long-term durability and sustainability of decentralized wastewater plants, proper maintenance after installation is just as important as technical analysis during design of the plant.

Keywords Decentralized wastewater treatment · India · Peri-urban
Feasibility study · ASP · FAB · DEWATS · Packaged treatment plant

1 Introduction

The current urban wastewater management process typically includes transportation of wastewater over long distances from its source of generation to the central municipal treatment plant through sewers and finally ends at disposal of the effluent in a nearby water body. With only 51% of the total sewage generated in metropolitan cities of India, most of the effluent is disposed of untreated which directly contributes to the contamination of freshwater source [1]. It has been stated in many studies that inflow of untreated effluent into water bodies affects aquatic life due to imbalance in nutrients, temperature and salinity [2]. However, with the increasing awareness of the potential of wastewater, the government of India has implemented

S. Hazarika (✉) · D. Pandit
Faculty of Technology, CEPT University, Ahmedabad 38009, Gujarat, India
e-mail: shrutaswinee.hazarika.mtech16@cept.ac.in

D. Deb et al. (eds.), *Innovations in Infrastructure*, Advances in Intelligent Systems and Computing 757, https://doi.org/10.1007/978-981-13-1966-2_18

various schemes and policies which encourage recycling and reuse of water. More than 30% of water used in residences are used for outdoor uses such as irrigation and 28% of water is used in toilet flushing which is both potential reuse opportunities for treated water [3]. In a conventional centralized wastewater treatment system to maximize the use of reclaimed water, additional water distribution pipelines must be constructed so that the benefit reaches the customers which are dispersed in the city area. This expansion of centralized water network proves to be very expensive with the ever-increasing urban sprawl. Therefore, to alleviate this pressure on existing municipal services decentralized system of wastewater management proves to be the most appropriate solution. Further, government of India is encouraging decentralized approach of wastewater treatment through the integrated township policy which mandates all townships which are generally constructed in the city outskirts to have their own decentralized water treatment plants and reutilize the treated wastewater for multiple uses to prevent exploitation of city resources and preserve freshwater sources [4].

Therefore, there is a need to evaluate existing decentralized wastewater treatment plants to identify sustainability and applicability factors so that such systems can be replicated in future.

2 Literature Review

Decentralized wastewater treatment systems treat waste at or near the source of generation. They can serve individual houses, industries, business complexes and/or small communities of residential houses. They can be large capacity septic tanks or package plants which work similar to a centralized treatment system, typically using small diameter sewers and a small-scale treatment facility.

The treatment process of domestic wastewater generally consists of four stages, namely pre-treatment, primary treatment, secondary treatment and tertiary treatment. The first two stages of treatment are common in all the different treatment technologies. The following wastewater treatment methods are basically modification of carrying out secondary stage of treatment.

Activated Sludge Process (ASP) [5] (Technology n.d.): This is one of the most tested wastewater treatment technologies which are being used all over the world since decades. Generally, all the conventional centralized sewage treatment plant works on this principle. After the process of sedimentation, wastewater is continuously mixed and aerated in an aeration tank where it is mixed with oxygen. The aerated effluent is then sent to a clarifier where the flocs get settled at the bottom of the tank which is recirculated to the aeration tank. The main advantage of using this technology is that it can be operated at a range of organic and hydraulic loading. Disadvantages include the requirement of constant supply of power, highly skilled operators, expert design and construction.

Fluidized Aerobic Bio-Reactor (FAB): This system consists of floating media of different shapes and sizes, which provides large surface area for the bacteria to

Table 1 Details of WWTPs

Case Study	Start-up year	Design capacity (m^3/day)	Technology type
IFFCO	2002	500	Fluidized bio-reactor
IIT Gandhinagar	2015	600	DEWATS
Godrej Garden City	2014	1000	Activated sludge process

grow on it. Air is provided in the FAB tanks to create an optimum environment for bacterial growth. The main advantage of this system is very small amount of sludge is produced and the space requirement of the plant is also very less.

Decentralized wastewater treatment system (DEWATS) [6] (Sasse 1998): DEWATS is developed by Bremen Overseas Research and Development Association (BORDA), Germany. In India, this system is being promoted by CDD Bangalore. In DEWATS, secondary treatment is carried out in anaerobic baffled reactor and the tertiary treatment is carried out through a bed of planted gravel filter. Anaerobic baffled reactor is a chamber consisting of a number of baffles through which water moves from top to bottom. Planted gravel filter is a filter media consisting of gravel and water-resistant plants.

3 Research Methodology

A case study-based analysis is carried out where three WWTPs representing different technologies installed by institutional and residential organizations were selected (Table 1). During site visits, data related to treatment technology, site condition, financial and managerial requirements of the plant are captured. A questionnaire survey was done to interview developers, vendors, consultants, users and city officials to gain knowledge of the market, the motivation and incentives behind installing decentralized wastewater treatment plants and its limitations.

4 Case Study

4.1 Fluidized Aerobic Bioreactor (500 m^3/day)

This FAB treatment plant is situated in IFFCO Kasturinagar in Kalol district of Gujarat. The township of Kasturinagar was developed on 51 acres of land as a residential colony for its staff. This township consists of 315 residential quarters, offices, post office, recreational facilities for the young and old, water treatment plant and a

Fig. 1 IFFCO wastewater treatment plant

Table 2 IFFCO wastewater effluent quality

Parameter	Outlet value	GPCB standard
pH	7.56	6.5–8.5
TDS (mg/L)	1218	–
TSS (mg/L)	10	30
BOD (mg/L)	8	20
COD (mg/L)	22	–
Oil and grease	Nil	–
Residual chlorine	0.55	>0.5

Source Weekly Wastewater Analysis Report by Metro Enviro-Chem Associates

sewage treatment plant [7] (Kumar n.d.). The distance between the township and the factory is around 5 kms.

Project Design: As per CPHEEO manual sewage generated is 280 m^3/day and the sewage treatment plant installed is designed for a capacity of 500 m^3/day keeping in mind the future population demand as shown in Fig. 1. The plant is designed on the principal of Fluidized Aerobic Bio-reactor technology and achieves the quality of final treated water as follows (Table 2).

Process Description:

Fluidized Aerobic Bio-reactors system: The system comprises of two FAB reactors containing plastic media as shown in Fig. 2. Bio growth takes place in the plastic media which is kept in fluidized condition by blowing air through SS grids. No sludge recycle is required to maintain liquor suspended solids and micro-organisms ratio.

1. The raw sewage from the township is collected in the well and it is pumped to FAB-I. The sewage feed pumps to FAB are provided with strainers to prevent floating materials from entering FAB-I.
2. The effluent flows to FAB-I to FAB-II. Both the reactors are provided with air distribution grids and wire mesh box to arrest the media.

Fig. 2 Fluidized aerobic bio-reactor I and II

3. Effluent from FAB-II is transferred to tube settler which acts as clarifier.
4. The sludge from bottom of tube settler is removed and taken to sludge drying bed for dewatering and drying. The treated water from tube settler is taken to chlorination contract tank and sodium hypo-chloride solution is added for disinfection of treated water.
5. The treated water is again taken to dual media filter for removing suspended solids. The filtered water is then passed through activated carbon filter to remove residual organic matter and stored in clear water collection tank of capacity 100 m^3.
6. From the clear water tank, treated effluent is pumped to distribution network in township for horticulture purpose.

Finance, Operations and Management: Kasturinagar Township is the first township to receive ISO-14001 certificate in May 2001. One of the major requirements of ISO-14001 is to manage its wastewater and thus the township installed a decentralized wastewater treatment plant. The entire project is funded by IFFCO including capital investment and operations and management. Thermax Company is the current agency which is responsible for operating the treatment plant and is on a contract basis. The plant requires one supervisor and three operators. According to the supervisor, they have not encountered any major problem till date.

The total cost of the treatment plant is about 60 lakhs which were recovered within 5 years as per primary survey. The operations and management cost includes the salaries of operators and cost of electricity consumption. With the installment of the treatment plant, a savings of Rs. 25/m^3 of water is done.

Feedback: The whole project was community oriented. The residents were made aware about the environmental issues threatening life existence. The core group members included housewives to workers to officers to achieve comprehensive involvement. To encourage reuse of water, intensive awareness generation programs were done.

The treatment technology was selected based on its level of maturity, level of maintenance and space required. As per the views of the Joint General Manager (EPC), they have successfully achieved 100% reuse of wastewater.

4.2 DEWATS (600 m^3/day)

The DEWATS treatment plant is situated in a public engineering institution, established in 2008 by the Ministry of Human Resource Development. Currently, the institution is developing its new campus at a 400 acres site in Palaj, Gandhinagar. The site is located on the eastern side of river Sabarmati, across from the Gandhinagar city. The campus will be a 100% residential campus for staff and students and it is developed in a phased manner where phase I will have 1200 students, phase II will have 2400 students and phase III will have 4800 students. At present, phase I of the development is completed and construction for phase II has begun (masterplan).

Project Design: A wastewater treatment plant is proposed with a capacity of 2.44 MLD against 2.35 sewage generation (theoretical) at the end of phase III. This treatment plant is also developed phase wise and the present capacity of the treatment plant is kept at 600 m^3/day as per the demand of phase I.

The primary and secondary treatments are based on DEWATS and tertiary treatment includes pressure sand filter and ultrafiltration system as shown in Figs. 3 and 4. The treated water which meets the quality standards is used for non-direct human contact purposes such as toilet flushing and horticulture (Table 3).

Process Description: The DEWATS installed in the campus is a combination of the following treatment systems:

1. Primary treatment or sedimentation in settlers: the settler consists of two compartments where the first compartment occupies 2/3rd of the space and most of the sludge gets deposited in the first compartment. From the outlet of the settler, supernatant enters the anaerobic baffled reactor.

Fig. 3 Tertiary treatment

Fig. 4 DEWATS (plant gravel filter)

Table 3 IIT Gandhinagar wastewater effluent quality

Parameter	Outlet value	Standard
pH	7.56	6.5–8.5
TSS (mg/L)	2	30
BOD (mg/L)	<10	20
COD (mg/L)	40	–

Source Interview with operators and engineers

2. Secondary treatment in baffled reactors and anaerobic filters where anaerobic degradation of suspended and dissolved solid takes place. In this phase, 70% of BOD removal is achieved. The effluent from baffled reactor is directed to anaerobic filter where non-settleable and dissolved solids are removed. 40–50% of remaining BOD is removed in this phase.
3. Tertiary treatment in planted gravel filter: the treated wastewater from anaerobic filter is transferred to PGF in such a way that it allows uniform distribution of the flow over the filter area. Pressure sand filter and Ultrafiltration system: This stage further reduces suspended solids, turbidity and organics. The final treated water is then stored in a tank of capacity 2.4 MLD.

Finance, Operations and Management:

The motivation behind installing a decentralized wastewater treatment system is that the institution is a green campus with a five-star GRIHA and LEED rating. The entire project is funded by CPWD under Ministry of Human Resource Development. The estimated cost of the treatment system was 1% of the total construction cost of phase I of the campus. The operations and maintenance is done by Swastic Electricals on yearly contract basis. The system requires one skilled operator and one inspector for primary treatment and a helper to look after the PGF. In-house team and the operating agency do regular inspections of effluent quality. This system requires very less maintenance.

The total cost of the system is around 1.8 crores where the cost of tertiary treatment itself is 1.4 crores. The yearly operations and management cost are 8.5 lakhs.

Feedback:

The treatment plant has proved to be a success in the case of IIT Gandhinagar and 100% of the treated water is used for landscaping to keep the campus lush green without any use of freshwater and the campus has achieved to be a zero-liquid discharge campus.

But in phase II of the development, the institution has decided on dispensing the tertiary treatment of PSF and UF. It was observed that treated water from effluent is needed to be used immediately in the form of irrigation/horticulture. When water is kept stagnant in toilet flushes, the BOD increased. Most of the residential apartments and guest houses have low occupancy and thus the toilets are not used frequently, that is why the flush system acts as a breeding pool for germs.

4.3 *Activated Sludge Process (1000 m^3/day)*

Godrej Garden City is one of the many township projects built on the concept of water conservation. It is an integrated township and has numerous housing concepts within its premises. More than 13,000 apartments are proposed including schools, colleges, playgrounds, supermarkets, entertainment halls, offices, etc. At present, 2500 flats are occupied and the rest of the infrastructure of the township is under construction.

It has become mandatory for every township to be a self-sustained one. Moreover, the area of Jagatpur where the township lies has no water supply network or sewage network. The source of water for drinking water is bore well.

The team of Godrej garden city believes in the principle of sustainable development and with this aim, they have achieved to get platinum rating for this township. The wastewater treatment plant installed in the township works on the principle of ASP.

Project Design: Based on the present occupancy rate and future demand, four WWTPs are currently operating in the township with 1000, 600, 250 and 100 KLD, respectively. In this study, the plant with 1000 KLD capacity is studied. This plant is a decentralized version of the conventional method of activated sludge process. The treated water is within the prescribed standards and is used for horticulture purposes (Table 4).

Process Description: The township has an underground network of sewer pipes which connects all the apartment building and transport the collected wastewater into the treatment plant. Activated sludge process technology consists of a number of chambers through which wastewater flows clearing out impurities as follows.

1. The wastewater is collected in a screening chamber through a bar screen to prevent rags, polyethene, bottles, etc., from entering the sedimentation tank.
2. The effluent is transferred to a horizontal tank where sedimentation occurs, and the settled sludge is removed periodically.

Table 4 Godrej Garden City wastewater effluent quality

Parameter	Outlet value	Standard
pH	8.2	6.5–8.5
TSS (mg/L)	0.55	<30
BOD (mg/L)	18	<20
COD (mg/L)	70	–
Odour	Unobjectionable	Unobjectionable

Source Monthly Inspection Report

Fig. 5 Aeration chamber

Fig. 6 PSF and ACF

3. To achieve reduced organic loading of the effluent, if put in a large aeration tank where air is pumped to enhance the breakdown of organic particles as shown in Fig. 5. The effluent is transferred to a clarifier-settler. The sludge at the bottom of clarifier-settler is sent back to the aeration tank.
4. For final treatment, the water is passed through three filter chambers as shown in Fig. 6.

5. The supernatant obtained is then stored in tank and further treated with chlorine from where it is used for irrigation purposes.

Finance, Operations and Management:

It has become mandatory for every township to be a self-sustained one. Moreover, the area of Jagatpur where the township lies has no water supply network or sewage network. The source of water for drinking water is bore well.

The team of Godrej garden city believes in the principle of sustainable development and with this aim they have achieved to get platinum rating for this township. The Operation and management are done by Prabhu Enviro Tech Private Ltd. with one supervisor, three operators and one helper.

According to the General Manager, the total cost of installing the treatment plant along with sewer network was 6 crores which were recovered within 6 years of installation from the advance maintenance fees of all the buyers of property. The operations and management cost per head is negligible. The Godrej garden city is free from paying water/sewer tax under Ahmedabad Property tax.

Feedback:

From the interviews and site visit, it is clear that the township faces shortage of water and this reuse of water for non- human purposes has proved to be a success in terms of conserving water. The developer also gets benefitted in terms of increased FSI and the residents are free from paying water and sewerage charges.

5 Summary

The key parameters studied for assessing the treatment plants are given in Table 5 where the annual O & M cost includes electricity cost, salary of staff, admin charges, cost of chemicals, repairs and effluent inspection charges.

Based on the study, the following key factors for the success of any plant is obtained as follows.

The design capacity of the plant must be in accordance to the anticipated future demand. Most of the treatment plant fails because of running at a capacity much lower than it is designed for. In such cases, effluent also does not meet the standards of CPCB.

It is very important to consider the land area requirements of different treatment plants in situations where land is scarce. Private developers generally opts for package plants because of lesser land requirements.

The affordability of the treatment plant by the users should be carefully considered. Package treatment plants prove to be very expensive for single households but when it is for a community or township it is a viable option. DEWATS system has the lowest capital investment required compared to on-site packaged plants. Ability to pay for the O & M cost is also important to obtain the expected treatment efficiency out of the treatment plant.

Table 5 Summary of treatment plants studied

Case study	IFFCO Kasturinagar	IIT Gandhinagar	Godrej Garden City
Total area	51 acres	400 acres	200 acres
Treatment technology	Fluidized aerobic bio-reactor	DEWATS	Activated sludge process
Maturity of technology	Matured	Maturing	Matured
Year of commissioning	2002	2015	2014
Design capacity	500 m^3/day	600 m^3/day	1000 m^3/day
Operating capacity	280 m^3/day	400 m^3/day	900 m^3/day
Area occupied by the treatment plant	1497 m^2	4046 m^2	1057 m^2
Outlet parameter			
pH	7.56	6.5–8.5	8.2
BOD	8	3	18
COD	22	45	70
Uses of recycled water	Horticulture	Horticulture, toilet flushing	Horticulture
Uses of sludge	Manure	Manure	Manure
Ease of upgradation	Easy	Easy	Difficult
Number of operators	3	3	5
Total capital cost	₹60,00,000	₹165,000,000	₹60,000,000
Total O & M cost	₹1,020,000	₹875,000	₹4,200,000
Capital cost (Rs Lakh/m^3)	0.12/m^3	0.3/m^3	0.6/m^3
Operation cost (Rs/m^3)	Rs 2206/m^3	Rs 1416/m^3	Rs 4200/m^3
Land area required (m^2/KLD)	0.25 excluding sludge dying bed	6.74	1.057
Source of fund for capital cost	Self-funded	Ministry of human resource and development	Self-funded
Sources of fund for O & M cost	Self-funded	Ministry of human resource and development	Self-funded

The result of user perspective survey conveyed that residents prefer treatment plants that occupy less space and are aesthetically appealing. The plant should not obstruct view and should be located at the back for safety reasons.

6 Conclusion and Scope for Future Work

This project work gives only an overview of the treatment technologies. It is important to include life cycle cost assessment of the decentralized wastewater treatment plant to understand financial feasibility in terms of Indian context. During the analysis, it has been observed that inflation rate for CAPEX and OPEX is to be considered because the treatment plants have different market conditions with respect to the year of commissioning.

It is important to state that decentralized WWTP for individual household prove to be uneconomical because of high power consumption with respect to lesser design capacity. Based on personal interview, it was observed that middle or high-income group of people were aware about the environmental benefits of installing decentralized wastewater treatment system and were even ready to incur additional cost. But it is impractical to assume the same for lower income group people.

It was seen that most of the WWTPs have private effluent quality inspection. There is a requirement for stringent law enforcement for the proper monitoring of the operation of the WWTP. There should be random visits from government officials and penalty should be given to those who do not comply with the standards.

WWTP can never be a revenue-generating model as the product is treated water and a small fraction of sludge. It is not possible to create a business out of it. But in the long run, recycling and reusing wastewater can solve water issues to a great extent. Therefore, there is a need for innovations in the field of wastewater treatment and the government must encourage pilot projects on decentralized treatment plant so that efficient and low-cost technologies can be developed.

References

1. Poll, R. a L., Co, L., Board, O.: Performance Evaluation of Sewage Treatment Plants Under NRCD (2013)
2. Chen, Z., Ngo, H.H., Guo, W.: A critical review on the end uses of recycled water. Crit. Rev. Environ. Sci. Technol. **43**(14), 1446–1516 (2013). https://www.tandfonline.com/doi/abs/10.1080/10643389.2011.647788
3. Babcock, R.W., Asce, M., Mcnair, D.A., Edling, L.A., Nagato, H.: Evaluation of a system for residential treatment and reuse of wastewater. J. Environ. Eng. **130**(7), 766–773 (2004). https://doi.org/10.1061/(ASCE)0733-9372(2004)130:7(766)
4. Department of Urban Development and Urban Housing: Gujarat Integrated Township Policy, pp. 1–65 (2007)
5. Sustainable Sanitation and Water Management Toolbox

6. Sasse, L.: DEWATS: Decentralised Wastewater Treatment in Developing Countries, pp. 1–161 (1998). https://www.sswm.info/sites/default/files/reference_attachments/DEWATS_Guidebook_small.pdf
7. Kumar, D.: Environmental Management System-An exemplary for Urban Local Bodies

Experimental and Analytical Investigation on Shear Strength of Concrete Containing Slag Considering Sustainable Development and Waste Management Concept

Damyanti Badagha, C. D. Modhera and Sandip A. Vasanwala

Abstract There is a demand for sustainable infrastructure development due to the rapid growth of urbanization considering the waste management concept. To achieve the unmet need in this area, different waste and by-products from different industries are in use in the field of civil engineering. There is a need for concrete to give good workability, long-term high performance, durability, and sustainability for special structures. There is a requirement of better performance of concrete not only in compressive behavior but also in different behavior like shear performance. This implemented experimental work has been introduced to find out the analytical and experimental behavior of concrete with 50% cement replacement by steel industry waste for compressive and shear strength under different exposures. The experimental results show that waste exhibits better failure pattern compared to concrete without waste powder, higher ultimate strength in addition to much improved strength in different exposure conditions such as acidic and alkaline, for better infrastructure development.

Keywords High-performance concrete · Industrial waste · Compressive strength Shear test · Alkaline exposure · Acidic exposure · Sustainability

1 Introduction

The infrastructure development creates the urge to need to find the innovative sustainable technique in a different era. The innovation in waste management considering sustainability is one of the fundamental requirements for the development of different types of infrastructures. The utilization of cement concrete is tremendous in the civil industry as a construction material [1, 2] in the entire world after water, which is an important part of infrastructure development. This concrete manufacturing process required a large amount of raw materials. The key ingredients of concrete are solely

D. Badagha (✉) · C. D. Modhera · S. A. Vasanwala
Applied Mechanics Department, SVNIT, Surat 395007, Gujarat, India
e-mail: damyantibadagha@yahoo.in

D. Deb et al. (eds.), *Innovations in Infrastructure*, Advances in Intelligent Systems and Computing 757, https://doi.org/10.1007/978-981-13-1966-2_19

responsible for the 7% of the worldwide generated CO_2 emissions [3]. In this rapid infrastructure development, the waste disposal [4] is also a pinpoint for all industries. There are a number of waste such as steel industry waste, different slags; different types of ash, construction–demolition waste, plastic waste, e-waste, etc., which are used to manufacture special concrete for the infrastructure development considering waste management solution and sustainability. The use of metakaoline and micro silica also give better performance in strength point of view [5] and durability of concrete [6]. For the waste management [7] and concrete production, there are advanced software programs that are utilized for the modeling and strength prediction and mix proportioning of concrete [8].

The research investigation reveals that the by-products having cementitious properties strengthen harden and fresh properties of concrete [9–11]. These mineral admixtures contributed major role in the permeability of concrete considering pore size distribution and its effect due to admixtures in cement matrix [12–16]. The reduced median pore size is observed in the mature cement matrices with the comparison of Portland cement paste without admixtures. However, the utilization of mineral admixtures like GGBFS and fly ash influence the hydration process because of their fineness [17, 18].

This innovative concept of waste materials in high-performance concrete has been modified as per requirement and need of time, as earlier it was compared to the concrete falls under the high strength concrete category, which has some beneficial facts, but it is unable to give an exact idea considering different aspects. HPC has been defined differently by various authors, and high-performance concrete could be a concrete that created with suited materials, properly combined as per elite mix design, transported, placed, compacted and cured to urge a wonderful performance in the structure within the surroundings to that it is exposed and under subjected loads for design. Thus, high-performance concrete is directly related to durability of concretes which can be determined by conducting different experimental test series. The water penetration phenomena and aggressive ion attacks including chloride in concrete are fundamentally revealing the physical and chemical performance of concrete considering durability [17]. The fictitious crack model (FCM) was introduced by Hillerborg for the fracture behavior of concrete [18] including tension softening curve and fracture energy of concrete.

The concrete structures are having a complex behavior because of the structural system including boundary conditions, loading, and others. Therefore, it is the fundamental requirement to understand the model for concrete structures. However, the theoretical limit analysis methods are not giving the deformation solution, which demands experiment based shear fracture behavior of concrete.

From this rigorous research survey, the objective of this study was set to prepare concrete using waste from steel industry considering the durability under different exposure conditions. Compressive strength and shear strength were measured for conventional concrete (CC) and HPC. Shear strength of concrete was measured by two different methods as well as analytically, as there is not plenty of research work on the shear strength of concrete has been conducted like introduced comparative study.

2 Experimental Program

Concrete is considered as a material similar to stone having the cementitious property within which aggregates are embedded. In cement concrete, water and hydraulic cement are composed of the mixture. For the execution of this research work, as per the BIS 12269:2013 [19], 53 grade ordinary Portland cement was utilized and samples were prepared as per IS 3535:1986 [20]. As per BIS 383:1970 [21], coarse and fine aggregates were used. Cement replacement was done by the waste collected from local steel industry which was ground in a powder form. The physical and chemical components for that waste powder are achieved [7] as per BS 6699:1992 [22] and found similar like cement. The slag was used as cement replacement from 10 to 90% with the interval of 10% for preparation of mortar mixes. As per ASTM C989, the slag activity index was investigated for different percentage of waste in the waste cement mortar which is listed in Table 1.

$$\text{Slag-activity-index} = (\text{SP}/P) \times 100$$

SP Average compressive strength of slag cement mortar cubes in MPa, and
P Average compressive strength of cement mortar cubes in MPa.

The slag activity index for 50% replacement of cement by waste is 118 for 7 days and 124 for 28 days which indicates the grade of waste material is Grade 120. It was found that slag gives maximum increased strength for 50% of cement replacement.

Mix Design and Curing
The mix design for concrete with 50% cement replacement was achieved by taking the numbers of trails. The compressive strength varies with variation in water/binder ration as shown in Fig. 1. The water/binder ratio of 0.40 and 0.36 gives better results in compressive strength for conventional concrete (CC) and HPC, respectively.
The mix proportion comparison of CC was done with the concrete containing 50% cement replacement with waste powder, which is shown in Table 2. The slump value suggests good workability for HPC. The prepared specimens were cast and cured in tap water for 28 days and cured in water with 5% concentration of hydrochloric acid (HCl) for aggressive acidic environmental effect and in water with 3.5% concentration of sodium hhloride (NaCl) for aggressive alkaline environmental effect on it.

Table 1 Slag activity index for percentage variation of waste in waste cement mortar

Slag activity index	Percentage of waste in waste cement mortar								
	10	20	30	40	50	60	70	80	90
7 days	105	109	113	114	118	108	93	85	70
28 days	103	105	112	115	124	97	89	80	62

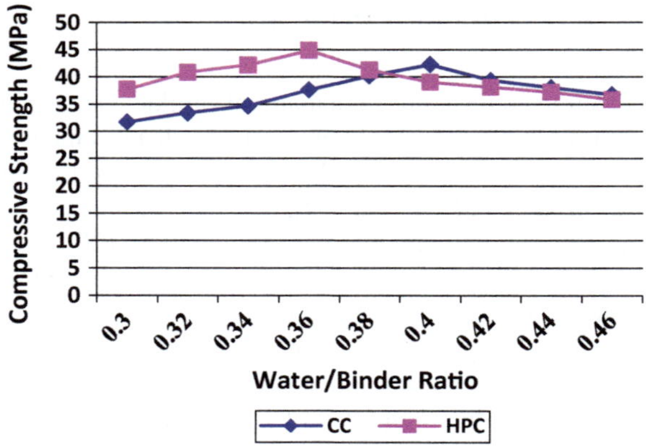

Fig. 1 Effect of water/binder ration on compressive strength of conventional concrete and HPC containing 50% waste at the age of 28 days

Table 2 Mix proportions

Material	Conventional concrete (CC)	HPC with waste as 50% cement replacement
Cement (kg)	465	235
Waste powder (kg)	–	235
Water (liter)	186	169
Sand (kg)	531	680
Coarse aggregate (10 mm) (kg)	423	428
Coarse aggregate (20 mm) (kg)	728	740
Admixture (kg)	–	3.1
Water cement ratio	0.4	0.36
Initial slump (mm)	140	165
Slump at 30 min	–	140 mm
Slump at 60 min	–	120 mm

Testing

The compression test was performed on a cube specimen with the edge of 150 mm as per IS 516:1959 [23]. The tests for shear strength of concrete have been performed as per JSCE-SF6 standard test method [24] and the method suggested by Bairagi and Modhera [25]. As per JSCE-SF6 standard test, the shear load is applied to the beam supported on a rigid block over a pair of 105 mm apart knife edges with the help of by a loading block with two sharp loading knife edges 100 mm apart as shown in Fig. 2. Thus, a narrow, 2.5-mm-wide region of the beam in between the supporting knife edges and the loading is subjected to a concentrated shear stress. The shear strength of concrete is evaluated with load *P* divided by shear area which is twice the width multiply bye depth.

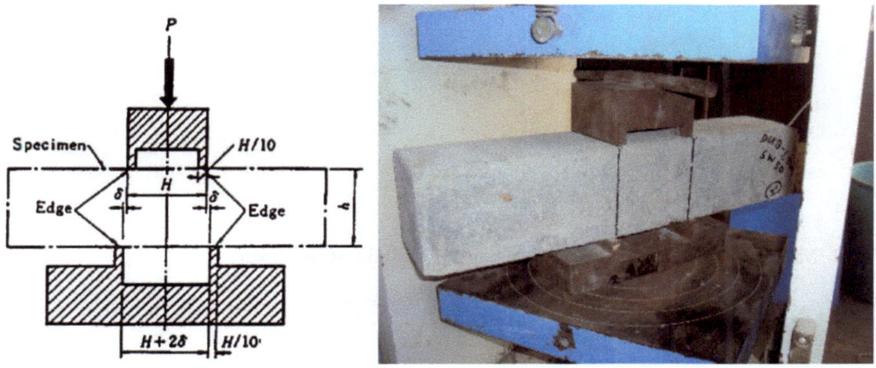

Fig. 2 Schematic diagram, actual specimen arrangement as per JSCE-SF6 method

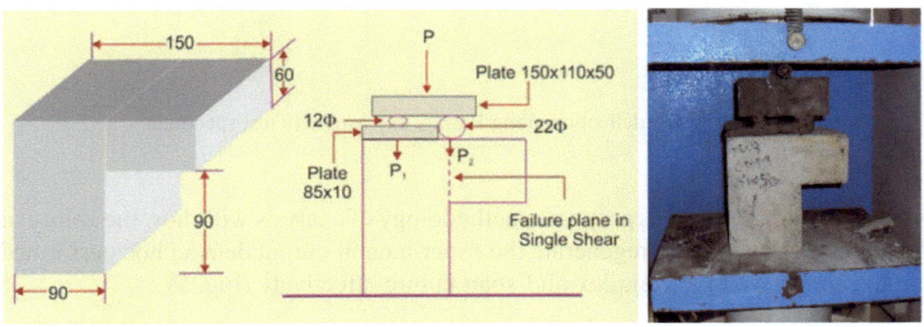

Fig. 3 Schematic diagram, actual specimen arrangement as per Bairagi and Modhera method

In the method, suggested by Dr. Bairagi and Dr. Modhera, specimen were cast and tested as per the arrangement shown in Fig. 3. Until the crack observed on the failure plane in single shear, load P was applied on the upper plate as shown in Fig. 3. The failure load P_2 is taken as a half of the total load P. The load P_2 divided by shear area (60 mm × 150 mm) in the unit N/mm^2 is considered as the shear stress for the given concrete specimen.

All specimens were air dried before the testing of specimen. Test was conducted in a universal compression testing machine. As per JSCE-SF6, the load has given without any impact continuously at a shearing stress rate of 0.06–0.1 MPa/s.

Analytic Investigation

The research experience concluded that the exhaustive experimental work and testing is a time-consuming process. For the betterment of research, advanced soft computing techniques, analysis, and modeling have been done to compare the experimental results with the software utilizing results (Fig. 4).

In this research work, ATENA software was used. In this section, three-dimensional nonlinear finite element analysis of concrete specimens has been conducted until the failure occurs in the crack form in order to similarly model interaction among

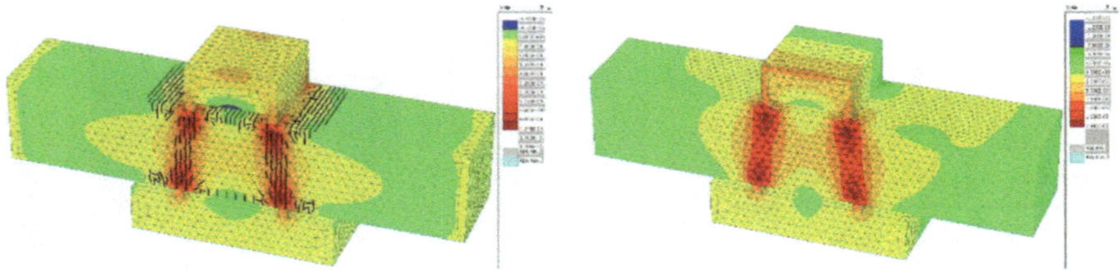

Fig. 4 Variation of crack and deflection along JSCE-SF6 test specimen

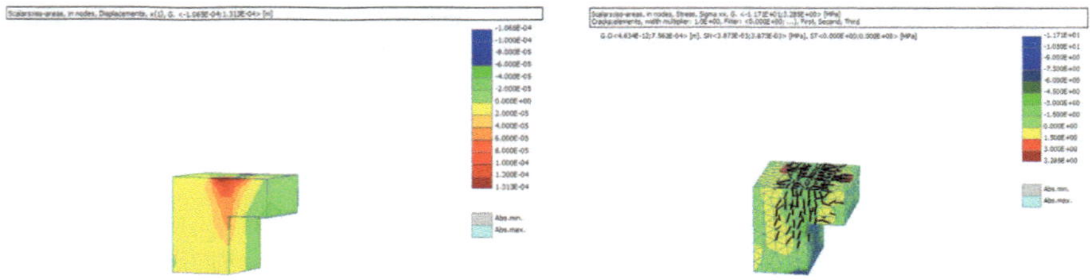

Fig. 5 Variation of crack and deflection along Bairagi and Modhera test specimen

concrete components. This proposed methodology of analysis will show the ability of the model to numerically regenerate the experimental output data ad hoc performed and to capture well the complex mid-span failure effectively (Fig. 5).

3 Results and Discussion

In this present study, the utilization of steel industry waste has been studied for concrete to reduce the consumption of cement. Compressive test and two different shear test methods were performed on conventional and waste utilized concrete and the results are plotted against the age of concrete as shown in Figs. 6, 7 and 8.

Figure 6 indicates the relation of compressive strength of CC and HPC in different exposures with the age of concrete. The compressive strength of HPC is higher than that of CC. Figures 7 and 8 indicate the relation of shear strength of CC and HPC in different exposures with the age of concrete using the JSCE-SF6 method and Bairagi and Modhera method, respectively.

In alkaline exposure, the compressive strengths of HPC and CC are reducing by 1 and 5%, respectively, where in acidic exposure, the compressive strength of concrete has been reduced by 3 and 10%, respectively. For HPC, in alkaline exposure and acidic exposure, the shear strength is reduced by 3 and 7% in JSCE-SF6 method and 4 and 7% in Bairagi and Modhera method, respectively. So the compressive strength results of concrete which are exposed to aggressive environment decrease in the comparison with of normal water cured concrete, but the aggressive exposure influence more on CC is more than that of high-performance concrete.

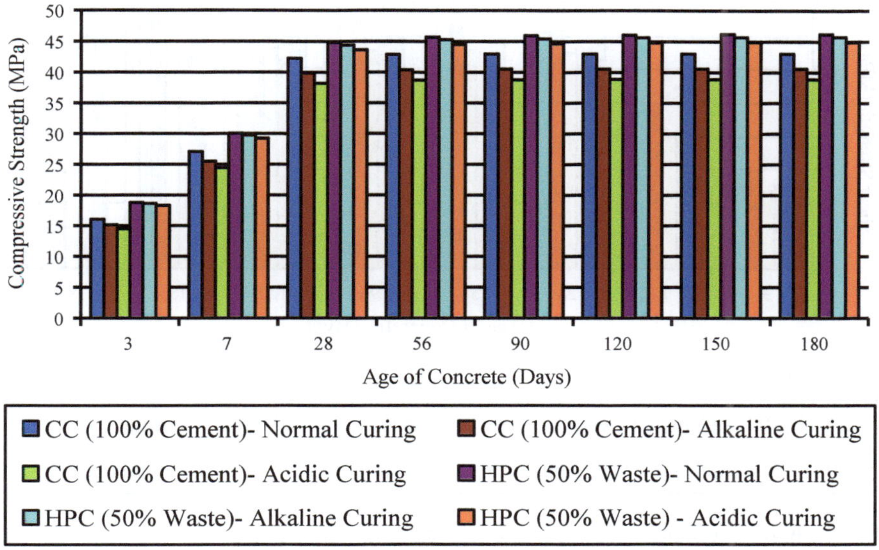

Fig. 6 Relation of compressive strength with age of concrete

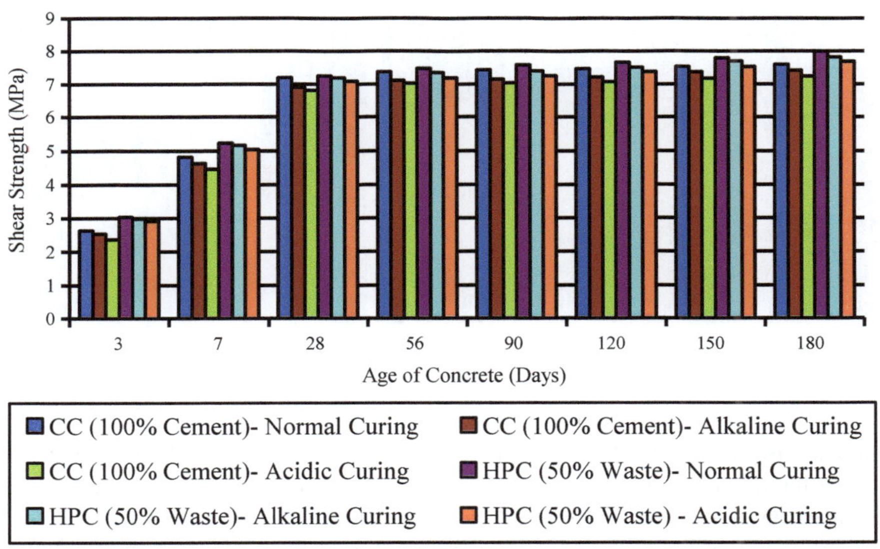

Fig. 7 Shear strength of concrete as per JSCE-SF6 at different age

This research indicated that in JSCE-SF6 procedure, the failure plane often deviated from the narrow region, prescribed plane for failure, under concentrated shear as shown in Fig. 9. The observed outputs of shear tests reveal that shear strength

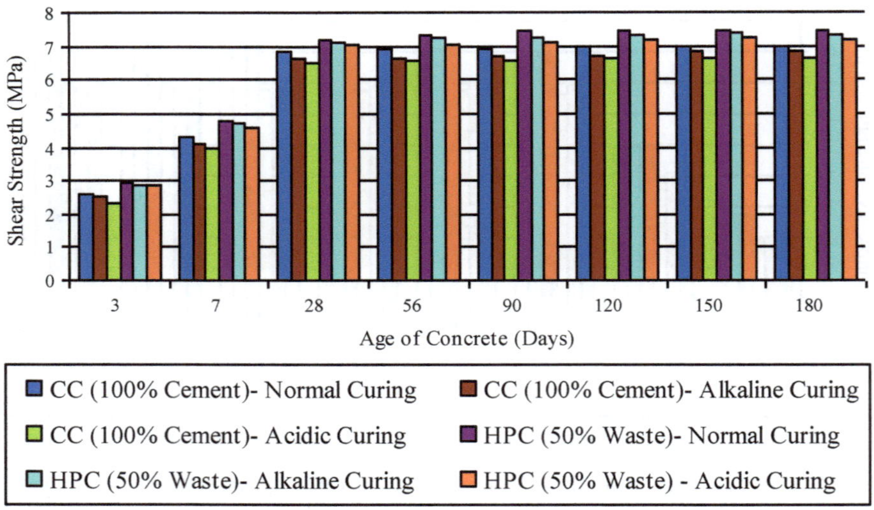

Fig. 8 Shear strength of concrete as per Bairagi and Modhera at different ages

Fig. 9 Crack pattern of the specimens for JSFC-SF6 and Bairagi and Modhera method

obtained from JSCE SF-6 gives a higher value than Bairagi and Modhera method. The difference in results is approximately up to 15% which is acceptable. The detailed comparison of results of FE modeling and experiments is tabulated in Table 3.

The results of FE modeling using ATENA software provide the best simulation for specimens of both tests. The variation for first crack load and ultimate load using FE modeling is less than 5% of the experimental results.

Table 3 Comparison of experimental results and analytical results of conventional concrete and HPC (50% replacement with waste) at 28 days in normal curing

Concrete	Test method	Experimental results		Analytic results	
		Load at 1st crack (KN)	Ultimate load (KN)	Load at 1st crack (KN)	Ultimate load (KN)
Conventional (100% cement)	JSCE-SF6	115	143.8	119.2	148.4
	Bairagi and Modhera	100.2	123.6	107	131.6
HPC (with 50% waste)	JSCE-SF6	120.2	145.0	123.5	152.0
	Bairagi and Modhera	105.6	129.2	113.6	136.5

4 Conclusion

Use of steel industry waste as the cement replacement in HPC is highly desirable and feasible because of its nature as an environment friendly, recycled material and the economic advantages by reducing cement consumption. It is the obvious fact from the study that the waste powder produced from steelmaking industry is giving desired compressive strength and shear strength for exactly half replacement of cement in concrete. Shear strength is very important in frame structure where beam column junction plays a critical role in the time of the earthquake. So it is suggested to use this HPC having industrial waste in place of CC to increase the shear strength of beam column junction. Slump value for HPC with industrial waste is observed good even after 60 min, so this mix design is suggested to be used in ready-mix concrete plant as the workability of HPC is better than CC. It is directly benefited as economical point of view to use industrial waste as cement replacement. The HPC has better performance on aggressive exposure which makes it durable in all kind of conditions. The effect of alkaline exposure is less on HPC so it is advisable to use this concrete in marine structures like docks, harbors, sea bridges, etc.

The analytic investigation gives results similar to the conducted experiment. So this model can also be used for the prediction of shear strength without laboratory test for the same. There are limited scope and research for load-bearing structure using concrete and it is a very important parameter for such structures, so this investigation found suitable to use this concrete where load-bearing structure have to construct.

For use of steel industry waste in concrete as a structural material, it is very important to find out the performance of reinforced HPC under different strength like flexure, shear, torsion, and compression. Some experimental and analytical tests focusing on different durability aspects considering the corrosion, chloride ion penetration, water permeability of steel reinforcement need investigation in a broad manner to understand this area. This study can be expanded considering different durability and physical parameter to make its use possible in practical field.

Acknowledgements The author would like to acknowledge Gujarat Council on Science and Technology (GujCOST) DST for funding this research work.

References

1. Prabhu, G.G., Hyun, J.H., Kim, Y.Y.: Effects of foundry sand as a fine aggregate in concrete production. Constr. Build. Mater. **70**, 514–521 (2014)
2. Meyer, C.: The greening of the concrete industry. Cement Concr. Comp. **31**, 199 (2009)
3. Malhotra, V.M.: Role of supplementary cementing materials in reducing greenhouse gas emissions. Concrete technology for a sustainable development in the 21st century. London, 226–35 (2000)
4. Kumar, S., Gupta, R.C., Thomas, B.S., Mehra, P.: Aggregate replacement and its usefulness in cement concrete for sustainable development-a study on rubber, jarosite and sandstone aggregates, advances in human factors and sustainable infrastructure. Adv. Intell. Syst. **493**, 13–25
5. Murgul, V., Popovic, Z.: Research on long-term strength of glass-fiber reinforced concrete. In: International Scientific Conference Energy Management of Municipal Transportation Facilities and Transport EMMFT 2017, Adv. Intell. Syst. vol. 692, pp. 640–646
6. Szewczyk, R., Kaliczyńska, M.: Study of the durability of reinforced concrete structures of engineering buildings. Recent Adv. Syst. Control Inf. Technol. Adv. Intell. Syst. **543**, 659–663 (2016)
7. Badagha, D.G., Modhera, C.D., Desai, S.N.: Waste material utilization as a cementitious material to understand tensile behaviour of concrete under alkaline and acidic exposure. In: Advances in Concrete, Structural & Geotechnical Engineering, 1st edn. Bloomsbury Publishing India Pvt Ltd., pp. 135–139 (2018)
8. Badagha, D.G., Modhera, C.D., Vasanwala, S.A.: Mix proportioning and strength prediction of high performance concrete including waste using artificial neural network. World Acad. Sci. Eng. Technol. Int. J. Civil Environ. Eng. **12**(2), 124–127 (2018)
9. Mehta, P.K., Malhotra, V.M.: Pozzolanic and Cementitious by Products as Mineral Admixtures For Concrete—A Critical Review. ACI Special Publication SP-79, pp. 1–46 (1983)
10. Mehta, P.K., Malhotra, V.M. (eds.).: Pozzolanic and cementitious by products in concretes—another look, vol. 1. ACI Special Publication SP-114, 1–45 (1989)
11. Mehta, P.K.: Malhotra V.M. (ed.).: Durability of Concrete—Fifty Years of Progress? ACI Special Publication SP-126, pp. 1–31 (1991)
12. Giorv, O.E., Vennesland, O.: Diffusion of chloride ion from sea water. Cement Concr. Res. **9**(2), 229–238 (1979)
13. Manmohan, D., Mehta, P.K.: Influence of pozzolanic, slag and chemical admixture on pore size distribution and permeability of hardened cement paste. Cement Concr. Aggr. **3**(1), 63–69 (1981)
14. Kumar, A., Roy, D.M.: Pore structure and ionic diffusion in admixture blended cement system. In: Proceedings of the 8th Congress Chem Cement, Brazil, pp. 73–81 (1986)
15. Hooton, R.D., Frohnsdorff, G. (ed.): Blended cement. ASTM STP **897**, 128–143 (1986)
16. Roy, D.M.: Mechanism of cement paste degradation due to chemical and physical factors. In: Proceedings of the 8th International Congress Chem Cement, Brazil, 362–80 (1986)
17. Ulrik, N.A., Monteiro, J.M.: Concrete: a three phase material. Cement Concr. Res. **23**(1), 147–151 (1993)
18. Hillerborg, A., Modeer, M., Petersson, P.E.: Analysis of crack formation and crack growth in concrete by means of fracture mechanics and finite elements. Cem. Concr. Res. **6**(6), 773–782 (1976)
19. Indian Standard, IS 12269:2013.: Ordinary Portland Cement, 53 Grade—Specifications. Bureau of Indian Standard (2013)

20. Indian Standard, IS 3535:1986.: Methods of Sampling Hydraulic Cement. Bureau of Indian Standard (2004)
21. Indian Standard, IS 383:1970.: Specification for Coarse and Fine Aggregate from Natural Sources for concrete. Bureau of Indian Standard (2002)
22. British Standards, BS 6699:1992.: Specification for ground granulated blast furnace slag for use with Portland cement (1992)
23. Indian Standard, IS 516:1959.: Methods of Tests for Strength of Concrete. Bureau of Indian Standard (2004)
24. Japan Society of Civil Engineers, JSCE-SF6: Method of Test for Shear Strength of Steel Fiber Reinforced Concrete. Tokyo, 67–69 (1990)
25. Bairagi, N.K., Modhera, C.D.: Shear strength of fiber reinforced concrete. ICI J. Chennai, 47–52 (2001)

Manufacturing of Lithium Cobalt Oxide from Spent Lithium-Ion Batteries: A Cathode Material

Ravi Methekar and Sandeep Anwani

Abstract Disposal of large amount of spent lithium-ion batteries will cause adverse environment impact as well as will create waste management problem. An eco-friendly disposal method needs to be devised. Recycling is one of the best options to address the disposal problem of the spent batteries. Use of Extracted metals from recycling for manufacturing of cathode active material, i.e., lithium cobalt oxide will help in reducing the rate of depletion of these valuable metals. Recycling process consists of peeling, dissolution, and extraction steps. Variations of parameters for enhancing the peeling and the dissolution steps along with extraction of cobalt oxalate and manufacturing of lithium cobalt oxide have been studied here. Optimal parametric values have been identified for higher dissolution efficiency. The dissolution efficiency was found to be 95.15% at the optimal operating condition. The extraction efficiency of cobalt oxalate is found to be 91% with purity of 90%. The purity of manufactured lithium cobalt oxide is found to be 91%.

Keywords Lithium-ion batteries · Recycling · Spent batteries
Dissolution · Lithium cobalt oxide

1 Introduction

Lithium-ion batteries (LIB) are considered to be one of the best power sources for many portable devices as well as for the transport applications that can operate at higher voltage and higher energy density. In the last decade (2000–2010), the production of LIB has increased at an alarming rate of 800% worldwide [1]. It is expected that there will be a huge growth of LIB industry close $32 billion by 2020 [2]. However, due to huge demand and usages, there are a large amount of LIB to be disposed off after their useful life. These batteries are called as spent lithium-

R. Methekar (✉) · S. Anwani
Department of Chemical Engineering, Visvesvaraya National Institute of Technology, Nagpur 440010, India
e-mail: methekar.vnit@gmail.com

D. Deb et al. (eds.), *Innovations in Infrastructure*, Advances in Intelligent Systems and Computing 757, https://doi.org/10.1007/978-981-13-1966-2_20

ion batteries and cannot be disposed off into the landfills because of safety and economic reasons. Also spent LIB are rich source of valuable metals like lithium, cobalt, manganese, nickel, aluminum, copper, etc. On the other hand, lithium and cobalt shortage may inevitable in few decades. Hence, recycling/or recovering of the major valuable metal contents are critical to prevent adverse environmental impacts and also to ensure sustainable material supply.

Several researchers are presently working in the field of recycling of spent lithium-ion batteries. To efficiently recover the valuable metals, various physical and chemical processes have been reported which includes crushing, dismantling, sieving, chemical precipitation and electrochemical methods. The recycling process can be broadly categorized into pyrometallurgy, hydrometallurgy, and bio-hydrometallurgy processes. Pyro-metallurigical process has the disadvantages like emission of toxic gases, higher cost of recycling. A bio-hydrometallurgical process takes longer time for treatment and also the microbes are very difficult to incubate [3]. Hydrometallurgical process is considered to be more suitable among all the available processes. In hydrometallurgical process, acidic or alkaline leaching is carried out for extraction of valuable metals by using chemical precipitation and separation.

The strong acids like sulfuric, hydrochloric, nitric or phosphoric [4–8] or weak acids like citric, formic, tartaric, succinic, lactic, or oxalic [1, 9–12] are used in the hydrometallurgical process. It is reported in the literature that the leaching efficiency of the valuable metals using the strong acids lies in the range of 85–99% [13, 14] whereas for weak acids lies in the range of 70–99% [15, 16]. Leaching efficiency consists of efficiencies of all the steps incurred in extraction of the valuable metals. However, very few literature reports the efficiency of each step incurred. In general, extractions of valuable metals from the spent lithium-ion battery consist of three main steps, viz., (1) peeling of active cathode materials from the cathode electrode (2) dissolution of active materials into the acid and (3) extraction of valuable metals from the dissolved active materials using chemical precipitation. Efficiency of each of the abovementioned step will have a great impact on the overall efficiency of the recycling process. Hence, it is extremely important to enhance the efficiencies of all these steps. The high efficiency of these steps ultimately results in high recovery of valuable metals. If high purity of extracted metals is achieved then those metals in the form of carbonates or oxalates can be combined to form the active cathode material for manufacturing of cathode of the lithium-ion battery and hence depletion rate from global reservoir of these metals can be reduced to some extent.

In this study, the dissolution efficiency is enhanced by optimizing the parameters like molarity of the acid, particle size of the active materials, operating temperature, and solid-to-liquid ratio. The cobalt oxalate is extracted using precipitation method at the optimal operating condition with efficient method with high purity. Lithium cobalt oxide is manufactured using extracted cobalt oxalate and procured lithium carbonate. The analysis of the extracted components is carried out using standard analytical method like XRD, XRF, and ICP AES for confirming the metal phase and also to calculate the purity of the extracted metals.

2 Materials and Methods

The cathode and anode electrodes of cell phone batteries are mainly made from $LiCoO_2$ amorphous powder and carbon black powder layered around aluminum and copper foil, respectively, with the help of binder.

First, a battery is fully discharged manually and then dismantled to separate out various components like cathode, anode, separator, current collectors, and connectors from the casing. The cathode strip is weighed (6.4 g) and cut into small pieces. Then, the pieces of cathode are dipped into 40 ml N-Methyl-2-pyrollidone (NMP) solvent. The conical flask containing the NMP solvent and cathode pieces are then subjected to ultrasonic cleaning for about 5 min. Ultrasound cleaning helps in breaking the bond between binder and active cathode material layer from the aluminum foil. Then the aluminum pieces which were separated from cathode materials due to ultrasonic cleaning are collected and separated out. The solution deprived off the aluminum is centrifuged to separate out the solid cathode materials from NMP solution. The solid untreated cathode material is then dried at 700 °C for 5 h in muffle furnace to burn off any binder as well as organic volatile matter. Then this material is ball milled to reduce the particle size less than 25 μm. This material is called as active material from spent lithium-ion batteries.

In dissolution, appropriate amount of sulfuric acid with optimal molarity is added in a three-neck round bottom flask with stoichiometric quantity of hydrogen peroxide. Then the three-neck round bottom flask is subjected to magnetic stirring for 2 h at 90 °C. Once the dissolution is over, solution is filtered out. The gouge materials are separated, dried and weighed. The dissolution efficiency is then calculated as the ratio of initial sample weight minus weight of gouge materials to the initial sample weight. To optimize the dissolution process, the parameters like molarity of acid, particle size, operating temperature, and solid-to-liquid ratio are varied and various runs of experiments are carried out to confirm the repeatability of the results.

For extraction of cobalt in the form of cobalt oxalate, the pH of the solution needs to maintain at the value near to 2. This is done by adding the appropriate amount of 5M NaOH solution to the mixture. Once the pH of the solution is reached near to 2, oxalic acid at the stoichiometric ratio is added drop by drop to the contents at a constant temperature of 50 °C with stirring speed of 300 rpm for 1 h. Pale pink precipitate will be formed. This precipitate is filtered out. Then, the obtained precipitate is thoroughly washed to remove any water dissolved impurities like sodium hydroxide and then dried in the oven.

To manufacture lithium cobalt oxide, appropriate amount of cobalt oxalate and lithium carbonate is well mixed in the mortar and pester and then calcined at 800 °C in muffle furnace for 5 h. The black powder obtained is cooled down to room temperature and sent for analysis.

2.1 Methods of Analysis

Various phases of the valuable metals are present in the active cathode materials obtained from the spent LIB. This information is very helpful in choosing the right acid for the dissolution of the active materials. This can be done using X-ray Diffraction (XRD). Using XRD analysis, we will able to find the phases of the valuable metals present in the mixture. Once XRD analysis is carried out then sample is subjected to either inductive coupled plasma atomic emission spectroscopy (ICP-AES) or X-ray fluorescence (XRF) which tells us the weight percentage of various valuable metal ions present in the sample. Using this information, we can find extraction efficiency as well as the purity percentage of the valuable metals.

3 Results and Discussions

3.1 Composition and Phases of Cathode Active Material

The XRF analysis of the untreated cathode material presented in Table 1 shows that the cobalt (45.1 weight %), manganese (11.8 weight %), and lithium (6.3 weight %) metals are present in the significant amount as transition metals. Calcium, sodium, magnesium, iron, etc. are present in small quantities. Once we got to know the metal contents, an XRD analysis is carried out for knowing the phase of the valuable metals. XRD analysis shows that the lithium cobalt oxide is present in a significant amount along with cobalt (II, III) oxide. As the Co (II) oxide is unstable, we need to add some reducing agent while the dissolution of the materials. Magnesium aluminum silicate, phosphorus pentoxide, and ferroferric oxide are also present in small quantity. The XRD graph is shown in Fig. 1.

3.2 Dissolution of Active Material and Parametric Study

Dissolution of active material is carried out in 500 ml three-neck round bottom flask. Condenser assembly is provided to avoid the evaporation losses of solvent during

Table 1 Composition of active cathode material using XRF analysis

Metal	Weight %	Metal	Weight %	Metal	Weight %	Metal	Weight %	Metal	Weight %
Li	6.3	Cu	0.6	K	0.02	Si	0.09	Fe	0.11
Co	45.1	Al	0.67	Ca	0.03	Cl	0.02	Zr	0.01
Mn	11.8	P	0.32	Na	0.23	Ti	0.02	Nb	0.01
Ni	0.30	S	0.16	Mg	0.41	Cr	0.09	O2	Rest

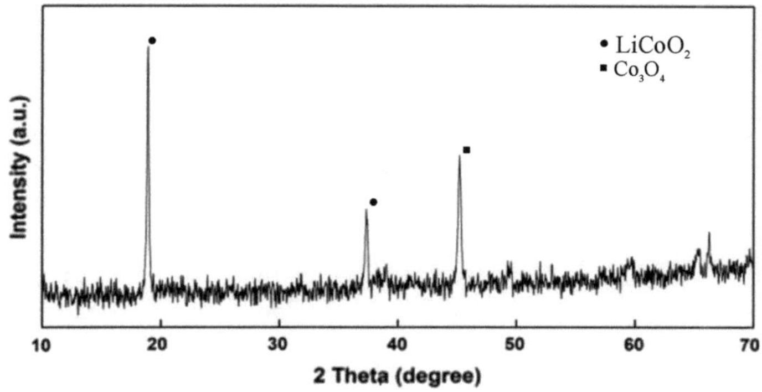

Fig. 1 Phase analysis of active cathode material using XRD analysis

the dissolution step. 4 gm of active cathode material is mixed in the solution of 2M sulfuric acid with solid-to-liquid ratio of 10 g/l. The sample of active cathode material is added into the acid media with hydrogen peroxide at the thermal equilibrium. This mixture is heated at constant temperature of 90 °C for 3 h with constant mixing speed of 300 rpm. Once the dissolution is over, the solution is filtered. The gouge materials are separated, dried and weighed. The dissolution efficiency can be calculated as follows:

$$D\% = ((m^0 - m^f)/m^0) \times 100$$

where $D\%$ is the percent dissolution efficiency; m^0 is the initial mass of the active material and m^f is the mass that remains undissolved solid after the leaching.

To enhance the dissolution efficiency, the operating parameters are optimized. The operating parameters like temperature, particles size, molarity of the acid used, and solid-to-liquid ratio are varied. The range of variations is fixed from the literatures. Table 2 shows the operating parameters matrix for optimization as well as corresponding dissolution efficiencies. The maximum dissolution efficiency is achieved equal to 95.15% at the temperature of 90 °C, particle size less than 25 μm, 2 molar sulfuric acid and solid-to-liquid ratio of 10 g/l. We found that the dissolution efficiency increases with increase in temperature, decrease in particle size, increase in molarity of the acid with certain upper level, and decreases with solid-to-liquid ratio. The results of parameter optimization are in well agreement with literature [1, 5–8].

3.3 Cobalt Oxalate Extraction and Characterization

Clear solution from the dissolution step is taken and 1M oxalic acid is added dropwise stoichiometrically with cobalt ions present in the solution. The stoichiometric amount of cobalt ions is calculated using the initial percentage of cobalt metal present in the

Table 2 Operating condition matrix for optimization of dissolution efficiency

Molarity (M)	Temperature (°C)	Particle size < (μm)	Solid-to-liquid ratio (g/l)	Dissolution efficiency %
2	90	25	10	95.15
2	90	90	10	82.15
2	60	25	10	79.50
2	75	25	10	87.00
1	90	25	10	93.50
2	90	90	20	46
2	90	25	20	53.5
1.5	90	25	10	94.02

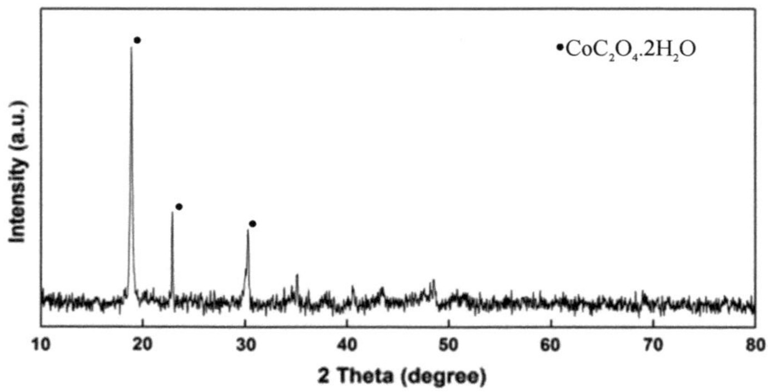

Fig. 2 Phase analysis of cobalt oxalate using XRD analysis

sample and the dissolution efficiency. The temperature is maintained at 60 °C with 300 rpm stirring speed for 2 h. The pH of the solution is maintained at 2 with the help of 10M NaOH solution. The pink pale precipitate is formed, which filtered and separated out. The precipitate of cobalt oxalate is dried and weighed for the calculation of extraction efficiency and then sent for characterization. 91% cobalt extraction efficiency is achieved. To reproduce the consistency in results, three runs of experiment have been carried out and average value is reported.

The formation of cobalt oxalate is confirmed using XRD analysis. The XRD peaks at particular position are in well agreement with the reported literature for cobalt oxalate [17]. The result of XRD analysis is shown in Fig. 2.

Then this sample is analyzed for purity of the cobalt oxalate using ICP-AES technique which is shown in Table 3. It is found that cobalt and carbon are present in significant amount. It is also found that Mn and Ni are also present in the sample along with Fe and Na in very small quantities. From ICP-AES analysis, the purity of cobalt oxalate is found to be 90%.

Table 3 Composition of cobalt oxalate using ICP-AES analysis

Metal ions	Weight percentage
Co	28.96
C	10.03
Li	0.02
Mn	1.07
Ni	1.01
Na	0.128
Fe	0.137

Table 4 Composition of lithium cobalt oxide using ICP-AES analysis

Metal ions	Weight percentage
Co	55.17
C	10.03
Li	7.22
Mn	2.23
Ni	2.32
Na	0.231
Fe	0.22

3.4 Manufacturing of Lithium Cobalt Oxide and Its Analysis

To manufacture lithium cobalt oxide (a cathode battery material), the extracted cobalt oxalate and procured lithium carbonate are mixed in the ratio of 1:1.1 on mass basis in the mortar and pester assembly. The well-mixed solid mixture is then subjected to heat treatment in the muffle furnace at 800 °C for 5 h. Then black colored powder is cooled down to room temperature and sent for analysis.

The XRD analysis of the sample confirmed that black powder mainly consists of lithium cobalt oxide. The ICP-AES data tabulated in Table 4 shows that the purity of lithium cobalt oxide is 91% with small traces of manganese and nickel along with sodium metal.

4 Conclusions

The recycling of secondary lithium-ion battery is necessary from the environment point of view as well as from the sustainable metal supply. In this study, recycling of mobile phone's lithium-ion secondary batteries is carried out using sulfuric acid as the leachate. The dissolution efficiency of the recycling process is enhanced using various optimal operating conditions. The dissolution efficiency is optimum at the operating condition of temperature equal to 90 °C, particle size less than 25 μm, 2 molar sulfuric acid and solid-to-liquid ratio of 10 g/l. The study shows that the maximum dissolution efficiency achieved is 95.15% at the optimal condition. The extraction efficiency of cobalt is 91% with 90% purity of cobalt oxalate.

The battery grade lithium cobalt oxide is manufactured from the extracted cobalt oxalate and procured lithium carbonate (Loba Chemicals, India). It is found that the purity of lithium cobalt oxide is 91%. However, the battery grade cathode material should have the purity of 99.5% and hence further research is going to improve the purity of cobalt oxalate and also to extract lithium carbonate from the spent lithium-ion batteries, so that cathode active material can be purely manufactured from metals extracted from spent lithium-ion batteries.

Acknowledgements for Financial Support The authors acknowledge the department of science and technology, Government of India for financial support for carrying out this work under the DST-SERB-ECR project (ECR/2016/000422).

References

1. Li, L., Fan, E., Guan, Y., Zhang, X., Xue, Q., Wei, L., Wu, F., Chen, R.: Sustainable recovery of cathodic materials from spent li-ion batteries using lactic acid leaching system. ACS Sustain. Chem. Eng. **5**(6), 5224–5233 (2017)
2. Zhang, X., Cao, H., Xie, Y., Ning, P., An, H., You, H., Nawaz, F.: A closed process for recycling $LiNi_{1/3}Co_{1/3}Mn_{1/3}O_2$ from the cathode scraps of lithium-ion batteries: process optimization and kinetics analysis. Sep. Puri. Technol. **150**, 186–195 (2015)
3. Jha, M.K., Kumari, A., Jha, A.K., Kumar, V., Hait, J., Pandey, B.D.: Recovery of lithium and cobalt from waste lithium ion batteries of mobile phone. Waste Manag. **33**(9), 1890–1897 (2013)
4. Zhang, P., Yokoyama, T., Itabashi, O., Suzuki, M.T., Inoue, K.: Hydrometallurgical process for the recovery of metal values from spent lithium ion spent batteries. Hydrometallurgy **47**, 259–271 (1998)
5. Castillo, S., Ansart, F., Laberty-Robert, C., Portal, J.: Advances in the recovering of spent lithium battery compounds. J. Power Sour. **112**, 247–254 (2002)
6. Lee, C., Rhee, K.: Reductive leaching of cathodic active materials from lithium ion battery wastes. Hydrometallurgy **68**, 5–10 (2003)
7. Shin, M.S., Kim, N., Sohn, J., Yang, H.D., Kim, H.Y.: Development of a metal recovery process from Li-ion battery wastes. Hydrometallurgy **79**, 172–181 (2005)
8. Meng, Q., Zhang, Y., Dong, P.: Use of glucose as reductant to recover Co from spent lithium ions batteries. Waste Manag. **64**, 214–218 (2017)
9. He, L.-P., Sun, S.-Y., Mu, Y.-Y., Song, X.-F., Yu, J.-G.: Recovery of lithium, nickel, cobalt, and manganese from spent lithium-ion batteries using L-tartaric acid as a leachant. ACS Sustain. Chem. Eng. **5**(1), 714–721 (2017)
10. Li, L., Wenjie, Q., Zhang, X., Lu, J., Chen, R., Wu, F., Amine, K.: Succinic acid-based leaching system: a sustainable process for recovery of valuable metals from spent Li-ion batteries. J. Power Sour. **282**, 544–551 (2015)
11. Chen, X., Luo, C., Zhang, Z., Kong, J., Zhou, T.: Sustainable recovery of metals from spent lithium-ion batteries: a green process. ACS Sustain. Chem. Eng. **3**(12), 3104–3113 (2015)
12. Nayaka, P.G., Pai, V.K., Santhosh, G., Manjanna, J.: Recovery of cobalt as cobalt oxalate from spent lithium ion batteries by using glycerin as leaching agent. J. Envn. Chem. Eng. **4**, 2378–2383 (2016)
13. Aktas, S., Fray, D.J., Burheim, O., Fenstad, J., Acma, E.: Recovery of metallic values from spent lithium ion secondary batteries. Miner. Process. Ext. Metal **115**(2), 95–100 (2006)

14. Swain, B., Jeong, J., Lee, J.C., Lee, G.H., Sohn, J.: S, Hydrometallurgical process for recovery of cobalt from waste cathodic active material generated during manufacturing of lithium ion batteries. J. Power Sour. **167**, 536–544 (2007)
15. Li, L., Jing, G., Renjie, C., Feng, W., Shi, C., Xiaoxiao, Z.: Environmental friendly leaching reagent for cobalt and lithium recovery from spent lithium ion batteries. Waste Mange. **30**, 2615–2621 (2010)
16. Chen, L., Tang, X., Zhang, Y., Li, L.: Process for recovery of cobalt oxalate from spent lithium ion batteries. Hydrometallurgy **108**, 80–86 (2011)
17. Zhu, S.G., He, W.Z., Li, G.M.: Recovery of Co and Li from spent lithium-ion batteries by combination method of acid leaching and chemical precipitation. Trans. Nonferrous Met. Soc. China **22**, 2274–2281 (2012)

Flexural Strengthening of RC Beams Using NSM Technique

Janki R. Patel, Mitali R. Patel, Tejendra G. Tank, Sandip A. Vasanwala and C. D. Modhera

Abstract Reinforced Concrete (RC) beam elements use bending action to resist the loads imposed on them. In critical situations, ideally, an RC beam must permit ample time to the occupants to evacuate the structure prior to collapse. Structures become weak over time, owing to environmental action, change in occupancy conditions, excessive loading, etc. An effort has been made here, for flexural strengthening of pre-designed beam element, using NSM technique by adopting three types of bars as variables, and by keeping all other parameters constant. The study includes the use of Extruded Glass Fibre Reinforced Polymer (GFRP) bars, Fe-415 bars, and Pultruded Basalt Fibre Rock Polymer (BFRP) bars for strengthening. End de-bonding was observed in all three cases. However, amongst all three, the best one was selected for end confinement using 50 mm wide CFRP wrap, using U-wrap technique in order to ascertain the performance enhancement. Here, since end confinement played its role, no de-bonding was observed. Conclusions are drawn based on observations pertaining to load, deflection, strain in steel, and compressive strain in concrete and de-bonding.

Keywords RC beams · Flexural strengthening · NSM · De-bonding
End confinement · Performance

J. R. Patel (✉) · T. G. Tank
Department of Civil Engineering, Indus Institute of Technology and Engineering,
Indus University, Rancharda, Via: Thaltej, Ahmedabad 382 115, Gujarat, India
e-mail: jankipatel.cvl@indusuni.ac.in

M. R. Patel
Institute of Research & Development, Gujarat Forensic Sciences University,
Nr. DFS Head Quarters, National Highway 8C, Sector 9, Gandhinagar 382007, Gujarat, India

S. A. Vasanwala · C. D. Modhera
Applied Mechanics Department, Sardar Vallabhbhai National Institute of Technology,
Ichchanath, Surat 395007, Gujarat, India

D. Deb et al. (eds.), *Innovations in Infrastructure*, Advances in Intelligent Systems and Computing 757, https://doi.org/10.1007/978-981-13-1966-2_21

1 Introduction

Strengthening of existing structural elements has gained tremendous popularity, owing to the availability of a variety of strengthening materials in the growing era of globalization. Structures undergo deterioration owing to many factors, the environmental action being predominant one. Again, poor quality control on filed results in early deterioration of structural elements [8]. There is always an option of demolishing an old structure and constructing a fresh one. Although there is an increase in the percentage of CO_2 emissions, however an awareness towards reduction of cement in construction practice has also increased. This awareness leads us to emphasize on a simple fact, that if we strengthen an existing structure, and make it last for few more years, human race is on dual gain. Primarily, we are pushing back the concrete debris generation, resulting from demolition of an old structure, by few years. Secondly, CO_2 emission levels of a newly constructed structure are much higher than that generated by an old structure strengthened.

Beam elements ideally must resist the loads by bending action, and should show the signs of failure, prior to collapse. In several cases, it has been observed that beam elements undergo deterioration to an extreme extent of becoming weak in flexure, and thereby may not be able to resist working loads for their designed life span. Eventually, a need is felt here to strengthen the element so as to ensure its performance for years to come. The use of composites, for external strengthening, involves various bonding mechanisms for strengthening, such as wrapping, side bonding, U-jacketing and Near Surface Mounting (NSM) techniques. While applying NSM bars to strengthen concrete beam in flexure, a grove is made at soffit of the member and thereafter the groove is filled with epoxy paste. The bar then is introduced in the groove ensuring that the groove is completely covered and the surface is levelled [6]. In addition to this, the practical advantage of NSM includes less risk to de-bonding, better protection against accidental impacts and unchanged aesthetic features [2]. The research carried out till date has indicated that the tensile strength of the externally bonded strengthening material is not completely utilized owing to their premature de-bonding [1, 5, 7, 9].

An attempt has been made in this manuscript to understand the adaptability of NSM technique for flexural strengthening of RC beams using Fe-415, GFRP, and BFRP rods. Behaviour is studied based on the loads, deflections, strains in main reinforcement, and compressive strain in concrete. Looking into the use of NSM technique adopted by researchers till date, it has been observed that confinement of bar at the ends is missing in majority of the strengthened structural elements, owing to which de-bonding of the bar occurs, either from concrete substrate, or from adhesive concrete interface, or from adhesive-bar interface [10]. This manuscript is inclined towards strengthening of beam element using NSM technique, also towards prevention of de-bonding failure by incorporation of CFRP (Carbon Fibre Reinforced Polymer) strip confinement using U-wrap technique at both ends of the NSM groove.

2 Problem Description

Fifteen pre-designed beam elements (200 mm × 270 mm × 2000 mm) were considered here for strengthening. The beams were intentionally designed to be weak in flexure, so that when tested, all non-strengthened beams fail in flexure only. Thereafter, NSM technique was adopted here as the mechanism to strengthen beams in flexure, using three different types of rods for strengthening. Parameters listed in Table 1 were kept constant for all beam specimen.

Apart from abovementioned parameters, two anchor bars of 10 mm diameter (Fe-415 grade) are provided in all beams to hold shear reinforcement. The details of beams tested (Strengthened as well as non-strengthened) are as mentioned in Table 2. There were three beams tested for each material used for strengthening, and since each of them showed more or less similar results and behaviour, the results of the best among the tested three are considered in this manuscript.

Initially, non-strengthened beams were tested, so as to evaluate its performance. Eventually, for each strengthening material adopted using NSM technique, beam specimens were tested so as to ascertain the performance of the strengthened specimen. For strengthening of SF-1, SF-2, and SF-3, Extruded GFRP, Fe-415, and pultruded BFRP bars were used, respectively, and no confinement at ends was provided in any of the cases. However, for SF-4, pultruded BFRP bars were used, and confinement was provided at the ends using 5 cm wide CFRP wrap, using U-wrap technique,

Table 1 Parameters kept constant

$L = 2000$ mm	$d = 239$ mm	$F_{ck} = 25$ N/mm^2	$B = 200$ mm
$l = 1800$ mm	$S = 200$ mm	$F_y = 415$ N/mm^2	$D = 270$ mm
$f_y = 250$ N/mm^2	$d' = 25$ mm	$A_{st} = 339.12$ mm^2 (3 Nos of 12 mm diameter bars)	

Table 2 Nomenclature of beams

S. No.	Beam nomenclature	Bar used for NSM strengthening	Diameter of bar	Remarks
1	NSF	–	–	Non-strengthened
3	SF-1	Extruded GFRP rod	12 mm ø	Strengthened
2	SF-2	Fe-415 rod	12 mm ø	Strengthened
4	SF-3	Pultruded BFRP rod	12 mm ø	Strengthened
5	SF-4	Pultruded BFRP rod and CFRP wrap	12 mm ø Rod, 50 mm wide CFRP Wrap	Strengthened

with an objective to evaluate the change in behaviour and performance of the beam element.

3 Material Properties and Detailing of NSM Groove

Grooves of 20 mm × 20 mm size were used for NSM, and epoxy adhesive comprising of base and hardener having properties mentioned in Table 3 was used for experimental work carried out. Properties of extruded GFRP bars, Fe-415 bars and pultruded BFRP bars used for the experimental program, are included in Table 4. Properties of CFRP wrap used for end confinement in SF-4 beam are as mentioned in Table 5.

It is evident that in actual beam element, is it practically impossible to introduce a groove beneath the supports, hence groove for NSM at the soffit of the beam for flexural strengthening, was introduced for a length of 1800 mm, leaving a distance

Table 3 Properties of epoxy adhesive

Density		1.30 kg/l
Elastic modulus	Flexural	3800 MPa
	Tensile	4500 MPa
Bond strength		Concrete fracture (>4 MPa)
Tensile strength		30 MPa
Elongation at break		0.9%

Table 4 Properties of bars used for strengthening

Property	Fe-415 rod	GFRP rod	BFRP rod
Density (g/cm^3)	7.8	1.9	1.95
Modulus of elasticity (MPa)	2 × 105	30,000	55,000
Poisson's ratio	0.3	0.3	–
Yield tensile strength (MPa)	432.5	200	–
Ultimate tensile strength (MPa)	639.3	506	1160

Table 5 Properties of CFRP warp

Density	1.82 g/cm^3
Wrap thickness	0.129
Tensile modulus	220 kN/mm^2
Tensile strength	3200 MPa

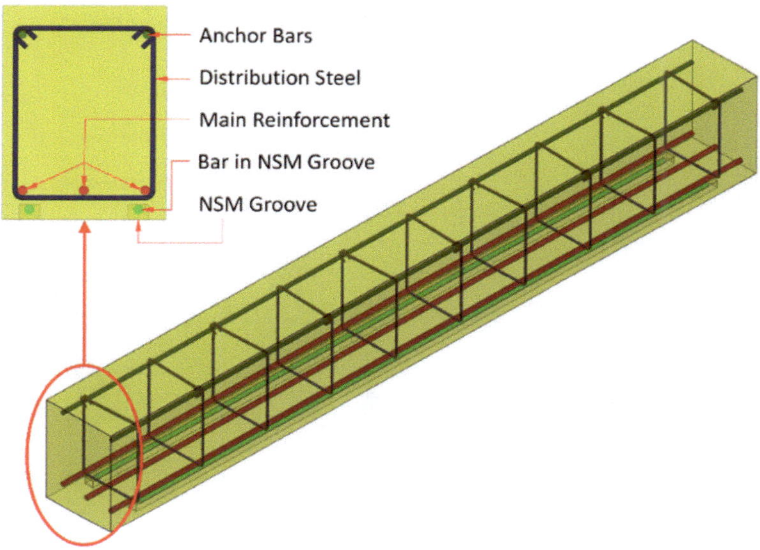

Fig. 1 Sketch of beam detailing including NSM

Fig. 2 Beam soffit marked for NSM

Fig. 3 Beam soffit strengthened using NSM technique

of 100 mm from both the ends. Here for NSM groove cut near the supports, bars introduced in the groove were of 1760 mm length (again leaving 20 mm from ends of groove) for better bonding of bars. A typical sketch of the same along with actual beam marked for NSM, and actual beam strengthened using NSM technique are as shown in Figs. 1, 2 and 3.

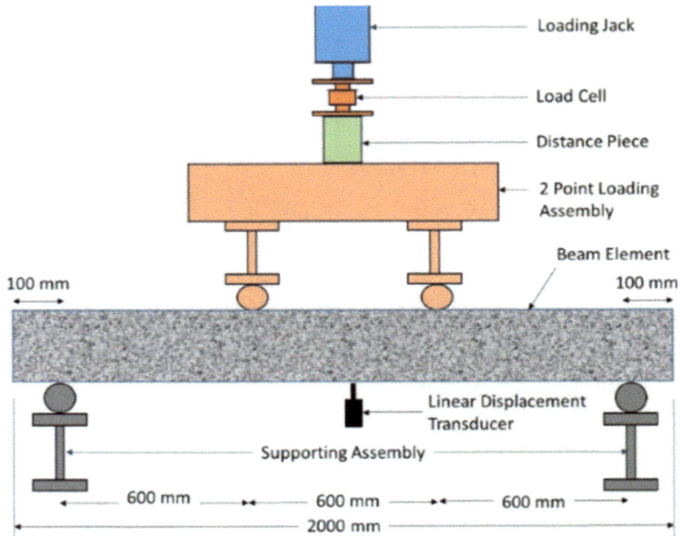

Fig. 4 Sketch of experimental test assembly

4 Experimental Test Setup

Four-point load pattern was adopted for testing of all specimen. A structural test loading frame was used along with a motorized hydraulic jack of 400 kN capacity for load application. Load cell of 300 kN capacity was used to transfer real-time load values to an eight-channel data logger. A linear displacement transducer capable to measure deflections up to 100 mm with an accuracy of 0.01 mm was used to transfer real-time deflection values to the logger. High-precision strain gauges capable to give precise strain values were used to measure strain in central steel bar and concrete compression. Selection of strain gauges for steel, concrete was as per the recommendations of manufacturer. Strains in steel were measured using 120-Ω strain gauge of 5 mm gauge length, fixed on central steel bar of each specimen. Strains in compression in concrete were measured using gauge having 30 mm gauge length, fixed at the centre of concrete top surface, on all specimen. A typical sketch of the experimental test setup assembly and actual test setup assembly, are as shown in Figs. 4 and 5 respectively.

Real-time measurements for load, deflection, and strain values were recorded using an eight-channel data logger with preconfigured and pre-calibrated channels for corresponding parameters to be recorded. Utmost care has been taken while selecting the gauge type, and during gauge application so as to maintain the accuracy of the work.

Fig. 5 Actual experimental test assembly

5 Results and Discussions

All Beam specimen considered for research were tested until failure. Load, deflection, strain in steel, and compressive strain in concrete were recorded at first crack as well as at ultimate load value. Observations were made pertaining to the behaviour of NSF for each parameter mentioned above and compared to those of SF-1, SF-2, SF-3, and SF-4.

5.1 Load v/s Deflection

NSF when tested showed pure flexural failure. The first crack in the beam was induced at a load of 50.87 kN, when deflection was 11.5 mm. The ultimate load carrying capacity of the beam was 130.23 kN at an ultimate deflection of 25.5 mm. Figure 6 shows combined load v/s deflection for all beam specimen tested.

It is evident that NSF is more ductile than any of the strengthened beam specimen, this is purely dependent on the stiffness of the material used for external strengthening of RC beam specimen. Stiffer the strengthening material, less ductile it is.

SF-1 showed the first crack at a load value 15.8% more than NSF, however at first crack, 72.3% reduction in deflection as compared to NSF was observed. Looking into the ultimate load capacity, there was a marginal increase of 4.7% as compared to NSF, however reduction in ultimate deflection was 73.9%. Among all four strengthening techniques, SF-1 showed the least deflection, this is owing to the rigidity of extruded GFRP bar used for NSM strengthening. There was no direct de-bonding of GFRP rod from parent concrete, this is attributed to strong bond between GFRP rod and adhesive, and between adhesive and parent concrete. However, failure of SF-1 was

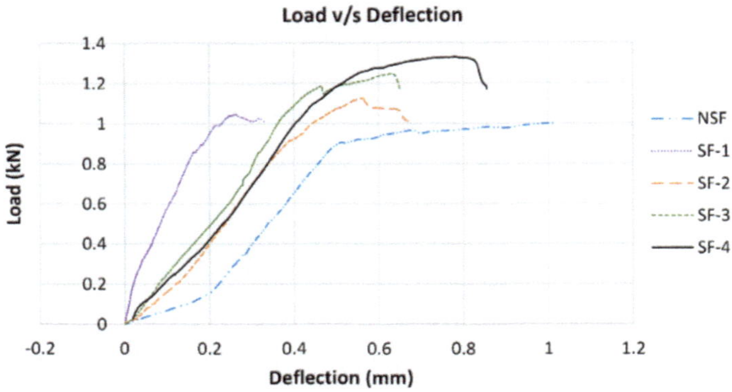

Fig. 6 Load v/s deflection for all beam elements

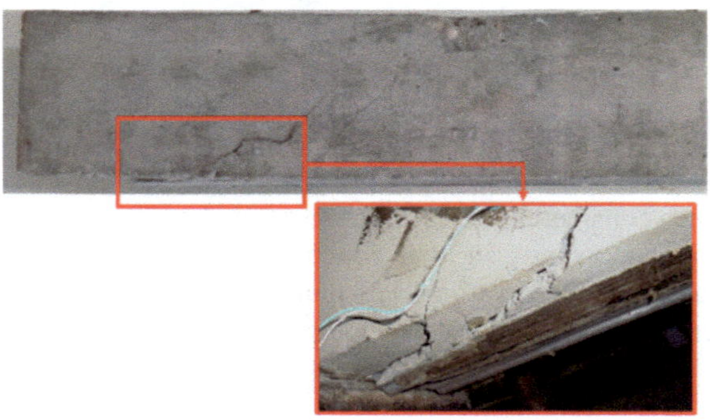

Fig. 7 Failure of SF-1

due to gradual de-bonding of concrete cover substrate from main steel. Figure 7 shows failure of SF-1.

At first crack, it was observed that SF-2 showed 6.4% increase in load carrying capacity as compared to NSF, and 28.8% reduction in deflection. At ultimate load value, 12.7% increase in load carrying capacity was recorded with 44.1% reduction in ultimate deflection value in comparison to NSF. Increase in ultimate load carrying capacity as collated with NSF and SF-1 was owing to the increased stiffness imparted to the beam by steel—adhesive bond. However, parallel increase in deflection is due to more ductility of steel when correlated with GFRP rod in SF-1. Failure of SF-2 was due to gradual de-bonding of concrete cover substrate from main steel, same as that observed in SF-1. Figure 8 shows the failure of SF-2.

When compared to NSF, the first crack in SF-3 showed 18.7% improvement in load carrying capacity with reduction in deflection value by 31.8%. Ultimate load carrying capacity improved by 24.8% with reduction in ultimate deflection up to 36.6%. The ductility of BFRP rod is more than that of steel and GFRP. This played a vital role in imparting ductility to SF-3. Mode of failure in SF-3 was de-bonding

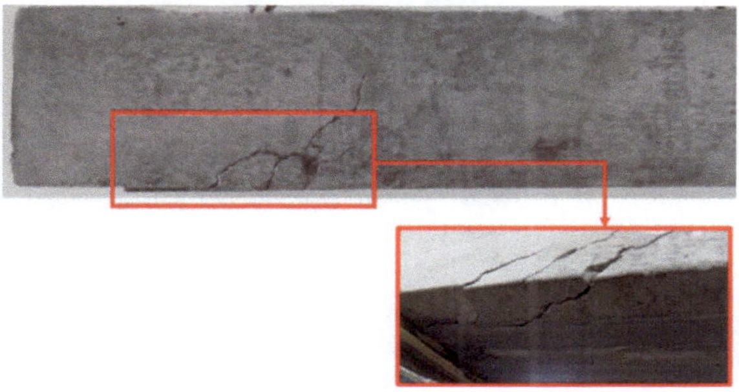

Fig. 8 Failure of SF-2

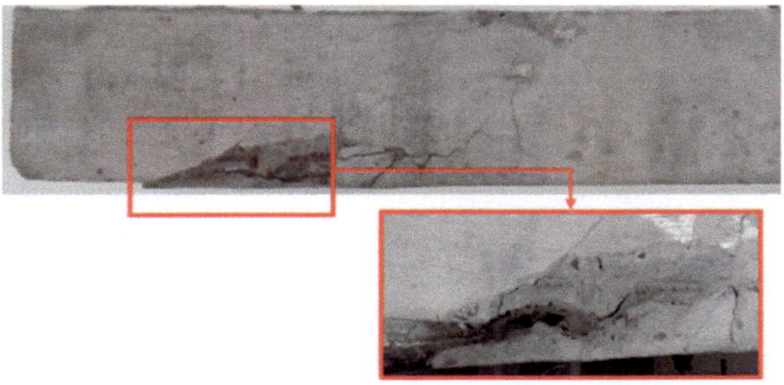

Fig. 9 Failure of SF-3

failure from one of the ends, wherein concrete cover substrate had de-bonded from main reinforcement. However, it was observed that SF-3 showed sudden failure and gave no warning prior to its sudden de-bonding failure. Figure 9 shows the failure of SF-3, highlighting the sudden failure due to de-bonding.

It was observed that in all NSM strengthened specimen, de-bonding had occurred, however since SF-3 was the best among NSF, SF-1 and SF-2, SF-4 was specifically considered to address the de-bonding problem. The only change made was that ends of the pultruded BFRP bars were confined using 5 cm wide CFRP strip and U-wrap technique, so as to prevent de-bonding of ends. SF-4 showed 28.7% increase in load carrying capacity at first crack, with 18% reduction in deflection as compared with NSF. At ultimate load, it showed 33.1% increase in load carrying capacity with 21.7% reduction in ultimate deflection in correlation to NSF. The ultimate load carrying capacity of SF-4 is the highest among all mechanisms. The specimen although has gained some stiffness after strengthening, it has not lost its ductility as much as other specimens have. Eventually, SF-4 showed a non-catastrophic failure. Figure 10 shows the failure of SF-4, wherein confinement has played a vital role in preventing de-bonding.

Talking about load and deflection, there are two types of cases to be emphasized on. The first case is of beam elements where an increase in load carrying capacity is observed, the second case is of beam elements where decrement in deflection is observed. Figure 11 shows ultimate load carrying capacity of each specimen with their corresponding ultimate deflection values.

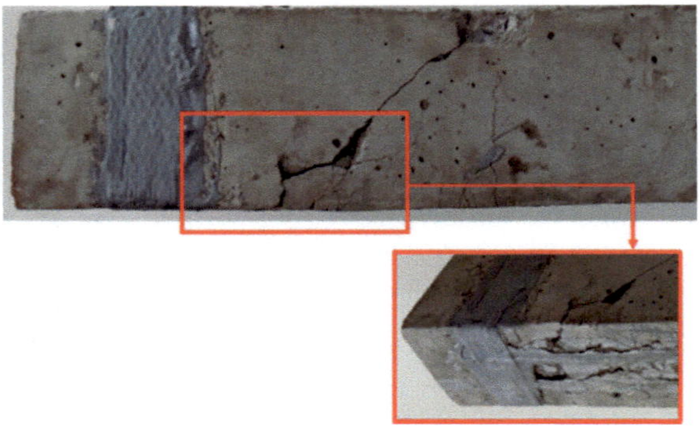

Fig. 10 Failure of SF-4

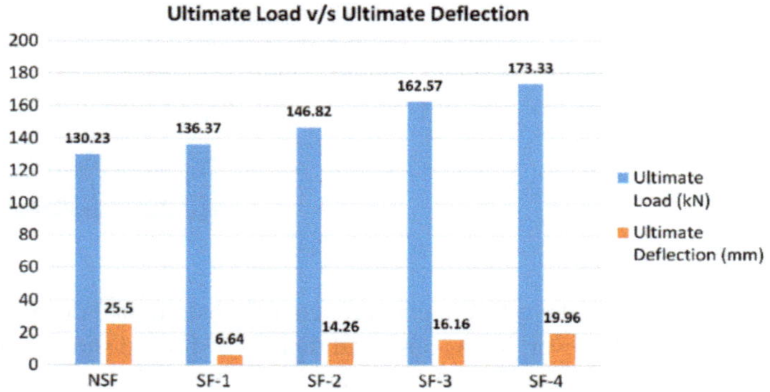

Fig. 11 Ultimate load v/s ultimate deflection for all specimen

5.2 Load v/s Strain in Main Steel

Strains were recorded for steel bar embedded in concrete for all specimen. When reinforcing bars embedded in concrete were stressed in tension, concrete assisted steel in carrying the tensile force, this continued even after cracks went severe. This concrete contribution is termed as tension stiffening by concrete [3]. For a bare reinforcement bar, the behaviour seems to be predictable, however for a bar embedded in concrete, the strain values decrease, which is owing to assistance received by steel from concrete in carrying tensile forces. Again, additional resistance to tensile force offered by strengthening material to concrete, and thereby to steel, in resisting tension also plays a vital role in strain in steel. Stiffer the strengthening material, resistive it gets. Bond between strengthening material to adhesive, and adhesive to parent concrete also offers resistance to bending, and thereby contributing to a decrease in strain in steel [4]. Figure 12 shows the behaviour of strain in steel for each specimen along with their corresponding load values. Normalized values are shown for clear understanding.

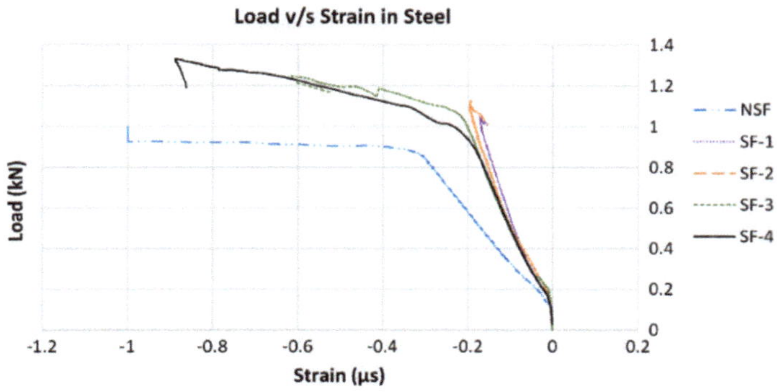

Fig. 12 Load v/s strain in main steel for all specimen

It can be observed that for all specimen, after the ultimate load, strain relaxation takes place. This is owing to the fact that bond between concrete and steel is disturbed and thereby deformations continue, but strain values in steel step back. Again, as mentioned earlier, the de-bonding in SF-1, SF-2, and SF-3 have occurred due to rapture between concrete cover and main steel interface, hence leading to relaxation of main steel after initiation of de-bonding. In case of SF-4, CFRP strips did not permit for end de-bonding and thereby the failure was gradual, but even here, the failure occurred as the bottom concrete had given up. The bond between strengthening material and adhesive, as well as the adhesive and parent concrete was stronger than the bond between main steel and parent concrete as a result of which the de-bonding had occurred from the main steel to parent concrete interface.

5.3 Load v/s Compressive Strain in Concrete

Values for strain in compression in concrete were measured for all specimen, wherein the gauge was placed at the centre of top surface of each beam specimen. It is observed that for all specimen tested, strain in concrete shows nonlinear behaviour right from the beginning. Strain in compression in concrete for NSF, SF-1, and SF-4 show increase in strain post failure. This is owing to a reason that even though the specimen has failed, after ultimate load capacity, deflections go on. There is no slip in concrete compression, and thereby no strain relaxation for concrete. However, in SF-2 and SF-3 strains are seen to decrease as we approach towards the ultimate load value, this was due to resistance offered by the strengthening technique, wherein Fe-415 rod and BFRP rod offered a pre-failure resistance, which was not offered in specimen strengthened using extruded GFRP bars owing to the stiffness of the GFRP rod. In SF-3, the resistance offered by the BFRP rod lasted only up to 162.57 kN wherein the specimen failed due to de-bonding. BFRP rod could have offered more resistance if de-bonding would not have occurred. This was observed in SF-4, wherein the CFRP

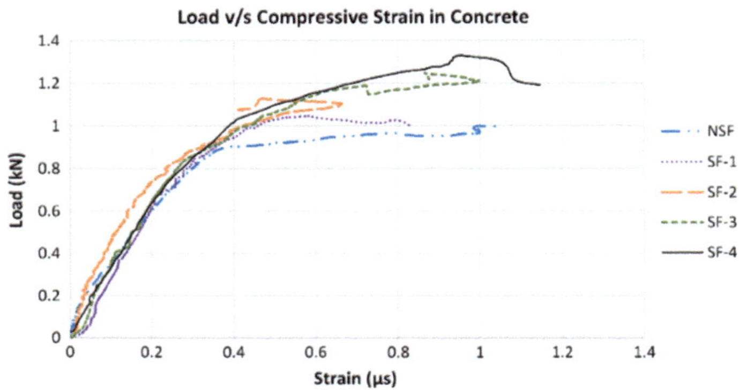

Fig. 13 Load v/s compressive strain in concrete for all specimen

wrap confinement at ends played its role and permitted the BFRP rod to offer more resistance. Although this was 6–7% more, however, failure pattern had changed to gradual rather than sudden. Figure 13 shows the behaviour of compressive strain in concrete for each of the specimen along with their corresponding load values. Normalized curves are indicated for better understanding and clear comparisons.

6 Conclusions

Strengthening of deficient beam elements needs to be done with utmost care. Here, all specimens were weak in flexure. NSM strengthening was adopted to strengthen all specimen, the only difference was the strengthening material used. Following conclusions are drawn from the experimental work carried out.

1. Since all specimen were strengthened in flexure, all have inclined towards shear failure. SF-1 and SF-2 exhibited gradual failure, however, SF-3 showed catastrophic failure. In all cases, namely SF-1, SF-2, and SF-3, the failure was owing to de-bonding of concrete cover interface from main reinforcement.
2. The reason attributed to sudden failure of SF-3 is the ultimate tensile strength of pultruded BFRP bar, which is almost 2–2.25 times more than Fe-415 and extruded GFRP bars, respectively. More the ultimate tensile strength, more is the bending resistance, and thereby more is its ultimate load carrying capacity.
3. In all cases, the bond between the strengthening material and adhesive, and between adhesive and parent concrete, was stronger than the bond between main steel and concrete. Hence, all de-bonding failures have occurred in the interface zone between concrete cover and main steel. However, for SF-4, confinement was from both the ends, and thereby failure was due to concrete crushing and not due to de-bonding.

4. End confinement in SF-4 has resulted into prevention of de-bonding of BFRP rods. NSM although is one of the oldest techniques, it is practically impossible to provide grooves beneath the supports and thereby inserting rod for NSM strengthening under the supports too becomes unrealistic.
5. As deflections increase, strain values in main steel and concrete compression increase. Controlled strain values can be achieved with controlled deflections, which purely depends on the mechanism and material used.

References

1. Barros, J.A.O., Fortes, A.S.: Flexural strengthening of concrete beams with CFRP laminates bonded into slits. Cem. Concr. Comp. **27**, 471–480 (2005)
2. Barros, J.A.O., Ferreira, D.R.S.M., Fortes, A.S., Dias, S.J.E.: Assessing the effectiveness of embedding CFRP laminates in the near surface for structural strengthening. Constr. Build. Mater. **20**, 478–491 (2006)
3. Carreira, D.J., Chu, K.H.: Stress-strain relationship for reinforced concrete in tension. ACI J. **83**(3), 21–28 (1986)
4. Chen, S., Kabeyasawa, T.: Average stress-strain relationship of steel bars embedded in concrete. In: Proceedings of 13th World Conference on Earthquake Engineering, Vancouver, B.C., Canada, Paper No. 3290 (2004)
5. Cruz, J.S., Barros, J.: Modeling of bond between near-surface mounted CFRP laminate strips and concrete. Comp. Struct. **82**, 1513–1521 (2004)
6. De Lorenzis, L., Nanni, A.: Characterization of FRP Rods as near-surface mounted reinforcement. J. Compos. Constr. 114–121 (2001)
7. Hollaway, L.C., Teng, J.G.: Strengthening and Rehabilitation of Civil Infrastructures Using Fibre-Reinforced Polymer (FRP) Composites. Woodhead Publ, Boca Raton (2008)
8. Laraba, A., Merdas, A., Chikh, N.E.: Structural performance of RC beams strengthened with NSMCFRP. In: Proceeding of the World Congress on Engineering, vol. II, London, 978–988 (2014)
9. Lorenzis, L.D., Rizzo, A., Tegola, A.L.: A modified pull-out test for bond of near-surface mounted FRP rods in concrete. Comp.: Part B **33**, 589–603 (2002)
10. Morsy, A.M., El-Tony, M., El-Naggar, M.: Flexural repair/strengthening of pre-damaged R.C. beams using embedded CFRP rods. Alexandria Eng. J. **54**, 1175–1179 (2015)

Usage of Crushed EAF Slag in Granular Sub-base Layer Construction

Radha J. Gonawala, Rakesh Kumar and Krupesh A. Chauhan

Abstract An attempt has been to study the optimum mix of crushed Electric Arc Furnace (EAF) slag with natural aggregate in Granular Sub-base (GSB) layer. Detailed experimental investigation was carried out for strength, durability and drainage properties. The dynamic CBR (CBR_d) using Light Weight Deflectometer (LWD) was conducted on a mix of EAF slag with natural aggregate. Totally 30 combinations of mix proportion were examined for six gradations of GSB to find an optimum mix based on strength and permeability criteria. Results show that mixture of 50% EAF slags with natural aggregate was found to be optimum mix. The closed graded permeability was estimated in the range between 380.96 m/day and 406.44 m/day, which falls under Indian specification. Similarly, static-soaked CBR for open-graded gradation was found around 150 and 160% from dynamic CBR. Closed graded dynamic CBR is varied in the range of 180 $\pm$ 10% for different closed graded. Therefore, up to 50% of EAF slag mixed with natural aggregate most satisfactory for construction GSB layer.

Keywords EAF slag · GSB layer · Permeability · Dynamic CBR · LWD

1 Introduction

Granular sub-base (GSB) layer acts as one of the structural and drainage layers of pavements [1]. The aggregates in sub-base layers serve a variety of purposes, namely, reducing the stress applied to the subgrade layer and providing adequate drainage for the pavement structure [2]. Natural materials being exhaustible, its quantity is declining gradually, whereas road construction work is growing [3]. The industrial waste product can be the alternate materials in highway construction. It is also stated

R. J. Gonawala (✉)
S. V. National Institute of Technology, Surat, Gujarat, India
e-mail: radhagonawala@gmail.com

R. Kumar · K. A. Chauhan
CED, S. V. National Institute of Technology, Surat, Gujarat, India

D. Deb et al. (eds.), *Innovations in Infrastructure*, Advances in Intelligent Systems and Computing 757, https://doi.org/10.1007/978-981-13-1966-2_22

that if industrial waste materials can be appropriately used in highway construction, pollution, and disposal problems may be reduced to a great extent [4]. Electric Arc Furnace (EAF) slag is a waste material produced during the manufacturing of steel. In steel production practice, about 150 kg of EAF slag produced per tonne of steel formed [5]. The original EAF slag comes out like a melted liquid which is a compound mixture of silicates and oxides that solidifies upon cooling [6].

The comparative study of recycled and natural materials did not show significant differences in the physical characterizations in petrography analysis during the durability and permeability tests [3]. The research indicates the strength property, binder adhesion and resistance to frictional and abrasion and concludes that it is possible to use steel slag aggregate in bituminous road surface and its base layers, including unbound bases and sub-base layer of pavement [4].

Researchers in Egypt also study the effect of quantity of steel slag (up to 70%) that could be blended with crushed limestone aggregates (30%), to be used as a bound sub-base layer of flexible pavement. It shows that increasing the steel slag percentage to the limestone in the blended mix rises its maximum dry density (MDD) value, California Bearing Ratio (CBR) and resilient modulus of the blend bound material layer [7]. The EAF slags also show acceptable chemical and physical characteristics in the limited replacement of 30–70% of natural aggregates to form appropriate and durable cement-treated materials [6]. Further, as a base course material, the physical and chemical properties of the slag show better results in the comparison of the natural aggregates like resilient modulus and permanent distortion characteristics (durability). Subsequently, the addition of lime and cement binders improve these features even more [8].

It is known that water intrusion is the primary reasons for premature failures in pavement layers. The required specifications repeatedly match the requirement of the slag in GSB and base layer as a structural member. A little study is reported about whether drainage properties and fractional gradation fulfils the drainage requirements of EAF slag [2]. It is said that the excess moisture could be drained out quickly if the GSB layer is more permeable in the flexible pavement [9]. The permeability in the range of 0.2 cm/s (500 ft/d) or higher is considered to be satisfactory. The permeability range of less than 3.5×10^{-4} cm/s (1 ft/d) would be regarded as too slow, with intermediate values being marginal and may hamper the drainage requirements of pavement layer [10].

Therefore, in this study apart from strength analysis (using LWD also), permeability is studied experimentally in the laboratory. The experiments were designed for the construction of GSB layer using all six different gradations of aggregate. Further, the optimized mix of EAF slag with natural aggregate for construction of the granular sub-base (GSB) layer of flexible pavement.

2 Materials and Methods

In this study, EAF slag was obtained from the M/s Essar steel Ltd, Hazira, Surat, Gujarat, India. EAF slag is the solid material of greyish colour and odour free. As per IS 2386 (Part 1)-1963, sieve analysis of EAF slag and natural aggregate were performed. EAF slag found as well-graded aggregate. The natural aggregate was collected from the quarry near Chikhli town, Surat, Gujarat, India. The nominal maximum aggregate size used as 40, 20, 10 and 6 mm for all six gradations as per MORTH (fifth revision, 2013) (Table 1). All the engineering properties of the EAF slag and aggregate found according to the Indian specification for construction of flexible pavement [11]. Further the chemical composition of EAF slag was examined with X-ray fluorescence (XRF) test.

CBR test obtained the strength parameter at Optimum Moisture Content (OMC), and MDD derived from Modified Proctor test. In this study, both static and dynamic CBR test was conducted on the soaked GSB specimens in the laboratory.

The dynamic CBR (CBR_d) laboratory test was designed as an alternative to the static CBR test, as easy to perform and the short period required due to the elimination of loading frame. The dynamic CBR (CBR_d) also feasible to do an in situ test on prepared sub-base layer. Dynamic CBR (CBR_d) test, in the laboratory, was conducted using LWD instrument in standard CBR mould [12].

3 Experimental Studies

EAF slag and aggregate were blended according to the gradation requirement. The moisture content and specific gravity test were performed according to the IS: 2386

Table 1 Gradation for GSB layer

IS sieve designation (mm)	Percent by weight passing the IS sieve					
	Grading I	Grading II	Grading III	Grading IV	Grading V	Grading VI
75.1	100	–	–	–	100	–
53.0	80–100	100	100	100	80–100	100
26.5	55–90	70–100	55–95	50–80	55–90	75–100
9.50	35–65	50–80	–	–	35–65	55–75
4.75	25–55	40–65	10–30	15–35	25–50	30–55
2.36	20–40	30–50	–	–	10–20	10–25
0.85	–	–	–	–	2–10	–
0.425	10–15	10–15	–	–	0–5	0–8
0.075	<5	<5	<5	<5	–	0–3

(part 3)-1963. The EAF slag and aggregate crushing value were checked according to IS: 2386 (part 4)-1963, and impact test carried was out according to IS: 2386 (part 4)-1963 as shown in Table 2. A maximum value of 40 percent is allowed for sub-base course in India. The water absorption of slag was reported 2.89 which is greater than 2% as per MORTH specification. Therefore, the slag was tested for soundness test (IS: 2386 (part 5)-1963).

Due to higher porosity of slag, water absorption is high. The higher porosity compared to that of natural aggregates was confirmed by visual examination of the surfaces of the grains of EAF slag. This property is due to the production methods of the slag. Slag from the steel industry comes from the rapid cooling of the oxidized and superficial liquid phase present in electric arc furnaces. The trapping of tiny air bubbles in melted mass during cooling leads to characteristic microporosity of the material [6]. Soundness test on the EAF slag indicates a loss by weight 5.072% with magnesium sulphate reagent. The weight loss after soundness test with magnesium sulphate is permitted up to 18%. It is worth stating that the sound mechanical characteristics of EAF slag, justifies its usage as GSB layer material.

Impact value and a crushing value obtained for EAF slag are inferior as compared to the natural aggregate. The molten EAF slag gets cool and hardened in the open air or to quicken the process by a jet of water. During the setting process of molten slag, the hydration and carbonation of the calcium oxide may take place. It could be the reason for irregular expansion and destruction of the EAF slag [6]. However, it fulfils the MORTH requirement for GSB layer.

The permeability of the GSB material found in the laboratory by constant head permeability test. The strength and permeability tests conducted on natural aggregate, EAF slag and combination of natural aggregate and EAF slag. EAF slag was blended with natural aggregate by the 0, 30, 50, 70, and 100% by the weight.

The OMC and MDD found according to the IS 2720 (part 8)-1983 for all aggregate-slag mixes. CBR tests conducted on combinations as per IS 2720 (part 16)-1987, at the OMC. After compaction, the CBR samples soaked in water for four days before testing. In CBR_d the loading applied by LWD, containing falling weight to use a defined force on the CBR specimen. A force of 2.44 kPa applied to the all samples subjected to CBR_d [12] (Fig. 1). CBR_d is estimated based on empirical formula [13]. As:

Table 2 Physical properties of natural aggregate and EAF slag

	Natural aggregate	EAF slag
Aggregate impact value (%)	7.91	16.13
Aggregate crushing value (%)	18.37	28.34
Liquid limit and plastic limit (%)	Non-plastic	Non-plastic
Specific gravity	2.79	3.09
Water absorption (%)	1.60	2.89

Fig. 1 Performing of dynamic CBR test using LWD

$$\text{CBR}_d = \frac{87.3}{s^{0.59}}\ \% \tag{1}$$

where the '87.3' value is dynamic loading coefficient including empirical coefficient, and '*s*' is the settlement (mm).

The coefficient of permeability test was determined by constant head permeability method for the flow of water through the granular aggregates as per IS 2720 (part 17)-1986. During testing, granular aggregates containing more than 10% of aggregate passing through 75-μm sieve are restricted, which is suitable for sub-base used as a drainage layer [9]. The coefficient of permeability, k, was calculated as per Darcy's law as follows:

$$k = \frac{q \times L}{A \times h} \tag{2}$$

k coefficient of Hydraulic conductivity (cm/sec),
q quantity of water discharged (cm^3/sec),
L mean length of specimen (cm),
A cross-sectional area of specimen (cm^2),
h head difference (cm).

4 Results and Discussion

The chemical composition of EAF slag mainly consists of CaO (27.59%), FeO (25.70%), Al_2O_3 (21.18%), SiO_2 (13.54%), and MgO (1.03%) as per XRF analysis. Further, XRF analysis results show that more than 85% of the total mass was represented by the CaO, FeO, Al_2O_3, SiO_2, and MgO, which is by standards EAF slag [6]. From the gradation of EAF slag, it can be classified as an 'all in aggregate' as per IS 383-2016. The EAF slag and aggregate were mixed in a manner to get its gradation curve closer to that specified grading. The results of Grading I blending for all five combinations plotted in Fig. 2.

As shown in Fig. 3, with the increase of the EAF slag up to 50% the MDD increases and then decreases after 50% replacement of the EAF slag. EAF slag is a crushed material, and while increasing a crushed material up to some extent, interlocking is increased, and hence higher MDD is observed. From the result of different combination for OMC shown in Fig. 4, the lowest water content is found for the 50% aggregate 50% EAF slag combination, for which MDD is highest. The difference in OMC depends upon the gradation of the blend. For densely graded mix higher, OMC observed as compared to the open-graded gradation, i.e. Grade 3 and 4. This trend is followed by all the mixtures of natural aggregate and EAF slag.

The highest value of CBR is observed for the 50% aggregate—50% slag combination as in Fig. 5 in range of 149–163%. It indicates maximum EAF slag percentage that can replace the natural aggregate. CBR results show that, as slag content increases, the strength increases due to better interlocking in-between crushed slag and natural aggregate. However, as slag rises beyond 50%, CBR starts decreasing due to inferior mechanical properties of the EAF slag as compare to natural aggregate. However, the CBR found is more than 90% for all open grading for all combination.

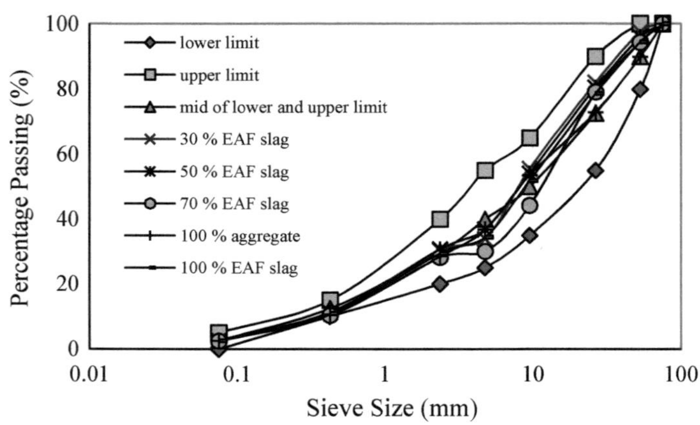

Fig. 2 Gradation of Grading I for a different combination of EAF slag and natural aggregate

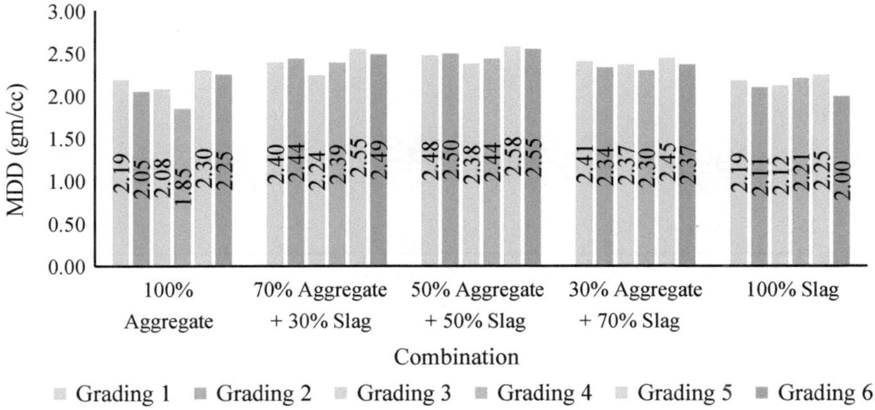

Fig. 3 Maximum dry density for different aggregate-EAF slag mix and grading

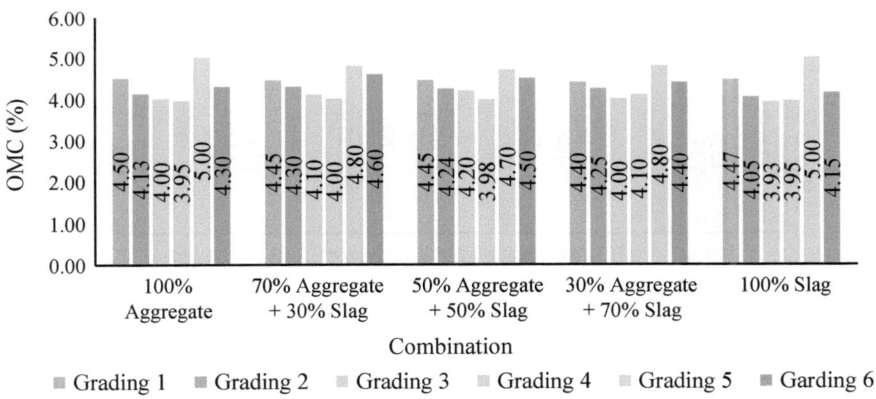

Fig. 4 OMC for different aggregate-EAF slag combination (mixtures) and grading

It shows the suitability to use EAF as a GSB material in accordance to the strength requirement.

Later, after levelling off the surface of the same GSB specimen mould and removing of the basis of the mould, dynamic CBR was carried out. CBR_d results tests are shown in Fig. 6. These results of CBR_d follow the same trend of the conventional CBR. Grading 3 and 4 are open-graded course and have inferior interlocking with adjacent grain shows comparatively low CBR values for all natural aggregate and EAF slag combination. Grading 1, 2, 5, and 6 are densely graded course and having all aggrade size in grading. It shows better interlocking in-between the aggregate to give better strength. This phenomenon observed in both conventional CBR and CBR_d.

Among six gradations, grading 3 and 4 are open-graded gradations. The other four grades 1, 2, 5, and 6 are a densely graded gradation. Grading 3 and 4 are highly

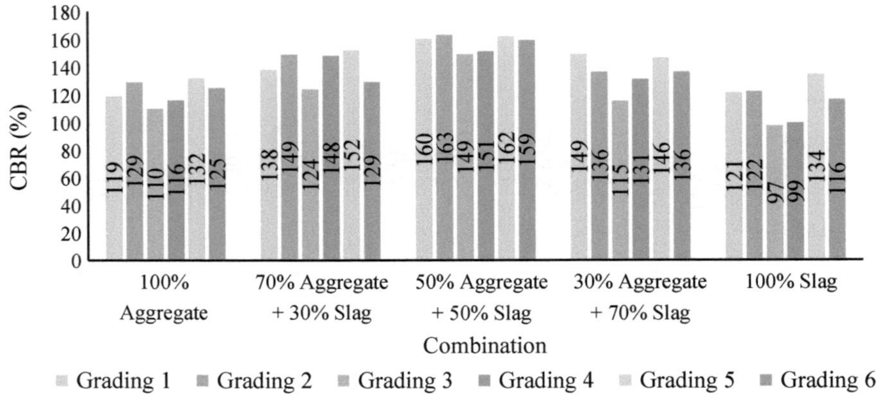

Fig. 5 CBR for different aggregate-EAF slag combination (blends) and grading

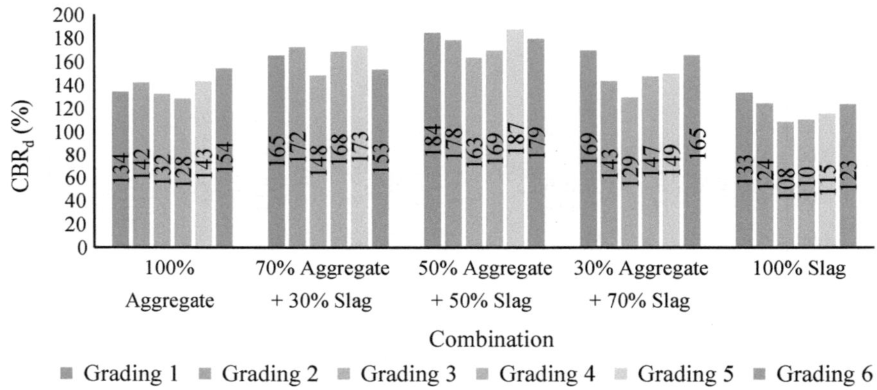

Fig. 6 CBR_d for different aggregate-EAF slag combination (mixtures) and grading

permeable grading developed to be used as a drainage layer as per MORTH specification. For grading 3 and 4, the permeability observed was high for all aggregate-slag combination as they are open-graded. Figure 7 shows maximum permeability saw for the 100% aggregate combination because of least interlocking between the particles, which allow unobstructed flow of water. The next higher permeability is found for the 70% aggregate—30% slag combination because of increment of interlocking due to crushed slag and its texture, and then permeability goes on decreasing as the slag content increases. All combinations satisfy the IRC: 37—2012 requirements of the granular sub-base permeability value of 300 m/day.

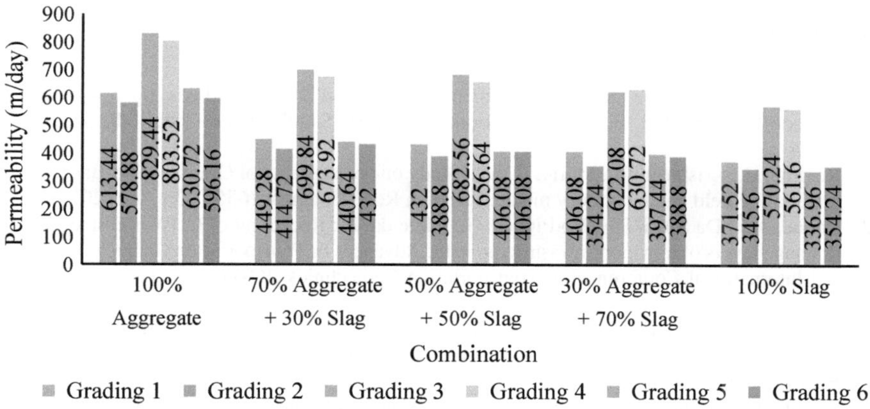

Fig. 7 Permeability for different aggregate-EAF slag combination (mixtures) and grading

5 Conclusion

In this study, partial replacement of natural aggregate with EAF steel slag was studied. The physical and chemical parameters of EAF slag were studied and found its compliance with all Indian specifications except moisture contents. XRF analysis shown that EAF slag contain CaO (27.59%), FeO (25.70%), Al_2O_3 (21.18%), SiO_2 (13.54%), and MgO (1.03%). Water absorption was reported as 2.89% which come under high side, and further, the durability test was performed. The results found satisfactory. Hence, EAF slag can be used for GSB layer. EAF slag was blended with natural aggregate by the 0, 30, 5070 and 100% by the weight. Results show that in almost all the combination, maximum MDD value observed is at 50% Aggregate 50% slag combination. Therefore, for strength and durability parameters optimum slag and aggregate with 50% mix were used in this study for GSB layer. EAF and natural aggregate mix gradation specimen CBR and dynamic CBR were reported in the range of 97–163% and 108–187%, respectively. The results or CBR and CBR_d for 50% aggregate and 50% EAF slag are $150 \pm 10\%$ for open gradation and $180 \pm 10\%$ for close gradation was found. The permeability of all mixes was observed in the range of 345.6–829.44 m/day. For 50% aggregate and 50% EAF slag, the permeability found in a range of 400 ± 30 m/day for closely graded gradation and more than 650 m/day for open-graded. Hence, from the experimental study, it can be concluded that, for all grading mentioned in MORTH specifications for granular sub-base layer, the replacement of natural aggregates with EAF slag is possible up to 50%.

Acknowledgements The authors would like to thank M/s Essar steel Ltd, Hazira, Surat, Gujarat, India for supplying EAF slag. The authors acknowledge Mr. Kalyan Kumar and Mr. Ghanshyam Kela M.Tech. students of Transportation Engineering and Planning Department of S.V. National Institute of Technology, Surat, India for their support and assistance in the laboratory testing work.

References

1. Huang, Y.H.: Drainage Design in Pavement Analysis and Design, 2nd edn, pp. 334–335. NJ, Pearson Prentice Hall, Pearson Education, Inc. Upper Saddle River, Chap. 8th, Sect. 8.1.1 (2004)
2. Kavitha, G., Krishnamurti, Murty, B.R.S.: Study on permeability of Granular Sub-Base (GSB) material by field and laboratory model. J. Indian Road Congress (655), 333–342 (2016)
3. Roque, A.J., Da Silva, P.F., Rodrigues, G., Almeida, R.: Recycling of CDW and steel slag in drainage layers of transport infrastructures. In: Advances in Transportation Geotechnics 3, The 3rd International Conference on Transportation Geotechnics (ICTG), vol. 143, pp. 196–203 (2016)
4. Yi, H., Xu, G., Cheng, H., Wang, J., Wan, Y., Hui, C.: An overview of utilization of steel slag. In: The 7th International Conference on Waste Management and Technology. Procedia Environ. Sci. **16**, 791–801 (2012)
5. Autelitano, F., Giuliani F.: Electric arc furnace slags in cement-treated materials for road construction: mechanical and durability properties. Constr. Build. Mater. **113**, 280–289 (Elsevier) (2016)
6. Pasetto, M., Baldo, N.: Experimental evaluation of high-performance base course and road base asphalt concrete with electric arc furnace steel slags. J. Hazard. Mater. **181**, pp. 938–948 (Elsevier) (2010)
7. Ahmed Ebrahim, A.E.M.B.: Evaluation of steel slag and crushed limestone mixtures as subbase material in flexible pavement. Ain Shams Eng. J. **4**, 43–53 (2013)
8. Byers, C.C., Saleh, M.F., Pidwerbesky, B.D.: Evaluation of melter slag as a base course material. Int. J. Pavement Eng. **5**(4), 193–199 (2004) (Taylor & Francis)
9. Elsayed, A.S., Lindly, J.K.: Estimating permeability of untreated roadway bases. Transp. Res. Rec. (1519), 11–18 (1996)
10. Richardson D.N.: Drainability characteristics of granular pavement base material. J. Transp. Eng. ASCE **123**(5), 385–392 (1997)
11. Ministry of Road Transport and Highways (MORTH): Specifications for Road and Bridge Works, 5th Revision. Published by Indian Roads Congress, New Delhi, India (2013)
12. Zabielska, K.A., Sulewska, M.J.: Dynamic CBR as a method of embankment compaction assessment. In: Proceedings of the 18th International Conference on Soil Mechanics and Geotechnical Engineering, pp. 1–4, Paris (2013)
13. Zabielska, K.A., Sulewska, M.J.: Dynamic CBR test to assess the soil compaction. J. Test. Eval. **43**(5), 1028–1036 (2015)

Categorization of Drought During Twentieth Century Using Precipitation in Banaskantha District, Gujarat, India

Pranav B. Mistry and T. M. V. Suryanarayana

Abstract Drought is a natural hazard due to adverse effects in climate change in earth's environment. Drought assessment is very important to manage water resources in lean period. In the present study, drought years and degree of deficit of annual rainfall are determined by use of Standardized Precipitation Index (SPI-3 to 12 Months), and Percentage Departure from mean (PD) methods. For this study, monthly rainfall data of 102 years (1901–2002), of Banaskantha District, were used. The months of January, February, March, April, May, November, and December have been identified 68, 73, 78, 71, 68, 81, and 79 times as drought months, respectively, in the twentieth century, indicating that these months must be provided with assured Irrigation. From the annual rainfall departure analysis, the drought years have been identified and it is observed that 1901, 1904, 1911,1915,1923,1939,1969,1987, and 2002 are affected by severe drought and 1974 is affected by extreme drought condition. The study also reveals that 6, 15, and 15% of extreme dry years, severe dry years, and moderate dry years occur among drought years considered, which means 36% years are categorized into moderate to extreme drought years out of the total drought years.

Keywords Drought · Standardized precipitation index · Percentage departure from mean · Drought frequency

1 Introduction

The flood and droughts are the main two aspects of hydrological hazard. Among this drought is a potential threat with a destructive damage to agricultural as well as natural production. Drought in India occurs due to irregular monsoon rainfall in space

P. B. Mistry · T. M. V. Suryanarayana (✉)
Faculty of Technology and Engineering, Water Resources Engineering and Management Institute (WREMI), The Maharaja Sayajirao University of Baroda, Samiala 391410, India
e-mail: drsurya-wremi@msubaroda.ac.in

D. Deb et al. (eds.), *Innovations in Infrastructure*, Advances in Intelligent Systems and Computing 757, https://doi.org/10.1007/978-981-13-1966-2_23

and time. For the assessment of drought severity, various indices are used. Global warming will have an adverse effect on climatic conditions and water resources. So it is necessary to determine level of severity and trends to help in effective decision-making process in reducing the impact of drought. Patel et al. [4] calculated SPI for shorter and longer duration in Surat district and determined the dry and wet periods during twentieth century. Das et al. [1] showed the severity of drought in Kutch district using LANDSAT ETM+ and SPI method for drought assessment. Malakiya and Suryanarayana [2] developed SPI and RDI for assessment of the drought for Amreli district, wherein they concluded that it is better to use SPI and RDI of 12 months' index for analysis of drought for this study.

In rain-fed agriculture, rainfall has a crucial role to play for suitable crop planning. Twenty-eight years (1983–2010) daily rainfall data has been analyzed by Ray et al. [5] found various drought scenario based on weekly, monthly, seasonally and yearly at Barapani station in Ri-Bhoi district. They have also concluded that maximum drought frequency as 11 times in 28th week and as 14 times in December month. Shah et al. [6] showed in their results that SPI method showing as 50% variation of normal to wet and normal to dry conditions, respectively.

2 Study Area and Data Collection

Banaskantha is one among the 33 districts of the Gujarat state of India. The district is situated between $23^{\circ}33'$ and $24^{\circ}45'$ north latitude and 72°15' and 73°87' east longitude which shown in Fig. 1.

The annual precipitation variation between 214 and 1801 mm and annual average precipitation is 863.01 mm. For the analysis of drought monitoring, precipitation data has been collected for 102 years (i.e., 1901–2002) from Indian Meteorological Department (IMD), Pune.

3 Methodology

3.1 Standardized Precipitation Index (SPI)

The SPI method is of meteorological drought assessment and is formulated by McKee et al. in [3]. The SPI index is calculated for various shorter (3, 6, 9 months) and Longer (12, 24, 48 months). For calculation of SPI, only rainfall data is required and calculated using Drin C Software. Shahabfar and Eitzinger computed SPI with different time steps (e.g., one month, three months, and 24 months).

The Standardized Precipitation Index is calculated using the below formula and the range of different drought condition is also given in Table 1.

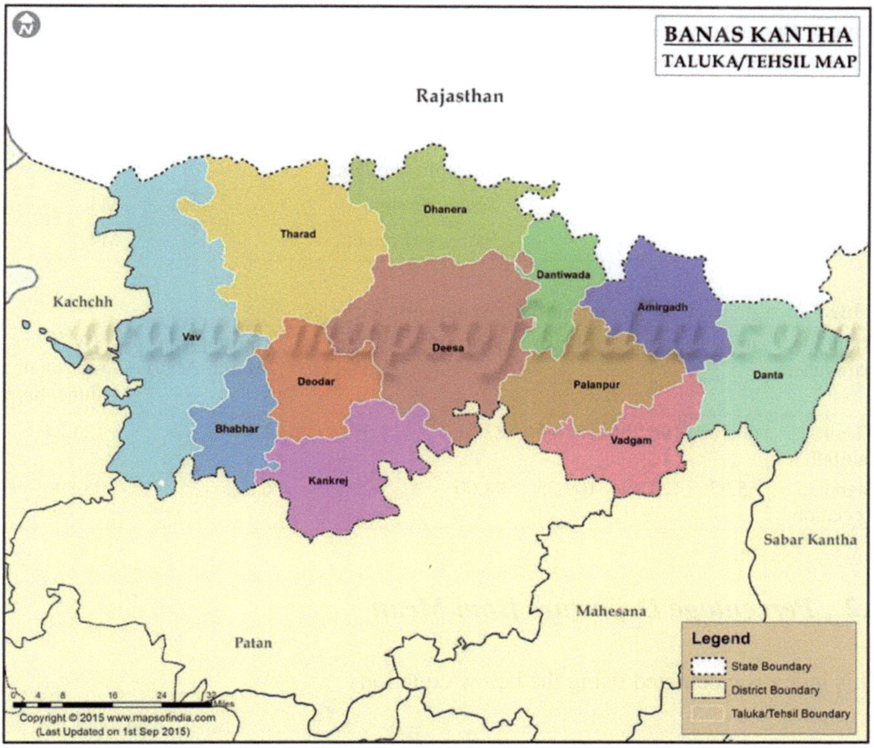

Fig. 1 Taluka map of Banaskantha

Table 1 Drought classification using SPI (McKee et al. [3])

SPI values	Drought condition
2.0 +	Extreme wet
1.5–1.99	Very wet
1.0–1.49	Moderate wet
−0.99 to +0.99	Near normal
−1.0 to −1.49	Moderate dry
−1.5 to −1.99	Severe dry
−2.0 to less	Extreme dry

$$\text{SPI} = \frac{x_i - x}{\sigma}$$

where

x the mean annual rainfall,

x_i the annual rainfall at any year and

σ the standard variation.

Table 2 Drought classification on percentage departure from normal (IMD 1971)

Percentage departure from normal	Intensity of drought	Code
>0.0	No drought	M0
0 to −25	Mild drought	M1
−25 to −50	Moderate drought	M2
−50 to −75	Severe drought	M3
−75 or less	Extreme drought	M4

Table 3 Seasonal rainfall in mm for Banaskantha district

Rainfall parameters	June	July	August	Sept	Winter (Jan–Feb)	Pre-monsoon (March–May)	SW monsoon (June–Sept)
Mean rainfall	60.70	196.75	152.91	82.61	1.51	2.90	123.24
Standard deviation	55.73	124.49	109.35	94.00	3.23	6.89	95.89

3.2 *Percentage Departure from Mean*

This index is calculated using the below equation:

$$P_d = \left[\frac{P_i - \overline{P}}{\overline{P}}\right] \times 100$$

where P_d is the percentage of departure from mean; P_i is monthly rainfall at any time, and P is the mean rainfall. Also, various codes assigned to each drought condition based on its value are given in Table 2.

4 Results and Discussion

The statistical parameters for the rainfall data for Banaskantha for 102 years were determined and are given in Table 3, which indicates that there are high fluctuations in occurrence of rain ranging from maximum rainfall 1072.68 mm in the year of 1917 and minimum rainfall 115.74 mm in the year of 1974 in the study area.

The monthly analysis of rainfall for Banaskantha district is given in Table 4. The same is based on assumption that a month is with drought when the rainfall for the month is less than half of the average rainfall for this month. The percentage of drought in a month was calculated as (number of drought months for a particular month/total number of drought months) multiplied by 100, for example, for January, the percentage of drought month is equal to 9.379%. Also, the highest average rainfall

Table 4 Drought analysis based on monthly rainfall for Banaskantha

Year	Average rainfall	Half of average rainfall	No. of drought months	Percentage of drought months
Jan	1.554	0.777	68	9.379
Feb	1.473	0.737	73	10.069
Mar	2.528	1.264	78	10.759
Apr	0.641	0.321	71	9.793
May	5.532	2.766	68	9.379
Jun	60.703	30.351	43	5.931
Jul	196.74	98.373	23	3.172
Aug	152.90	76.455	25	3.448
Sep	82.608	41.304	51	7.034
Oct	8.800	4.400	65	8.966
Nov	5.088	2.544	81	11.172
Dec	1.113	0.556	79	10.897

was observed as 196.75 mm in the month of July and lowest average rainfall was observed in the month of April as 0.641 mm in the span of 102 years. The months of February, March, April, November, and December have 73, 78, 71, 81, and 79 times drought months, respectively, indicating that these months must be provided with assured irrigation. The months of July and August had minimum number of drought months, i.e., 23 and 25, respectively.

4.1 Results of PD

The percentage of annual rainfall departures in Banaskantha district is computed. Also, from the annual rainfall departure analysis, the drought years have been identified and it is observed that 1901, 1904, 1911,1915,1923,1939,1969,1987, and 2002 years are affected by severe drought and 1974 year affected by extreme drought condition. The precipitation data of Banaskantha which is collected from IMD, Pune shows maximum Precipitation as 1073 mm in 1917 year and minimum precipitation as 116 mm in 1974 year. The percentage departure from normal is given in depicted in Fig. 2.

4.2 Results of SPI

The study of Standardized Precipitation Index produced the drought severity at 3, 4, 5, 6, 7, 8, 9, 10, 11, and 12 months' time steps, in the Banaskantha district of North

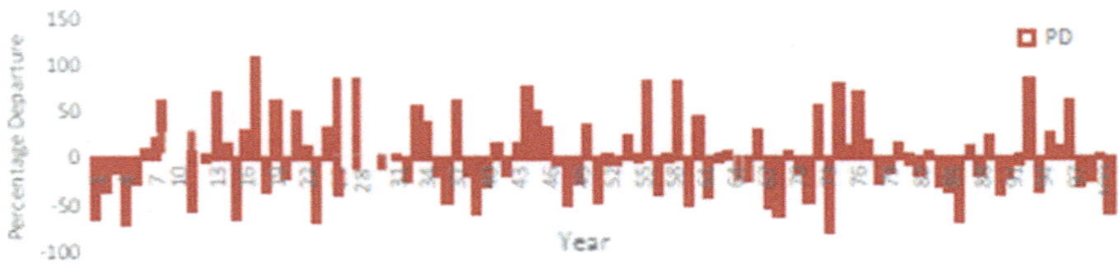

Fig. 2 Percentage of departure from mean in Banaskantha in twentieth century

Table 5 Total number of years affected by drought in diverse conditions for altered SPI time scale

SPI time scale/conditions	SPI 3	SPI 4	SPI 5	SPI 6	SPI 7	SPI 8	SPI 9	SPI 10	SPI 11	SPI 12
Extreme dry	0	0	2	3	4	3	2	4	2	3
Severe dry	9	8	4	4	3	4	7	4	10	7
Moderate dry	12	12	16	11	10	8	5	8	5	7
Near normal	63	62	60	63	65	69	67	69	74	66
Moderate wet	10	11	11	12	11	13	15	11	3	11
Very wet	8	7	6	9	9	4	6	4	6	7
Extreme wet	0	2	3	0	0	1	0	2	2	1
Total years	102	102	102	102	102	102	102	102	102	102

Gujarat, India. The total number of years affected by drought in different conditions for various altered SPI timescales are given in Table 5 and also shown in Fig. 3.

From Fig. 4, it is shown that only 1% year is obtained as the extreme wet conditions in the last 102 years because the average annual rainfall of Banaskantha is only 520 mm. Also from Fig. 5, it reveals that out of the drought years, 36% years are categorized into moderate to extreme drought years, which includes 6% extreme dry, 15% severe dry, and 15% moderate dry years.

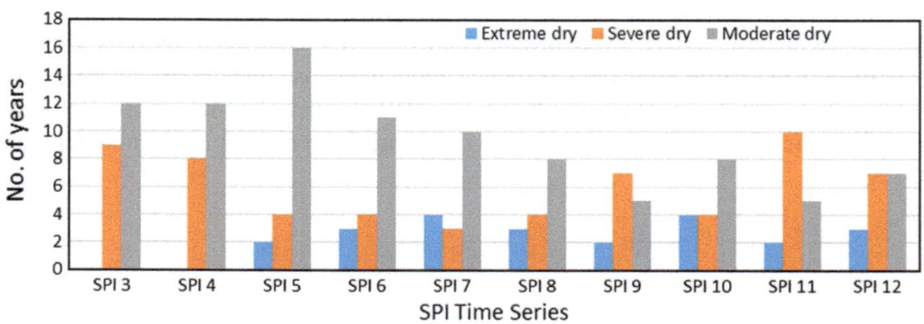

Fig. 3 Drought conditions in various SPI timescales

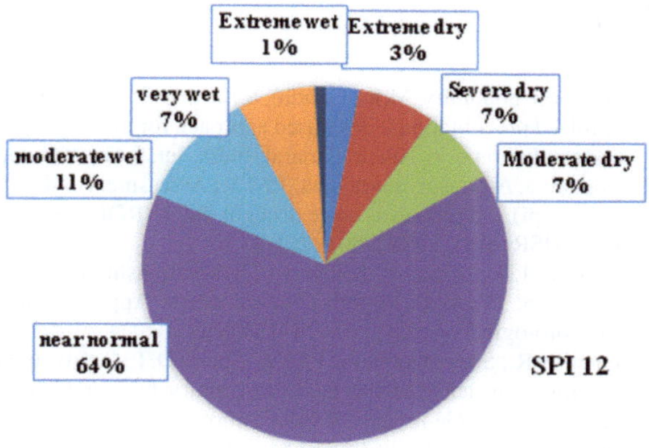

Fig. 4 Distribution of wet and dry periods (SPI 12 months)

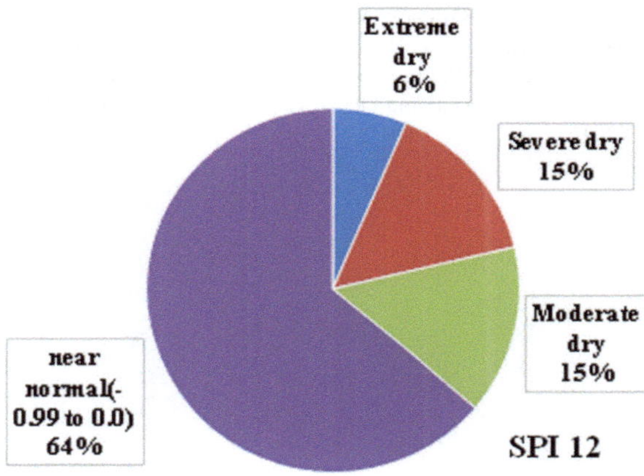

Fig. 5 Frequency of drought condition

5 Conclusions

The highest annual average rainfall was observed in Banaskantha as 196.75 mm in July and lowest in the month of April as 0.641 mm in the span of 102 years. In the present study, drought months are February, March, April, November, and December with total number of drought months with 73, 78, 71, 81, and 79, respectively, which indicatesthat these months must be provided with assured Irrigation. Similarly, months of July and August had minimum number of drought months, i.e., 23 and 25, respectively.

From the results, it is concluded that 1901, 1904, 1911, 1915, 1923, 1939, 1969, 1987, and 2002 are affected by severe drought and 1974 is affected by extreme drought condition. The results of SPI 12 show it is shown that only 1% year considered as extreme wet conditions and 6, 15 and 15% of extreme dry years, severe dry years, and moderate dry years for 102 years which means 36% years are categorized into moderate to extreme drought years out of the total drought years.

References

1. Das, S., Choudhury, M.R., Gandhi, S., Joshi, V.: Application of earth observation data and standardized precipitation index based approach for meteorological drought monitoring, assessment and prediction over Kutch, Gujarat. India. Int. J. Environ. Geoinformatics **3**(2), 24–37 (2016)
2. Malakiya, A.D., Suryanarayana, T.M.V.: Assessment of drought using standardized precipitation index (spi) and reconnaissance drought index (RDI): a case study of Amreli District. Int. J. Sci. Res. (IJSR) **5**(8), 1995–2002 (2016)
3. McKee, T.B., Nolan, J., Kleist, J.: The relationship of drought frequency and duration to time scales. In: Preprints, Eighth Conference on Applied Climatology, Anaheim, CA, American Meteorological Society, pp. 179–184 (1993)
4. Patel, N.R., Suryanarayana, T.M.V., Shete, D.T.: Analyzing extreme events using standardized precipitation index during the 20th century for Surat District, India. In: Majumder, M. (ed.) Application of Geographical Information Systems and Soft Computation Techniques in Water and Water Based Renewable Energy Problems, pp. 41–50. Springer Verlag, Singapore (2017)
5. Ray, L.I.P., Bora, P.K., Ram, V., Singh, A.K., Singh, R., Feroze, S.M.: Meteorological drought assessment in Barapani, Meghalaya. J. Indian Water Resour. Soc. **32**(1–2), 56–61 (2012)
6. Shah, R, Bharadiya, N., Manekar, V.: Drought index computation using standardized precipitation index (SPI) method for Surat District, Gujarat. In: International Conference on Water Resources, Coastal and Ocean Engineering (ICWRCOE 2015). Aquat. Procedia **4,** 1243–1249 (2015)

Effect of Moment Capacity Ratio on RC Framed Structure

Ram Arjun Sargar and Jyoti Pushan Bhusari

Abstract In RC framed structure, the beam column joint design is crucial. The behaviour of beam column joint is governed by the moment capacity ratio which is generally greater than one. There is a large variation in this value among various codes. The IS: 13920 clearly mentions that the moment capacity ratio at joint to be taken 1.4. Pushover analysis is performed on a frame whose moment capacity ratio (MCR) is increased increased with an increase in column dimensions as well as reinforcements to ensure strong column–weak beam. Its effect on the lateral displacement, base shear, storey drift, ductility and formation of hinges are studied and optimum value of moment capacity ratio is calculated. It is concluded that for better ductility and attaining plastic hinges at end of beams rather than in column, the MCR should not be less than 1.4. To achieve this, increasing the reinforcement in column proves to be more effective rather than increasing column dimensions.

Keywords Beam column joint · Ductility · Moment capacity ratio · Strong column–weak beam

1 Introduction

1.1 General

A beam column joint is a very crucial zone in RC framed structure. Where the elements of the frame interconnect. The transfer of forces and moments between the elements like beams and column take place effectively through the joint of the frame thereby ensuring continuity of a structure [9]. The beam column joint is vulnera-

R. A. Sargar (✉) · J. P. Bhusari
Sinhgad College of Engineering, Pune, Maharashtra, India
e-mail: ramsargar143@gmail.com

J. P. Bhusari
e-mail: jpbhusari.scoe@sinhgad.edu

D. Deb et al. (eds.), *Innovations in Infrastructure*, Advances in Intelligent Systems and Computing 757, https://doi.org/10.1007/978-981-13-1966-2_24

ble to failure since there is a sudden change in the geometry leading the development of complex state of stresses at the joint, thereby displaying critical behaviour. Research undertaken to study the joint behaviour has revealed the factors for the critical behaviour of joints, viz., cross-sectional details, reinforcement provided, type of concrete, and loading. Ductility of the structure is the prime important factor in case design of structure for earthquake forces. Increase in ductility of the structure increases the capacity of the structure to deform even in the plastic range without collapse, resulting in a reduction in effective earthquake forces.

1.2 Strong Column–Weak Beam Design (SCWB)

Murthy has emphasise the strong column–weak beam philosophy, wherein it is expected that the plastic hinge be formed near the end of the beam. This can be achieved by adopting ductile detailing for the structure ensuring complete energy dissipation. It is expected that before the collapse of structure occurs, the maximum number of plastics hinges should be formed [3].

The hierarchy of formation of plastic hinges is first on beams, then at the bottom of columns in the first storey. This can be achieved in design by maintaining proper ratio of the moment carrying capacities of beam and column meeting at the joint. The strength hierarchy in capacity based design procedure is first set at the local member level and later at the global level. This requires adjustment of column strength (M_c) to be greater than the strength of beams (M_b) at the junction. Thereby $M_c \geq M_b$.

Adopting the SCWB design, the beam is designed to be the weakest. This results in flexure failure of beams, avoiding the failure of the column and the joint, thereby preventing global failure of the entire structure. In addition, even if the flexural failure of the beam takes place, the beams will be able to carry the gravity loads since its shear capacity will remain intact. If SCWB design is not adopted, it may lead to formation of soft storey as the columns will become weaker than beams.

2 Literature Review

Moment capacity ratio is the ratio of sum of the moment carrying capacities of all columns to the sum of moment carrying capacities of all beams meeting at joint. Mathematically it can be expressed as:

$$\text{Moment capacity ratio} = \frac{\Sigma M_c}{\Sigma M_b} \tag{1}$$

where ΣM_c and ΣM_b are sum of the moment carrying capacities of the column and beam, respectively, connected at the joint in a particular direction as given in Fig. 1.

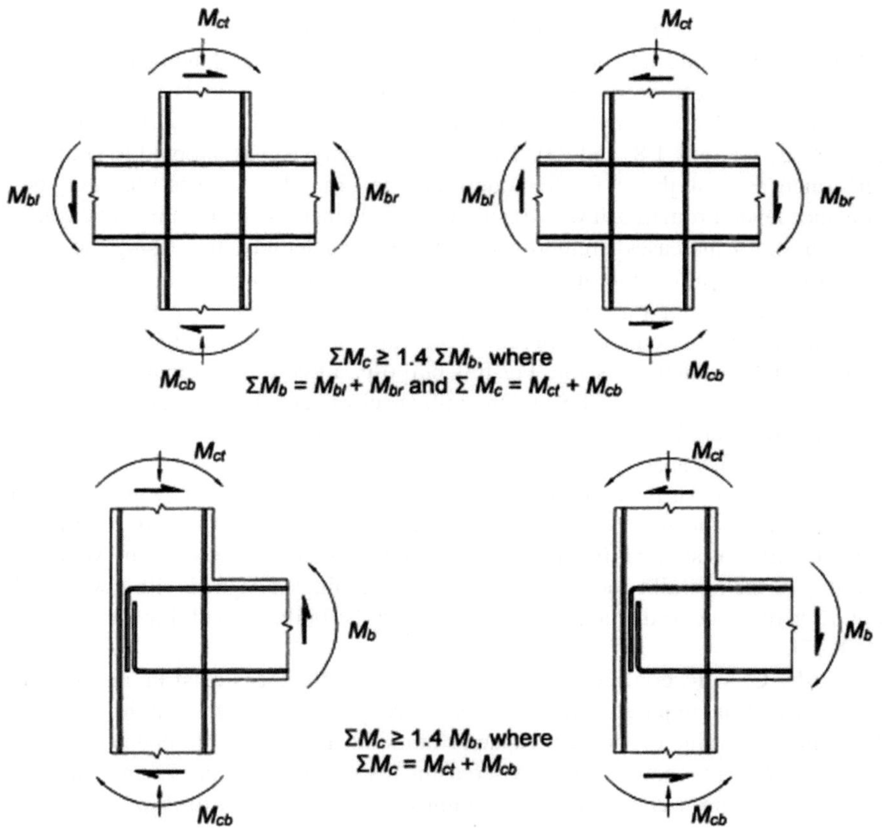

Fig. 1 Moment capacity ratio [5]

Table 1 Codal provisions on moment capacity ratio

Code	ACI318M-02	NZS3101:1995	EN1998-1:2003	IS13920:2014 (Draft)	IS13920:2016
Moment capacity ratio	$\Sigma M_c \geq 1.2\ \Sigma M_b$	$\Sigma M_c \geq 1.4\ \Sigma \Phi_0 \Sigma M_b$	$\Sigma M_c \geq 1.3\ \Sigma M_b$	$\Sigma M_c \geq 1.7\ \Sigma M_b$	$\Sigma M_c \geq 1.4\ \Sigma M_b$

Literature shows considerable research done on effect of moment capacity ratio on RC framed structure. Review of different codes has been carried out and codal provisions related to moment capacity ratio is given in Table 1.

Φ_0 is the overstrength factor for beams equal to 1.47.

With reference to above codal provisions, research work was carried out by Uma and Jain [11]. For an RC moment resisting frame subjected to horizontal seismic loads at joints, the summation of moment capacity of columns should be greater than 1.1 times summation of moment capacity of beams; i.e. $\Sigma M_c \geq 1.1\ \Sigma M_b$.

Ketut Sudarsana et al. investigated the influence of moment capacity ratio on behaviour of ductile RC buildings. Two building categories having five and ten storeys, respectively, were selected and 14 interior frame models of these building were modelled and analysed. The moment capacity ratio was varied and taken as 1.0, 1.2, 1.4, 1.6, 1.8, and 2.0. The pushover analysis was adopted to study the performances of all the 14 frame models. All the 14 models showed life safety performance level. From this, it was concluded that the increase in the moment capacity ratio up to 1.4 increases strength and ductility considerably. Increasing the MCR further does not affect the ductility [10]. Shivakumara Swamy et al. also adopted pushover analysis for various stiffness ratios by changing the cross section of column rather of increase in column reinforcement. For all the cases, the cross section of the beam was kept the same. For this purpose, a two-dimensional RC concrete frame was modelled in ETABS and the influences of seismic zones and category of soil was considered. From this, it was concluded that structures with lesser MCR have low seismic capacity [8].

Pushover analysis by adopting greater values of MCR was carried out by Praveen Kumar Parasa. The effect of application of higher values of MCR on the ductility of the structure was studied. It was concluded that with increase in value of MCR, the ductility of the overall structure enhances. The buildings designed with lower MCR values were prone to damage substantially than the building with higher values of MCR [7].

Murthy C. V. R. et al. found that considerable differences exist in current code provisions on the required column-to-beam strength ratio. In addition, it is clarified that minimum value of moment capacity ratio 1.2–1.4 adopted in codes is insufficient. A simple procedure was proposed using results of linear elastic static analysis, to strong column–weak beam design philosophy, which lead to desired inelastic behaviour of moment frame buildings during strong seismic action. It is clear that the design moment capacity of column should be 2.4–2.8 times the design moment capacity of beam [12].

3 System Development

Present research work is carried out to study the performance of the structure for increasing moment capacity ratio by increasing column dimensions as well as longitudinal reinforcement of column to ensure strong column–weak beam using pushover analysis in SAP2000 v19.

3.1 Pushover Analysis

In the present work, pushover analysis with displacement controlled method is adopted for the analysis. In this method, the lateral forces on the structure are

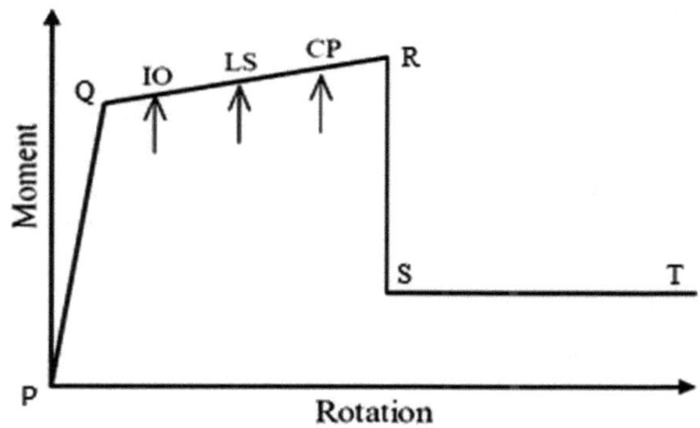

Fig. 2 Typical force–displacement curve with performance levels

increased till a target displacement defined by the designer is achieved. The total target displacement used is 4% of total height of building.

As shown in Fig. 2, performance level points are labelled as P, Q, R, S, and T. These points express the moment rotation performance of the hinge formed and further points IO, LS, and CP denoting immediate occupancy, life safety, and collapse prevention, respectively, define the acceptance criteria for the hinge.

3.2 Building Design and Modelling

The RC multi-storeyed frame used in present work has the following data given in Tables 2, 3, 4, and 5.

As per specified in Sect. 2, three models A, B, and C are finalised, the details of which are given in Table 5.

Model A corresponds to MCR value less than 1.4. Model B corresponds to increasing the MCR with increase in column dimensions and model C corresponds to increase in the MCR at beam-column joints with increase in the longitudinal reinforcement of column.

Table 2 General building details

Type of frame	SMRF
Seismic zone	V
Exposure conditions	Mild
Soil category	Medium
Damping factor	0.05
No. of storey	(G+6)
Storey height (m)	3
No. of bay	4
Bay width (m)	4

Table 3 Details of materials

Concrete	Grade = M25 Poisons Ratio = 0.3 Modulus of Elasticity = 25,000 MPa Density = 25 N/mm^2
Reinforcement	Grade = Fe415 Modulus of Elasticity = 200 GPa

Table 4 Design loads

Lateral loads (EQ)	As per IS 1893 i. Seismic Zone V ii. Importance factor-1.5 iii. Soil category-II iv. S_a/g-2.5 v. Factor for response reduction-5
Dead load	20 kN/m
Live load	10 kN/m

Table 5 Details of sections

Model	A	B	C
MCR	MCR < 1.4	MCR > 1.4	MCR > 1.4
Column	270 mm × 300 mm External	350 mm × 350 mm External	270 mm × 300 mm External
	330 mm × 350 mm Internal	380 mm × 380 mm Internal	330 mm × 350 mm Internal
Beam	Cross section of 300 mm × 300 mm for all the models A, B and C		

Two-dimensional analysis is performed by considering an intermediate plane frame of the building subjected to only unidirectional lateral loading. While developing the model, the following assumptions were made:

a. All column supports have fixed end condition at the bottom.
b. Effect of non-structural elements is not considered.
c. The location of plastic hinges assigned is at a distance 0.05L and 0.95L for both column as well as beam. Axial force and biaxial moment hinges (PM_2M_3 hinges) are provided at the ends of the columns. For the beams, only bending moment hinges (M_3 hinges) are provided.
d. The total target displacement used is 4% of total height of building.

4 Performance Analysis

4.1 Pushover Curve

Pushover curve (capacity curve) depicts the maximum lateral displacement of the structure as a function of the base shear force applied to the base of structure.

Figure 3 shows these curves for model A, B and C respectively and Table 6 shows the respective values of pushover curve.

The comparison of pushover curves obtained from pushover analysis for model A, model B and model C are shown in Table 6 and Fig. 3.

Figure 3 shows that the increase in moment capacity ratio caused a rise in the maximum lateral load carrying capacity.

Table 6 shows the displacement and base force corresponding to collapse condition for each model.

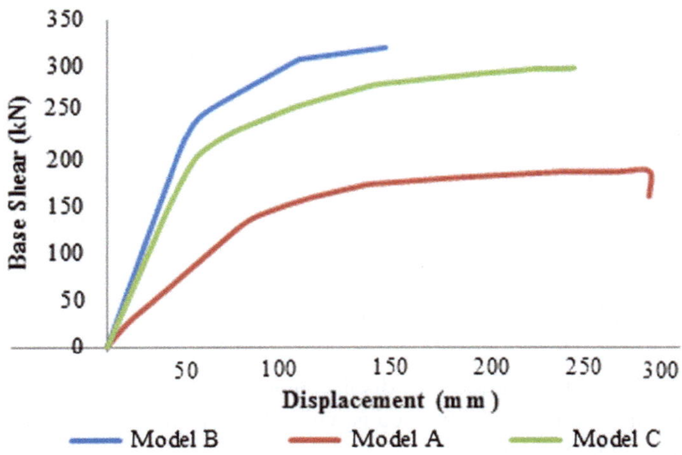

Fig. 3 Comparison of pushover curves

Table 6 Comparison of pushover curve parameter

	Model A	Model B	Model C
Base force (kN)	187.79	319.93	298.78
Displacement (mm)	282	146	245

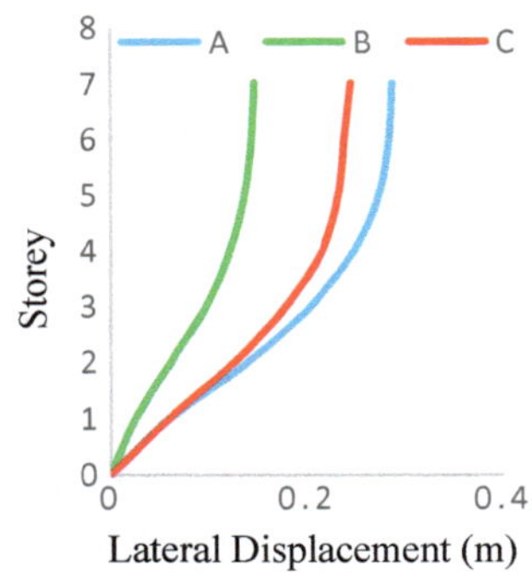

Fig. 4 Lateral displacement comparisons

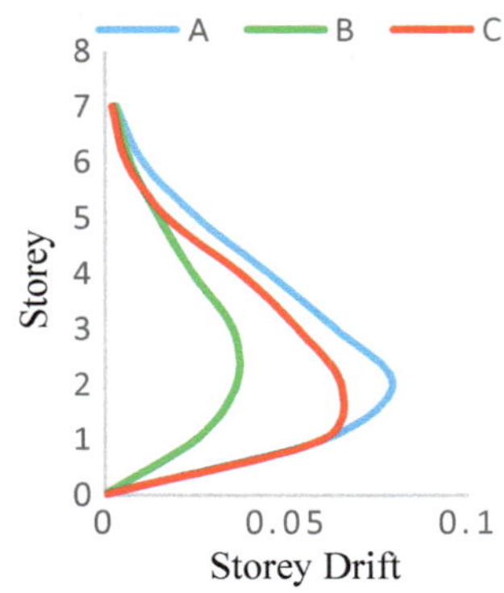

Fig. 5 Storey drifts comparison

4.2 Lateral Displacement and Storey Drift

Figures 4 and 5 show the lateral storey displacement and storey drift for all the models A, B, and C. Inter-storey drift obtained is within 0.4% of the total height as per the IS1893:2016 codal requirement.

4.3 Plastic Hinges Results

Figure 6 shows the plastic hinge formation, details of which have been shown in Table 7. Model A has 11 hinges going beyond collapse prevention range, whereas model B and model C have no hinge beyond collapse prevention range. So model B and model C are safe for collapse prevention performance level.

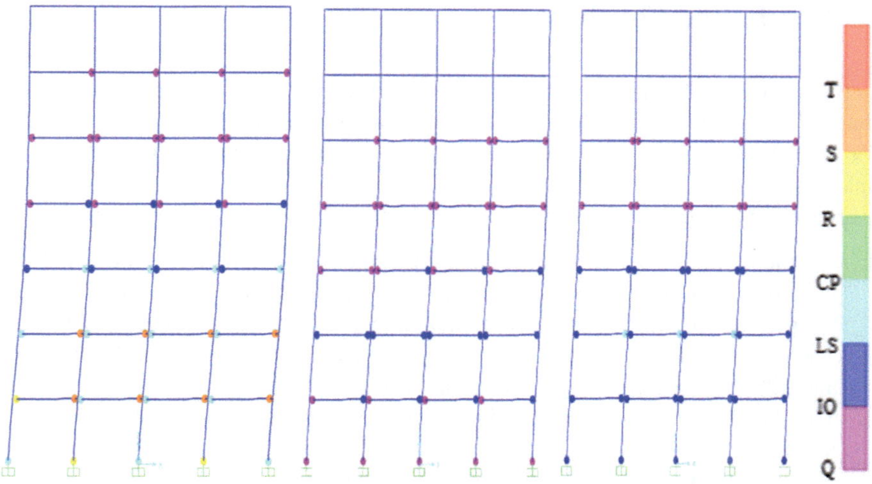

Fig. 6 Plastic hinge formation

Table 7 Plastic hinge mechanism

Model	P–Q	Q–IO	IO–LS	LS–CP	CP–R	R–S	S–T	>T	Total
A	77	16	8	14	0	3	8	0	126
B	84	27	15	0	0	0	0	0	126
C	84	13	26	3	0	0	0	0	126

4.4 Comparison of Various Parameters from Pushover Curve

Table 8 shows the comparison of stiffness, strength, and ductility of model A, B, and C respectively.

Results show that model B has higher strength and stiffness than model A and model C whereas model C has the highest ductility.

Table 8 Comparison of various parameters

Name	Stiffness (kN/m)	Strength (kN)	Yield displacement (mm)	Ultimate displacement (mm)	Ductility ratio
Model A	2086.55	187.79	88	282	3.09
Model B	5244.75	319.93	61	146	2.43
Model C	4979.66	298.78	60	245	4.08

5 Results and Conclusions

In the present study three models, viz., model A having moment capacity ratio less than 1.4 and model B and C having moment capacity ratio more than 1.4 are designed for Zone V. After design, the performance of all models is checked using pushover analysis.

Based on the observations of this study the following conclusions are made:

1. For the model A, lateral displacement of frame is found to be 282 mm. With the increase the column sizes in model B, the displacement reduces to 146 mm. With the increase in reinforcement in model C, the displacement reduces to 245 mm. The corresponding lateral load carrying capacity for model B and C are 319.9 and 298.7 kN which is higher than that of model A (187.8 kN).
2. The ductility of the frame is reduced from 3.09 to 2.43 when the column size increases, whereas it increases to 4.08 with the increase the reinforcement.
3. Model B and C, both have moment capacity ratio more than 1.4 and both models show better performance than Model A having moment capacity ratio less than 1.4.
4. Though Model B shows substantial increase in strength and stiffness by increase in column sizes, the ductility of structure decreases substantially. Model C proves more effective to ensure ductility by an increase in the column reinforcement without any compromise in the strength and stiffness.

Thereby, it is recommended that MCR > 1.4 should be maintained and to achieve this, increasing the reinforcement in column proves to be more effective rather than increasing column dimensions.

References

1. ACI 318-02: Building code requirements for structural concrete (ACI 318 M-02) and commentary. American Concrete Institute (2002)
2. EN 1998-1-3: Design provisions for earthquake resistant structures-part 1: general rules, seismic actions and rules for building. Brussels (2003)
3. Goswami, R., Murthy, C.V.R., Vijayanarayanan, A.R., Mehta, V.V.: Some Concepts in Earthquake Behavior of Buildings. Gujarat State Disaster Management Authority, Government of Gujarat (2012)
4. IS 13920 Draft: Indian Standard code of practice for ductile detailing of concrete structures subjected to seismic forces. BIS, New Delhi (2014)
5. IS: 13920: Indian Standard code of practice for ductile detailing of concrete structures subjected to seismic forces. BIS, New Delhi (2016)
6. NZS 3101: Part 1: Concrete structures standard, part 1: the design of concrete structures. New Zealand Standard, New Zealand (1995)
7. Parasa P.K.: In a Rectangular RC framed building moment capacity ratio at beam column joints. Int. J. Adv. Res. Sci. Eng. (2015)
8. Shivakumara Swamy, B., Prasad, S.K., Sunil, N.: Influence of strong column and weak beam concept, soil type and seismic zone on seismic performance of RC frames from pushover analysis. Int. J. Res. Eng. Technol. (2015)

9. Subramanian, N., Prakash Rao, D.S.: Design of joints in RC structures with particular reference to seismic conditions. The Indian Concr. J. (2003)
10. Sudarsana, I.K., Budiwati, A.M., Aditya P.W.: Effect of column to beam strength ratio on performance of reinforced concrete frames. ICETIA (2014)
11. Uma, S.R., Jain, S.K.: Seismic design of beam-column joints in RC moment resisting frames—review of codes. Struct. Eng. Mech. **23**(5) (2006)
12. Vijayanarayanan, A.R., Goswami, R., Murty, C.V.R.: Simple linear elastic static analysis procedure to attain desired collapse mechanism for moment resisting frames. In: 16th World Conference on Earthquake, Santiago Chile (2017)

Printed Circuit Board Modelling for Thermal Analysis of Electronics Packages

Jayesh Mahitkar, Mayank Chauhan, Hardik Gediya and Vivek K. Singh

Abstract Thermal design of electronics subsystems containing printed circuit boards (PCBs) for space applications is quite a challenging and time-consuming job due to lack of convective environment and no possibility of repair in case of failure once subsystem is launched in the mission. Thermal designer tries to reduce modelling and computational effort by modelling multilayer PCBs in equivalent lesser number of layers. This paper deals with case study of a typical electronics subsystems having one multilayered PCB thermal modelling. Two cases are considered for thermal analysis. In one case, all layers of PCB are modelled as they are and in second case multilayered PCB (10 layers) is modelled in two equivalent layers.

Keywords PCB · Thermal modelling · Electronics package · Thermal design

List of Symbols

$k_{p,e}$ Effective parallel thermal conductivity
$k_{n,e}$ Effective normal thermal conductivity

1 Introduction

Thermal subsystems of electronics subsystems used for nuclear, aerospace, defense and space industries, etc., are very critical and challenging task. Typically, an electronic subsystem used for space applications consists of single or multiple PCBs

J. Mahitkar
Department of Mechanical Engineering, PDPU, Gandhinagar 382421, Gujarat, India

M. Chauhan · H. Gediya · V. K. Singh (✉)
Space Applications Centre, Indian Space Research Organisation, Ahmedabad 380015, Gujarat, India
e-mail: singhvk2006@gmail.com

D. Deb et al. (eds.), *Innovations in Infrastructure*, Advances in Intelligent Systems and Computing 757, https://doi.org/10.1007/978-981-13-1966-2_25

kept inside the metallic box. Day-by-day requirement of miniaturization of electronics subsystem is increasing, which leads to increase in heat flux/power density [1]. Various components like capacitors, diodes, resistors, transistors, etc. are mounted over PCB and they need to be maintained below a particular temperature for safe operation. Moreover, life of these components strongly depends on their junction temperature. Failure of one or more electronics components may lead to failure of complete subsystem [2].

Electronics subsystems used for ground applications thermal management are achieved by means of a combination of conduction, convection, and radiation. In the recent past, few other technologies like spray cooling, jet cooling, etc. also have been tried for thermal management of electronics subsystem for ground applications [3].

Due to the absence of convective environment and difficulty in using nonconventional techniques like spray cooling, air cooling, etc., thermal management of spaceborne electronics subsystem purely depends on conductive and radiative mode of heat transfer.

PCB is typically a multilayer board made of copper layers and dielectric materials like FR4, polyimide, poly-tetra-fluoro-ethylene, etc. Copper layer thickness is optimized for desired signal transfer and minimization of losses. In PCBs, few thermal planes are also provided for enhancing in-plane heat transfer. Various layers of PCBs are also connected by printed Through Holes (PTH) filled with metals [4].

2 Literature Review

1.6 mm overall thickness PCB modelling with the uniform 35 μm single layer of copper thickness is studied by Lemczyk et al. [5]. They observed that property of a material to conduct the heat is predicted by harmonic mean $2k_{p,e}k_{n,e}/\left(k_{p,e}+k_{n,e}\right)$ is more accurate rather than using parallel and normal effective thermal conductivity (ETC).

The geometric mean of parallel and normal ETC $\left[\sqrt{k_{p,e}k_{n,e}}\right]$ is more closely observed with the experimental study carried out by Lohan et al. [3]. The sensitivity analysis is carried out for PCB ETC in order to investigate copper content in a single layer, amount of copper in internal layers and its nearness to the case or surfaces.

The dependency of parameters in order to investigate ETC for uniform layer modelling of PCB is studied by Culham and Yovanovich [6]. They observed that the size of power dissipating component, its placement location as well as constraints are the major influencing parameter for ETC of PCB.

To save computational time and human efforts typically PCBs for electronic subsystems, multilayer PCBs for electronic subsystem is modelled in two layers. This paper is an attempt to present case study of modelling a PCB in all layer as original and two layer. A typical electronic subsystem is considered for the study having several resistors, thermistor, etc. mounded on both sides of PCB in which dissipation of components ranges from 0.02 to 2.0 W. Thickness of PCB is 2.4 mm. Section 2 of the paper deals with thermal modelling, Sect. 3 deals with results and discussion followed by conclusions and future scope of work in Sect. 4.

3 Thermal Modelling

Heat transfer from PCB and electronics box is modelled as three-dimensional heat conduction and radiation problem. Commercial software NX 8.5 Space Systems Thermal is used for thermal analysis.

A typical PCB tray under study is shown in Fig. 1. For both cases (i.e. PCB modelled into actual ten layers and two layers), the mechanical tray is modelled with structured hexahedral grid and the same tray is used for both cases. When PCB is modelled in actual ten layers, thermal conductivity of copper and dielectric material is provided as per their physical presence. When PCB is modelled into two layers in-plane and z-plane effective thermal conductivity for the first five and last five layers out of ten actual layers is computed.

Choice of appropriate grid size is important for capturing desired thermal phenomenon accurately within a reasonable timeframe. A grid independent study is carried out but not mentioned in this paper.

When PCB is modelled in two layers approximately 20,000 mesh are found adequate for the analysis while for 10 layers number of element for different layers are varying approximately from 1100 to 11,000. Thermal coupling is used to model the effective heat transfer between all joining surfaces [7] (Fig. 2).

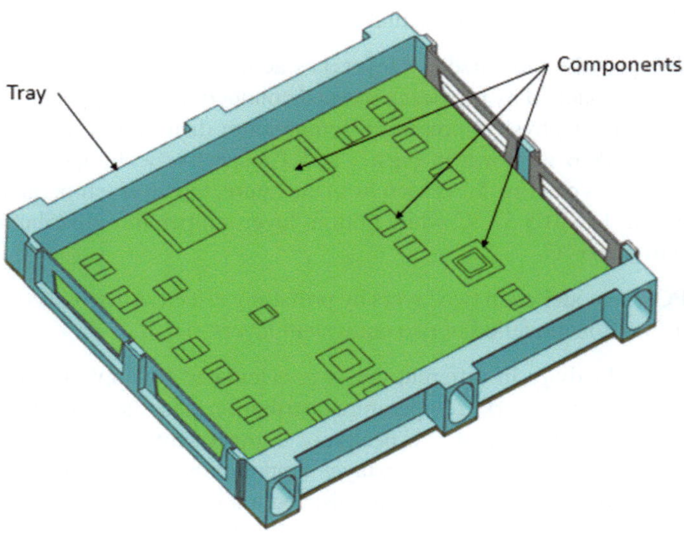

Fig. 1 Model of PCB

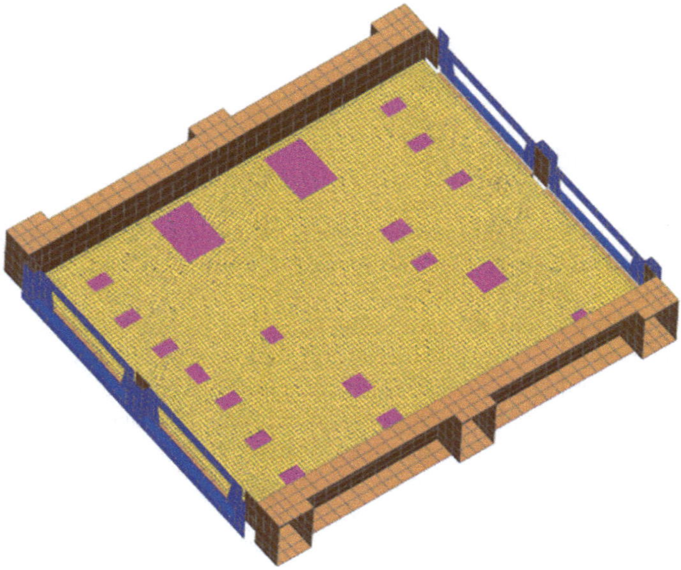

Fig. 2 PCB in two layers along with grid

4 Results and Discussions

A comparative study of PCB for both the cases is presented in this section.

Case (1): PCB is modelled in ten actual layers

All PCB layers are modelled (10 layers) as actual. Heat flow from one layer to another layer is modelled by providing thermal conductivity of copper and dielectric material as per their physical presence. Temperature distribution in top layer and bottom layer of PCB is shown in Figs. 3 and 4 respectively. Temperature of PCB components is shown in Fig. 5. As seen from the figure, top layer temperature in PCB varies from 80.71 to 105.57 °C while bottom layer temperature distribution varies from 77.19 to 97.65 °C.

Case (2): PCB is modelled in two layers by providing effective thermal conductivity. All layers of PCB is modelled in two equivalent layers.

Thermal conductivity in different zones is calculated by applying parallel and series heat resistance network. Temperature distribution in top layer and bottom layer of PCB is shown in Figs. 6 and 7 respectively. Temperature of PCB components is shown in Fig. 8. As shown in Fig. 6, top layer temperature in PCB varies from 74.7 to 105.1 °C while bottom layer temperature distribution is varies from 71.6 to 97.8 °C.

Difference in temperature of components and PCB at the same location is shown due to thermal resistance between component and PCB. In both layers of PCB, hot spot is seen but its value is well below the desired range of 110 °C. It is seen that there

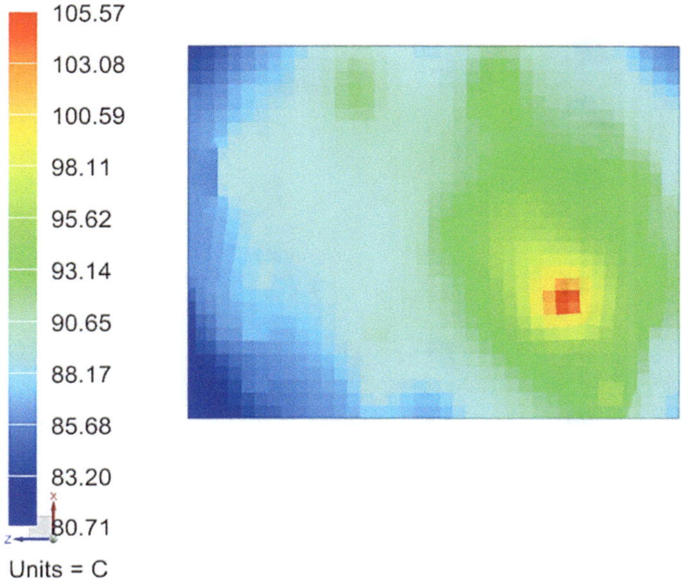

Fig. 3 Top layer temperature contour for case (1)

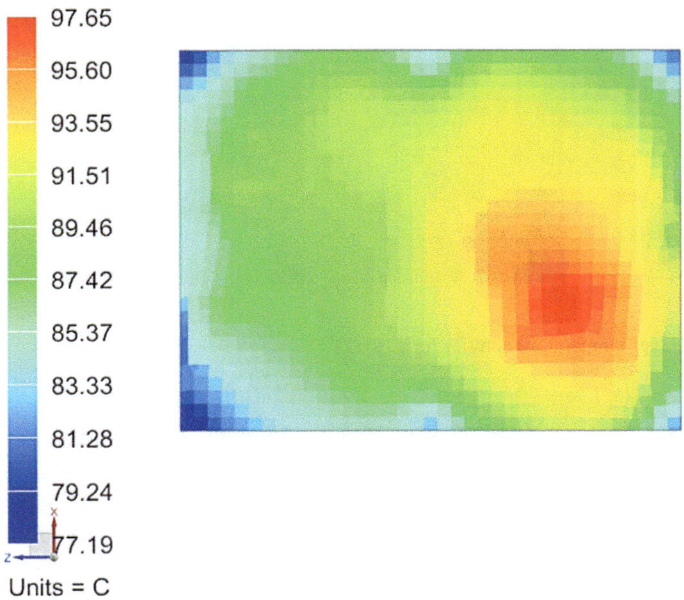

Fig. 4 Bottom layer temperature contour for case (1)

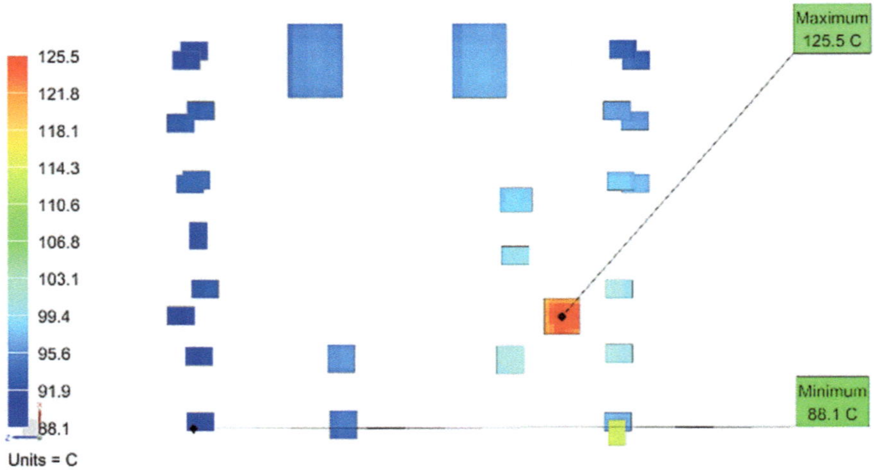

Fig. 5 Temperature contour of components for case (1)

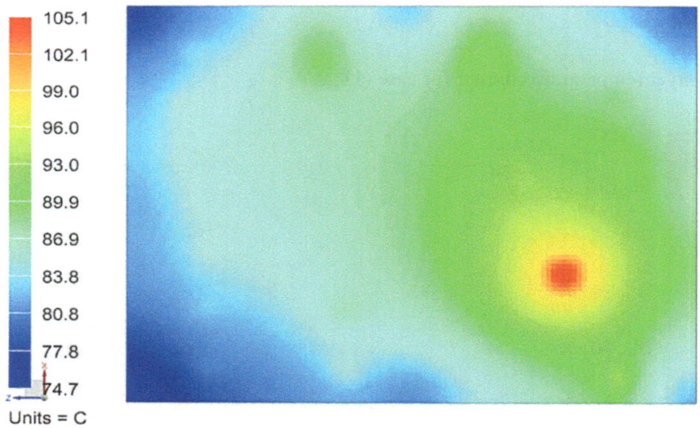

Fig. 6 Top layer temperature contour for case (2)

is small difference in temperature distribution when PCB is modelled in two layers and ten layers. Interest of PCB design is to ensure that temperature of all devices are below thermal acceptable limit.

Here for both case difference in temperature of devices is not very large, however modelling effort is significantly reduced. Hence, whenever there is almost uniform Copper distribution in different layers of PCB, it can be modelled in equivalent two layers without much compromise on temperature prediction of devices.

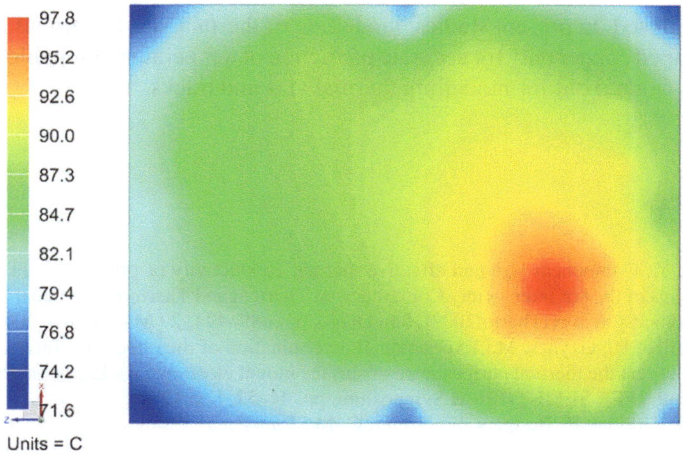

Fig. 7 Bottom layer temperature contours for case (2)

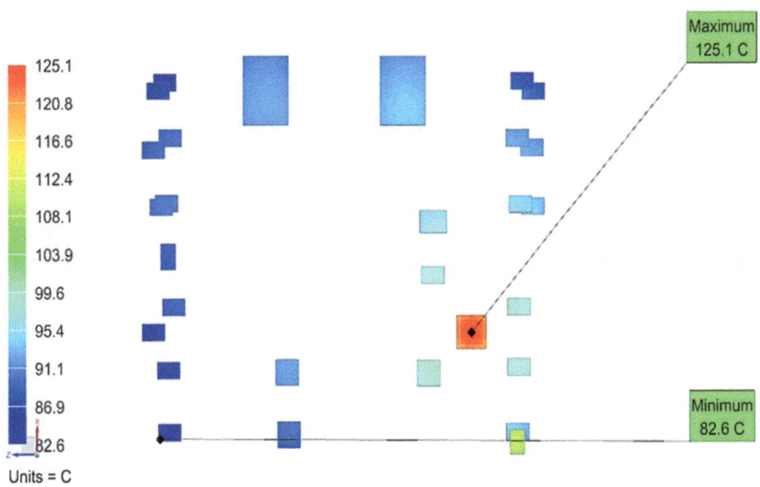

Fig. 8 Temperature contour of components for case (2)

5 Conclusions and Future Scope

PCB under consideration for present study has either significantly uniform distribution of copper in layers or very less amount of copper in layer in forms of tracks, thermal vias, etc. Hence, significant difference in temperature was not noticed when PCB was modelled in equivalent two layers. When PCB consists of evenly distributed layers modelling PCB in equivalent two layers may lead to significant inaccuracy in temperature of the same appropriate thermal modelling strategies need to be worked out and analysed. In present studies, modelling of thermal vias and tracks is considered as lumped copper only for accurate prediction of temperature a suitable approach need to be worked out for modelling thermal vias and tracks.

References

1. Shabany, Y.: Component size and effective thermal conductivity of printed circuit boards. In: Proceedings of the 8th Intersociety Conference on Thermal and Thermomechanical Phenomena in Electronic Systems (ITherm 2002), San Diego, pp. 489–494, 29 May–1 June 2002
2. Shaukatullah, H., Gaynes, M.: Experimental determination of the effect of printed circuit card conductivity on the thermal performance of surface mount electronic packages. In: Proceedings of the Tenth IEEE SEMI-THERM Symposium, pp. 44–52, Feb 1994
3. Lohan, J., Tiilikka, P., Rodgers, E'., Fager, C.-M., Rantala, J.: Using experimental analysis to evaluate the influence of printed circuit board construction on the thermal performance of four package types in both natural and forced convection. In: Proceedings of the Seventh Intersociety Conference on Thermal and Thermomechanical Phenomena in Electronic Systems (ITherm 2000), vol. 2, pp. 213–225, May 2000
4. Azar, K., Gralebner, J.E.: Experimental determination of thermal conductivity of printed wiring boards. In: Proceedings of the Twelfth IEEE SEMITHERM Symposium, pp. 169–182, March 1996
5. Lemczyk, T.F., Mack, B., Culham, J.R., Yovanovich, M.M.: PCB trace thermal analysis and effective conductivity. In: Proceedings of the Seventh IEEE SEMI-THERM Symposium, pp. 15–22 (1991)
6. Culham, J.R., Yovanovich, M.M.: Factors affecting the calculation of effective conductivity in printed circuit boards. In: Proceedings of the Sixth Intersociety Conference on Thermal and Thermomechanical Phenomena in Electronic Systems (ITherm'98), pp. 460–467 (1998)
7. Berggren, A.: Design of Thermal Control System for the Spacecraft MIST. KTH Royal Institute of Technology (2015)

Numerical Investigation of Cylindrical Heat Pipe Performance

Aishwarya Chaudhari, Mangesh Borkar, Arvind Deshpande, Mandar Tendolkar and Vivek K. Singh

Abstract The thermophysical properties of wick microstructures and heat interactions across the wick play a central role in heat pipe performance. It is therefore essential to analyse the wick microstructures. The performance of a cylindrical heat pipe with a screen mesh wick structure is analysed numerically using two-dimensional, axisymmetric, and transient model. A fully implicit, structured collocated finite-volume scheme is used in conjunction with the SIMPLE algorithm to solve Navier–Stokes equations for both, liquid and vapour phases. Present model assumes the wick to be saturated with the liquid and a porous medium formulation is employed for the wick. The evaporation and condensation rates at the diphasic interface are determined using kinetic theory. The model is tested for time step independency and a parametric study is performed by varying the permeability of wick and heat input. The results presented are in good agreement with the data from the previous literature.

Keywords Cylindrical heat pipe · Permeability · Evaporation · Condensation

Nomenclature

C_p	Specific heat
k	Thermal conductivity
K	Wick permeability
$\dot{m}''$	Interface mass flux
M	Molecular weight
N	Number of nodes

A. Chaudhari · M. Borkar (✉) · A. Deshpande · M. Tendolkar
Department of Mechanical Engineering,
Veermata Jijabai Technological Institute, Mumbai, India
e-mail: mdborkar_m16@me.vjti.ac.in

V. K. Singh
Thermal Division, Space Application Centre, ISRO, Ahmedabad, India

D. Deb et al. (eds.), *Innovations in Infrastructure*, Advances in Intelligent Systems and Computing 757, https://doi.org/10.1007/978-981-13-1966-2_26

T	Temperature
P	Pressure
$\vec{u}$	Velocity vector
Q	Heat input
T_{∞}	Coolant temperature
h_{∞}	Heat transfer coefficient
$\hat{P}$	Hydrodynamic pressure
L_E	Length of evaporator
L_C	Length of condenser
L_A	Length of adiabatic

Greek

σ	Accommodation coefficient
$\emptyset$	Porosity of the wick

Subscript

eff	Effective
int	Interface
op	Operating
0	Reference
v	Vapour
l	Liquid
r	Radial direction
x	Axial direction

1 Introduction

Heat pipes are considered to be the most effective passive cooling devices and find large applications in the fields of electronic cooling, thermal control system for spacecrafts, and nuclear reactors. Heat pipes use wick as a porous medium for returning the working fluid from the condenser to the evaporator section. The interstitial volumes of these porous structures are filled with the liquid due to which a solid–liquid—vapour contact line is formed between the solid particles. This gives rise to a large capillary force due to which the liquid can be pumped through the porous medium. At this diphasic interface, the phenomenon of thin-film evaporation takes place which is believed to be the effective mode of heat transfer. The interface shape formed in the region of solid particles governs evaporative heat transfer phenomenon.

With miniaturization of electronic devices and simultaneous increase in the dissipation of heat fluxes, it has become mandatory to improve the performance of the heat pipes to meet these changing needs. For the efficient working of the heat pipes, wick structure must be optimized. Increasing the capillary forces and decreasing the interface thermal resistance lead to increase in heat pipe efficiency.

Carbajal et al. [1–3] theoretically, numerically, and experimentally studied the transient heat transfer and fluid flow in flat heat pipes. In the 90s, many authors have examined the effects of permeability, particularly on sintered porous media [4] and screen wick [5, 6]. In order to illustrate the heat transport behaviour of sintered wick, thickness is altered by removing the layers of sintered copper particles in the experimental work reported by Hanlon et al. [7]. Singh et al. [8] investigated that the thermal performance of biporous wick is superior at higher heat loads than monoporous. Iverson et al. [9] analysed that the mass transfer due to vapourization of working fluid is an important aspect for heat dissipation from wick.

Cao and Faghri [10] investigated the performance of high-temperature cylindrical heat pipes. Two-dimensional, steady-state performance of flat and cylindrical heat pipes with different heating and cooling conditions is analysed by Aghvami et al. [11, 12]. Faghri and Buchko [13] analysed the performance of cylindrical heat pipes with multiple heat sources, both experimentally and numerically.

The objective of the present work is to predict the performance of cylindrical heat pipes. A two-dimensional, axisymmetric, transient model is used to solve the continuity, momentum and energy equations along with the mass transfer. First, the methodology is checked for time step independence. Then the effects of varying permeability on screen mesh wick are investigated. Also, the mass transfer balance at the diphasic interface is determined and analysed for different heat inputs in order to estimate the heat pipe performance.

2 Model Description

A schematic diagram of a two-dimensional axisymmetric cylindrical heat pipe with a screen mesh wick structure modelled in the present study is shown in Fig. 1. It is 370 mm wide and 7 mm long with 100 mm wide evaporator section and 180 mm condenser section. The wick and the wall materials are chosen to be copper and the working fluid is water. A vapour core thickness of 5.5 mm, a wick thickness of 0.5 mm and a solid wall thickness of 1 mm are selected. The permeability (K), porosity ($\emptyset$) and thermal conductivity (k) of screen mesh copper wick are 2.89×10^{-9} m^2, 0.7, 61.8 W/m K, respectively. Ergun's coefficient (C_{E}) is 0.24 [14]. The thermophysical properties of the heat pipe material and the working fluid are given in Table 1. Heat flux of 150 W (Q) is applied over the evaporator region. On the condenser side, the heat transfer coefficient and the coolant water temperature are 900 W/m^2 K and 294 K, respectively. Initially, heat pipe is at 294 K and vapour is assumed to be saturated. The commercial code ANSYS Fluent 17.2 is used for

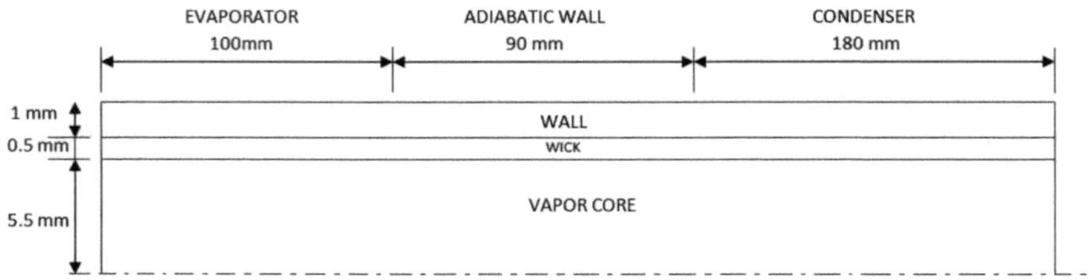

Fig. 1 Schematic model of 2-D axisymmetric cylindrical heat pipe

Table 1 Thermophysical properties of material and working fluid [14]

Domain	Properties	Values
Copper wall/wick	Thermal conductivity Specific heat Density Thermal conductivity of wick	387.6 W/m K 381 J/kg K 8978 kg/m^3 61.8 W/m K
Water vapour	Thermal conductivity Specific heat Density Viscosity	0.026 W/m K 2014 J/kg K 0.01 kg/m^3 8.49×10^{-6} N s/m^2
Water	Thermal conductivity Specific heat Density Viscosity	0.6 W/m K 4182 J/kg K 1000 kg/m^3 8×10^{-4} N s/m^2
Water/vapour	Latent heat of vapourization	2446.36 kJ/kg

the analysis with Volume of Fluid (VOF) model and the model is validated from numerical work [15].

3 Governing Equations

Since a two-dimensional axisymmetric model is used in the present work, governing equations in a cylindrical coordinate system are utilized. Fluid flows in wick and vapour domains are considered to be laminar and incompressible [16] as well as two-dimensional. Gravity effects are neglected and constant material properties are assumed except vapour and liquid density [14]. Both wick and vapour domains are assumed to be at equilibrium state throughout the process. Under these assumptions, the governing equations may be written on the similar line as reported in [15].

For the wick and the vapour core, the continuity equation is

$$\frac{\partial \rho}{\partial t} + \nabla.(\rho \vec{u}) = 0 \tag{1}$$

where ρ and t are density and time, respectively, and $\frac{\partial \rho}{\partial t}$ represents the mass addition or depletion in the wick and vapour domains.

The two-dimensional axisymmetric momentum equations in the wick and the vapour core are

$$\frac{\partial \rho \vec{u}}{\partial t} + \nabla.(\rho \vec{u}\ \vec{u}) = -\nabla.p + \mu \nabla^2 \vec{u} - \frac{\mu \emptyset}{K}\vec{u} - \frac{C_E \emptyset \rho |\overline{u}|}{K^{\frac{1}{2}}}\vec{u} \tag{2}$$

where K and C_E are the permeability and the Ergun's coefficient of the wick domain respectively. The last two terms on the right-hand side of Eq. 2 represent axial and radial component source terms. In the vapour domain, $K = \infty$ and $\emptyset = 1$.

The energy equation for wall, wick and vapour core is

$$\frac{\partial ((\rho C_p)_m T)}{\partial t} + \nabla.(([\rho C_p]_l \vec{u} T)) = k_{\text{eff}} \nabla^2 T \tag{3}$$

where k_{eff} is the effective conductivity of the porous medium. The term ρC_p assumes suitable values in their respective domains [14]. The work presents the effective conductivity value of 61.8 W/m K which is based on the thermal conductivities of water and copper [17].

3.1 Boundary Conditions

Diphasic interface affects the performance of heat pipe significantly. Thus, boundary conditions are the most critical part of the present model. The different boundary conditions imposed on the domains are as follows.

Outside Wall:

Evaporator Section: $-k\frac{\partial T}{\partial r} = q''\ \ 0 \le x \le L_E$
Adiabatic Section: $-k\frac{\partial T}{\partial r} = 0\ \ L_E \le x \le L_E + L_A$
Condenser Section: $-k\frac{\partial T}{\partial r} = h_\infty (T - T_\infty)\ \ L_E + L_A \le x \le L_E + L_A + L_C$

Wick-wall interface:

$$u_x = u_r = 0$$

where u_r and u_x are the radial and axial components of the velocity, respectively.

Wick-vapour interface:

The interface mass flux is given by kinetic theory as

$$\dot{m}'' = \left(\frac{2\sigma}{2-\sigma}\right)\frac{1}{(2\pi R)^{\frac{1}{2}}}\left(\frac{P_v}{(T_v)^{\frac{1}{2}}} - \frac{P_{\text{int}}}{(T_{\text{int}})^{\frac{1}{2}}}\right) \tag{4}$$

Here, $\dot{m}'' < 0$ denotes evaporation and $\dot{m}'' > 0$ denotes condensation. The flow is assumed to be normal to the interface while considering for momentum transport due to evaporation or condensation.

Lateral walls:

All the lateral walls are assumed to be adiabatic and thus the temperature gradients normal to the walls are zero. Also, assuming no-slip boundary condition, all the velocity components are zero.

$$\frac{\partial T}{\partial x} = u_x = u_r = 0 \tag{5}$$

Axisymmetric Line:

Because of the axisymmetric model considered in the present study, all the derivatives normal to the axisymmetric line are zero and the radial velocity is also zero. Thus, the boundary conditions for the axisymmetric line are written as

$$\frac{\partial T}{\partial r} = \frac{\partial u_x}{\partial r} = u_r = 0 \tag{6}$$

3.2 *Initial Conditions*

Wick and vapour domains are assumed to be at equilibrium and saturated state initially. The initial conditions are summarized as

$$T(x, r) = T_0 = T_\infty \tag{7}$$

$$P_v(x, r) = P_{\text{op}} = P_0 = P_{\text{sat}}(T = T_\infty) \tag{8}$$

$$\begin{aligned} u_r(x, r) &= u_x(x, r) = P_l(x, r) \\ &= \hat{P}(x, r) = \dot{m}(x) = \dot{m}''(x) = 0 \end{aligned} \tag{9}$$

4 Solution Procedure

The governing Eqs. (1)–(3) are discretized using the finite-volume method described in [15] with a fully implicit scheme. Pressure–velocity coupling is achieved using the SIMPLE algorithm. The resulting discretized equations are solved using line-by-line TDMA with a two-level algebraic multigrid procedure for acceleration. The transient terms are discretized as fully implicit leading to a stable solution. Central differencing scheme is used to discretize the diffusion terms while second-order upwind method is used for discretization of convective terms. The User-Defined Functions (UDF) are developed to model the mass transfer phenomenon, temperature and pressure at the diphasic interface as well as the densities of liquid and vapour.

5 Results and Discussion

5.1 Flow and Temperature Field

Figure 2 shows the temperature contour in an axisymmetric model of heat pipe with accommodation coefficient $(\sigma)=0.3$, porosity $(\varphi)=0.7$, permeability $(K_1)=2.89\times10^{-9}$ m^2 for heat input of 150 W. The working fluid evaporates at the liquid—vapour interface below the evaporator inside the heat pipe and the vapour condenses on the right side of diphasic interface. Figure 3 depicts the flow field in the vapour region.

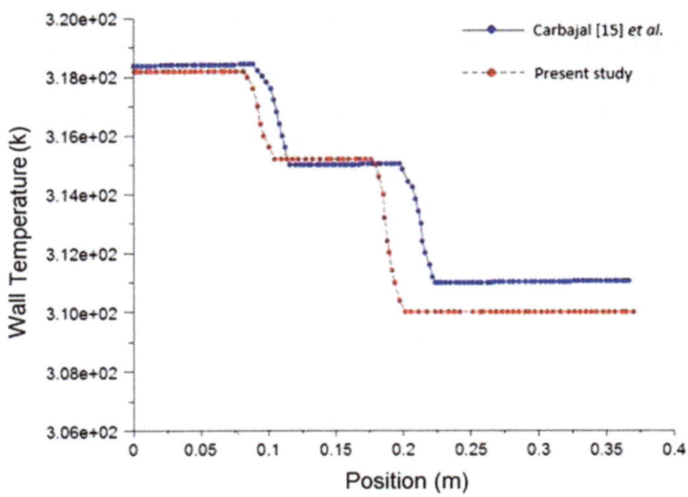

Fig. 2 Comparison of wall temperature distribution with Ref. [15]

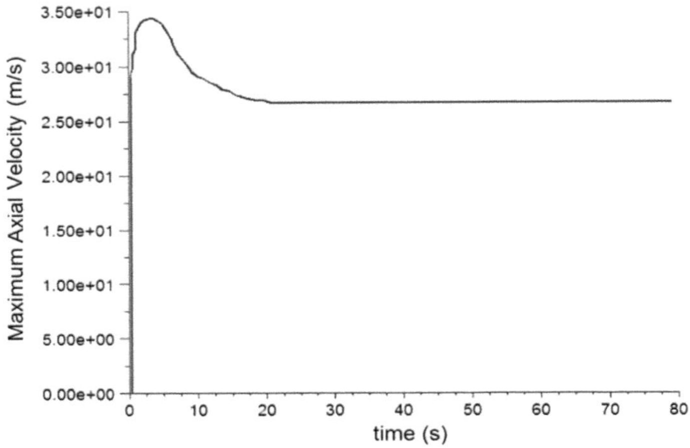

Fig. 3 Maximum axial velocity versus time

5.2 *Time Step and Grid Independence*

The proposed model is analysed for time step and grid independence. Figure 4a reflects three different time steps (0.1 s, 0.01 s, 0.001 s) compared for two different times 1 and 5 s. The solution starts with a time step of 0.001 s and increases 1.2 times each time step until it reaches the cap of 0.1 s. The results reveal that there are only 0.05 and 0.08% differences between time steps reported for wall temperatures with respect to 0.001 s after 1 and 5 s, respectively. Also for variable time step (0.001–0.1), less computational time is required than that for the smallest time step (0.001 s) and the results are presented for variable time step. Figure 4b depicts the wall temperature at time $t = 5$ s for coarse and very fine grid. For various grid sizes, wall temperature agrees well in such a way that a maximum difference of 0.06% is reported. The coarsest grid ($N_r = [20, 8, 8]$, $N_x = 74$) is chosen to perform the numerical simulations. Its results agree very well with the corresponding results from the finest grid ($N_r = [20, 8, 8]$, $N_x = 296$).

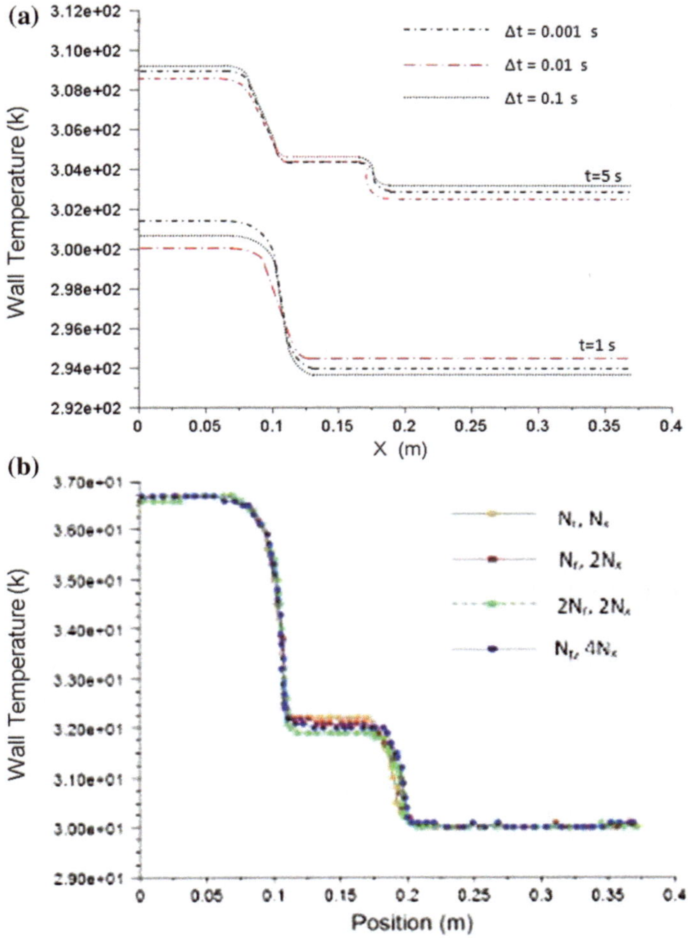

Fig. 4 **a** Wall temperature distribution for two cases with three different time steps.**b** Temperature distribution for various grid sizes in axial (N_x) and radial (N_r) direction

5.3 Operating Pressure

Operating pressure [14] is strongly influenced by the temperature in the vapour core. As shown in Fig. 5, operating pressure increases with increase in the temperature of vapour core as well as with time and reaches a steady value.

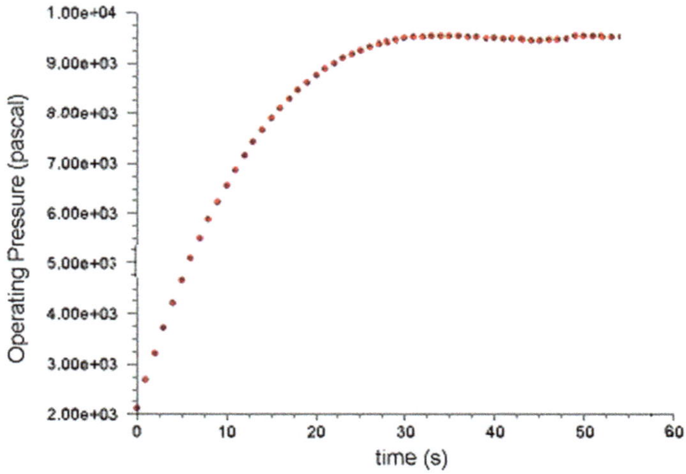

Fig. 5 Operating pressure variation with time with heat input of 150 W

5.4 *Effect of Permeability*

In order to study the effects of permeability, three cases of different permeabilities (viz., $K_1 = 2.89 \times 10^{-9}$, $K_2 = 9.011 \times 10^{-10}$, $K_3 = 3 \times 10^{-10}$) are considered for the wick. The analysis reveals that the liquid flow through the screen mesh increases with decrease in permeability. The transient pressure drops in wick for different cases of permeabilities are presented in Fig. 6. This variation plays a vital role in capillary

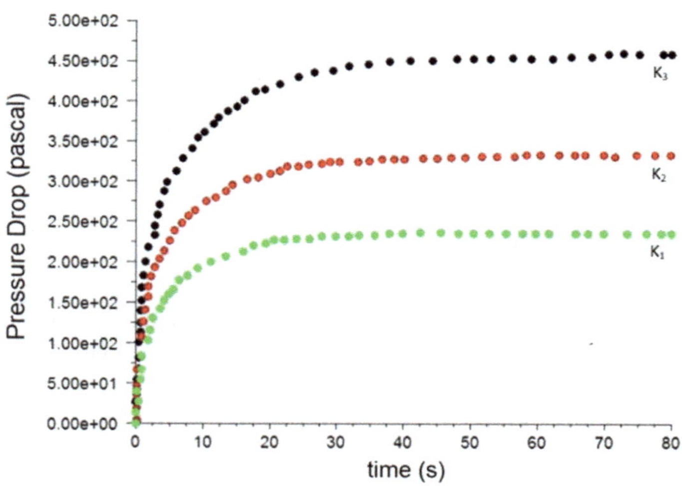

Fig. 6 Pressure drop variation for variable permeability versus time

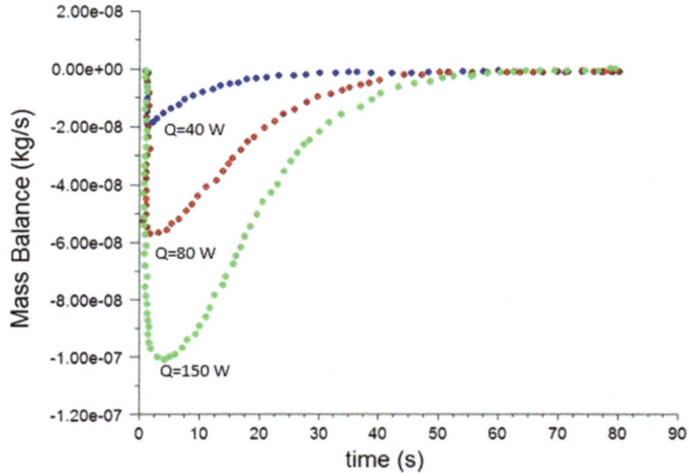

Fig. 7 Pressure drop variation for variable permeability versus time

limit calculations. Thus, the viscous properties of wick structure have to be specified accurately.

5.5 *Effect of Heat Input*

Figure 7 depicts the mass transfer balance at the liquid–vapour interface with respect to time for different heat inputs. It is observed that the mass transfer balance value first reduces sharply and then gradually increases to zero with time. This is attributed to the higher evaporation mass transfer rate than condensation mass transfer rate at the beginning and both reach the same values at the steady state. Also, this difference increases towards the high heat input and limit to zero with time. Also, the thermal resistance decreases (0.11645–0.08908 °C/W) with heat input.

6 Conclusions

In order to enhance the thermal performance of heat pipes, it is imperative to recognize and analyse the underlying phenomena that affect the performance. An axisymmetric, two-dimensional model is developed to analyse the performance of cylindrical heat pipe. The model accounts for coupling of continuity, energy and momentum equations in wick and vapour regions. A stable numerical approach is developed to solve the set of governing equations. This work also provides insights into the effects of variable permeability on pressure drop, optimization of time step and variation of mass transfer balance with different heat inputs. Thus, overall work is an attempt to investigate the performance of cylindrical heat pipe.

The basic methodology developed in this work can be extended to incorporate other types of operating conditions, microstructures and design model of device.

References

1. Carbajal, G., Sobhan, C.B., Peterson, G.P.: Numerical study of heat pipe heat spreaders with large periodic heat input. J. Thermophys. Heat Transf. **20**(4) (2006)
2. Carbajal, G., Sobhan, C.B., Peterson, G.P., Queheillalt, D.T., Wadley, H.N.G.: Thermal response of a flat heat pipe sandwich structure to a localized heat flux. Int. J. Heat Mass Transf. **49**(21–22), 4070–4081 (2006)
3. Carbajal, G., Sobhan, C.B., Peterson, G.P.B., Queheillalt, D.T., Wadley, H.N.G.: A quasi-3D analysis of the thermal performance of a flat heat pipe. Int. J. Heat Mass Transf. **50**(21–22), 4286–4296 (2007)
4. Guyon, E., Oger, L., Plona, T.J.: Transport properties in sintered porous media composed of two particle sizes. J. Phys. D. **20**(12), 1637–1644 (1987)
5. Noda, H., Yoshioka, K., Hamatake, H.: Experimental study on the permeability of screen wicks. JSME Int. J. **36**, 357–363 (1993)
6. Adkins, R., Moss, T.: Measuring flow properties of wicks for heat pipe solar receivers. In: Presented at 12th Annual International Solar Energy Conference, American Society of Mechanical Engineers, New York, pp. 103–108 (1990)
7. Hanlon, A., Ma, H.B.: Evaporation heat transfer in sintered porous media. J. Heat Transf. **125**, 644–652 (2003)
8. Singh, R., Akbarzadeh, A., Mochizuki, M.: Effect of Wick characteristics on the thermal performance of the miniature loop heat pipes. ASME J. Heat Transf. **131**, 1–10 (2009)
9. Iverson, D., Davis, T.W., Garimella, S.V., Kang, S.S.: Heat and mass transport in heat pipe Wick structures. J. Thermophys. Heat Transf. **21**(2), 392–404 (2007)
10. Cao, Y., Faghri, A.: Transient two-dimensional compressible analysis for high-temperature heat pipes with pulsed heat input. Numer. Heat Transf. Part A: Appl. **18**(4), 483–502 (1991)
11. Aghvami, M., Faghri, A.: Analysis of flat heat pipes with various heating and cooling configurations. Appl. Thermal Eng. **31**(14–15), 2645–2655 (2011)
12. Famouri, M., Carbajal, G., Li, C.: Transient analysis of heat transfer and fluid flow in a polymer-based Micro Flat Heat Pipe with hybrid Wicks. Int. J. Heat Mass Transf. **70**, 545–555 (2014)
13. Faghri, A., Buchko, M.: Experimental and numerical analysis of low-temperature heat pipes with multiple heat sources. J. Heat Transf. **113**(3), 728–734 (1991)
14. Vadakkan, U., Murthy, J.Y., Garimella, S.V.: Transient analysis of flat heat pipes. In: Proceedings of the ASME Summer Heat Transfer Conference, Las Vegas, Nevada (2003)
15. Carbajal, G., Li, C.: Transient analysis of heat transfer and fluid flow in a cylindrical heat pipe with hybrid wicks. Int. J. Heat Mass Transf. **70** (2016)
16. Ooijen, H.v., Hoogendoorn, C.: Vapor flow calculations in a flat-plate heat pipe. AIAA J. **17**(11), 1251–1259 (1979)
17. Rayleigh, L.: LVI. On the influence of obstacles arranged in rectangular order upon the properties of a medium. Lond. Edinb. Dublin Philos. Mag. J. Sci. **34**(211), 481–502 (1892)

Simulations of Exhaust Gas Recirculation and Its Impact on NO_x

Umang Makadia, Pulkit Kumar, Ajit Kumar Parwani and Dipankar Deb

Abstract Emission control are focused area in engine development. One of the important methods to reduce the nitrogen oxide (NO_x) is recirculating the exhaust gas into intake of engine. In the present work the numerical simulation is carried out using commercial software ANSYS with focus on the effect of NO_x emissions in the exhaust of CI (diesel) engine with exhaust gas recirculation (EGR). For the analysis purpose we assume CO_2 as exhaust gas, which is to be recirculated. This study is performed with and without EGR for the comparison. Simulation result shows that at the start of combustion the formation of nitrogen mono oxide (NO) are greater than nitrogen dioxide (NO_2) but later on since most of NO get oxidized to NO_2 its emissions get reduced. Results also show that with EGR both NO and NO_2 significantly reduces. However, there is slight reduction of power and slight increase in CO and CO_2 emissions are observed.

Keywords ANSYS · Numerical simulation · EGR · Emission · CI engine

1 Introduction

CI engines are gaining popularity because they use diesel as a fuel which is cheaper than gasoline for SI engine. CI Engine has higher thermal efficiency as well as it emits low carbon dioxide [1]. Advancement in technology is making diesel engine favorable for light-duty vehicle.

But the biggest disadvantage of diesel engine is the presence of NO_x emissions in the exhaust gas. NO_x is the main cause of acid rain, photochemical smog and

U. Makadia · P. Kumar (✉) · A. K. Parwani
Department of Mechanical Engineering, IITRAM, Ahmedabad, Gujarat, India
e-mail: pulkit.kumar.17pm@iitram.ac.in

U. Makadia
e-mail: Umang.makadia.16mm@iitram.ac.in

D. Deb
Department of Electrical Engineering, IITRAM, Ahemdabad, Gujarat, India

D. Deb et al. (eds.), *Innovations in Infrastructure*, Advances in Intelligent Systems and Computing 757, https://doi.org/10.1007/978-981-13-1966-2_27

environmental degradation [2] formation of NO_x takes place because of the reaction nitrogen and oxygen as shown in reaction given below [3]

$$O^* + N_2 = NO + N^*$$
$$N^* + O_2 = NO + O^*$$
$$N^* + OH = NO + H^*$$

The favorable conditions for the formation of NO_x are the presence of higher flame temperature and abundant amount of oxygen and nitrogen during combustion [4, 5].

One of the best methods to reduce the harmful emission of NO_x is Exhaust Gas Recirculation (EGR) where some portion of exhaust gas is recirculated back to the intake manifold of engine. The exhaust gas has very large amount of CO_2 so with an increase in EGR, CO_2 concentration at intake increases. The formation of CO_2 takes place by following reaction during combustion:

$$C + \tfrac{1}{2}O_2 = CO \quad \Delta H = -111\,\text{kJ}$$
$$CO + \tfrac{1}{2}O_2 = CO_2 \quad \Delta H = -394\,\text{kJ}$$

It will not only dilutes the oxygen and nitrogen concentration but also reduces the flame temperature during the combustion because CO_2 has high heat capacity [6].

The simultaneous reduction of NO_x and CO_2 from the exhaust of internal combustion engine and diesel generator set is proposed by Kumar et al. [7]. They used EGR for the reduction of NO_x and carbon capture and storage unit to capture the CO_2.

Selim et al. [8] performed an experiment on Ricardo E6 single cylinder, variable compression, indirect injection diesel engine running on a dual fuel of diesel, and compressed natural gas. They presented combustion noise and thermal efficiency data at different EGR ratios, engine speeds, loads, temperatures of EGR, and compression ratios. They concluded that 5% EGR ratio may be favorable in terms of improved thermal efficiency, reduced combustion noise, and reduced NO_x emission.

Torregrosa et al. [9] experimented on four strokes, four-cylinder diesel engine and concluded about EGR influence on fire deck temperatures and heat flux. He found that CO_2 formation increased due to EGR but NO_x formation decreased. These findings differ from few earlier reports in the literature which found no significant effect of EGR on those variables.

Praveen et al. [10] experimented on single cylinder four stroke water cooled diesel engine and studied the effect of multiwalled carbon nanotubes on EGR. They have mixed multiwalled carbon nanotubes in *Calophyllum inophyllum* biodiesel B20 blend and concluded that NO_x emissions are reduced by 25.6% for B20 with 20% EGR.

Recent advancements in computational fluid dynamics software make it easier to measure pollutants details at the time of combustion and post-combustion. Pratheeba et al. [11] generated Matlab code and studied the effect of EGR on homogeneous charge compression ignition engine (HCCI) engine and also developed numerical model which can predict NO_x formation at different crank angles for different EGR ratios. They found reduction of NO_x with increases in the percentage of EGR and validated the proposed model with experimental work.

Sharma et al. [12] developed an algorithm for diesel engine using commercial code AVL FIRE and KIVA-3V. This algorithm was able to map flow structures, temperature, pressure, and emission-related data. The multiple injections were applied for the constant speed, turbocharged engine and were found to reduce unburnt HC and CO with minimal NO_x penalty.

Many CFD packages are available but for this analysis, ANSYS is used for simulating the combustion process of diesel engine and predicting the emissions. In the current study, the emissions and temperature are determined at various crank angles and comparison is made for these emissions with and without EGR which as per author's knowledge has not been taken earlier in the literature.

2 Methodology

ANSYS is used for simulating the combustion process of diesel engine and predicting the emissions. This work strictly focuses on combustion in the engine. The details of computational domain, mesh, mathematical model, and boundary condition used in this work are described below.

2.1 Computational Domain

Combustion chamber has been selected as computational domain as shown in Fig. 1. The domain shown in the figure represents about 1/6th portion of the whole engine as end faces behave as periodic walls. The domain mainly includes swept volume, piston head, and all end walls. The zone representing spray cone is responsible for diesel spray and its propagation during combustion. As shown in Fig. 1 only section of engine is chosen to minimize computational time. Domain is divided into different zones for proper mesh generation. Also, swept volumes are discretized as moving mesh or dynamic mesh.

For a spray angle of 70°, connecting rod of 165 mm length, crank radius of 55 mm, cylinder radius of 45 mm, and stroke length of 110 mm is chosen. Clearance volume is calculated as 4.56 mm^3 for compression ratio of 15.6. Only compression and power stroke are taken into consideration.

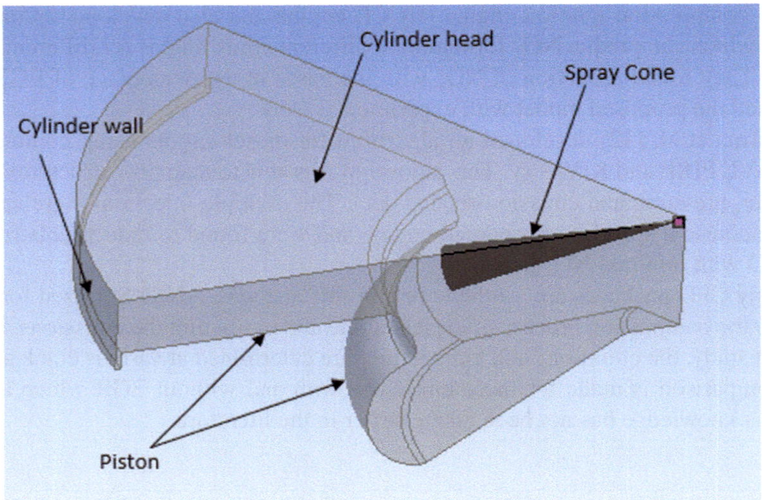

Fig. 1 Computational domain

2.2 Mesh Generation

Figure 2 shows the meshing model generated during simulation. Moving mesh boundary condition was given at piston because of its reciprocating nature and periodic mesh boundary condition was given to surface A and B.

Finite volume meshing method is used for discretizing the domain. Both structured and unstructured grids are taken into consideration for generation of mesh. In this case, multi-zone meshing is used in which different element sizes are given for different zones of domain. As a result, the prismatic mesh is generated near the walls and moving mesh (dynamic mesh) is generated for swept volume. Smoothing, layering and re-meshing were set for dynamic condition of meshing.

2.3 Model Selection

As the combustion is a complex phenomenon, so it is difficult to define whole combustion process into only one model. Therefore, various mathematical models are used to determine the combustion process.

For determination of flame speed, "power law" is used which correlates the laminar flame speed to pressure, temperature, and equivalence ratio. Equation of power law is given below [13]

$$S_L^0 = S_{L,\mathrm{ref}}^0 (T/T_{\mathrm{ref}})^{\alpha} (P/P_{\mathrm{ref}})^{\beta} F_{\mathrm{dil}} \tag{1}$$

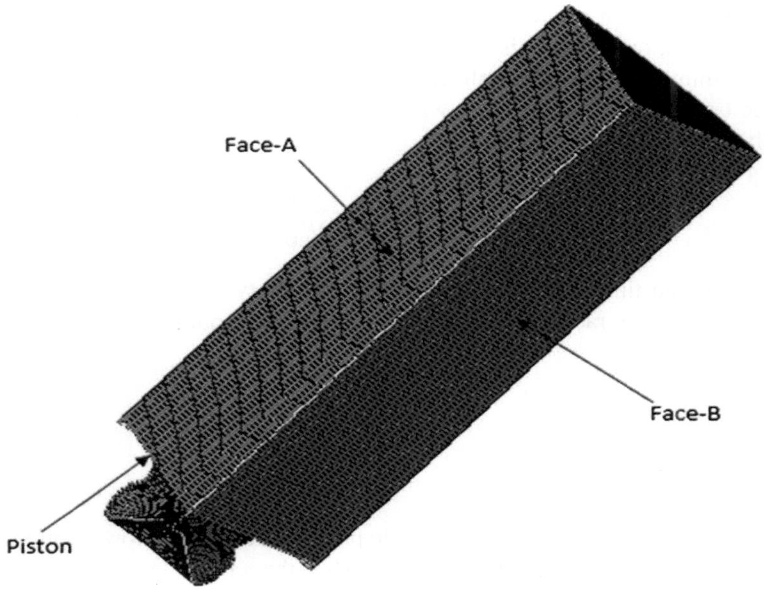

Fig. 2 Meshing model

where reference condition is 298 K temperature and 1 atm pressure. Factor of diluent effect is denoted by F_{dil} and speed of flame is denoted by S_L. Here, lean mixture is used in diesel engine. So Gülder expression was used to fulfill the purpose [14].

$$S^0_{L,\text{ref}} = \omega\phi^{\eta}\mathrm{e}^{-\xi(\varphi-\sigma)\wedge 2} \tag{2}$$

where ω, φ, η, ξ are data fitting coefficients. Also, the factor of diluent effect can be written as

$$F_{\text{dil}} = 1 - f_{x1} X_{\text{dill}}^{fx2} \tag{3}$$

where f_{x1} and f_{x2} are the empirical constant and X_{dil} is mass fraction diluent.

For turbulence, k-ε RNG model is used which gives better turbulence effect of fluid inside the combustion chamber. The governing equations of RNG model is [15]

$$\frac{\partial(\rho\varepsilon)}{\partial t} + \frac{\partial(\rho U j\,\varepsilon)}{\partial x_j} = \frac{\partial}{\partial x_j}\left[\left(\mu + \frac{\mu_t}{\sigma_{\varepsilon\text{RNG}}}\right)\frac{\partial\varepsilon}{\partial x_j}\right] + \frac{\varepsilon}{k}\left(C_{\varepsilon 1\text{RNG}} P_k - C_{\varepsilon 2\text{RNG}}\rho\varepsilon + C_{\varepsilon 1\text{RNG}} P_{\varepsilon b}\right) \tag{4}$$

Here, k is turbulent kinetic energy and ε is turbulence dissipation rate and $C_{\varepsilon 1\text{RNG}}$ and $C_{\varepsilon 2\text{RNG}}$ are the RNG model constant.

For demonstration of diesel spray, "Radius of Influence Model (ROI)" is used. In ROI method, one particle is allowed to collide with another only if this particle resides within in the radius of influence of the other. With the help of ROI model, distance traveled by spray particle within the given time step is calculated by the formula given below [16, 17].

$$\Delta t \times U_d = C_{\text{col}} \frac{R_{\text{inf}}}{N_{\text{ROI}}^{1/3}}. \tag{5}$$

where Δt is the time step, U_d is the average velocity of droplet and C_{col} is the Constant. R_{inf} is the radius of Influence and N_{ROI} is the number of particles in the sphere of influence.

2.4 Boundary Conditions

Moving wall boundary condition is given to piston wall which follows motion as per crank angle (i.e., slider-crank mechanism). Also, the temperature of piston wall is kept at 500 K and liner wall temperature is taken as 440 K. The cylinder head wall temperature is kept at 480 K. No slip conditions are given to all solid walls.

Two faces A and B as shown in Fig. 2 are periodic faces. Initially Pressure is taken as 344,737.5 Pa and temperature as 404 K for the working fluid (air). The simulation starts from the end of suction stroke at 570° crank angle and ends at the end of power stroke with 833° crank angles.

Atmospheric air is chosen as oxidizer along with certain amount of exhaust gas and *n*-heptane (diesel) is taken as fuel. Atmospheric air contained $O_2 = 0.13$, $N_2 = 0.74$, $H_2O = 0.03$ and $CO_2 = 0.07$ by mass. All the properties are taken from CHEMKIN database files available with current software. Species model is used for chemical processes. The injection of fluid takes place just before end of compression stroke. Here, injection starts at 712° crank angle and ends at 738.2° crank angle. Overall 40.2 mg of diesel is being injected.

3 Result and Discussion

Simulation was done for both the case, i.e., without EGR and with EGR. Simulation was done in dell inspiron laptop with 1.6 GHz CPU, 6 GB RAM, 2 GB graphic card. It requires approximately 36 h for the final result. Figure 3 represents maximum temperature inside the combustion chamber. The sudden rise in temperature was observed after 712° crank angle due to formation of flame takes place initially which fully develops at a maximum temperature around 2361 K and 721° crank angle in case of without EGR while it gets reduced to 2317 K with EGR which is negligible. This temperature drop occurs due to high specific heat of CO_2 compared to oxygen.

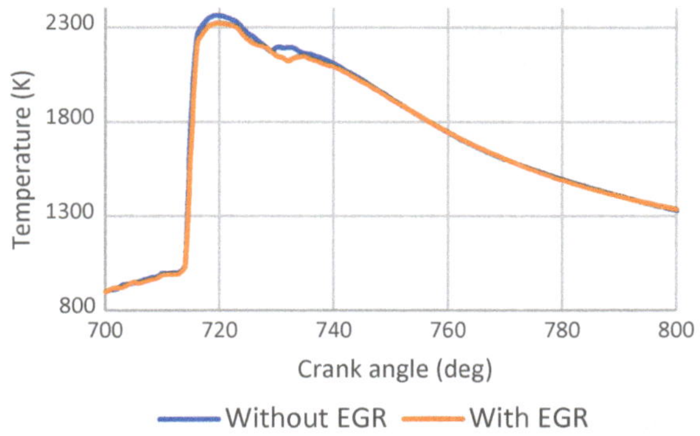

Fig. 3 Maximum temperature inside cylinder at different crank angle

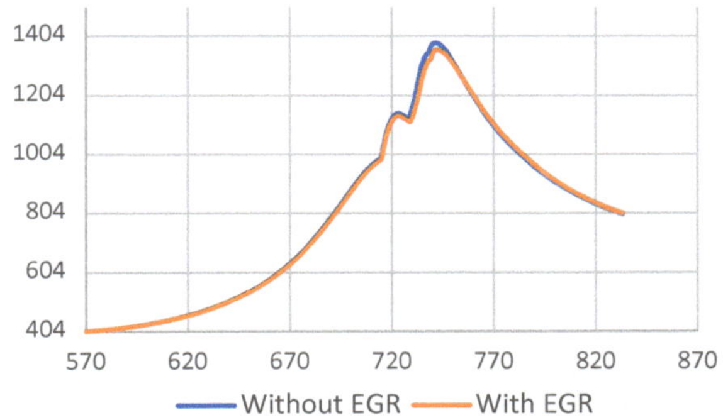

Fig. 4 Average temperature inside cylinder at different crank angles

Figure 4 represents the average temperature inside the cylinder during combustion at different crank angles. Combustion started with an injection of diesel at 712° crank angle. When the piston moves to top dead center (TDC), temperature reached to its peak value. Again, peak temperature inside combustion chamber in case of without EGR was lower as compared to with EGR. This also happened because of less oxygen availability for combustion. Similar trend of temperature decrease with EGR was observed by Torregrosa et al. [10] in their experimentations.

Figures 5 and 6 represent the formation of nitrogen monoxide and nitrogen dioxide at different angles, respectively. Due to the high temperature at 712° crank angle, more nitrogen oxide is formed. After achieving maximum temperature, large amount of oxygen was remained inside the combustion chamber. This leads to the formation of NO_2 from NO due to oxidation. After 740° crank angle NO_2 formation increases suddenly due to the conversation of NO into NO_2.

Figures 5 and 6 also represent that, at maximum temperature, the formation of NO and NO_2 gets decreased due to less availability of oxygen in case of with EGR system when compared to without EGR system. But at end of power stroke, the amount of NO is more with EGR system because of less oxygen present in the chamber. But overall NO_x production gets reduced to 32.28% with EGR. Similar trends of EGR effect on NO_x was observed by Torregrosa et al. [10] in their experimentations.

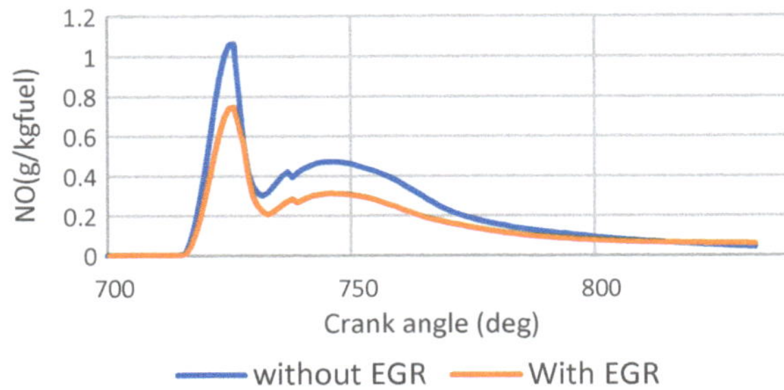

Fig. 5 NO formation at different crank angle

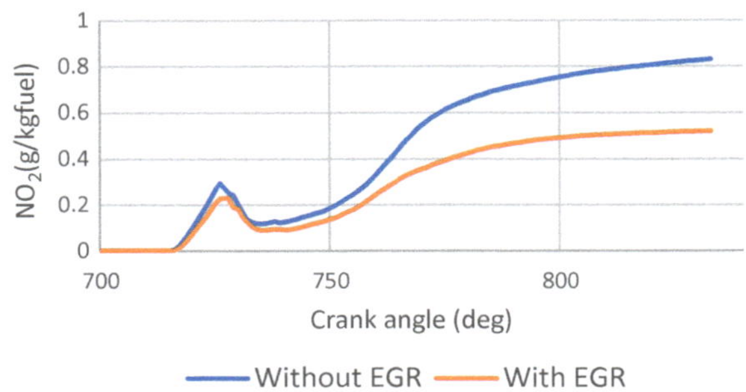

Fig. 6 NO_2 formation at different crank angle

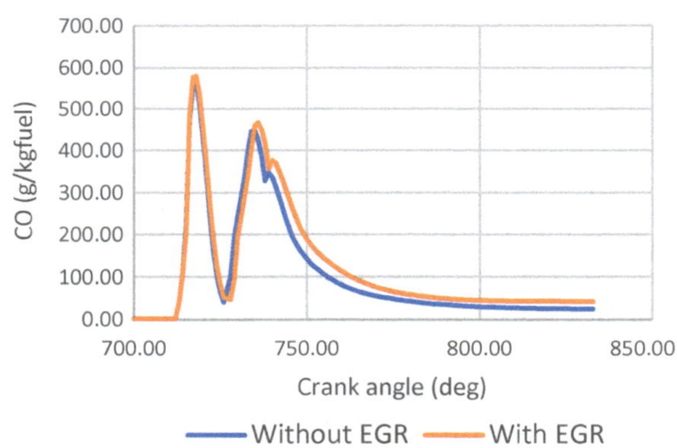

Fig. 7 CO formation at different crank angle

Due to less quantity of oxygen present at the time of combustion in case of with EGR system, mass fraction of CO increased, and CO_2 concentration in exhaust gas also increased is shown in Figs. 7 and 8. CO concentration increased from 21.48 to 41.58 g/kg of fuel. CO_2 concentration increased because of the addition of CO_2 present in exhaust gas due to combustion with recirculated exhaust gas. Similar trends of EGR effect on CO and CO_2 were observed by Torregrosa et al. [9] in their experimentations.

Fig. 8 CO_2 Formation at different crank angle

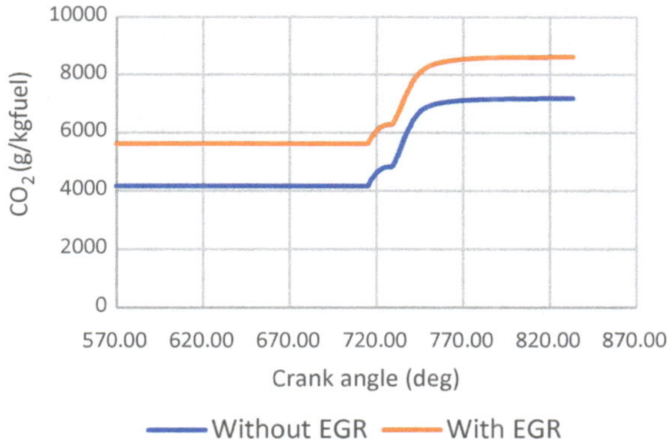

4 Conclusions

The current paper focusses on numerical analysis of diesel engine using ANSYS. The temperature and emissions are determined at different crank angles with EGR and without EGR. Following are the outcomes of this research.

1. Maximum temperature (flame temperature) decreases by approximately 44 K in case of with EGR system. There was a drop of 23 K temperature in average temperature of combustion chamber with EGR system. Pressure remained same in both the cases.
2. There was reduction of 32.28% in NO_x with EGR compare to without EGR.
3. CO production increased from 21.48 to 41.58 g/kg of fuel in case of with EGR system.
4. CO_2 at exhaust increased in case of with EGR system but there is no change in CO_2 after combustion.
5. Exhausted Gas Recirculation (EGR) was found to be the most promising solution for reducing NO_x, however CO emissions increase with EGR.

References

1. Kim, H., Lee, D., Park, S., Choi, K., Wang, H.: An experimental study on heat exchange effectiveness in the diesel engine EGR coolers, **22**, 361–366 (2008)
2. Jang, S., Park, S., Choi, K., Kim, H.: Experimental investigation of the influences of shape and surface area on the EGR cooler efficiency, pp. 621–628 (2011)
3. Heywood, J.: *Internal Combustion Engine Fundamentals* (1998)
4. Zelenka, P.: A key technology for future efficient HD Diesels. SAE (1998)
5. Hoard, J., Abarham, M., Styles, D., Giuliano, J.M., Sluder, C.S., Storey, J.M.E.: Diesel EGR Cooler Fouling, vol. 1(1) (2008)
6. Jafarmadar, S., Nemati, P.: Analysis of exhaust gas recirculation (EGR) effects on exergy terms in an engine operating with diesel oil and hydrogen. Energy **126**, 746–755 (2017)

7. Kumar, P., Makadia, U., Parwani, A., Deb, D.: Simultaneous reduction of NO_x and CO_2 using exhaust gas recirculation and carbon capture. 201821006612 (2018)
8. Selim, M.Y.E.: Effect of exhaust gas recirculation on some combustion characteristics of dual fuel engine. Energy Convers. Manag. **44**, 707–721 (2003)
9. Torregrosa, A.J., Broatch, A., Olmeda, P., Salvador-Iborra, J., Warey, A.: Experimental study of the influence of exhaust gas recirculation on heat transfer in the firedeck of a direct injection diesel engine. Energy Convers. Manag. **153**, 304–312 (2017)
10. Praveen, A., Rao, G.L.N., Balakrishna, B.: The combined effect of multiwalled carbon nanotubes and exhaust gas recirculation on the performance and emission characteristics of a diesel engine. Int. J. Ambient Energy **0**(0), 1–32 (2018)
11. Pratheeba, C.N., Aghalayam, P.: Effect of exhaust gas recirculation in NO_x control for compression ignition and homogeneous charge compression ignition engines. Phys. Procedia **66**, 25–28 (2015)
12. Sharma, C.S., Anand, T.N.C., Ravikrishna, R.V.: A methodology for analysis of diesel engine in-cylinder flow and combustion, vol. x(x), 157–167 (2010)
13. Metghalchi, M., Keck, J.C.: Burning velocities of methanol, ethanol and iso-octane-air mixtures. Symp. Combust. Combust. Inst. **19**, 275 (1983)
14. Gülder, O.: Correlations of Laminar Combustion Data for Alternative S.I. Engine Fuels. SAE Technical Paper Series 841000 (1984)
15. Yakhot, V., Orszag, S.A., Thangam, S., Gatski, T.B., Speziale, C.G.: Development of turbulence models for shear flows by a double expansion technique. Phys. Fluids A Fluid Dyn. **4**(7), 1510–1520 (1992)
16. Munnannur, A., Reitz, R.D.: A comprehensive collision model for multi-dimensional engine spray computations. At. Sprays **19**, 597–619 (2009)
17. Munnannur, A.: Droplet Collision Modeling in Multi-Dimensional Engine Spray Computations. University of Wisconsin-Madison (2007)

Thermo-Magneto-Electric Generator—An Approximate Theoretical Approach

Sanjay Kumar Sinha, Brajesh Tiwari and Dipankar Deb

Abstract Thermo-magneto-electric-generator can be a very efficient system of heat extraction and electricity production with the help of phase transition property of soft magnets and piezoelectric property of some elastic materials. It works due to oscillation of soft magnet between heat source and heat sink which cause vibration of piezoelectric cantilever. It is better than other heat recovery ventilations because it does not waste the heat but also uses that energy to produce electricity. The main source of potential for oscillation of soft magnet is linear magnetic field produced by hard magnet which in return causes magnetic potential energy in soft magnet. The hard magnet also acts as an unlimited fuel for the system until considerable amount of heat is extracted.

Keywords Phase transition · Soft magnet · Hard magnet · Piezoelectric material Heat extraction

1 Introduction

In our day to day life, there is plenty of heat wastage through small electronic devices. The investigation of wastage of heat energy can also provide alternative energy sources and yield significant energy savings in the industrial and commercial building sectors. Waste heat recovery technologies such as cogeneration have the potential to produce electricity below the price charged by the local electricity provider. These

S. K. Sinha (✉) · B. Tiwari · D. Deb
Institute of Infrastructure Technology Research and Management,
Ahmedabad 380026, Gujarat, India
e-mail: sanjay.sinha.14e@iitram.ac.in

B. Tiwari
e-mail: brajeshtiwari@iitram.ac.in

D. Deb
e-mail: dipankardeb@iitram.ac.in

D. Deb et al. (eds.), *Innovations in Infrastructure*, Advances in Intelligent Systems and Computing 757, https://doi.org/10.1007/978-981-13-1966-2_28

technologies reduce dependency on the electrical grid [1]. A system can be developed which can use that waste heat to generate electricity. We intend to accomplish the dual task of removal of heat and generation of electricity.

This system will consist of three main constituents: Hard magnets, soft magnets, and piezoelectric material. The phase transition of soft magnet, i.e., Gd from ferromagnet to paramagnet due to application of heat has been employed in this system [2]. A constant temperature difference of around 80 °C has been produced to carry out the project. The hard magnet attached to heat source which has a temperature of 70 °C is always above its Curie temperature. The Gd is connected to a spring and initially in contact with cold side, i.e., heat sink. Since the curie temperature of Gd is very close to room temperature which is around 21 °C [3], the thermal gradient will produce mechanical vibration of PVDF due to phase change of Gadolinium [4]. PVDF(Polyvinyl-Difluoride) is attached with Gd and is a piezoelectric material, and deformation in its structure will produce electrical output. Also, Gd will extract heat from Nd (Heat source) to carry out its phase change from ferro to para and transfer it to colder side (Heat sink) which again induces phase change from para to ferro. In this way, we did the dual task of electricity generation and heat removal.

The thermo-magneto electric generator and energy harvester gives a completely new and reliable technology to effectively use the waste heat and sufficient energy production for small electronic devices like UAVs [5]. The efficiency of power degrades at high temperature in solar module so we need a heat extracting system [6] (Fig. 1).

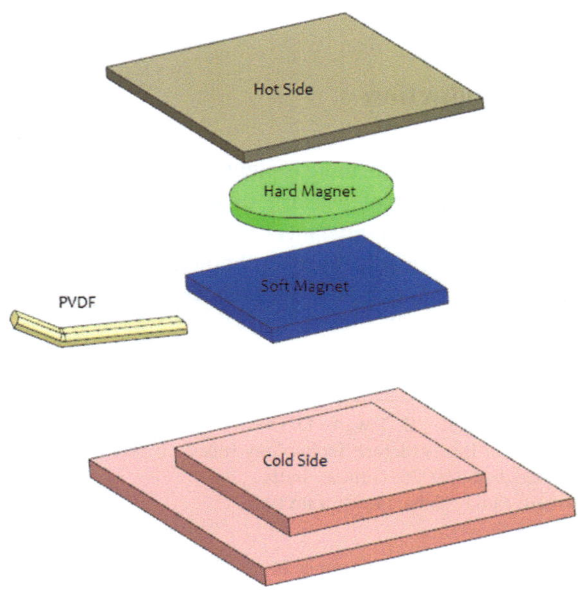

Fig. 1 Split view of PVDF model. Different materials depicted in the system are in regular progression according to their use, at hot side we have Nd, at cold side we have Gd, from Gd PVDF is attached

2 Working Principle

We have assumed that initially Gd is attached with PVDF and lying flat on colder side at temperature $T_{cold} = 263$ K with no tension force in piezoelectric material (PVDF). As magnetic field is created by Nd, the electromagnetic energy will develop in Gd and it will be in motion going towards hot side and then it will be attached to Nd until its phase changes because of change in temperature. So, it will be pulled back by the elastic force of PVDF (cantilever). Above process takes place due to ferromagnetic to paramagnetic phase transition of Gd. Now, Gd will be back on cold side hence it will regain its ferromagnetic strength because soft magnet has phase transition on temperature below Curie point. Thus, the oscillation of Gd between hot and cold side will continue until considerable amount of heat is extracted. Due to oscillation stress will develop in PVDF which is the cause of production of electricity (Fig. 2).

The hot side or source contains hard magnet such as Neodymium and colder side or heat sink have soft magnets such as Gadolinium initially, temperature difference must be at least 80 K. This oscillation will continue until the temperature gradient between source and sink is maintained. We are using two properties of materials to calculate different parameters associated with TMEG. The two properties are heat transfer, which is for the calculation of frequency and electromagnetic attraction, which is for the calculation of energy and forces. There will be mainly two assumptions for this model: (i) time elapse in motion of Gd is negligible. (ii) height of Gd is negligible.

2.1 Heat Transfer

In order to know the frequency of oscillating gadolinium, we have to calculate the rate of heat transfer, when it goes to hot side with initial temperature of that of cold side. We will have to obtain the time in which gadolinium will be attached to hotter as well as colder side.

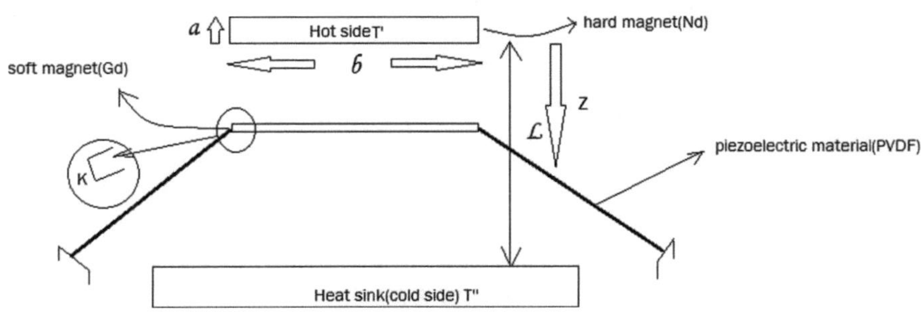

Fig. 2 Schematic of TMEG. At t = 0, PVDF is placed horizontally unelongated(normal sized, without stress), attached with Gd who is sitting on the cold side at 263 K. Then field is produced by Nd due to this motion will start and stress will start developing in PVDF

Let heat be Q, so required mathematical expression is $\frac{dQ}{dt}$. The Ohm's law analogous for heat transfer equation is:

$$\frac{dQ}{dt} = \Delta T \times \frac{KA}{k}, \tag{1}$$

where K is thermal conductivity, A is area of Gadolinium, and k is height of Gadolinium. where T is temperature. We will analyze first at hotter side and temperature at hotter side is T′. Let the temperature at other side is T.
So now,

$$\frac{dQ}{dt} = (T' - T) \times \frac{KA}{k}, \tag{2}$$

Now with the help of heat capacity equation which is $dQ = m \times C \times dT$, we will construct our heat transfer model after blending these two equations. We will get

$$m \times CdT = (T' - T) \times \frac{KA}{k} dt, \tag{3}$$

where m is the mass of material and C is heat capacity. Which implies, $\frac{mC}{T' - T}.dT = \frac{KA}{k}.dt$, From here, our mathematics for the calculation of time in which attachment between Gadolinium and Neodymium is no more, to achieve this the temperature of the side which is not in contact should reach curie temperature which ensures that the whole Gadolinium has temperature more than curie value of temperature, which is T_c. So, T will carry from T'' to T_c. Finally we have to solve

$$\int_{T''}^{T_c} \frac{m \times C}{T' - T} dT = \int_0^{t_0} \frac{KA}{k} dt. \tag{4}$$

In order to get approximated time t_0 of attachment of Gd and Nd.

Similarly at colder side T will vary 343–293 K and fixed temperature at colder side is 263 K.

$$\int_{T'}^{T_c} \frac{m \times C}{T'' - T} dT = \int_0^{t'_0} \frac{KA}{k} dt. \tag{5}$$

After solving above equation, find out time of attachment at cold surface, i.e., t'_0.

Now we put values temperatures, heat constant, conductivity, and mass, $T' = 343$ K, $T'' = 263$ K, $T_c = 293$ K, $C = 240$ J/g K, $K = 11$ W/m K and $m = 0.141$ gm [7] to get $t_0 = 0.18$ s, $t'_0 = 0.37$ s and net time of attachment is $t_{net} = t_0 + t'_0 = 0.55$ s. So, net frequency (f) of oscillation is 1.81 Hz.

2.2 Electromagnetic Energy

We will start modeling magnetic part of the array with an assumptions. The distance between the magnets, i.e., L is much larger than height of Gd, i.e., k.

Now, in this array, we have two magnetic material soft magnet (Gd) and hard magnet (Nd). Here, the main magnetic field which is the cause of all the movement is generated by Nd is denoted by B_{mag}. Now from (Furlani 2001)the magnetic field (B_{mag}) generated by Nd(only magnitude of magnetic field is considered) is [8]

$$B_{mag}(z) = \frac{B_r}{2} \tan^{-1}\left(\frac{(z+l)(a^2+b^2+(z+l)^2)^{1/2}}{a \times b} - \tan^{-1}\frac{(z(a^2+b^2+(z)^2)^{1/2})}{a \times b}\right), \tag{6}$$

where remanence and dimensions are as follows: B_r = 1.2 T, 2a (length) = 20 mm, 2b (width) = 20 mm, l (height) = 1.58 mm, and z is the distance from hotter side of the array. The variation of magnetic field is shown in Fig. 3.

The attraction between Gd and Nd will also be due to magnetization of Gd, which will necessarily increase or decrease due temperature difference because Gd is a soft magnet. The magnetization curve of Gd is given in Fig. 4 [4].

From the given figure, it is clear that around room temperature magnetization of Gd will become zero. And at temperature below 263 K, Gd will be having considerable magnetization. The force on Gd will be solely dependent on both magnetization of Gd

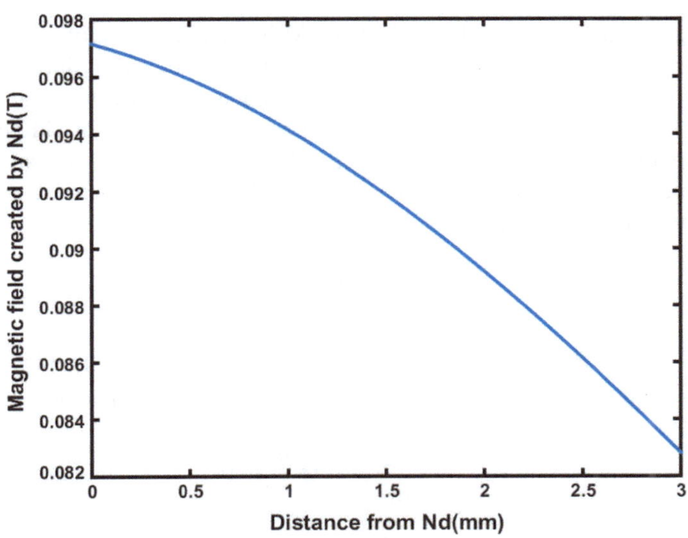

Fig. 3 Variation of magnetic field with respect to distance from hotter side. This figure shows field at different point relative to hot side, it is inevitable that it will decrease as we go far from Nd because according to Biot–Savart law B (magnetic field) is inversely proportional to distance (z)

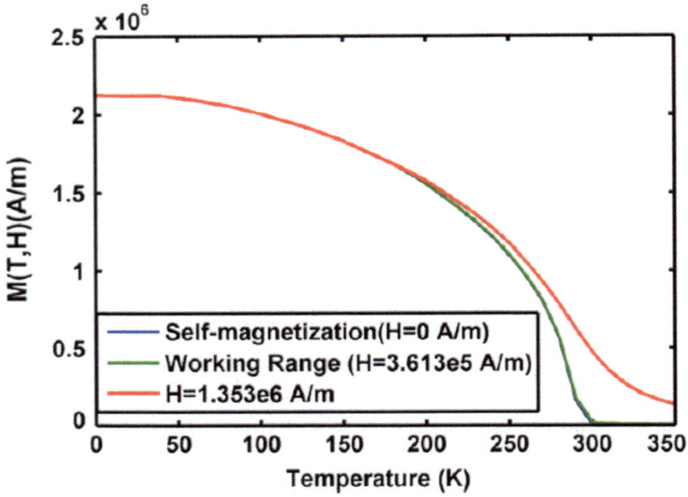

Fig. 4 Variation of magnetization with respect to temperature [4] (adopted by [2])

and Magnetic field generated by Nd. Due external magnetic field and magnetization magneto-static potential is developed in Gd. We have the formulae for magneto-static potential energy [8],

$$E_{mag} = \int_V MB_{mag} dV, \tag{7}$$

where B_{mag} is magnetic field due to ND and M is the magnitude of magnetization of Gd. In our work, we will ignore any change in magnetization during traveling or moving phase of Gd because it will be sudden and there will be no appreciable change in magnetization of Gd. B_{mag} is independent of volume of Gd because of our assumption $L \gg k$. So, from the given graph of magnetization, we can know the approximate magnetization of Gd when it is at colder side of the array and during traveling period, it will be constant. Hence, we get

$$E_{mag} = MB_{mag} V, \tag{8}$$

where M = 1.2 × 10^6 A/m and $V = 20 \times 20 \times 0.127$ mm^3. As, we know $F = -\frac{dE}{dz}$, i.e., the magnitude of potential gradient is force applied on that material [4].

$$F = -\frac{dE}{dz} = -M \frac{B_{mag}}{dz} V. \tag{9}$$

The variation of force is given in Fig. 5. In this way, we have calculated the applied force on Gd.

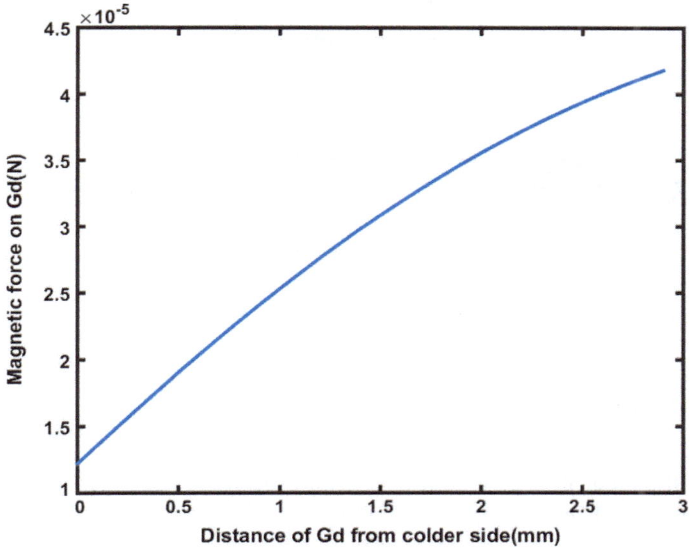

Fig. 5 Variation of force on Gd with respect to distance from colder side. The figure is showing that as the distance increases the force on Gd increases because the source of magnetic field, i.e., Nd is at the other side so Gd is going towards Nd

2.3 *Stress on PVDF (Piezoelectric Material)*

As we all know stress is the total force applied by material to regain its original stature divided by cross-sectional area. Here from Fig. 6, we can see T is the net force applied by PVDF to regain its original shape.

By analyzing whole FBD, we can say

$$T = F_{mag} \times \cos\theta, \tag{10}$$

where $\cos\theta$ is ratio of displacement of Gd by height of PVDF and F_{mag} is force calculated in Eq. 9.

After all this, we can write Stress (S) with cross-sectional area A on PVDF as follows:

$$S = \frac{T}{A}, \tag{11}$$

where A is cross-sectional area = 0.3 m^2. Stress on piezoelectric material is shown in Fig. 7.

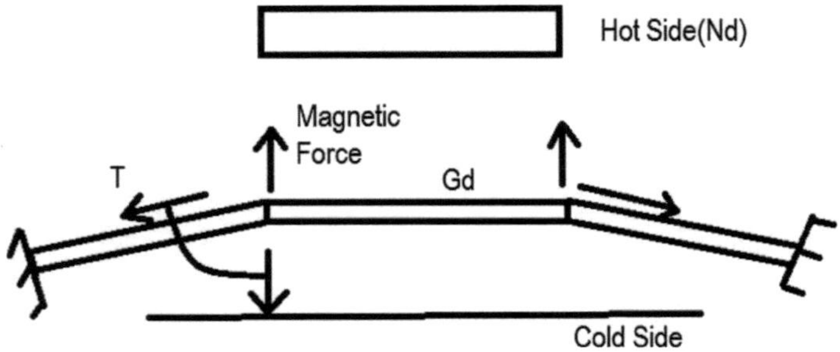

Fig. 6 Free body diagram of array

3 Results and Discussion

After calculation of stress, we have to apply piezoelectric equation with its different coefficients to find variation of voltage in PVDF. The required equations are [9, 10]

$$R = s^E T + dE, \tag{12}$$

and

$$D = dT + e^T E, \tag{13}$$

where R is strain, s is compliance when electric field is constant, T is applied stress, D is the electric charge density displacement, e is permittivity when temperature is constant, and E is electric field strength. In paramagnetic state, voltage will be decreasing because of decrease in stress while descending to the colder side. In this way, we will get periodic cyclic A.C. voltage. In the given Fig. 8, we have shown only ascending side when voltage value is reaching its peak. At the low dimensions of Gd and Nd, we obtain peak voltage per unit length 8.9 mV/m. Here, we consider the length of PVDF is 10 mm, so the net peak voltage produced is around 89 mv.

In order to calculate power we can use electromechanical transduction $d_{31} = 3.98$ pC N^{-1} [11], The variation of power obtained is given in Fig. 9. We can increase our output by adding more than one PVDF strip. If we go in further detail in mathematics of piezoelectric material then we can easily find that increase in frequency (which is not dealt in this work) of Gd will also increase the voltage output [2]. So, temperature gradient can also be increased to increase the voltage output. In this way, we are able satisfy duel usage of removal of heat from the devices and production of electricity. After proper mathematical analysis, we have planned our future venture on working real-time model of this array. So if we make an optimal model with proper dimensions then efficiency of system will definitely increase in great fashion because of heat removal along with production of extra power due to TMEG.

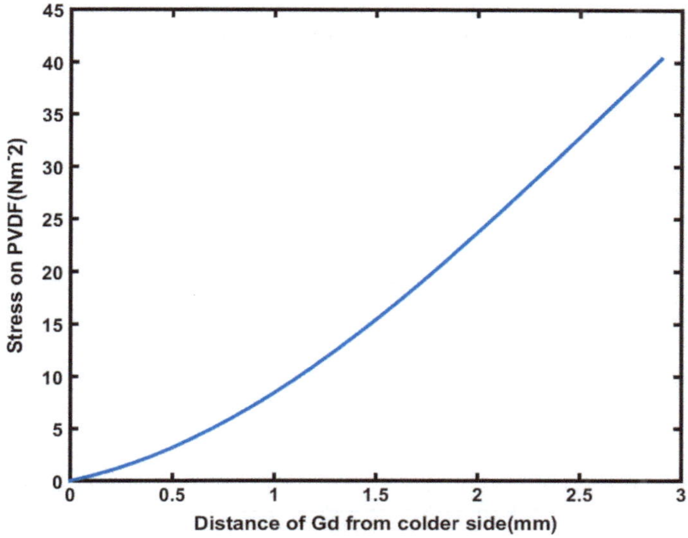

Fig. 7 Variation of stress in PVDF versus distance from colder side

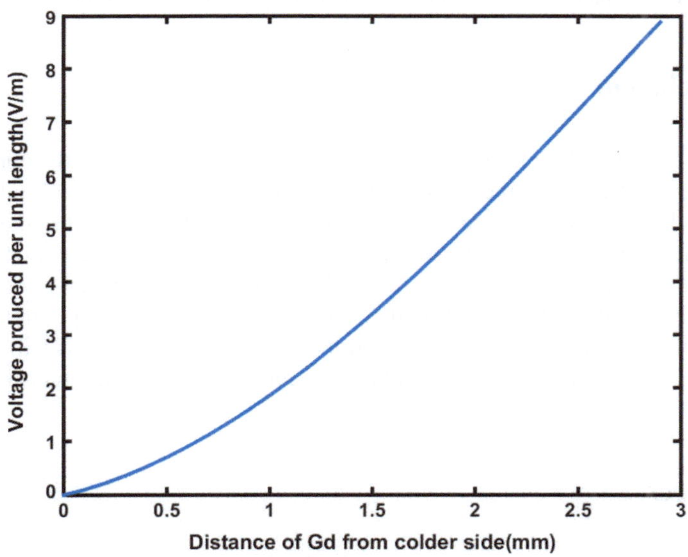

Fig. 8 Variation of voltage appeared on PVDF with respect of distance from colder side

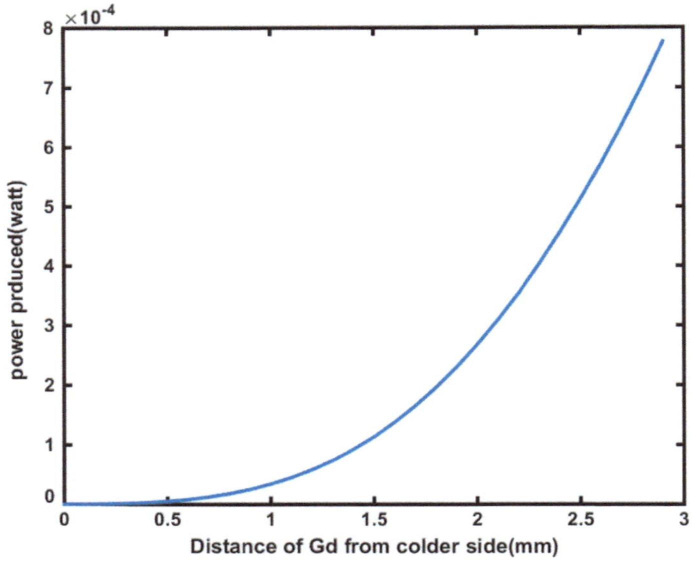

Fig. 9 Variation of power appeared on PVDF with respect of distance from colder side

4 Conclusion

The given model of thermo-magneto-electric generator has given us the opportunity of comparing different attributes of materials such as thermal, magnetic and electric, and blend those in approximate manner so as to get appropriate electrical output. In order to vary peak voltage output, we can increase or decrease the separation between source and sink depending upon the need. To increase the frequency of oscillation, we can decrease the temperature gradient up to a certain limit. To get more output practically, we can increase the overall dimension of the system. This will certainly increase mathematical complicacy and require more mathematical precision.

References

1. Waste heat minimisation and recovery — Energy EXchange [Internet]. Eex.gov.au. 2018 (cited 18 March 2018). Available from: https://www.eex.gov.au/technologies/waste-heat-minimisation-and-recovery
2. Chun, J., Song, H., Kang, M., Kang, H., Kishore, R., Priya, S.: Thermo-magneto-electric generator arrays for active heat recovery system. Sci. Rep. **7**, 41383 (2017)
3. Dankov, S., Tishin, A., Pecharsky, V., Gschneidner, K.: Magnetic phase transitions and the magnetothermal properties of gadolinium. Phys. Rev. B **57**, 3478–3490 (1998)
4. Joshi, K., Priya, S.: Multi-physics model of a thermo-magnetic energy harvester. Smart Mater. Struct. **22**, 055005 (2013)

5. Tang, X., Chen, P., Li, B.: Optimal air route flight conflict resolution based on receding horizon control. Aerosp. Sci. Technol. **50**, 77–87 (2016)
6. Shiau, Jaw-Kuen, Ma, Der-Ming, Yang, Pin-Ying, Wang, Geng-Feng, Gong, J.H.: Design of a solar power management system for an experimental UAV. IEEE Trans. Aerosp. Electr. Syst. **45**, 1350–1360 (2009)
7. Online Materials Information Resource - MatWeb [Internet]. Matweb.com. 2018 (cited 18 March 2018). Available from: http://matweb.com
8. Furlani, E.: Permanent Magnet and Electromechanical Devices. Academic, San Diego, CA (2001)
9. Lefki, K., Dormans, G.: Measurement of piezoelectric coefficients of ferroelectric thin films. J. Appl. Phys. **76**, 1764–1767 (1994)
10. Cha, S., Kim, S., Kim, H., Ku, J., Sohn, J., Park, Y., et al.: Porous PVDF as effective sonic wave driven nanogenerators. Nano Lett. **11**, 5142–5147 (2011)
11. Bernard, F., Gimeno, L., Viala, B., Gusarov, B., Cugat, O.: Direct piezoelectric coefficient measurements of PVDF and PLLA under controlled strain and stress. In: Proceedings, vol. 1, p. 335 (2017)

Estimation of Time-Varying Heat Source for One-Dimensional Heat Conduction by Conjugate Gradient Method

Sanil Shah and Ajit Kumar Parwani

Abstract Inverse analysis using Conjugate Gradient (CGM) with adjoint problem is carried out for estimating time-varying heat source in one-dimensional heat conduction. Three types of source terms are considered: constant, linearly increasing, and linearly decreasing. The direct problem is first solved with actual heat source values and temperature data is obtained for the domain. The inverse problem is then solved for the estimation of heat source using few of these temperature data. In actual practice, temperatures are measured experimentally and hence contain measurement errors. Therefore, the random error have been added in the temperature data for simulating temperature readings taken from experiments. The direct problem, inverse problem, sensitivity problem, and adjoint problem equations are discretized by Finite volume method (FVM) and solved by developing MATLAB code. For constant and linear source, the estimated source values have the same behavior with actual source values but for linearly decreasing source, there is large deviation of estimated values from actual values.

Keywords Inverse analysis · CGM · Sensitivity problem · Adjoint problem FVM

1 Introduction

Heat transfer is the most important phenomena in all thermal systems and processes. In normal heat transfer, the value of heat flux or temperature is known to boundaries and temperature distribution throughout the body estimated by governing mechanisms of heat transfer, i.e., by conduction, convection and radiation. Thus, in direct heat transfer problem, values of quantities at boundaries are known. But for some cases, it is difficult to measure flux or temperature at boundary directly. Inverse heat transfer deals with such problems where direct measurement of flux or source

S. Shah (✉) · A. K. Parwani
Department of Mechanical Engineering, IITRAM, Ahmedabad, Gujarat, India
e-mail: sanil.shah.mech10@gmail.com

D. Deb et al. (eds.), *Innovations in Infrastructure*, Advances in Intelligent Systems and Computing 757, https://doi.org/10.1007/978-981-13-1966-2_29

generation is not possible. By the knowledge of measured temperature within the boundary, inverse heat transfer estimates unknown heat flux, source, or thermophysical properties [1].

Inverse heat transfer has numerous applications in engineering and science such as mechanical, aerospace, astrophysics, nuclear science, mathematics, etc. In most inverse methods, the solutions of inverse heat transfer problems are obtained in the least square basis. Since the inverse heat transfer is ill-posed problem [1], it requires mathematical treatment to become well-posed problem. Tikhonov et al. [2] introduced regularization procedure in which the least squares equation modified by adding smoothing terms which reduces the unstable effects of the measurement errors. Alifanov et al. [3] introduced iterative regularization principle in which a sequential improvement of the solution takes place. The stopping criterion for iterative procedure is chosen so that the final solution is stabilized with respect to errors in the input data. Alifanow et al. [4] gave solution of nonlinear inverse heat transfer process for determining the heat loads. Beck et al. [5] introduced method of parameter estimation in inverse heat transfer. There are many applications of inverse heat transfer techniques. Bokar et al. [6] estimated radiation temperature source term in sphere, Huang et al. [7] estimated plane heat source inside a domain by inverse heat transfer.

In an inverse heat transfer method, the solution is obtained by minimizing the least square function or an objective function. An iterative approach is used for this process. There are many methods available for solving inverse heat transfer problems. Parwani et al. [8] used Differential Evolution (DE) algorithm for estimating strength and position of heat source in participating media. They got accurate prediction of strength and position of time-varying source. CGM is extensively used for inverse problems by various researchers [1]. In CGM for the estimation of a variable, no a priori information on the functional form of the unknown variable is available [9, 10]. Jarrny et al. [11] introduced parameter estimation in CGM. Later, on CGM with functional estimation was introduced in which no initial estimation of function is available [1]. Beckman [12] gave solution of linear inverse heat transfer problems by CGM. Parwani et al. [13–16] developed hybrid method by the combination of CGM and DE for estimation of boundary condition in various heat transfer cases. They concluded that the CGM is accurate method for prediction of strength and position of boundary heat flux. They also found that hybrid approach was more accurate and faster than CGM or DE approach. CGM is very fast method and is mostly used method for minimization [16]. The main advantage of using CGM is that it is stable method and it is self-regularized method. The current work focuses on estimation of time-varying source for one-dimensional heat conduction domain with CGM. Commercial software MATLAB R2016a is used for coding and running the algorithm. A desktop with 3.7 GHz processor with 4 GB RAM is used for running the code.

2 Estimation of Time-Varying Source by CGM

2.1 Direct Problem

Figure 1 shows a one-dimensional domain where estimated time-varying source $g_p(t)$ is given. Both ends of domain are insulated. The total length of domain is 1 m. The source is applied for the time duration of t_f sec. The main aim of CGM is to minimize the objective function of least squares given by [2]

$$S[g_p(t)] = \int\limits_{t=0}^{t=t_f} \{Y_m(t) - T[x_{\text{meas}}, t, g_p(t)]\}^2 \text{d}t \tag{1}$$

where $Y_m(t)$ measured temperature values and T are $[x_{\text{meas}}, t, g_p(t)]$ are the estimated values at that measuring point. And all values considered here for time interval of $t = 0$ to $t = t_f$ where t_f is the final time.

Here, $Y_m(t)$ values are the measured values containing some measurement errors as measuring process has some errors (i.e., instrumental error, measuring error, etc.) But for the current case, numerical experiment is carried out in which $Y_m(t)$ are obtained by solving direct heat transfer problem and adding measurement error term. Here, the time-varying source term $S(t)$ is known.

Direct problem is given by

$$\frac{\partial}{\partial x}\left(k\frac{\partial Y}{\partial x}\right) + S(t)\delta(x - 0.5) = \rho c\frac{\partial Y}{\partial t} \tag{2}$$

In the above equation, the value of source is given and according to it is behavior, the values of Y varies. Here, $S(t)$ is applied only on centre point (i.e., $x = 0.5$ m) which is indicated by Dirac Delta Function $\delta(x - 0.5)$. From the above equation, temperature field $Y(t)$ is obtained.

Also by considering measurement errors, the measured temperatures $Y_m(t)$ can be obtained by

$$Y_m = Y(t) + \sigma \text{e} \tag{3}$$

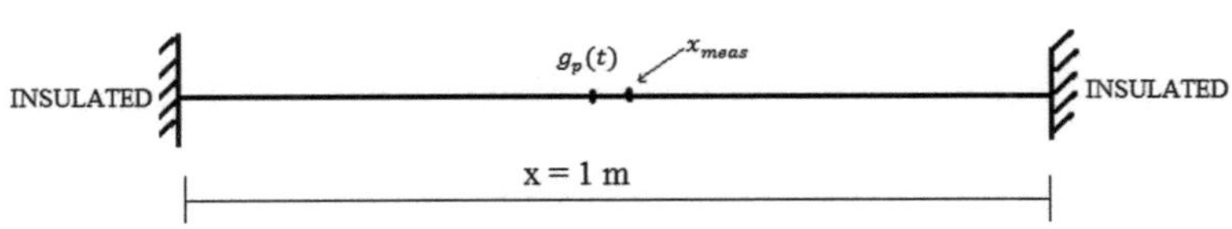

Fig. 1 Geometry of problem

where σ is the standard deviation of measurement errors which is assumed same for all measurements and e is normally distributed random error.

2.2 Inverse Problem

, the time-varying source $S(t)$ (Eq. 2) is unknown that is estimated by measured values $Y_m(x_{\text{meas}}, t)$ at any location near to source. It is an iterative approach in which $S(t)$ is estimated at the end of Kth iteration.

The formulation of heat conduction equation for inverse problem is the same as the direct problem

$$\frac{\partial}{\partial x}\left(k\frac{\partial T}{\partial x}\right) + g_p(t)\delta(x - 0.5) = \rho c\frac{\partial T}{\partial t} \tag{4}$$

where $g_p(t)$ is estimated source value. From the above equations, estimated temperature field $T(x, t)$ is obtained. For solution of above problem by CGM, two additional functions are required: Lagrange multiplier $\lambda(x, t)$ and sensitivity function $\Delta T(x_{\text{meas}}, t)$ [1]. The sensitivity equation can be obtained by replacing $g_p(t)$ by $g_p(t) + \Delta g_p(t)$ and $T(x, t)$ by $T(x, t) + \Delta T(x, t)$ in Eq. (4) and subtracting the resulting equation from main Eq. (4). The final result will be a sensitivity equation with boundary conditions

$$\frac{\partial^2 \Delta T(x,t)}{x^2} + \Delta g_p(t)\delta(x - 0.5) = \frac{\partial \Delta T(x,t)}{\partial t} \tag{5}$$

$$\frac{\partial(\Delta T(0,t))}{\partial x} = 0, \; t > 0 \tag{5a}$$

$$\frac{\partial(\Delta T(1,t))}{\partial x} = 0, \; t > 0 \tag{5b}$$

$$\Delta T(x,t) = 0 \text{ for } t = 0, \; 0 < x < 1 \tag{5c}$$

Inverse heat conduction problems are ill-posed problems and do not satisfy the existence and uniqueness conditions for the solution [1]. To make problem solvable, Lagrange Multiplier $\lambda(x, t)$ is required which is obtained by solving adjoint problem given by [1]

$$\frac{\partial \lambda}{\partial t} + \frac{\partial^2 \lambda}{\partial x^2} + 2[T(x,t) - Y_m(t)]\delta(x - x_{\text{meas}}) = 0 \tag{6}$$

$$\frac{\partial \lambda(0,t)}{\partial x} = 0, \; \text{for } t > 0 \tag{6a}$$

$$\frac{\partial \lambda(1,t)}{\partial x} = 0, \; \text{for } t > 0 \tag{6b}$$

$$\lambda(x, t_f) = 0, \; \text{for } 0 < x < 1 \tag{6c}$$

The CGM for the estimation of source term $g_p(t)$ is given by [1]

$$g_p^{k+1}(t) = g_p^k(t) - \beta^k d^k(t) \tag{7}$$

Here, $g_p^{k+1}(t)$ is source at $k+1$ iteration and depend upon previous values $g_p^k(t)$ and search-step size β^k and direction of decent $d^k(t)$. The direction of decent $d^k(t)$ at iteration k is given by [1]

$$d^k(t) = \nabla S\left[g_p^k(t)\right] + \gamma^k d^k(t) \tag{8}$$

where γ^k is called conjugate coefficient and $\nabla S\left[g_p^k(t)\right]$ is the gradient of $g_p^k(t)$.

The conjugate coefficient is obtained by Fletcher-Reeves [17] expression

$$\gamma^k = \frac{\int_{t=0}^{t=t_f} \left\{\nabla S\left[g_p^k(t)\right]\right\}^2 \mathrm{d}t}{\int_{t=0}^{t=t_f} \left\{\nabla S\left[g_p^{k-1}(t)\right]\right\}^2 \mathrm{d}t} \tag{9}$$

Here, $\gamma^0 = 0$ and gradient is equivalent to the value of Lagrange multiplier at mean point

$$\nabla S\left[g_p(t)\right] = \lambda(0.5, t) \tag{10}$$

Search step size is given by the following equation [1]:

$$\beta^k = \frac{\int_{t=0}^{t=t_f} \left\{T\left[x_{\mathrm{meas}}, t, g_p^k(t)\right] - Y_m(t)\right\} \Delta T\left[x_{\mathrm{meas}}, t, d_p^k(t)\right] \mathrm{d}t}{\int_{t=0}^{t=t_f} \left\{\Delta T\left[x_{\mathrm{meas}}, t, d_p^k(t)\right]\right\}^2 \mathrm{d}t} \tag{11}$$

From Eq. (11), values of search step β^k is obtained and from this value and value of decent $d^k(t)$, $g_p^{k+1}(t)$ is calculated from Eq. (7).

The following stopping criteria are used in the absence of measurement errors [1]:

$$S\left[g_p^{k+1}(t)\right] < \varepsilon, \tag{12}$$

But when measurement errors are present, the discrepancy principle is used as stopping criteria [1]

$$S\left[g_p^{k+1}(t)\right] < \sigma^2 t_f \tag{13}$$

Hence for this case, $\varepsilon = \sigma^2 t_f$

where ε is a small number chosen for stopping criteria.

2.3 The Solution Algorithm

The following steps are used for solving inverse heat transfer problem:

Step 1: Put any random initial value of unknown source $g_p^0(t)$ in Eq. (4). In the current paper, the initial guess for unknown source taken to be zero.
Step 2: Solve direct problem given by Eq. (4) and obtain temperature field $T(x, t)$.
Step 3: Knowing the estimated temperatures $T(x_{\text{meas}}, t)$ and measured temperatures $Y_m\ (x_{\text{meas}}, t)$, find Lagrange multiplier $\lambda(x, t)$ from Eq. (6).
Step 4: Knowing $\lambda(x, t)$ at measured location, find gradient $\nabla S\left[g_p(t)\right]$ by taking $\nabla S\left[g_p(t)\right] = \lambda(0.5, t)$.
Step 5: Compute conjugate-coefficient γ^k from Eq. (9).
Step 6: Calculate the direction of decent $d^k(t)$ from Eq. (8).
Step 7: Solve sensitivity problem by putting $\Delta g_p(t) = d^k(t)$ in Eq. (5) and obtain $\Delta T(x, t)$.
Step 8: Compute step size β^k from Eq. (11) in going from k to $k+1$.
Step 9: Compute $g_p^{k+1}(t)$ from the Eq. (7) and check the stopping criteria set by Eq. (12) or (13). Stop the iteration if a convergence criterion is satisfied, and if it does not satisfy the criteria then go to step 2.

3 Results and Discussions

Grid independency test should be performed on direct heat conduction to obtain optimum grid size. For this case, different grid sizes varying from $n = 100$ to $n = 210$ are taken and temperature distribution at centre point ($x = 0.5$ m) is taken. The source term $S(t)$ here is considered as constant value of 10,000,000 W/m^3 which is applied for 100 s.

As shown in Fig. 2, temperature distribution at different time changes with grid size but after 200 grid points, there is no significant change in temperatures and hence, 200 grid points were found optimum for this analysis. The total number of time steps (tn) are 100 for each case.

Also measurement errors in temperature are considered here. The measured values of temperature are given by Eq. (3). For current study, three cases of error are chosen: $\sigma = 0$, $\sigma = 0.001$, $\sigma = 0.002$. Also with this, three different cases of time-varying source $S(t)$ considered: (1) Constant. (2) Linearly increasing. and (3) Linearly decreasing.

For all cases, the measuring point is taken as $x = 0.505$ m, i.e., 5 mm away from source. For the convergence, the value of ε is chosen as 10^{-5} for $\sigma = 0$. Total numbers of iterations are set to 500,000 for each case. For checking the accuracy of estimation, Root Mean Square (RMS) error is calculated given by

$$\text{RMS error} = \frac{1}{\text{tn}}\sqrt{\sum\nolimits_{n=1}^{\text{tn}} \left[S(t) - g_P(t)\right]^2} \tag{14}$$

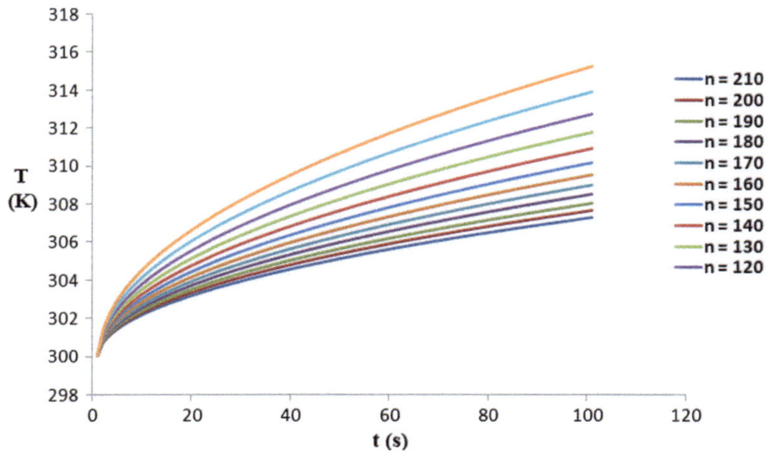

Fig. 2 Temperature distribution versus time for different grid sizes

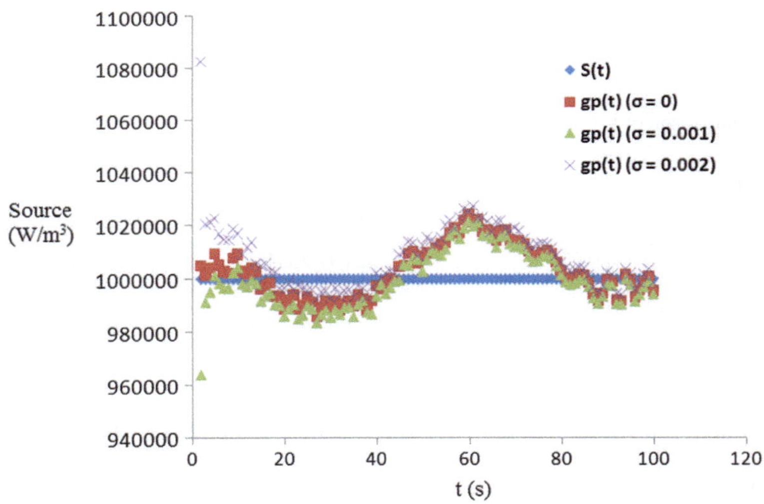

Fig. 3 Comparison of $S(t)$ and $g_p(t)$ for constant $S(t)$

Case 1: Constant Source

Constant value of source $S(t) = 1{,}000{,}000$ W/m^3 is taken and by algorithm, $g_p(t)$ is estimated. Value of ε reaches up to 0.0022 after 500,000 iterations and comparison of actual source $S(t)$ and estimated $g_p(t)$ is shown in Fig. 3.

As shown in Fig. 3, the estimated source $g_p(t)$ deviates around the actual value $S(t)$. This is due to the fact that CGM is a gradient-based method, in which gradient required for minimizing least square term. For constant source, it is difficult to obtain gradient, also sometimes CGM got struck in local minima [1]. From Table 1, it is

Table 1 Details of CPU time and RMS error for constant source

Sr. No.	σ	CPU time (s)	RMS error (W/m^3)
1	0.00	24,632.153965	9883.343835
2	0.001	26,712.411142	10,544.74906
3	0.002	26,545.213975	14,730.0952

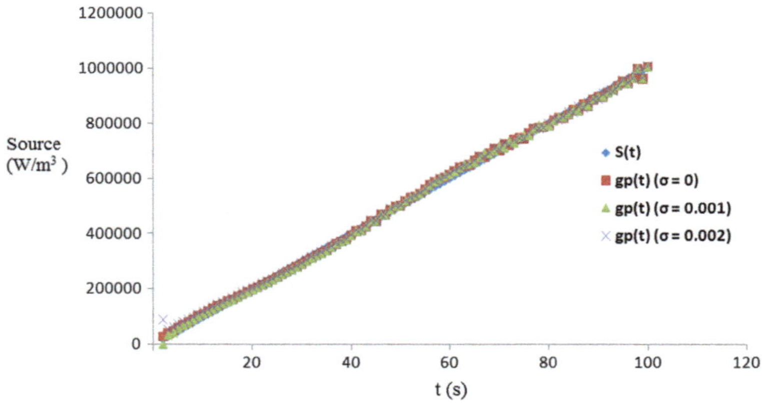

Fig. 4 Comparison of $S(t)$ and $g_p(t)$ for linearly increasing $S(t)$

Table 2 Details of CPU time and RMS error for linearly increasing source

Sr. No	σ	CPU time (s)	RMS error (W/m^3)
1	0.00	22,091.765482	10,438.97288
2	0.001	25,354.382623	10,189.95472
3	0.002	25,803.605136	14,063.07774

clear that CPU time increases with error, also RMS error increases with measurement errors.

CGM is a fast method in which about 246,500 s required for solving 500,000 iterations and accuracy can be increased by increasing number of iterations.

Case 2: Linearly Increasing Source

For the second case, linearly increasing source $S(t) = 10{,}000$ (tn) is applied at centre point and $g_p(t)$ is estimated from measured temperatures Y_m and CGM algorithm. Figure 4 shows comparison of both sources. Good convergence ($\varepsilon = 0.0022$) reached by system for 500,000 iterations for each case of measurement errors. Figure 4 shows good agreement of estimated source values with actual values (Table 2).

Here, CPU time and RMS error increases with measurement error same like constant source.

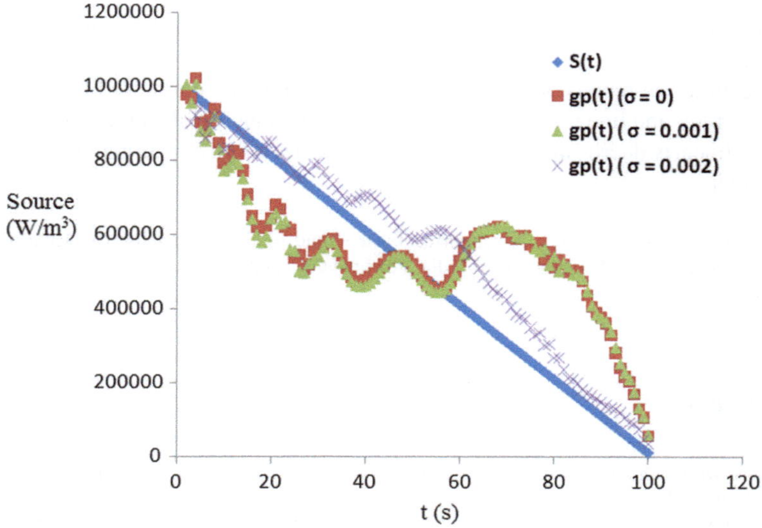

Fig. 5 Comparison of $S(t)$ and $g_p(t)$ for linearly decreasing $S(t)$

Table 3 Details of CPU time and RMS error for linearly decreasing source

Sr. No	σ	CPU time (s)	RMS error (W/m^3)
1	0.00	38,982.204589	193,262.0429
2	0.001	38,628.643461	198,948.3016
3	0.002	36,067.951481	75,855.46119

Case 3: Linearly Decreasing Source

For this case, linearly decreasing source $S(t) = 1{,}000{,}000–10{,}000(\text{tn})$ is applied to mid-point and comparison with estimated values $g_p(t)$ for different cases is shown in Fig. 5. The estimated values are diverged from $S(t)$ also ε has poor convergence values of 0.2752 ($\sigma = 0$), 0.2913 ($\sigma = 0.001$), and 0.0873($\sigma = 0.002$). The main reason behind poor convergence is that as value of source decreases with respect to time, least square terms reduces which give convergence at local minima and this will deteriorate the results. Table 3 shows the same thing about least square errors.

4 Conclusions

Estimation of time-varying source for one-dimensional heat conduction has been carried out using inverse heat transfer technique. For current study, CGM with functional estimation is chosen. Also, errors in temperature measurements considered. Following are the conclusions:

- CGM is very fast method for estimation and is accurate enough for gradient-based source.
- For accuracy of results, sufficient amount of iterations are required in CGM.
- For constant and linear source, the estimated data match with actual source values but for linearly decreasing source, there is large deviation of estimated values from actual values.

Acknowledgements This work is supported by the fund of SERB division of Department of Science and Technology, Government of India. The financial support towards this research is greatly appreciated.

References

1. Ozisik, M.N., Orlande, H.R.B: Inverse Heat Transfer: Fundamentals and Applications. Taylor and Francis, New York (2000)
2. Tikhonov, A.N., Arsenin, V.Y.: Solution of Ill-Posed Problems. Winston & Sons, Washington DC (1977)
3. Alifanow, O.M., Artyukhin, E.A.: Regularized numerical solution of nonlinear inverse heat conduction problem. Inzhenerno-Fizicheskii Zhurnal **29**, 159–164 (1975)
4. Alifanov, O.M.: Determination of heat loads from a solution of the nonlinear inverse problem. High Temp. **15**, 498–504 (1977)
5. Beck, J.V., Arnold, K.J.: Parameter Estimation in Engineering and Science. Wiley, New York (1977)
6. Bokar, J., Ozisik, M.N.: An inverse problem for the estimation of radiation temperature source term in a sphere. Inverse Prob. Eng. **1**, 191–205 (1995)
7. Huang, C.H., Ozisik, M.N.: Inverse problem of determining the unknown strength of an internal plane heat source. J. Franklin Inst. **75** (1992)
8. Parwani, A.K., Talukdar, P., Subbarao, P.M.V.: Simultaneous estimation of strength and position of heat source in participating medium using DE algorithm. J. Quant. Spectrosc. Radiat. Transfer **127**, 130–139 (2013)
9. Neto, A.G.S., Ozisik, M.N.: The estimation of space and time dependent strength of a volumetric heat source in a one-dimensional plate. Int. J. Heat Mass Transf. **37**, 909–915 (1994)
10. Neto, A.G.S., Ozisik, M.N.: Two-dimensional inverse heat conduction problem of estimating the time-varying strength of a line heat source. J. Appl. Phys. **71**, 5357–5362 (1992)
11. Jarny, Y., Ozisik, M.N., Bardon, J.P.: A general optimization method using adjoint equation for solving multidimensional inverse heat conduction. Int. J. Heat Mass Transf. **34**, 2911–2919 (1991)
12. Beckman, F.S.: The Solution of Linear Equation by Conjugate Gradient Method, Mathematical Methods for Digital Computer, A. Chap. 4. Wiley, New York (1960)
13. Parwani, A.K., Talukdar, P., Subbarao, P.M.V.: Estimation of inlet temperature of developing fluid flow in parallel plate channel. Int. J. Therm. Sci. **57**, 126–134 (2012)
14. Parwani, A.K., Talukdar, P., Subbarao, P.M.V.: Estimation of boundary heat flux using experimental temperature data in turbulent forced convection flow. Heat Mass Transf. **51**, 411–421 (2015)
15. Parwani, A.K., Talukdar, P., Subbarao, P.M.V.: A hybrid approach using CGM and DE algorithm for the estimation of boundary heat flux in parallel plate channel. Numer. Heat Transf. Part-A **65**, 461–481 (2014)

16. Parwani, A.K., Talukdar, P., Subbarao, P.M.V.: Performance evaluation of hybrid differential evolution approach for the estimation of strength of heat source in radiatively participating medium. Int. J. Heat Mass Transf. **56**, 552–560 (2013)
17. Fletcher, R., Reeves, C.M.: Function minimization by conjugate gradients. Comput. J. **7**, 149–154 (1964)

H_2S Absorption from Syngas in Physical Solvent DMEPEG

Ashok Dave, Bhumika Pathak, Medha Dave, Poonam Kashyap and Ye Huang

Abstract H_2S is removed from sour syngas upstream of CO_2 capture, thus producing sweeter syngas. Simulation results of H_2S absorption in a physical or chemical solvent may be unreliable due to assumptions like the equilibrium-based modelling, theoretical number of stages, tray efficiency, height equivalent of theoretical plate, as also due to the lack of prior experience and the lack of pilot plant data at appropriate scale of operation. Rate-based process simulations carried out using ProTreat software are described in this publication using process flow diagram (PFD) and simulation flow sheet—to describe the packed tower design and process design for H_2S Absorption from syngas in physical solvent DMEPEG. Consideration of physical model of packed tower (packing type, size and material) imparts credibility to the simulation results. Sensitivity analysis is conducted for process parameters. Performance of packer tower is analysed and compared with process requirement and recommended practice. Similar performing tower internals is suggested.

Keywords H_2S absorption · Packed tower · HETP · DMEPEG · PFD

A. Dave
Mechanical Engineering Department, School of Engineering, Indrashil University, Rajpur, Taluka Kadi, Mehsana, Gujarat 382740, India
e-mail: ashokddave@yahoo.com

B. Pathak (✉)
Pramukh Swami Medical College, Karamsad, India
e-mail: docbhum29@yahoo.com

M. Dave
California State University, Long Beach, CA, USA

P. Kashyap · Y. Huang
CST, Ulster University Jordanstown, Newtownabbey, Antrim, Northern Ireland BT37 0QB, UK

D. Deb et al. (eds.), *Innovations in Infrastructure*, Advances in Intelligent Systems and Computing 757, https://doi.org/10.1007/978-981-13-1966-2_30

1 Introduction

Among the variety of fossil fuels naturally available to mankind, coal is the most evenly [1] spread and abundantly available [2] fuel around the world. Hence, the coal has been the primary energy source of mankind for centuries and it promises the future energy security for years to come. Gasification of coal (by producing and pretreating syngas) enables the consumption of its thermal energy in a gas turbine followed by bottoming steam turbine. Syngas is also useful for the production of petrochemicals, urea, etc. This process route necessitates the desulphurization of syngas for plant protection against corrosion and the environment protection against harmful (and greenhouse) gas emission. Removal of sulphur-rich gas from the syngas by its absorption in physical solvent is the first step of the syngas desulphurization process. Knowledge of the detailed process model can help to lower down the risk in its implementation and the resulting cost. Sensitivity analysis of process parameters affecting the performance of H_2S Absorption (from syngas produced in a pre-combustion IGCC power plant into physical solvent DMEPEG) is described in this book chapter (based on the Ph.D. thesis [3] of the primary author). Research input and outcome are matched with literature throughout the chemical process development.

2 Process Condition for H_2S Capture

The composition of syngas fed to H_2S capture plant is described in Table 1. The packed tower of the AGR plant is designed to absorb H_2S from 450 Tonne per Hour (TPH) syngas to meet the 608 MW Th heat demand (607.793 to be precise) which is considered for a modern F class industrial gas turbine (generating 301 MWe). The large mass transfer area necessary for processing 450 Tonne per hour (TPH) syngas (and 2526.55 TPH solvent) is provided by a packed tower (H_2S Absorber). The composition of syngas fed to H_2S Absorber (in AGR process) is described in Table 1. The amount and composition of syngas and solvent fed to H_2S Absorber (boundary condition) are described in Tables 2, 3 and 4 in detail.

Table 1 Composition of syngas (588 TPH) fed to H_2S Absorber (as per industrial practice) [5]

Syngas constituent	Composition (mole %)
H_2	46.28%
CO_2	42.1%
H_2S	0.5075%
N_2	8.36%

Table 2 H_2S absorption sensitivity analysis

Equipment size	
Packing material	Random Packing IMTP 40
Packing height of H_2S absorber	50 m (130 segments of equal size)
Solvent feed	
DMEPEG and dissolved gases	1.2–3.5 kmol/s for sensitivity study at 36.40 bar-a, 35.0 °C
Composition	64.5% DMEPEG with dissolved gas including 34.80% CO_2 and other dissolved gas

3 Solvent Characteristics and Process Simulation Strategy

Molecules like H_2S, CO_2, H_2O, etc., are classified as polar molecule due to the nature of its atomic bond and the uneven resulting in uneven distribution of electric charge. Physical solvents such as poly glycol-ether, propylene carbonate, etc., and the chemical solvents such as those having amine group can attract such polar molecules. DMEPEG is the abbreviation of the variety of physical solvents classified as dimethylether of polyethyleneglycol [4].

Solubility of H_2S in various polar organic solvents such as various specific polymer compounds of DMEPEG is described in Fig. 1. Comparability of published data for solubility of CO_2 in DMEPEG solvent with the property model implemented in ProTreat software (Version 5.2, 2013) is described by Dave et al. [2, 3, 5], 2016. The property model for H_2S absorption by DMEPEG solvent and the strategy of ProTreat [6] software for the rate-based mass transfer [7] simulation is elaborately described by Dave et al. [2, 3, 5], 2016. Refer to the publications [5, 8] regarding: (a) ProTreat software and the rate-based mass Transfer, (b) Validation of ProTreat software and (c) Simulation.

Relative solubility of various syngas constituents [9] in DMEPEG solvent accordingly is reported by Barry Burr and Lili Lyddon. Various other researchers have published [10–13] the gas solubility data for various other blends [14] of DMEPEG solvent through the applicable range of temperature and pressure. Yuming Xu [15] et al. describe the falling solubility of H_2S in DMEPEG solvent with rising temperature from $K_H = 0.44$ (Henry's Number) at 298 K to $K_H = 1.01$ at 333 K. The selectivity of H_2S compared to CO_2 falls down from 8.11 at 298 K to 6.485 at 333 K. Gas solubility in DMEPEG solvent is also governed by the composition of the solvent. Solubility of CO_2 in the individual species of DMEPEG is reviewed [8] by Dave et al. 2016.

Table 3 Syngas and solvent fed to H_2S capture plant

		Syngas				Solvent				
Pressure		37 bar-a				38 bar-a (inlet to H_2S Absorber)				
Temperature		25 °C				35 °C				
Flow rate		6.178 k-mole/s (125 kg/s)				1.2–3.5 k-mole/s				
	Species and composition (1 ppm = 1×10^{-6} mol fraction = 1×10^{-4} %)									
	H_2O	H_2S	CO_2	DMEPEG	CH_4	H_2	He	CO	N_2	Ar
Syngas	0.17%	0.31%	38.2%	Nil	0.05%	53.6%	0.27%	1.73%	4.77%	0.94%
Solvent	11.66 ppm	11.6 ppm	34.8%	64.5%	17.65 ppm	0.4684%	39.63 ppm	0.029%	0.21%	119 ppm

Table 4 Composition of solvent fed to H_2S capture plant (final design)

H_2O	H_2S	CO_2	DMEPEG	CH_4	H_2	He	CO	N_2	Ar
17.4%	2.864 ppm	27.8%	54.2%	17.89 ppm	0.4709%	39.74 ppm	0.0294%	0.128%	119.9 ppm

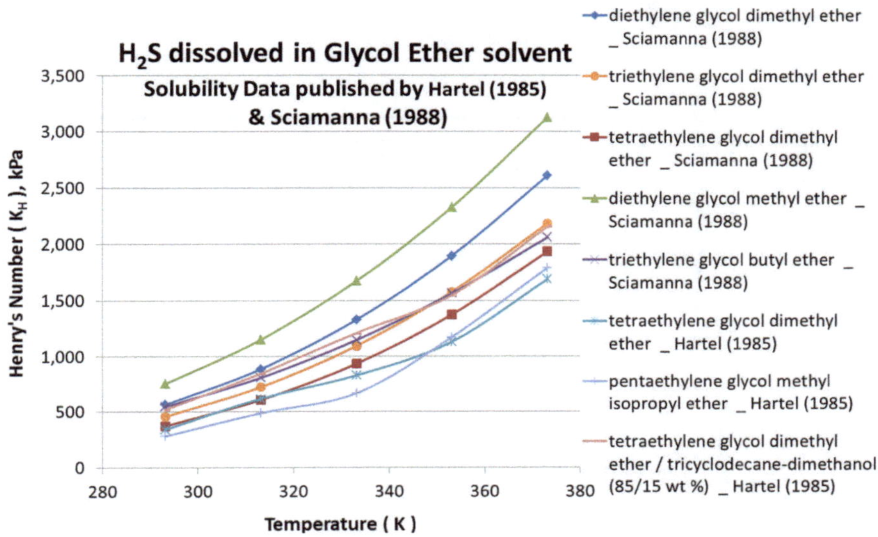

Fig. 1 Solubility of H_2S in various DMEPEG polymers [17, 18]

4 Process Description

Absorption of H_2S from syngas is inevitably in the beginning of CO_2 capture from syngas. H_2S is capture using DMEPEG solvent by a three-stage process. The first stage consists of the absorption of H_2S from plain or CO_2-loaded solvent.

Within the second stage, the solvent (having dissolved CO_2 and H_2S) is heated and passed through another packed tower (using N_2 as stripping media), thus liberating the dissolved CO_2 (enrichment of H_2S).

The third and last stage consists of the de-pressurization of the solvent and stripping of concentrated H_2S dissolved within the solvent in a thermal stripper using steam as stripping media. This publication focuses on H_2S Absorption. Process flow diagram is shown in Fig. 2. Flow sheet for modelling H_2S Capture Process is shown in Fig. 3.

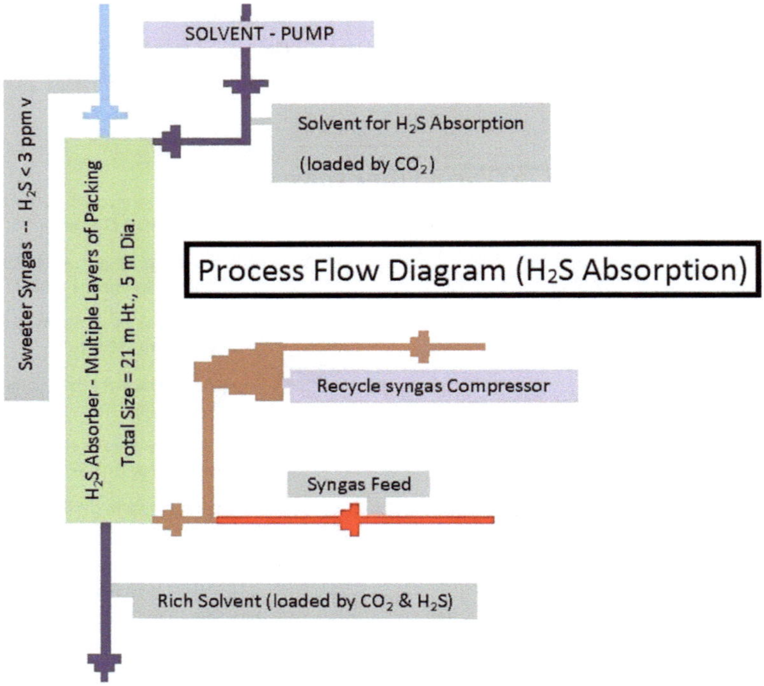

Fig. 2 Process flow diagram of H_2S absorption

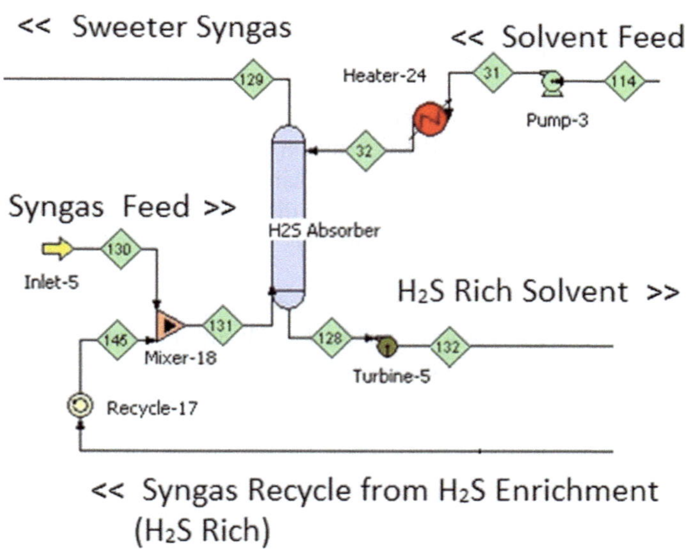

Fig. 3 Flowsheet for modelling of H_2S absorption using ProTreat software

5 Process Model [16]

The process of H_2S absorption is modelled as a steady-state simulation using version 5.2 (2013) of the ProTreat software. The non-ionic gas solubility model of the ProTreat software (applicable to DMEPEG solvent) is utilized for rate-based simulation of H_2S Absorption in a packed tower. Sensitivity study is conducted to assess the performance of various process configurations. Major process parameters are optimized for process design, basic engineering and economic assessment. H_2S Capture Plant can be designed for a particular amount of residual H_2S content within processed syngas. This is achieved by designing appropriate process configuration of 'H_2S Absorber' (height and diameter of packed sections, solvent composition, temperature and feed rate, discharge condition of syngas, etc.) of appropriate height.

6 Sensitivity of H_2S Absorption Depending on Solvent Circulation Rate

H_2S Absorption can be promoted by either additional solvent circulation or by lowering the temperature of the incoming solvent. The solvent coming out of CO_2 absorber is largely saturated by absorption of CO_2. Further cooling increases its overall gas dissolving capacity resulting in additional co-absorption of CO_2 while absorbing H_2S. Such additionally absorbed gas has to be desorbed prior to the H_2S Enrichment. The other technique of enhancing H_2S Absorption is to increase the solvent flowrate, i.e. diverting increasing proportion of solvent from the CO_2 Capture Plant to the H_2S Capture Plant. However, this increasing throughput of solvent results in dilution of composition of absorbed H_2S in solvent which is not desirable for H_2S Enrichment.

6.1 Specifications (Assumptions) for H_2S Absorber Equipment

Sensitivity study is conducted to investigate the effect of solvent stream on H_2S absorption. The equipment details of H_2S Absorber and the thermodynamic parameters of solvent stream is in Table 2. 125 kg/s syngas (6.178 kmol/s) is injected to the H_2S Absorber at 25 °C and 37 bar-a.

6.2 Specification (Assumptions) for Feed Condition

Composition of syngas and solvent for sensitivity analysis of H_2S Absorber size is described in Table 3.

6.3 Analysis and Results

Figure 4 shows the results of sensitivity analysis of rate-based mass transfer simulations for varying rate of solvent flow rate. Packing height of 50 m is divided into 130 segments of equal size. The segments from top to bottom of packed tower (H_2S Absorber) are numbered from 0 to 130 (X-axis in Fig. 4). The resulting molar (or volumetric) composition of H_2S in syngas emanating from each of the 130 stages is shown on Y-axis in Fig. 4. Solvent circulation at packed tower inlet is in the range of 1.2–3.5 kmol/s. H_2S composition of exit syngas (from H_2S Absorber) continues falling with rising solvent circulation rate. H_2S composition in exit syngas can be lowered by increasing the solvent circulation from 1.2 kmol/s up to 1.7 kmol/s. H_2S composition of syngas tends to stabilize between 2.5 and 3 ppm despite additional Incremental solvent circulation rate. Hencem for 50 m tower height, the rise in solvent circulation beyond 1.7 kmol/s is not much advantageous.

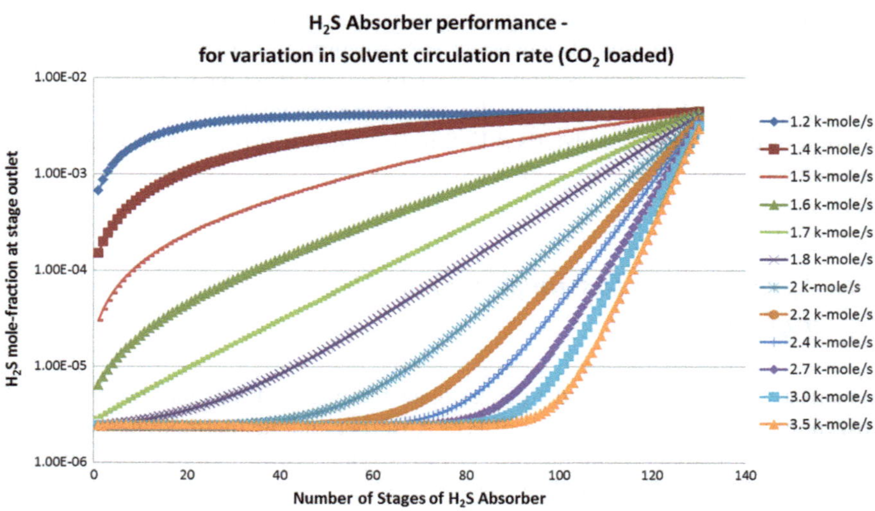

Fig. 4 Susceptibility of H_2S absorption with reference to the solvent circulation rate

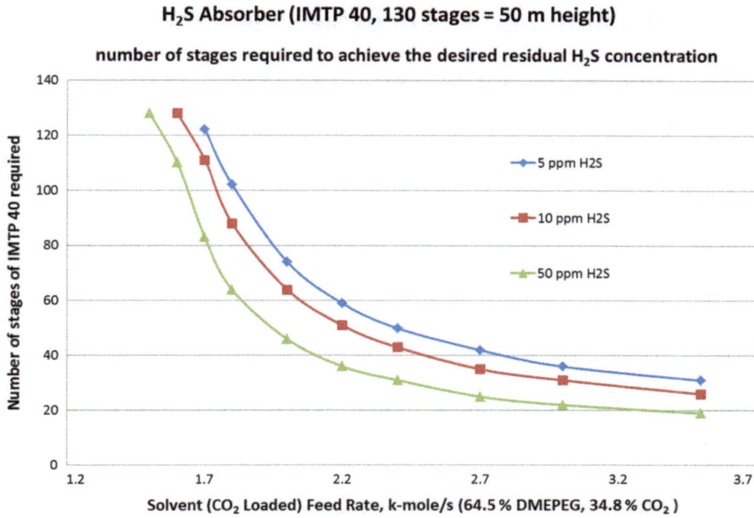

Fig. 5 Mass transfer area of H_2S Absorber with respect to solvent circulation rate as per the targeted concentration of residual H_2S in processed syngas

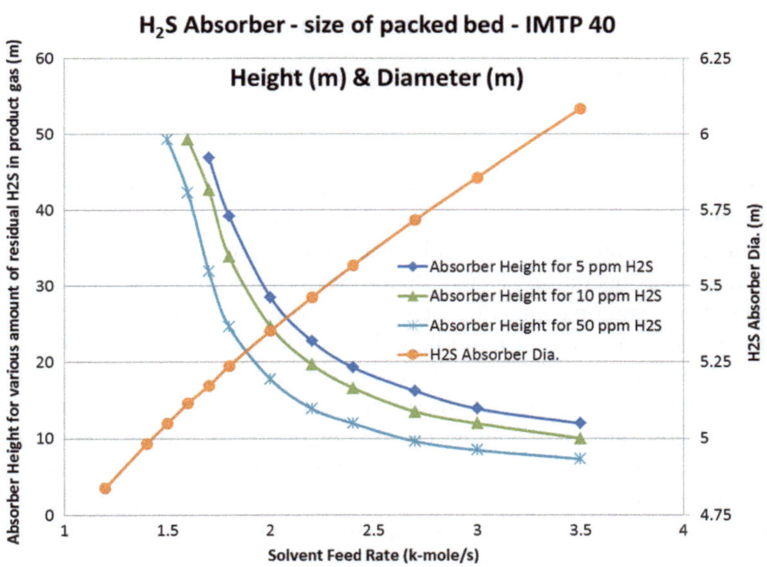

Fig. 6 Packed tower dimensions (height and diameter) with respect to solvent circulation rate (for 70% flood)

As per the anticipated H_2S composition in exit syngas (possibly 50 ppm, 10 ppm or 5 ppm), the solvent flow rate can be raised above 1.7 kmol/s to adjust packed tower height. Packed tower height necessary to achieve either 50 ppm or 10 ppm or

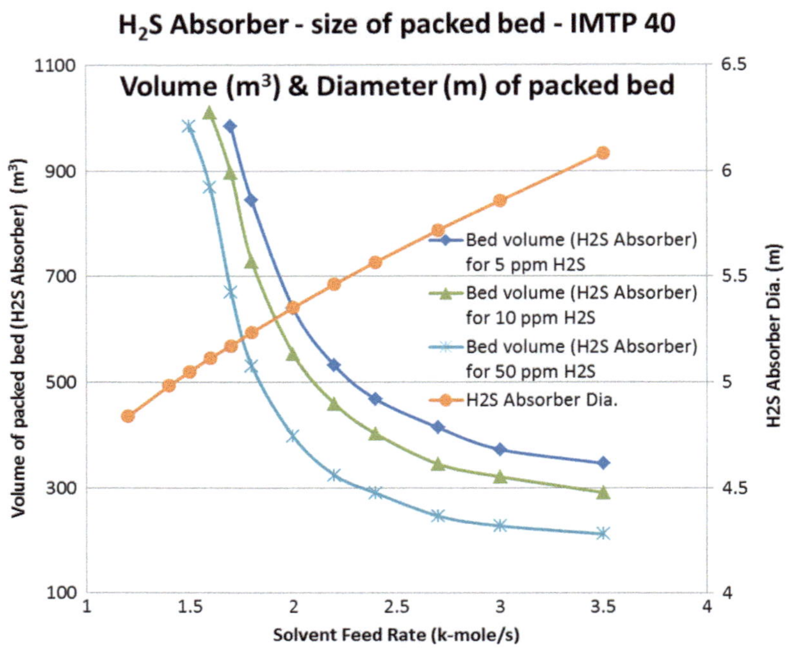

Fig. 7 Susceptibility of the diameter and volume of packed tower (H_2S absorber) with respect to solvent circulation rate

5 ppm H_2S in syngas exit is shown in Fig. 5. Solvent flow rate of 2 kmol/s fed to tower height of 30 m can achieve residual H_2S composition below 5 ppm.

Solvent circulation rate of 2 kmol/s necessitates packing diameter of 5.35 m to maintain tower operation at 70% flood condition. Further rise in solvent flow rate necessitates larger packing diameter (considering 70% flood) or more intensive tower operation (operation at more than 70% flood condition) as per Fig. 6. The product quality (composition of syngas) is governed by tower height, whereas the plant size (throughput of syngas stream and solvent stream) is governed by the packing diameter (and the resulting c/s area of the packed tower). Diameter and height of the packed tower determine the gross (bulk) volume of the packing material accommodated within the sections of the packed tower as shown in Fig. 7. The amount of packing material determines the capital cost of the packed tower, besides another major component of the cost of the containment (containment is the metallic shell enveloping the pressurized syngas and solvent in the packed tower).

CO_2 enters the H_2S Capture Plant as a compound dissolved in solvent (its amount is in direct proportion of the rate of solvent circulation). 34.8 mol% of the circulating solvent consists of dissolved CO_2. Recirculation of this CO_2 back to the H_2S Absorber (through H_2S Enrichment) causes additional power consumption by syngas recirculation compressor as described in Fig. 8.

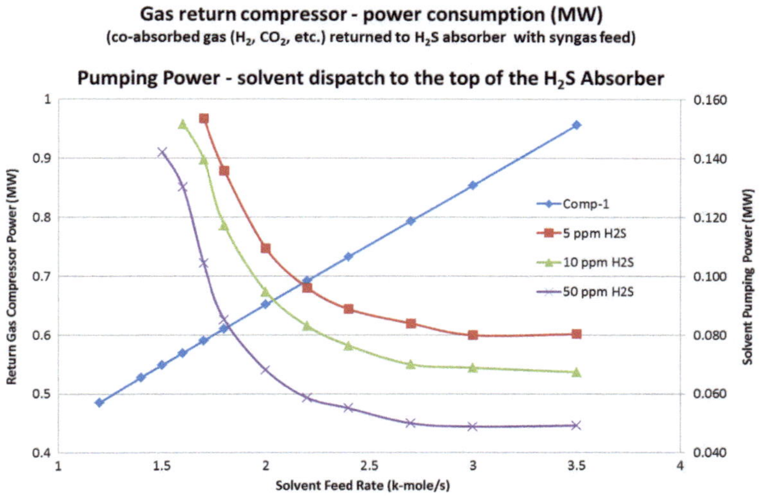

Fig. 8 Susceptibility of power demand (for circulation of solvent and recycle syngas and their total) with reference to the solvent circulation rate

Power consumption of solvent pump is in direct proportion of the rate of solvent circulation and the height of solvent inlet to packed tower (static head). Higher mass transfer from syngas to solvent (enabled by higher solvent circulation rate) results in the requirement of shorter height of the packed tower. The overall effect of variation of solvent circulation rate and pumping head results in variation of pumping power demand as described in Fig. 8. In addition to the height of the packing, the height of the tower also consists of the height of dead volume such as solvent distributor, re-distributor and collector, the space between packed sections for manhole, welding allowance, etc. The height of the tower due to these features would also contribute to the requirement of additional static head in solvent supply pipeline (which is neglected for the calculation of power consumption for solvent pumping in this analysis).

7 Final Design

RSR (Raschig Super-Ring) No. 1 is considered as packing material for the finalized process configuration designed for 70% flood condition. Diameter of packed section is 5 m and total height of packed section is 21 m. The syngas feed condition is as per Table 3. 2.1 k-mole/s (350.891 kg/s) solvent is fed at 25 °C at 40 bar-a pressure at pump discharge at ground level. Solvent composition is described in Table 4 (units are same as Table 3). The design and operating parameters of H_2S Absorber for the finalized process configuration are described in Table 5. H_2S concentration of 2.8 ppm

Table 5 Specification and performance of H_2S absorber

H_2S absorber	Specification	Performance achieved
Flood in the Packing	70% flood	69.1–69.3% flood along 21 m packing height
Packing pressure drop	Reasonable	3.5 kPa for packed height of 21 m, i.e. 166.6 Pa/m
Superficial Velocity Syngas (4.51–4.54 m^3/s)		0.236 m/s (849.6 m/h) 0.231 m/s (831.6 m/h)
Solvent (0.343 m^3/s)	10–200 m/h [19]	0.0174 m/s (62.64 m/h) 0.0176 m/s (63.36 m/h)
H_2S in processed sweeter syngas at outlet	<5 mg/Nm^3 [a] <5 ppmv	4.05 mg/Nm^3 2.8 ppmv
Solvent (DMEPEG) loss	Minimize	Negligible (as per simulation convergence tolerance)

[a] H_2S Specification for Natural Gas Pipeline [20, 21] is 5 mg/Nm^3

Table 6 Composition of processed sweeter syngas

H_2O	H_2S	CO_2	DMEPEG	CH_4	H_2	He	CO	N_2	Ar
0.031%	2.8 ppm	41.7%	0.016	0.0444%	47.1%	0.238%	1.52%	8.51%	0.828%

Table 7 Composition of H_2S-loaded solvent

Species	H_2O	H_2S	CO_2	DMEPEG	Total
Flow rate (kmol/s)	0.375	3.642 e−2	0.664	1.138	2.227
Composition (mole %)	16.84%	1.635%	29.816%	51.100%	100%

Table 8 Recovery of syngas constituents from H_2S capture

H_2S	H_2	CO_2	N_2
99.895% (19.13 mol/s)	Nearly 100% (3.310 kmol/s)	99.83% (2.355 kmol/s)	203.4% (0.6 kmol/s)

is achieved for the finalized configuration using packed height of 21 m for RSR No. 1 [for solvent feed rate of 2.1 kmol/s at 25 °C (including dissolved CO_2 and other gases)]. The performance of finalized H_2S Capture Process is compared against the industry standard specification in Table 5. Composition of processed (sweeter/H_2S lean) compounds of Syngas is and H_2S loaded solvent are as per Tables 6 and 7. Overall recovery of compounds of Syngas is as per Table 8.

Table 9 Options for H_2S Absorber packed tower

Option 1	IMTP of Koch-Glitch (Diameter = 5.2 m) IMTP 40: 21 m height (70.0% flood)
Option 2	I-Ring of Sulzer (Diameter = 5.2 m) I-Ring 40: 21 m height (70.0% flood)
Option 3	RSR of Raschig (Diameter = 4.95 m) RSR No. 1: 21 m height (70.0% flood)

8 Conclusion

Detailed process design is developed for H_2S Absorption from syngas (with estimation of equipment size). H_2S in processed sweeter syngas being 2.8 ppmv is less than 5 mg/Nm^3 which itself is less than 5 ppmv. Packed Tower Height of 21 m is sufficient if RSR 1 packing is used for tower diameter of 5 m to process 450 TPH syngas having 0.31% H_2S (Table 9). Various other options are suggested for packing material. Sizing and rating of various process equipment and machinery can be done based on this work. This publication can be used as model case to up/downscale similar process or to analyse various process configurations. The basic strategy for process design is illustrated.

Acknowledgements The Author acknowledges the support of DECARBit Project funded by European Union (FP7 Grant No. 2119171) for part of this work.

References

1. IEA flagship publication World Energy Outlook (WEO). http://www.iea.org/weo/
2. BP Statistical Review of World Energy. https://www.bp.com/en/global/corporate/energy-economics/statistical-review-of-world-energy.html
3. Dave, A.: Ph.D. thesis titled "Techno-Economic assessment of IGCC systems with CO_2 capture using novel absorption/ desorption technologies", ISNI 0000 0004 5372 1661, uk.bl.ethos.675929 (EThOS ID). http://ethos.bl.uk/OrderDetails.do?uin=uk.bl.ethos.675929
4. A webpage of Digital Refining website regarding acid gas removal by physical solvent DMEPEG. http://www.digitalrefining.com/data/digital_magazines/file/1767276991.pdf
5. Dave, A., Dave, M., Rezvani, S., Huang, Y., Hewitt, N.: Process design for H_2S Enrichment in physical solvent DMEPEG. Int. J. Greenhouse Gas Control **50**, 261–270 (2016). https://doi.org/10.1016/j.ijggc.2016.02.004
6. ProTreat software for rate-based mass-transfer simulation from Optimized Gas Treatment (OGT), USA. http://ogtrt.com
7. ProTreat Software Publications. http://ogtrt.com/information/publications
8. Dave, A., Dave, M., Rezvani, S., Huang, Y., Hewitt, N.: Process design for CO_2 absorption from syngas using physical solvent DMEPEG. Int. J. Greenhouse Gas Control **49**, 436–448 (2016). https://doi.org/10.1016/j.ijggc.2016.03.015
9. Burr, B., Lyddon, L.: A Comparison of Physical Solvents for Acid Gas Removal. Bryan Research & Engineering, Inc., Bryan, Texas, U.S.A. http://www.bre.com/portals/0/technicalarticles/a%20comparison%20of%20physical%20solvents%20for%20acid%20gas%20removal%20revised.pdf

10. Sciamanna, S.F., Lynn, S.: Solubility of hydrogen sulfide, sulfur dioxide, carbon dioxide, propane, and n-butane in poly(glycol ethers). Ind. Eng. Chem. Res. **27**, 492–499 (1988)
11. Greene, P.A., Kutsher, G.S.: Carbon dioxide removal and recovery new polyglycol ether processes. In: Proceedings of the 1965 Laurance Reid Gas Conditioning Conference, Norman, OK
12. Hartel, G.H.: Low-volatility polar organic solvents for sulfur dioxide, hydrogen sulfide, and carbonyl sulfide. J. Chem. Eng. Data **30**, 57–61 (1985)
13. Sweny, J.W.: High CO_2-high H_2S removal with SELEXOL solvent. In: Proceedings of the 59th Annual Gas Processors Association Convention, pp. 163–166
14. Aionicesei, E., Skerget, M., Knez, Z.: Measurement and modeling of the CO_2 solubility in poly(ethylene glycol) of different molecular weights. J. Chem. Eng. Data **53**, 185–188 (2008)
15. Xu, Y., Schutte, R.P., Hepler, L.: Solubilities of carbon dioxide, hydrogen sulfide and sulfur dioxide in physical solvents. Can. J. Chem. Eng. **70**, 569–573 (1992)
16. Webpage of Dr. Ashok Dave on Google Scholar and LinkedIn at https://scholar.google.co.in/citations?user=qZS4WHcAAAAJ&hl and https://www.linkedin.com/in/ashok-dave-2b348514/
17. Sciamanna, S.F., Lynn, S.: Solubility of hydrogen sulfide, sulfur dioxide, carbon dioxide, propane, and n-butane in poly(glycol ethers). Ind. Eng. Chem. Res. **27**, 492–499 (1988)
18. Hartel, G.H.: Low-volatility polar organic solvents for sulfur dioxide, hydrogen sulfide, and carbonyl sulfide. J. Chem. Eng. Data **30**, 57–61 (1985)
19. Kolev, N., Nakov, S., Ljutzkanov, L., Kolev, D.: Institute of Chemical Engineering, Bulgarian Academy of Sciences, Acad. G. Bonchev St., Bl.103, 1113 Sofia, Bulgaria. Comparison of the effective surface area of some highly effective random packings third and forth generation. Symposium Series No. 152, IChemE (2006) http://www.nt.ntnu.no/users/skoge/prost/proceedings/distillation06/CD-proceedings/paper073.pdf
20. Page No. 3 technical report. Natural Gas Specification Challenges In The Lng Industry, on http://www.kbr.com/Newsroom/Publications/Technical-Papers/Natural-Gas-Specification-Challenges-in-the-LNG-Industry.pdf
21. Website of National (Gas) Grid, UK. http://www2.nationalgrid.com/uk/industry-information/gas-transmission-system-operations/gas-quality/

Effect of Shoulder Diameter on Friction Stir Welding of Al 6061 to SS 304

Deepika M. Harwani and Vishvesh J. Badheka

Abstract Friction stir welding, a solid-state welding process, is an emerging technology for welding various dissimilar metals. It is characterized by rotation and translation of a non-consumable tool along the centerline of the weld. Welding of aluminium and steel alloys is gaining worldwide momentum as this combination is utilized prominently in fabrication of vehicles and shipbuilding industries. The present study investigated the effect of shoulder diameter on the tensile strength, microhardness and microstructure for the butt joint of Al 6061 to SS 304. It was observed that the shoulder diameter of 22 mm yielded maximum tensile strength than 18 and 20 mm shoulder diameters, respectively. The same effect was observed for percentage elongation for the joints. The maximum microhardness in the nugget region was found to be about 100 VHN.

Keywords Friction stir welding · Aluminium · Stainless steel

1 Introduction

Invented by Wayne Thomas and patented by TWI (The Welding Institute), Cambridge, United Kingdom in 1991, friction stir welding process has been researched upon extensively to improve and expand its range of applications. A conventional fusion method involves welding of the base materials in their molten state, which poses a great challenge when welding dissimilar metals, while friction stir welding (FSW) is a process that welds the metals in the solid state by causing localized plastic

D. M. Harwani (✉)
Department of Mechanical Engineering, R.C. Technical Institute,
Sola, Ahmedabad, Gujarat, India
e-mail: deeps2908@gmail.com

V. J. Badheka
Department of Mechanical Engineering, School of Technology,
Pandit Deendayal Petroleum University, Gandhinagar, Gujarat, India
e-mail: drvishvesh@gmail.com

D. Deb et al. (eds.), *Innovations in Infrastructure*, Advances in Intelligent Systems and Computing 757, https://doi.org/10.1007/978-981-13-1966-2_31

deformation. As the fusion welding of dissimilar systems is quite rigorous, FSW is now widely employed for welding dissimilar metals, owing to its low temperature which helps to prohibit the formation of secondary undesirable phases [1].

Demonstrating several advantages over fusion welding processes, like reduced concentration of defects, eco-friendliness and versatility, this welding technique has joined the league of other processes for welding dissimilar metal alloys. The process of combining lightweight materials such as aluminium and high strength ferrous alloys like stainless steel has found potential applications in automotive and other industrial sectors. The combination of steel and aluminium finds applications in many fields ranging from aquatic transportation machines to the automotive parts like tailor welded blanks, chassis module and windshield frames. Due to the large differences in the mechanical and physical properties, it has always been challenging to produce a sound weldment of aluminium and stainless steel possessing high strength. Generally, the solubility of Fe in Al matrix is very limited, which eventually leads to the formation of brittle intermetallic compounds (IMCs). With the help of many experimental studies and analysis of the binary phase diagram of Al–Fe, it has been reported that during welding of Al alloy and steel brittle IMCs of Fe_2Al_5, $FeAl_2$, FeAl and Fe_3Al easily form at the weld interface. Such IMCs result in crack propagation and poor strength of the weld and should be controlled with the optimization of the weld parameters [2–5].

An experimental investigation was carried out to butt weld 2-mm-thick sheets of Al 5083 and SS 400 to analyse the effects of pin rotation speed, pin diameter and position of pin axis on the tensile strength and microstructure of the joint. It was reported that it was impossible to weld aluminium alloy to steel by rotating the pin in counterclockwise direction, and the maximum tensile strength obtained was about 86% of the Al base metal [6, 7].

Chen [8] successfully achieved a butt joint between AA 6061 and SS 400 mild steel and employed ANOVA to analyse the experimental data considering four controllable parameters as tool rotation speed, traverse speed, tool tilt angle and pin tool diameter. It was found that lower traverse speed and rotation speed will give higher impact value, and the tensile strength of the joint obtained was 76% of the AA 6061 base material.

Another experimental study performed friction stir welding to join Al 5186 to mild steel (St 52) having thickness 3 mm and indicated that generation of IMCs in the weld nugget was prominent at lower welding speeds (14 mm/min). It was concluded that the joint strength increased with the decrease in tool plunge depth and simultaneous increase in weld speed [9]. Additional to the tool geometry and careful selection of process parameters, proper tool offsetting is also important in case of welding of dissimilar metals like aluminium and steel. As reported in earlier studies, it has been observed that better joint strength is obtained when the pin is kept at an optimum offset from the weld centre towards the aluminium side. It will also decrease the tool wear [2, 6, 9].

One recent study experimentally investigated the feasibility of dissimilar friction stir weld joint between AA5083-H321 (an alloy of aluminium and magnesium) and A316L austenitic stainless steel and led to the inference that welding traverse velocity has an essential controlling influence on the formation of intermetallic compounds as well as tunnelling defects. The peak temperature during the welding was unaffected by the tool traverse velocity [10].

Laser welding process has also been widely researched upon for joining dissimilar Al and steel alloys. Causing the fusion of the two metals at the interface, laser welding process results in thick IMCs which are undesirable and decrease the durability of the joints. Good quality weld joints can be produced by preheating one metal with respect to another in welding of dissimilar systems, especially when there is a large difference in the melting point of both metals. Some hybrid welding techniques like laser-brazing and laser-arc have also been experimented with and studied to decrease the IMC thickness and increase the mechanical strength of the joints. It has been found that laser welding of aluminium alloys was quite difficult. Cracks were formed due to the damage to the heat treatment of the alloys as the intense laser beam heats the metals beyond their annealing point [11–13].

This particular study aims to investigate the effect of various shoulder diameters on the strength of joint of Al 6061-T651 to SS 304L, keeping the rest parameters unchanged.

2 Experimental Work

2.1 Materials and Welding Conditions

Pre-sized small plates of Al 6061-T651 (chemical composition as 1.03% Mg, 0.56% Si and 0.12% Mn) and SS 304L (chemical composition as 18.78% Cr, 8.2% Ni and 1.69% Mn) having dimensions $100 \times 50 \times 3$ mm were butt welded at the Welding Research Laboratory of School of Technology, Pandit Deendayal Petroleum University, Gandhinagar. The plates were machined and cut from the sheets of $300 \times 50 \times 3$ mm dimensions with the help of a shearing machine. The ultimate tensile strength of parent metals was tested and found to be about 668 and 302 MPa for SS 304 and Al 6061-T651, respectively. Three coupons were welded by varying three different shoulder diameters, keeping the rest parameters unchanged.

2.2 Experimental Setup and Welding Conditions

A special purpose vertical milling machine has been established there for friction stir welding with ranges for rotational speed as 35–1500 RPM, welding speed as 20–800 mm/min and tilt angle as 0°–50° with motor power of 3 HP. The experimental setup is shown in Fig. 1.

Earlier investigations have focused on the parameters like tool rotation, welding speed, pin size, tool plunge depth and tool tilt angle. So, for this experimental work,

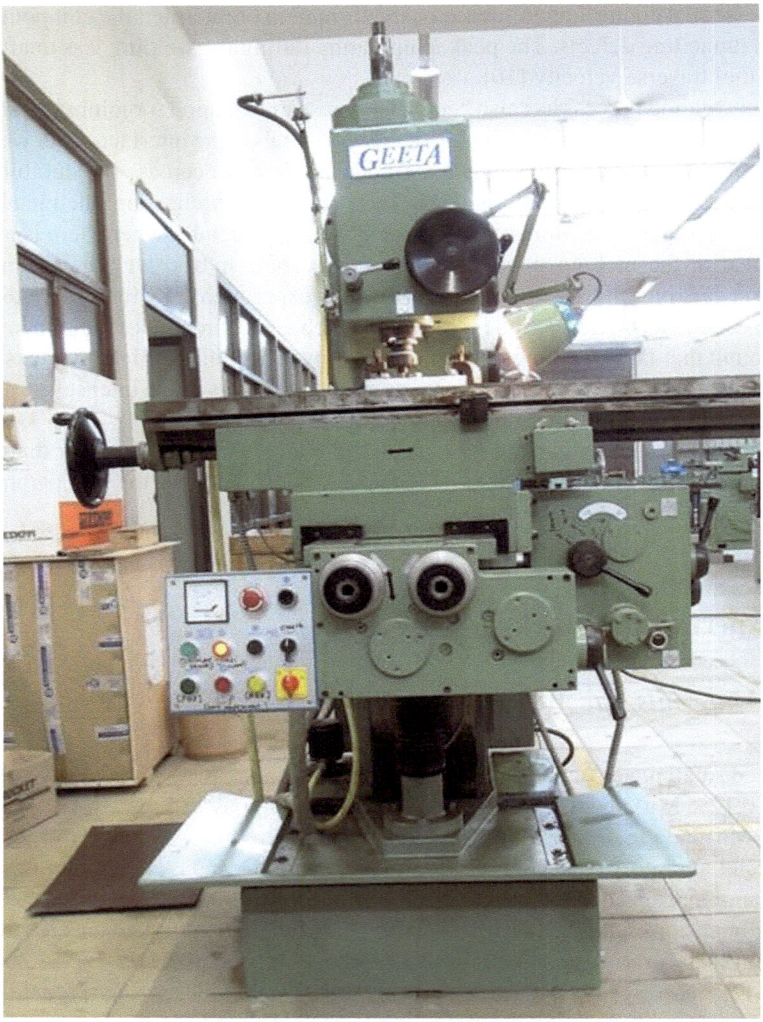

Fig. 1 Setup for FSW on special purpose VMC (available at PDPU Welding Research lab under the sponsored project of ISRO E33011/60/2010-V)

the variable parameter selected was tool shoulder diameter and the rest parameters like tool rotation speed, welding speed, tool tilt angle and tool pin offset were kept constant. The summary of the process parameters is shown in Table 1.

Table 1 Welding parameters selected for the experimental runs

Sample ID	Shoulder diameter (mm)	Tool rotational speed (RPM)	Welding speed (mm/min)	Tool tilt angle (°)	Tool pin offset (mm)
SD18	18	1500	31.5	2	2
SD20	20				
SD22	22				

Fig. 2 Tungsten carbide tool

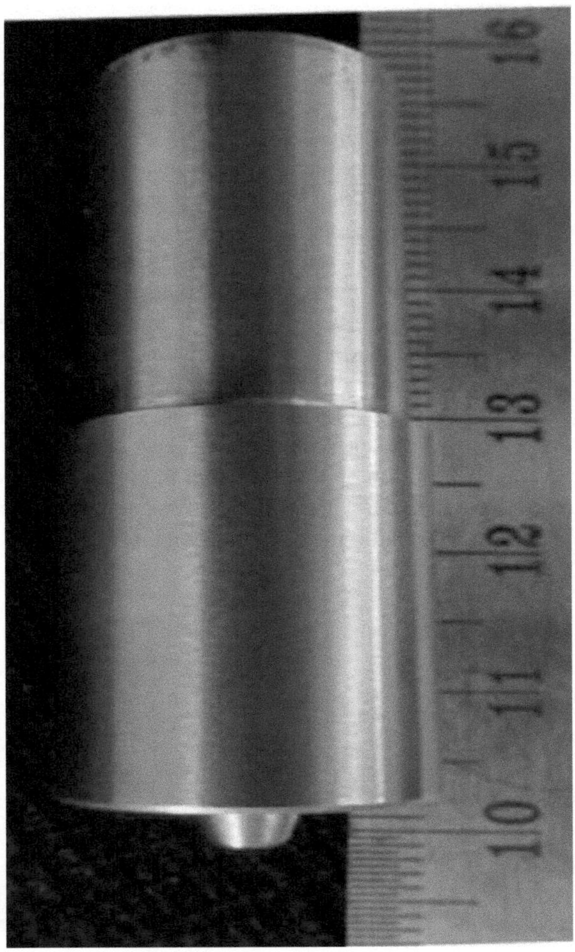

2.3 *Tool Material and Design*

A non-consumable rotational tool is used in friction stir welding which causes the friction between two metals to be joined by traversing through them. Its rotation applies the mechanical pressure and intermixes the softened metals at the place of the joint, causing localized deformation. For this study, an unthreaded tungsten carbide alloy tool (with 88% tungsten carbide + 12% cobalt by weight), having a tapered cylindrical profile, was employed for welding the aluminium to stainless steel (Fig. 2).

The tool material was procured from Sinter Sud Pvt. Ltd., Italy and was given cryogenic heat treatment after machining. The hardness of the tool was 92.8 HRA, and the torsional resistance was about 2100 N/mm^2. The pin had root diameter and tip diameter as 5 and 3 mm, respectively. The tool was specially designed and machined for welding of plates having thickness of 3 mm. A stainless steel fixture (having dimensions 230 × 195 × 25 mm) was specially designed for the friction stir welding of Al plates, as shown in Fig. 3.

With 18, 20 and 22 mm shoulder diameters, three dissimilar coupons were joined as shown in (Fig. 4). With the tool rotating in clockwise direction, SS 304 was kept on the advancing side and Al 6061 was kept on the retreating side during the welding. The earlier investigations for welding of aluminium to steel have suggested that it is not possible to carry out such dissimilar welding with aluminium on the advancing side and rotating the tool in counterclockwise direction [6]. It has also been observed that when Al is kept on advancing side, the weld is restricted to the top surface only and shows little bonding within the plate. Improved mechanical properties are obtained in the joint when the softer base material is kept on the retreating side while harder one is kept on the advancing side [1].

Fig. 3 Plates fixed in the fixture

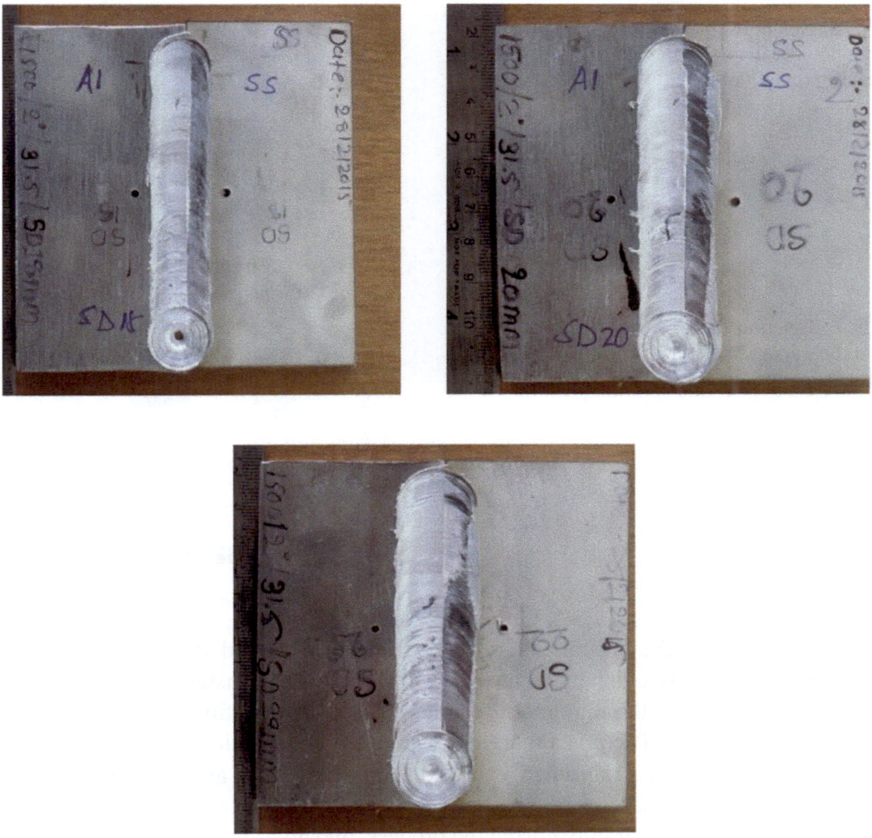

Fig. 4 Al–SS friction stir welded coupons employing tool with three different shoulder diameters

3 Results and Discussions

3.1 Tensile Test

The joint strength and ductility were tested at Electrical Research & Development Association (ERDA), Baroda with the help of a well-calibrated Universal Testing Machine. The tensile specimens were machined on a milling machine as per the standard ASTM E8M-04, tensile specimens having gauge length of 32 mm and shoulder of 10 mm. The tensile tests were carried out at room temperature with a constant load application rate of 1.5 mm/min. The relationship between the tensile strength of the joints and the shoulder diameters is shown in Fig. 5. Similar relationship between elongation and the shoulder diameters is illustrated in Fig. 6.

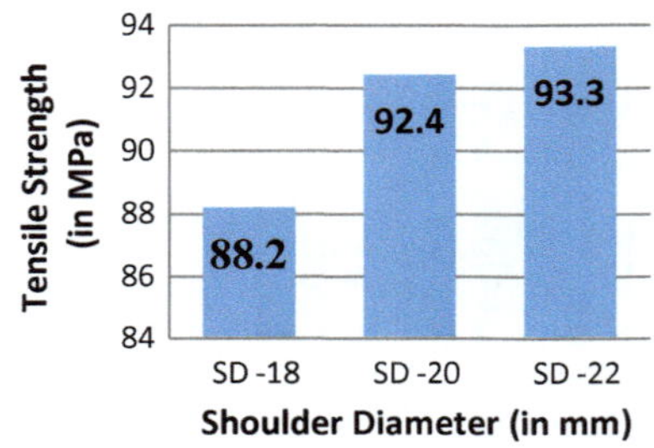

Fig. 5 Relationship between the shoulder diameter and tensile strength of joints

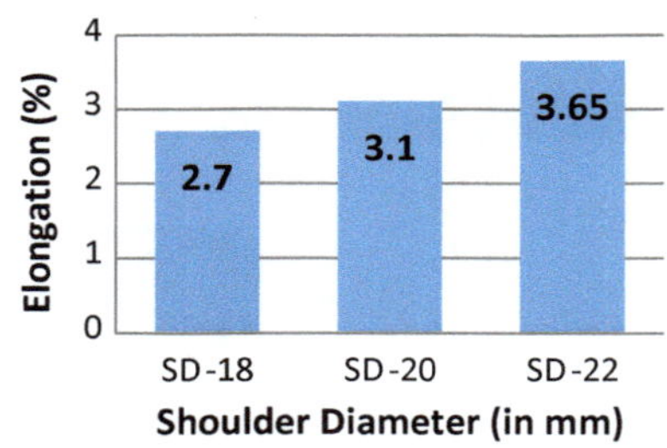

Fig. 6 Relationship between the shoulder diameter and elongation of joints

The fracture of the specimens was noted along the weld interface. It can be observed that the shoulder diameter of 22 mm with the constant offset of 2 mm on the Al side yields the maximum tensile strength and maximum elongation out of the three coupons. There is a marginal increase in the tensile strength along with elongation tendency of joints when observed from 18 to 20 mm and from 20 to 22 mm diameters of shoulders, respectively.

To measure the temperature during the process, two holes were drilled 2 mm away from the nugget zone and two K-type contact thermocouples were inserted in the holes, one on the advancing and one on the retreating side. As observed from Fig. 7, the temperature on the advancing side was more than the retreating side during the welding. This is because the heat generated during the process depends on the relative velocity of the tool, which is higher on the advancing side than the retreating side.

With the increase in the shoulder diameter of the welding tool, the peak temperature achieved during the welding increases and with this more SS particles are being scattered in the aluminium matrix which yields higher strength with higher shoulder diameter. The shoulder diameter plays an important role in the heat generation and the material flow during the welding process. So, with the increase in the shoulder diameter, more heat is generated which results in the higher peak temperature.

Out of the three coupons, shoulder diameter 22 has the temperature ranges, for both advancing and retreating sides, between shoulder diameter 18 and shoulder diameter 20. The maximum tensile strength of the friction stir welded joint comes out to be about 31% of the Al base metal. The percentage elongation of the joint also comes out to be maximum (3.65%) for 22-mm-diameter welding parameter.

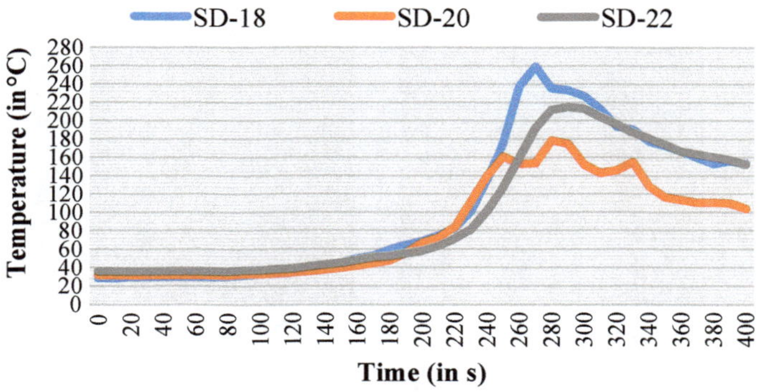

(a) Temperature Profile for SS 304 (Advancing Side)

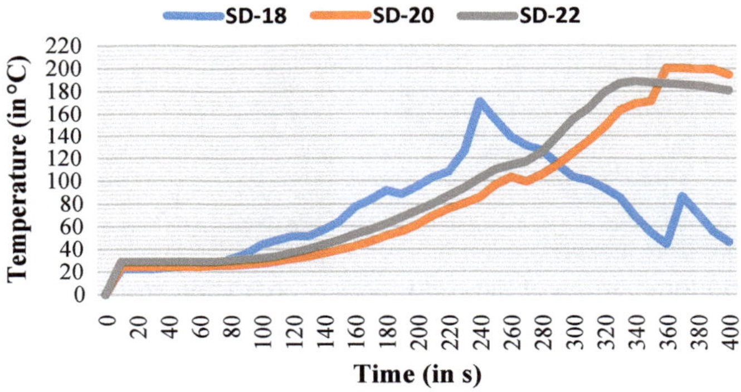

(b) Temperature Profile for Al 6061 (Retreating Side)

Fig. 7 Temperature profile of Al–SS joints

3.2 *Microstructure*

The maximum tensile strength was obtained for the joint welded with 22-mm-diameter shoulder. Optical micrographs of this joint were obtained under various magnifications to examine the microstructure and the formation of any intermetallic compounds. The etchant used for Al 6061 was the mixture of hydrochloric acid and nitric acid (having composition 60 ml of HCl + 40 ml of HNO_3). And the etchant used for SS304 region of the weld was having the composition of Aqua Regia 2 ml HNO_3 + 1 ml of HCl.

Figure 8 illustrates the scratching of SS particles in the Al matrix with the middle portion of the weld nugget having finer SS particles scattered than the top and bottom portion.

(a) Top portion of the weld

(b) Mid-portion of the weld

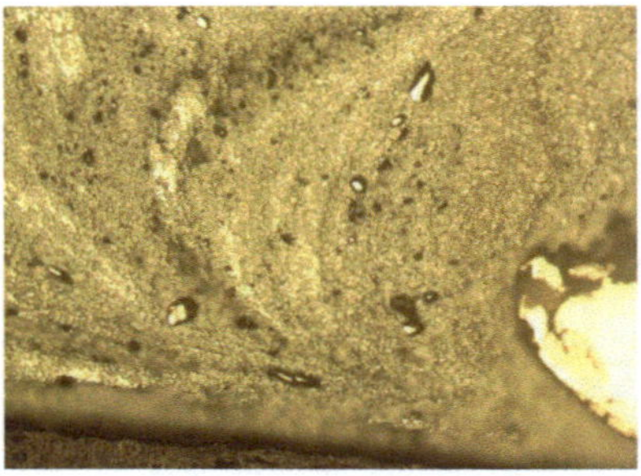

(c) Bottom portion of the weld

(d) Microstructure of Al 6061 base metal with 200× magnification

(e) Microstructure SS 304 base metal with 200× magnification

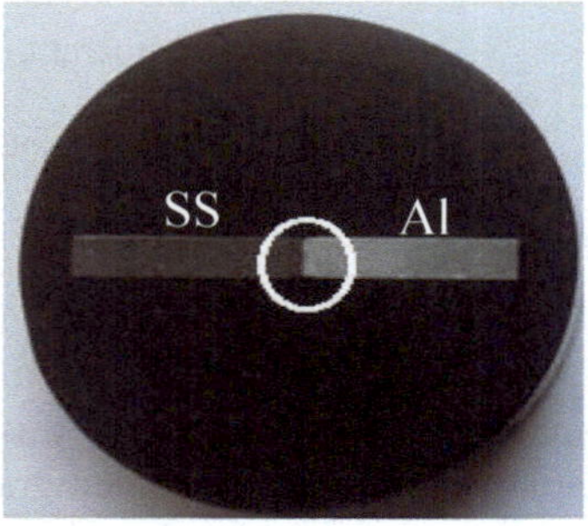

(f) Macrostructure of FSW joint welded with SD 22 tool

Fig. 8 Microstructure of the joint along weld interface with 200× magnification

3.3 *Microhardness Distribution*

Vickers microhardness distribution in the joint having maximum tensile strength (shoulder diameter 22 mm) is shown in Fig. 9. The load applied was 300 g at the spacing of 1 mm across the joint, and the dwell time was chosen as 15 s. The maximum

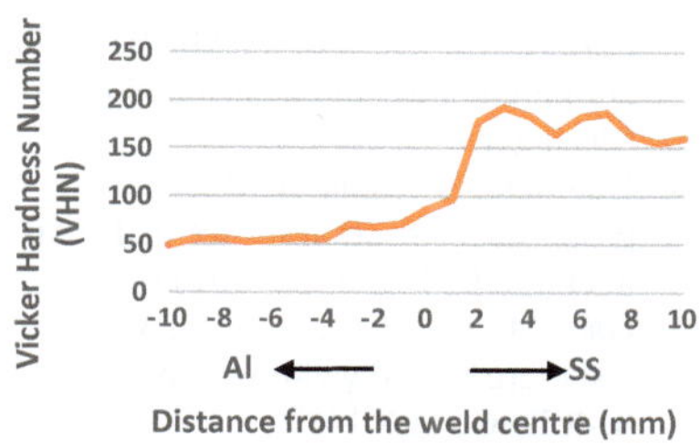

Fig. 9 Microhardness profile for FSW joint (VHN vs. Distance from the centre)

hardness achieved was about 192.5 VHN, which was at a distance of 3 mm from the nugget towards the SS side. In the Al side, the hardness remained more or less same and showed an increase while moving towards the nugget region because of the stirring action.

4 Conclusions

The dissimilar metals' butt joints for Al 6061 alloy and SS 304 alloy were successfully produced by the method of friction stir welding followed by the tensile tests and Vickers microhardness tests. The main results obtained from this study are summarized below:

- The tensile strength, percentage elongation and the joint efficiency increase marginally with the increase in the shoulder diameter, from 18 to 20 mm and from 20 to 22 mm, respectively, for friction stir welding process.
- Maximum tensile strength was achieved with the shoulder diameter 22 mm which was about 93 MPa, about 31% of the base Al metal.
- The microstructure study revealed that more SS particles scratched into the Al matrix for the joint produced with shoulder diameter 22 mm which resulted in better tensile strength as compared to the other joints.
- The hardness of the weld nugget was found significantly lower than that of base metal SS 304 and the maximum hardness obtained in the nugget was about 100 VHN.

References

1. Sadeesh, P., Venkatesh Kannan, M., Rajkumar, V., Avinash, P., Arivazhagan, N., Devendranath Ramkumar, K., Narayanan, S.: Studies on friction stir welding of AA 2024 and AA 6061 dissimilar metals. Procedia Eng. **75**, 145–149 (2014). (Elsevier Ltd.)
2. Sajan, S.G., Meshram, M., Pankaj, P.S., Dey, S.R.: Friction Stir Welding of aluminum 6082 with mild steel and its joint analyses. Department of Materials Science and Engineering, Indian Institute of Technology Hyderabad, Ordnance Factory Estate, Yeddumailaram (2013)

3. Lan, S., Liu, X., Ni, J.: Microstructural Evolution during friction stir welding of dissimilar aluminum alloy to advanced high-strength steel. Int. J. Adv. Manuf. Technol. **7531**(2) (2015) (Springer)
4. Habibnia, M., Shakeri, M., Nourouzi, S., Besharati Givi, M.K.: Microstructural and mechanical properties of friction stir welded 5050 Al alloy and 304 stainless steel plates. Int. J. Adv. Manuf. Technol. **014**(6306-5) (2014)
5. Gao, M., Chen, C., Mei, S., Wang, L., Zeng, X.: Parameter optimization and mechanism of laser–arc hybrid welding of dissimilar Al alloy and stainless steel. Int. J. Adv. Manuf. Technol. **014**(5996-z) (2014)
6. Kimapong, K., Watanabe, T.: Friction stir welding of aluminum alloy to steel. Weld. J. **83**, 277–282 (2004)
7. Watanabe, T., Takayama, H., Yanagisawa, A.: Joining of aluminium alloy to steel by friction stir welding. J. Mater. Process. Technol. **178**, 342–349 (2006)
8. Chen, T.: Process parameters study on FSW joint of dissimilar metals for aluminium-steel. J. Mater. Sci. **44**, 2573–2580 (2009)
9. Dehghani, M., Amadeh, A., Akbari, S.A.A.: Mousavi: investigations on the effects of friction stir welding parameters on intermetallic and defect formation in joining aluminum alloy to mild steel. Mater. Des. **49**, 433–441 (2013)
10. Rafiei, R., Ostovar iMoghaddam, A., Hatami, M.R., Khodabakhshi, F.: Microstructural characteristics and mechanical properties of the dissimilar friction-stir butt welds between an Al-Mg alloy and A316L stainless steel. Int. J. Adv. Manuf. Technol. **016**(9597-x) (2016)
11. Shah, L.H., Ishak, M.: Review of research progress on aluminium-steel dissimilar welding. Mater. Manuf. Process. (2014). https://doi.org/10.1080/10426914.2014.880461
12. Kah, P., Rajan, R., Martikainen, J., Suoranta, R.: Investigation of weld defects in friction-stir welding and fusion welding of aluminium alloys. Int. J. Mech. Mater. Eng. **10**(26), 1–10 (2015)
13. Meco, S., Pardal, G., Ganguly, S., Williams, S., McPherson, N.: Application of laser in seam welding of dissimilar steel to aluminium joints for thick structural components. Opt. Lasers Eng. **67**, 22–30 (2015) (Elsevier Ltd.)

Experimental Investigation of High-Speed Turning of INCONEL 718 Using PVD-Coated Carbide Tool Under Wet Condition

Ankur Chaurasia, Vishal Wankhede and Rakesh Chaudhari

Abstract With the advancement in the field of materials, the material removal processes with high surface accuracy have also become more and more challenging. This paper reports use of PVD-coated tungsten carbide tool CNC lathe machine under wet condition using cutting oil for high-speed turning of Inconel 718. Effect of process parameter—speed, feed, and depth of cut on response parameter change in surface roughness and tool wear—has been studied on INCONEL-718 using ANOVA. Also, the effect of given parameters on the tool wear and chip formation was investigating through SEM micrographs. It has been found that speed and feed have significant effect on percentage change in surface roughness. It is also cleared that at low depth of cut, low speed and moderate feed, the chip is a continuous type whereas at high speed, low feed and high depth of cut, formation of chip becomes sawtooth type.

Keywords PVD · SEM · ANOVA · INCONEL-718 · CNC

1 Introduction

The progress of recent industry tends to employ an advanced material which appreciates its technological properties, better performance in products, or enable new designs. Nickel-based alloys owing to their good mechanical properties in high tem-

A. Chaurasia (✉) · R. Chaudhari
Department of Mechanical Engineering, Pandit Deendayal Petroleum University, Gandhinagar, India
e-mail: ankur.chaurasia@sot.pdpu.ac.in

R. Chaudhari
e-mail: rakesh.chaudhari@sot.pdpu.ac.in

V. Wankhede
Department of Industrial Engineering, Pandit Deendayal Petroleum University, Gandhinagar, India
e-mail: vishal.wankhede@sot.pdpu.ac.in

D. Deb et al. (eds.), *Innovations in Infrastructure*, Advances in Intelligent Systems and Computing 757, https://doi.org/10.1007/978-981-13-1966-2_32

peratures are commonly used in the aerospace industry. The most popular types of heat resistant super alloys include Inconel-718, Waspaloy, and Inconel-625. Inconel alloys are corrosion resistant, and oxidation materials well suited for service in intense environments subjected to heat and pressure. When heated, Inconel forms a stable, thick, passivating oxide layer shielding the surface from attack. Nowadays, CNC turning process is an advanced process which used for material removal of advanced engineering materials. With the help of this process, internal as well as external area can be machined. A lot of work has been done on metal and alloys, but relatively less work has been made on superalloys. Recently, research work on nickel-based superalloys has rapidly increased due to its better properties and variety of industrial application. Industries are focusing on high-temperature strength and corrosion resistance material, and therefore Inconel-718, which has high-temperature strength and corrosion resistance, was chosen for study. CNC turning process is useful for material removal with very high surface finish and less tool wear for INCONEL-718. In CNC turning process, the material is removed by removing the chips using the specific cutting tool insert. The turning process has been in the presence of cutting oil. The PVD-coated tungsten carbide insert is used in wet condition for high-speed turning of INCONEL-718 in CNC lathe machine. The basic purpose of literature review is to know about the process. Here, with the help of literature review, we know about the process parameter and their relationship to each other. How the process parameter affects the surface roughness during finishing of Inconel-718 on the CNC lathe machine? From the observation of the literature review, it has been found that many researchers carried out the theoretical and experimental studies on Inconel 718. Share their experience about the effect of parameters and process performance. Kamata and Obikawa [1] reported that MQL carrier gas plays significant role in cooling of cutting tool using argon as the carrier gas in turning of Inconel 718 with different types of coated carbide tools. Joshi and Brahmankar [2] studied on finding out the effect of cutting edge geometry on the surface integrity of high-speed turning of Inconel 718. Thakur et al. [3] carried out an experimental study on the Inconel 718 during turning using tungsten carbide insert (K20) tool. They examine the effect of machining parameters on the cutting temperature, cutting force, specific cutting pressure, surface finish, and tool wear. Cantero et al. [4] performed tool wear patterns analysis in finishing turning of Inconel grade 718 (Ni alloys). Patil et al. [5] performed relative analysis of high-speed machining of Inconel grade 718 with coolant as compressed cold carbon dioxide gas in dry condition. Tebaldo et al. [6] focused on substitution or removal of conventional lubricants those results in the reduction of environmental impact, operating costs, filtration, and metalworking fluids disposal. Lima et al. [7] studied the tool wear and tool life of SiAlON, whisker, and mixed ceramic (ceramic tool) during turning of Inconel grade 751 in oxygen and argon atmospheres. Design of experiment has been used to carry out the experiments.

From the literature review, it has been observed that very less attempts have been made study of machining parameter effect on high-speed turning of Inconel 718 using PVD-coated tungsten carbide insert using aqua cut solution cutting oil. This type of cutting tool satisfied our main objective "economic machining of Inconel on CNC lathe machine" compared to CBN and diamond cutting tool.

2 Experimental Setup and Experimentation

2.1 Selection of Process Parameter

Based on literature review, recommended data by insert supplier, and available speed and feed range in available CNC lathe machine, value of the parameter has been selected. CNC lathe machine is used so that there is no any constraint of machine specification for selecting the process parameter. Experiment has been designed on Taguchi's $L_9(3^3)$ orthogonal array using the Minitab software version 17. The experiment is designed on the basis of $L_9(3^3)$ orthogonal array in which three factors at three level is to be taken. The orthogonal array is $L_9(3^3)$, having nine rows and each row has one trial condition with factors and level indicators in the number which shows the level. The process parameter and their levels have given below in Table 1.

Calculate the % improvement in the surface roughness given in Eq. (1).

$$\Delta \mathrm{Ra} = \left(\frac{\text{initial surface roughness} - \text{final surface roughness}}{\text{initial surface roughness}} \right) \times 100. \quad (1)$$

General regression model is used in this present study in order to check the effect of the various process parameters on percentage change in surface roughness. Minitab-17 is used for general regression model. ANOVA Table 2 is given below.

After analyzing analysis of variance table, it has been clear that speed and feed have significant effect on % change in surface roughness. The *P* value in the ANOVA table less than 0.05 for process parameter is considered to be significant.

Table 1 Process parameter and its level

Factor	Value		
	Level-1	Level-2	Level-3
Cutting speed (m/min)	20	25	30
Feed (mm/rev)	0.1	0.15	0.2
Depth of cut (mm)	0.05	0.075	0.1
Response: % (percentage) change in surface roughness (%ΔRa)			

Table 2 Analysis of variance (ANOVA) for percentage change in surface roughness

Source	DF	Adj SS	Adj MS	F value	P value	Contribution (%)
DOC	2	57.713	28.8567	102.08	0.010	16.427
Feed	2	139.595	69.7973	246.91	0.004	38.267
Speed	2	162.124	81.0622	286.76	0.003	45.145
Error	2	0.565	0.2827			0.1608
Total	8	351.329				100.0

3 Results and Discussion

The results obtained in the turning of INCONEL-718 on CNC lathe machine using the PVD-coated tungsten carbide insert under wet condition have been analyzed. This process is not much effective using the conventional cutting tool but with some coating of the cutting tool and some alteration in machining condition, the effectiveness of this process is improved. Additional coating on cutting tool increases the strength of cutting tip which results in better surface finish.

3.1 *Effect of the Process Parameter on the Percentage Change in Average Surface Roughness (%ΔRa)*

To verify the effect of process parameter on % change in surface roughness, every one of selected process parameters (speed, feed, and depth of cut) has been varied from lower level to higher level as given in Table 1 according to the orthogonal array. It has been observed that percentage change in surface roughness increased from 75.25% to 98.25% with the use of PVD-coated tungsten carbide cutting tool insert under wet condition. The main effect plot of effective parameter (speed, feed, and depth of cut) is shown in Fig. 1. It is observed from the graph that cutting speed and feed have significant effect on the percentage change in surface roughness. As cutting speed increases, tool wear rate rapidly increases resulting in decrease in the percentage change in surface roughness. At high-speed cutting operation, the strain rate is high in the shear zone, so more heat energy is generated. Also, it depicts higher temperature at the tool–chip interface. The time required for heat dissipation decreases and therefore temperature rises at the interface. Surface quality plays a critical role in performance and fatigue life of the element. It is obvious that increase in the depth of cut leads to decrease in the percentage change in surface roughness. For higher depth of cut, the material removal rate is higher which leads to increased temperature and wear of the cutting edge so that there is decrease in the surface roughness.

From the above graph (obtained from MINITAB software), it is clear that when depth of cut is low, moderate feed and low depth of cut give the optimum result of surface finish. The percentage contribution of process parameter on percentage change in surface roughness is calculated from ANOVA (Table 2) as shown in the pie chart shown in Fig. 2.

3.2 *Chip Analysis Using SEM*

During machining, chips produced in different forms depend on the tool, machined material, and the cutting conditions. The chip separation and chip formation mecha-

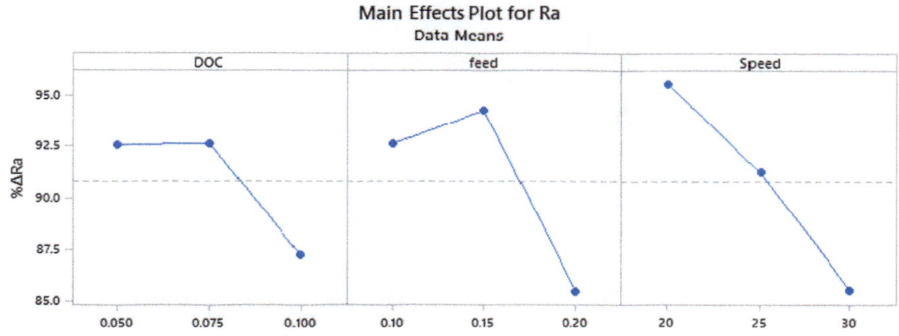

Fig. 1 Main effect plots for the effect of various process parameters on % change in average surface roughness during the turning operation

Fig. 2 Percentage contribution of effect of process parameter on % change in surface roughness

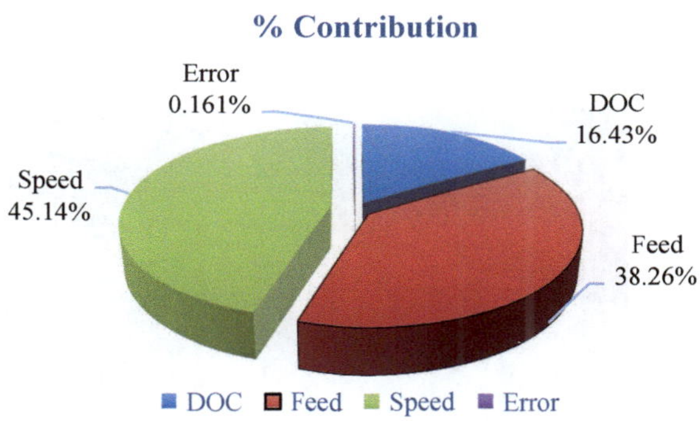

nism is by reason of severe strain rate occurs during the machining process. Because of the lower thermal conductivity of the material during Inconel 718 cutting, temperature can be very high locally in a few areas of the workpiece which results in thermal softening and reduces strain hardening capacity of the material. Consequently, in a narrowband of chips, the shear instability is observed. At low speed conditions, the chip shapes were observed as continuous, where in some cases sawtooth edged chips and segmented. The chip deformation is very high in a narrowband among the segments of chip and also inhomogeneous nature is observed. Abrasive sawtoothed edges were produced by shear localization in the chip. Continuous chips were observed at low cutting seed which changes to sawtooth at higher speed. The chips formed at different cutting speeds are collected and shown in Fig. 3. It shows direct relationship between speed and sawtooth type chips, i.e., the higher the speed, the higher the formation of sawtooth type chip. The critical speed accountable for the sawtooth chip formation in high-speed machining of Inconel 718 was found to be 25 m/min.

From above SEM image (Fig. 3), it is clear that at moderate feed, low depth of cut, and low speed, the segmented and sawtooth edged frequency in chip is very low.

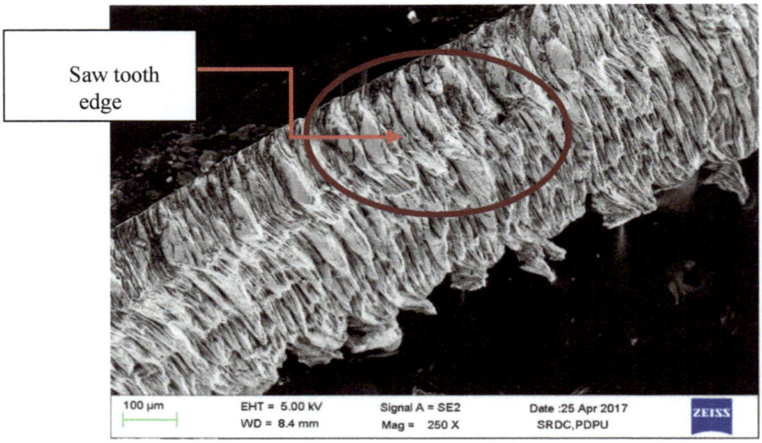

Fig. 3 SEM image of chip using speed 20 m/min, feed 0.15 mm/rev, DOC 0.05 mm

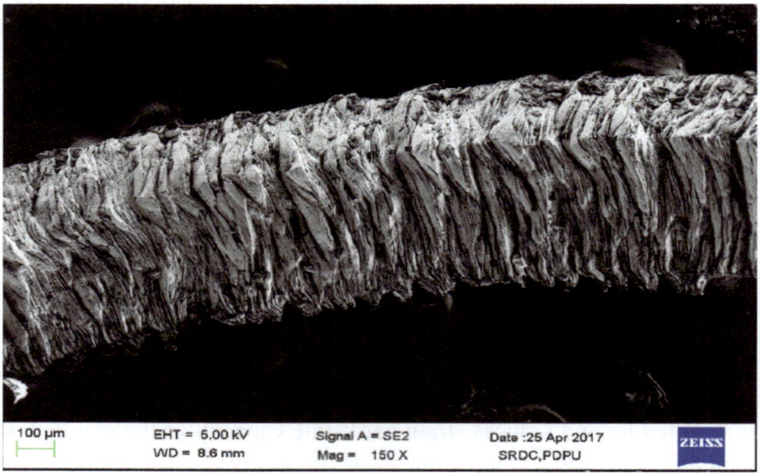

Fig. 4 SEM image of chip using speed 25 m/min, feed 0.2 mm/rev, DOC 0.05 mm

From above SEM image (Fig. 4), it is clear that at higher feed, low depth of cut, and moderate speed, the segmented and sawtooth edged frequency in chip is moderate.

At higher feed, high DOC, and high speed, sawtooth edged frequency in chip is very high (Fig. 5).

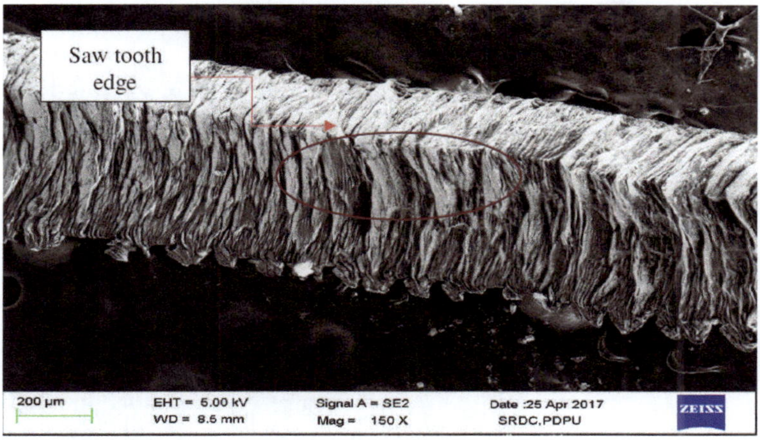

Fig. 5 SEM image of chip using speed 30 m/min, feed 0.2 mm/rev, DOC 0.1 mm

3.3 Effect on Tool Wear

Tool wear mechanisms are generally influenced by three phenomena, namely, diffusion, thermal softening, and notching at depth of cut and trailing edge. In case of superalloy, Inconel 718 a combination of large primary deformation zone heat source and a large secondary deformation zone shear stress gives rise to higher temperature and stress in the region. The SEM (Fig. 6) assessment depicts the existence of surface damage in the form of metal debris adhesion, side flow, smeared material, and feed marks of chips and wear of PVD coating due to scratch of chips.

4 Conclusions

The experimental setup was successfully developed for high-speed turning. The experiment has been successfully conducted; result obtained from the experiments analyzed to check the performance and efficiency of the high-speed turning process with PVD-coated cutting tool insert on INCONEL-718 under wet condition. The effects of process parameter on percentage change in surface roughness (%ΔRa) have been studied using L_9 array for INCONEL-718. Surface roughness has been successfully measured using stylus profilometer, and percentage change in surface roughness (%ΔRa) is found to be in range 75.25–98.25% and minimum surface roughness achieved up to 0.09 micron. Optimum value of the process parameter is 20 m/min speed, 0.15 mm/rev feed, and 0.05 mm depth of cut for the percentage change in surface roughness (%ΔRa). Percentage contribution of process parameter in %ΔRa is 45% for speed, 38% for feed, and 16% for depth of cut, near to 1% for error. The SEM assessment depicts the existence of surface damage in the form

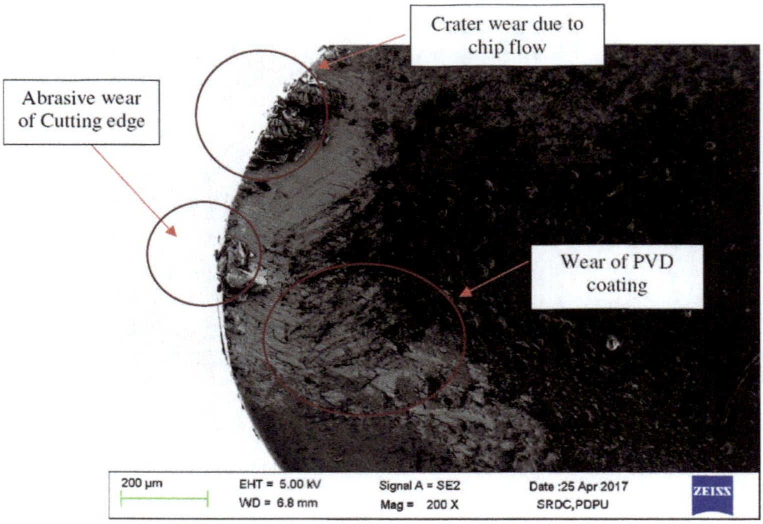

Fig. 6 Tool wear using SEM image

of metal debris adhesion, side flow, smeared material, and feed marks of chips and wear of PVD coating due to scratch of chips. At moderate feed, low speed, and low depth of cut, the continuous chips are observed while at increased speed, decrease feed, and increase depth of cut, it is changed to sawtooth type.

References

1. Kamata, Y., Obikawa, T.: High speed MQL finish-turning of Inconel 718 with different coated tools. J. Mater. Process. Technol. **192**, 281–286 (2007)
2. Pawade, R.S., Joshi, S.S., Brahmankar, P.K.: Effect of machining parameters and cutting edge geometry on surface integrity of high-speed turned Inconel 718. Int. J. Mach. Tools Manuf **48**(1), 15–28 (2008)
3. Thakur, D.G., Ramamoorthy, B., Vijayaraghavan, L.: Study on the machinability characteristics of superalloy Inconel 718 during high speed turning. Mater. Des. **30.5**, 1718–1725 (2009)
4. Cantero, J.L., et al.: Analysis of tool wear patterns in finishing turning of Inconel 718. Wear **297**(1–2), 885–894 (2013)
5. Pawade, R.S., et al.: An investigation of cutting forces and surface damage in high-speed turning of Inconel 718. J. Mater. Process. Technol. **192**, 139–146 (2007)
6. Tebaldo, V., di Confiengo, G.G., Faga, M.G.: Sustainability in machining: "Eco-friendly" turning of Inconel 718. Surface characterisation and economic analysis. J. Cleaner Prod. **140**, 1567–1577 (2017)
7. Lima, F.F., et al.: Wear of ceramic tools when machining Inconel 751 using argon and oxygen as lubri-cooling atmospheres. Ceram. Int. **43**(1), 677–685 (2017)

MADM Approach for Selection of Materials in Fused Deposition Modelling Process

PL. Ramkumar and Kumar Abhishek

Abstract When designing a new product, the manufacturer can efficiently predict the performance of the product by prototyping and using a real situation. Before freezing the final requirement of the design, prototype is built to understand the actual requirements of the product by the end users. Different prototyping techniques are available each with distinct strengths. Fused deposition modelling (FDM) is one of the most widely used techniques. Different materials are used in FDM. The problem of material selection in FDM is a concern for manufacturers for many years. This problem becomes more complicated in recent years, due to the increase in variety of materials and the properties they offer. In this paper, an objective-based multi-attribute decision-making (MADM) method is explored for the selection of material for FDM. The method follows fuzzy logic to transfigure the qualitative attributes into quantitative attributes. The scheme has been used to rank different materials to assist the manufacturer to select suitable material from long list of materials available for FDM. NYLON 12 has been identified as the suitable material based on the attributes considered. However, the final ranking may differ based on the attributes and alternatives considered.

Keywords Material selection · Fused deposition modelling · MADM · Rapid prototyping · Additive manufacturing

1 Introduction

Rapid prototyping is the technology that is used to construct physical models which replicates the actual product to be designed. Rapid prototyping technology is efficient in constructing parts of irregular geometries in the absence of tooling [1]. This has gained focus in product development in recent years. This technology comes under the field of additive manufacturing technology which is different from traditional

PL. Ramkumar · K. Abhishek (✉)
Mechanical Engineering, IITRAM, Ahmedabad, India
e-mail: krabhishek1987@gmail.com

D. Deb et al. (eds.), *Innovations in Infrastructure*, Advances in Intelligent Systems and Computing 757, https://doi.org/10.1007/978-981-13-1966-2_33

manufacturing methods like machining. In traditional manufacturing method, the raw materials are shaped by removing the materials. In contrary, during additive manufacturing, the parts are built by adding the materials [2].

Fused deposition modelling is one of the prevalent methods in additive manufacturing to produce small parts in the order of few cubic inches [3]. Variety of materials are used in fusion deposition modelling which includes ABS of different grades, polycarbonates, ULTEM, nylon and wax for making prototypes in different fields like aerospace, medical, automotive, electronics and other speciality applications.

In recent years, many researchers have concentrated on knowing the mechanical characteristics of the fused deposit modelling product. Fused deposit modelling specimen of two orientations (0°/90°) and crisscross (45°/−45°) has been considered and their mechanical properties are evaluated by Fatimatuzahraa et al. [1]. From his work, he concluded that the parameters like air gap and raster orientation significantly affect the mechanical properties of the FDM specimen. He also observed that orientation (0°/90°) has weak interface than (45°/−45°). The quality of FDM parts was assessed by Nur Saaidah et al. [4]. From his work, he determined the optimum process parameter which yields better quality of the FDM parts in terms of dimensional accuracy and surface roughness. According to Lee et al. [5], layer thickness, raster angle and air gap greatly affect the elastic performance of the FDM product. Van Weren et al. [6] had pointed that increase in neck growth affects the quality of the parts. There are few other researchers [7, 8] who have deeply addressed mechanical characteristics of the FDM parts.

From the review, it can be clearly understood that the researchers in the past have concentrated on knowing the optimum process parameter which can yield required mechanical properties of the FDM parts for different materials. Nevertheless, selecting the appropriate material considering different material attributes and alternative has not been investigated in detail yet.

For constructing fusion deposition modelling product, one of the important decisions taken during the design stage is the material selection. As there are variety of materials available and respective advantages and disadvantages are associated with each material, selection of suitable material for any specific application is a challenge. There are different methods for material selection suggested and adopted in the past. However, their precept is similar. While selecting materials for any specific applications, diverse number of characteristics or attributes are required to be considered. As the selection of suitable material for any specific application is complex, there is need for simple and systematic approach to guide the designers considering a number of selected attributes and their interrelations. For the above-described problems, multiple attribute/multi-criteria decision-making (MADM/MCDM) methodology is appropriate for the selection of a material based on a diverse set of characteristics/attributes [9]. MADM methodology is being applied in different areas which include electroplating system selection [10], supplier selection [11] and plant layout selection [12]. MADM and MCDM techniques were used for decision-making in various manufacturing situations. Maniya and Bhatt [13] used multi-criteria decision-making method for the appropriate selection of flexible manufacturing system. Also, recently Jain and Raj [14] proposed MADM for ranking different FMS systems.

Thirumalai and Senthilkumar [15] used this method in choosing the optimum machining parameter.

In this paper, an MADM procedure is adopted to assist the manufacturer to select the suitable material for FDM process. The procedure considers both quantitative and qualitative attributes. Fuzzy conversion scale is used to transmute the qualitative attributes into quantitative attributes. The correctness of the result has been verified by comparing the results with the result obtained from grey relation analysis.

2 Material Selection Using Objective-Based MADM Method

2.1 Description of Attributes in Material Selection of Fusion Deposition Modelling

Selection of a material in FDM process for a distinct situation is going to affect the overall efficiency of the product. Data related to every attribute governs the conduct of a product. It shows that description of attributes is critical in selection of material from number of different alternatives available. The present material selection problem includes 13 alternative materials and 8 different attributes to evaluate those alternatives. This matrix can be of the choice of the manufacturer/designer, as the attributes and alternative can be added or deleted depending upon their requirement. Thirteen alternative material considered for this work are ABS (acrylonitrile butadiene styrene) of different grades—ABSPLUS P-430, ABSi, ABS-ESD7, ABS-M30, ABS- M30i, ASA (Acrylonitrile Styrene Acrylate), PC-ABS (polycarbonate ABS), PC-ISO, Nylon 12, ULTEM-1010 (transparent polyetherimide), ULTEM-9085 and PPSF-PPSU (polyphenylsulfone). For the above materials, eight different attributes considered are tensile strength, elongation at break, flexural strength, Izod impact, heat deflection property, cost, colour available and number of layer thickness.

Multiple attribute decision-making is used to determine problems demanding selection from the midst of large number of preferences. The methodology subsisting of four steps for material selection [9, 16] is described below.

2.2 Four-Step Selection Procedure

Step 1: Decision Table

The material and their properties considered in the work as shown in Table 1 are overall common properties of the materials used in FDM, and it is not specific for any particular material manufacturer. In the decision table, the properties like tensile strength, flexural strength, etc. are referred as beneficial attributes (i.e. higher values are desired), whereas another category of attributes considered are cited as non-beneficial attributes whose lower values are desired like cost.

A quantitative or qualitative value can be designated to each attribute. The qualitative value is converted to quantitative value using a fuzzy conversion scale proposed by Chen and Hwang [17].

The values related to each attribute may be of different units (e.g. Impact strength is in J/m and elongation at break is in percentage). Thus, the normalized decision matrix is obtained using the following expression:

$$N_{ij}^{*} = \frac{N_{ij}}{\sum_{i=1}^{n} N_{ij}} \tag{1}$$

where N_{ij}^{*} is the normalized value of N_{ij} and $\sum_{i=1}^{n} N_{ij}$ gives the overall summation of jth attribute for n number of alternative. Table 2 shows the normalized decision matrix evaluated for the material selection problem based on Eq. (1).

Step 2: Objective Weight of Importance:

Statistical variance is used to evaluate the objective weight of priority of the criteria. Following relation implies the weight of priority of different criteria:

$$H_j = \left(\frac{1}{n}\right) \sum_{i=1}^{n} \left(N_{ij} - \left(N_{ij\text{mean}}\right)\right)2 \tag{2}$$

where H_j is the statistical variance related to the jth criteria. N_{ij} mean is the average value of the N_{ij}. Statistical variance of all the eight attributes is evaluated and presented in Table 3.

Objective weight O_j of particular criteria, i.e. jth attribute, can be obtained by finding the ratio between the statistical deviations of jth criteria with the sum of the statistical deviation of 'm' number of criteria. Therefore, O_j can be determined using the expression given below:

$$O_j = H_j / \sum_{j=1}^{m} H_j \tag{3}$$

The objective weight of importance of all the eight attributes is evaluated and presented in Table 4.

Step 3: Preference Index:

Preference index of every alternative is the grade of that alternative with reference to the other alternative. Preference index values P_i^0 for all the attributes are determined using Eq. (4) given below:

Table 1 Properties of candidate material and its respective properties

Alternatives/Attributes	Tensile strength in MPa	Elongation at break in %	Fleural strength in MPa	Izod Impact strength in J/m	Heat deflection (°c)	Cost in $	Colours available	No. of layer thickness
ABSPLUS-P430	33	6.00	58	106	96	350	EXH	VL
ABSi	37	4.40	62	96	87	350	VL	L
ABS-ESD7	36	3.00	61	111	96	395	EXL	EL
ABS-M30	32	7.00	60	128	96	350	A	L
ABS-M30i	32	7.00	61	128	96	350	EXL	L
ASA	33	9.00	60	64	0.45	350	EH	L
PC-ABS	34	5.00	59	235	110	395	EXL	L
PC-ISO	57	4.30	90	86	133	395	EL	VL
PC	68	4.80	104	53	138	395	EXL	VL
NYLON 12	48	30	69	200	82	395	EXL	VL
ULTEM-1010	81	3.30	0.04	41	216	685	EXL	EL
ULTEM-9085	71.64	5.90	115.1	106	153	685	EL	EL
PPSF-PPSU	55	3.00	110	58.73	189	685	EXL	EL

Table 2 Normalized decision matrix

Alternatives/Attributes	Tensile strength	Elongation at break in	Flexural strength	Izod Impact strength	Heat deflection	Cost	Colours Available.	No. of layer thickness
ABSPLUS-P430	0.053	0.065	0.064	0.075	0.064	0.061	0.195	0.076
ASBi	0.060	0.047	0.068	0.068	0.058	0.061	0.065	0.090
ABSESD7	0.058	0.032	0.067	0.079	0.064	0.068	0.039	0.062
ABS-M30	0.052	0.076	0.066	0.091	0.064	0.061	0.143	0.090
ABS-M30i	0.052	0.076	0.067	0.091	0.064	0.061	0.039	0.090
ASA	0.053	0.097	0.066	0.045	0.000	0.061	0.195	0.090
PC-ABS	0.055	0.054	0.065	0.166	0.074	0.068	0.039	0.090
PC-ISO	0.092	0.046	0.099	0.061	0.089	0.068	0.065	0.076
PC	0.110	0.052	0.114	0.038	0.092	0.068	0.039	0.076
NYLON 12	0.078	0.324	0.076	0.142	0.055	0.068	0.039	0.076
ULTEM-1010	0.131	0.036	0.000	0.029	0.145	0.119	0.039	0.062
ULTEM-9085	0.116	0.064	0.127	0.075	0.103	0.119	0.065	0.062
PPSF-PPSU	0.089	0.032	0.121	0.042	0.127	0.119	0.039	0.062

Table 3 Statistical variance for the eight attributes

Attributes	Tensile strength	Elongation at Break	Flexural Strength	Izod Impact strength	Heat deflection	COST	Colours available	No. of layer thickness
H_j	0.009	0.070	0.008	0.018	0.008	0.006	0.042	0.001

Table 4 Objective weight of importance for the eight attributes

Attributes	Tensile strength	Elongation at Break	Flexural strength	Izod Impact strength	Heat deflection	COST	Colours available	No. of layer thickness
O_j	0.056	0.420	0.052	0.113	0.048	0.041	0.255	0.010

$$P_i^o = \sum_{j=1}^{m} O_j N_{ij}^{**} \tag{4}$$

For beneficial and non-beneficial attributes, N_{ij}^{**} is given by $N_{ij}^{**} = \frac{N_{ij}^b}{\left(N_{ij}^b\right)\max}$ and $N_{ij}^{**} = \frac{\left(N_{ij}^{nb}\right)\min}{N_{ij}^{nb}}$.

N_{ij}^b and $\left(N_{ij}^{nb}\right)$ denotes the balanced amount of beneficial and non-beneficial criteria. $\left(N_{ij}^b\right)\max$ and $\left(N_{ij}^{nb}\right)\min$ state maximum and minimum values of jth beneficial or non-beneficial attributes.

Different alternates ranked based on the preference index score obtained. The preference index values and respective ranks of different alternatives are stated in Table 5.

Step 4: Final Selection

The final selection of the material for FDM process can be obtained from the preference index score obtained in the previous step. But the manufacturer/designer should keep in mind that the preference index/score obtained may change if the number of attributes used to evaluate the different alternatives is further added or deleted. Therefore, decision of including or eliminating any alternative or attribute plays major role in this decision-making process. Also, for the final decision, other practical constraints like material availability constraint, social constraints, management constraints, etc. can be considered.

Table 5 Preference index value and ranks for different alternatives

Alternatives/Attributes	P_i^o	Rank
ABSPLUS-P430	0.507034	3
ABSi	0.313167	9
ABS-ESD7	0.259843	13
ABS-M30	0.465178	4
ABS-M30i	0.329252	8
ASA	0.532662	2
PC-ABS	0.352133	6
PC-ISO	0.335262	7
PC	0.306162	10
NYLON 12	0.691525	1
ULTEM-1010	0.303589	11
ULTEM-9085	0.374274	5
PPSF-PPSU	0.27125	12

3 Conclusions

The paper presents a methodology for the selection of suitable material in fused deposition modelling process. The highlights of the present work are as follows:

1. Multi-attribute decision-making method is adopted for the material selection in FDM process. The approach is useful in comparing, ranking and selecting the suitable material from a long list of available alternative materials.
2. Using the adopted procedure NYLON 12 is identified as the suitable material based on the eight attributes considered. However, the final ranking may differ based on the attributes and alternatives considered.
3. The proposed methodology can handle both qualitative and quantitative attributes. Fuzzy conversion scale is used to convert qualitative to quantitative attributes.
4. The adopted method for material selection is comparatively simple and convenient. Only input required will be the decision matrix.
5. The proposed method evaluates the alternatives based on objective weight of importance which are determined using statistical variance. Therefore, manufacturer or designer does not require any past experience for deciding the subjective weights for different attributes.

References

1. Fatimatuzahraa, A.W., Farahaina, B., Yusoff W.A.Y.: The effect of employing different raster orientations on the mechanical properties and microstructure of fused deposition modeling

parts. In: Business, Engineering and Industrial Applications (ISBEIA), IEEE Symposium, pp. 1–9 (2011)
2. Kay, R.N.: Effect of raster orientation on the structural properties of components fabricated by fused deposition modeling, (thesis). Graduate School of Ohio State University (2014)
3. Singh, S.: Computer Aided Design and Manufacturing, 5th edn., p. 854 Khanna Publisher, India (2010)
4. Nur Saaidah, A., Alkahari, M.R., Boejang, H.: Analysis on fused deposition modelling performance. J. Zhejiang Univ. Sci. A. Appl. Phys. Eng. **11**, 972–977 (2010)
5. Lee, B.H., Abdullah, J., Khan, Z.A.: Optimization of rapid prototyping parameters for production of flexible ABS object. J. Mater. Process. Technol. **169**, 54–61 (2005)
6. Van Weren, R., Agarwala, M.K., Jamalabad, V.R.: Quality of parts processed by fused deposition. In: Proceeding of the Solid Freeform Fabrication Symposium, vol. 6, pp. 314–321. TX (1997)
7. Vega, V., Clements, J., Lam, T., Abad, A., Fritz, B., Ula, N., Es Said, O.S.: The effect of layer orientation on the mechanical properties and microstructure of a polymer. J. Mater. Eng. Perform. **20**, 978–988 (2011)
8. Mohamed, O.A., Masood, S.H., Bhowmik, J.L., Nikzad, M., Azadmanjiri, J.: Effect of process parameters on dynamic mechanical performance of FDM pc/abs printed parts through design of experiment. J. Mater. Eng. Perform. **25**, 2922–2935 (2016)
9. Ramkumar, P.L., Kulkarni, D.M.: Objective-based multiple attribute decision-making method for plastic manufacturing process selection. Int. J. Manuf. Technol. Manage. **28**, 184–199 (2014)
10. Abhishek, K., Agrawal, V.P.: Attribute based specification, comparison and selection of electroplating system using MADM approach. Expert Syst. Appl. **36**, 10815–10827 (2009)
11. Li, G.D., Yamaguchi, D., Nagai, M.A.: Grey-based decision making approach to the supplier selection problem. Math. Comput. Model. **46**, 573–581 (2007)
12. Yang, T., Hung, C.C.: Multiple-attribute decision making methods for plant layout design problem. Robot. Comput.- Integr. Manuf. **23**, 126–137 (2007)
13. Maniya, K.D., Bhatt, M.G.: The selection of flexible manufacturing system using preference selection index method. Int. J. Ind. Syst. Manage. **14**(3), 125–141 (2011)
14. Jain, V., Raj, T.: Ranking of flexibility in flexible manufacturing system by using a combined multiple attribute decision making method. J. Flex. Syst. (2013)
15. Thirumalai, R., Senthilkumaar, J.S.: Multi-criteia decision making in the selection of machining parameters for inconel 718. J. Mech. Sci. Technol. **27**, 1109–1116 (2013)
16. Rao, R.V., Patel, B.K.: A subjective and objective integrated multiple attribute decision making method for material selection. Mater. Des. **31**, 4738–4747 (2010)
17. Chen, S.J., Hwang, C.L.: Fuzzy multiple attribute decision making-methods and applications. In: Lecture notes in economics and mathematical systems, Springer, Berlin (1992)

Design and Simulation of Seat Handle Using Plastic Injection Molding Process

Shubham Jayswal, Harsh Jangade, Kumar Abhishek and PL. Ramkumar

Abstract Nowadays, the plastic industries are gaining more attention due to the tremendous increase in demand for plastic products mainly in household applications, space industry, automobile industry, sports industry, and so on. Therefore, manufacturers gave more emphasis on the plastic injection molding in order to manufacture intricate shape and complex geometry with good dimensional accuracy. In order to meet the demand, it is very essential to adopt various advanced technologies like CAD/CAM/CAE for the better understanding of injection molding process to gain quality products. Hence, this paper highlights the analysis of seat handle for automobile products used in injection molding process with help of software NX 9.0 and Moldex 3D. There are various types of software used to simulate the model of component and Moldex 3D is one of them. Moldex 3D is a software generally called Computer-Aided Engineering (CAE). It calculates the filling time, melt front temperature, cooling time, and pressure drop in the mold, and results depend on material used and technological parameters. In this work, ABS material has been considered. There are many parameters which affect the properties of plastic products. From the analysis, suitable gate position has been advised for the uniform flow of mold material. Injection gate visualizes the effect of temperature, pressure, and melt front time. The operating parameters have to be correct and can produce better product. There are changes occurred in product with change in parameters. Optimum parameter is related to the quality to produce product without defects such as improper surface finish, flow marks, hesitation, short shots, improper location of runner, over packing, and others.

Keywords Injection molding · Simulation · Seat handle

S. Jayswal · H. Jangade · K. Abhishek (✉) · PL. Ramkumar
Department of Mechanical Engineering, Institute of Infrastructure, Technology, Research and Management, Ahmedabad 380026, India
e-mail: krabhishek1987@gmail.com

D. Deb et al. (eds.), *Innovations in Infrastructure*, Advances in Intelligent Systems and Computing 757, https://doi.org/10.1007/978-981-13-1966-2_34

1 Introduction

The plastics are used to manufacture automotive products efficiently by reducing weight, together with providing durability, corrosion resistance, toughness, design flexibility, resiliency, and high performance at low cost. Therefore, this ongoing development of advanced, high-performance polymers has dramatically increased the application of tool and dies fabrication in plastic injection. However, the mold design and fabrication is very expensive and time-consuming process; therefore, simulations are required to analyze and design the complicated plastic part which enhances the productivity and process consistency. Due to the development of CAD/CAM/CAE technology especially Moldflow analysis, the number of trials on mold can be reduced to achieve good quality product. However, the major challenges using this software are to optimize the design and to understand the mechanism of hot runner in the injection molding process. Therefore, various researchers have attempted to analyze and design the mold to improve the quality of the finished product.

Longzhi et al. [1] assessed the effect of melt temperature, mold temperature, injection time, holding pressure, and cooling time on sink mark index in plastic injection molded component using Moldflow software. Shakkarwal and Yadav [2] presented a practical tool design procedure/methodology of an injection mold for connecting link. In this study, polyacetal has been chosen as a material for connecting link. The work also shows that the parameters such as sink marks, fill time, weld line, air traps, etc. have been affected the quality of the finished product. Kumar and Babu [3] worked on industrial helmet, its design parameters, and its analysis using 3D modeling module in Pro/Engineer software. Stanek et al. [4] had used the Cadmould Rubber to analyze filling time, speed, and vulcanization time. Cadmould Rubber is the software called Computer-Aided Engineering (CAE). Main objective for simulation analysis by Cadmould Rubber is to increase the injected rubber parts, cut the cost connected to the mold, repair, and accelerate the preparations of new production, especially new tool design. Martinsen et al. [5] described methods for fabrication and installation of sensors on surfaces of injection molding tools. In this study, the sensors were encapsulated in an Al_2O_3 layer for electrical insulation and wear resistance and data extracted from these sensors to calibrate the FEM sensors for calibration of FEM process simulation and optimization of conformal cooling channels on the production tool. Oktem et al. [6] established the relationship between parameters on both shrinkage and warpage during the simulation analysis on injection molded part using Moldflow plastic insight.

It is evident from the literature that a lot of researches are continuously carried out in the field of optimization procedures and technique for injection molding process but less work has been carried out to design and fabrication of seat handle. In this paper, mold design for seat handle which is a part of latch assembly used in automobile seat handle has been designed in high-end CAD software Siemens NX. Moldflow analysis was performed on the component in order to determine the molding defects and to reduce development time and expenses.

Table 1 Properties of ABS 633 Grade

Density	1.05 gm/cc
Linear mold shrinkage	0.004–0.006 cm/cm
Melt mold shrinkage	4.5 gm/10 min
Poisson ratio	0.39
Thermal conductivity	0.2 W/°C
Hold pressure	50–75% of injection pressure
Injection speed	High

2 Design for the Seat Handle

This paper describes the modeling and simulation of seat handle using plastic injection molding process. NX 9.0 software increases the simplicity for modeling of the component. This provides the support of analytical model using multi-CAD geometry using synchronous technology. After completing the modeling of component, it is required to simulate the model of component in Moldex 3D software by importing the file from NX 9.0. After completion of part simulation, Moldex 3D provides the important outputs such as filling time, maximum temperature, melt front temperature, pressure drop, etc.

2.1 Design for the Seat Handle

Proper material selection is important criterion for any design of the component. In this study, Acrylonitrile Butadiene Styrene 633 grade has been used because it is a thermoplastic material known for its high impact resistance and toughness. It also provides good resistance to scratching and rigidity with an excellent dimensional stability. ABS has low melting point and is also inexpensive material. Due to its thermoplastic properties, ABS is easily plastic injection molded. The properties of ABS 633 are listed in Table 1 as shown.

2.2 Process Parameter

It is essential to identify the important process parameter while designing any product as it affects the quality and efficiency of the product. It has been observed from the literature that melt front temperature, mold temperature, ejection temperature, and freezing temperature have been major effect in injection molding processes. Table 2 represents the process parameters and their description.

Table 2 Process parameters and their description [7]

Parameters	Description
Polymer	ABS
Material grade	ABS 633
Grade flow	High flow
Producer	Default
Comment	MVR (220, 10) = 45 cm^3/10 min, d = 1.05 g/cc
Melt temperature (minimum)	220 °C
Melt temperature (normal)	240 °C
Melt temperature (maximum)	260 °C
Mold temperature (minimum)	40 °C
Mold temperature (normal)	60 °C
Mold temperature (maximum)	80 °C
Ejection temperature	85 °C
Freeze temperature	115 °C
Shrinkage allowance	1.006%

2.3 Modeling of Seat Handle

A 3D model of the automobile component (Seat handle) is generated in Siemens NX 9.0 as it has been specifically required by the Company (Ashapura mold tech.) (Figs. 1 and 2).

3 Simulation Results

There are various technological parameters involved during injection molding process that have crucial impact on the final quality of the product which includes the location of injection gate, melt front temperature, pressure drop, and melt front temperature.

3.1 Position of Gate

Figure 3 indicates the optimum location of gate. The selection of the gate position is one of the important criteria as it reduces the filling time and controls the distribution of mold temperature during injection molding process. In order to select the optimum location for the gate, this paper adopted trial-and-error method by positioning the gate at different locations.

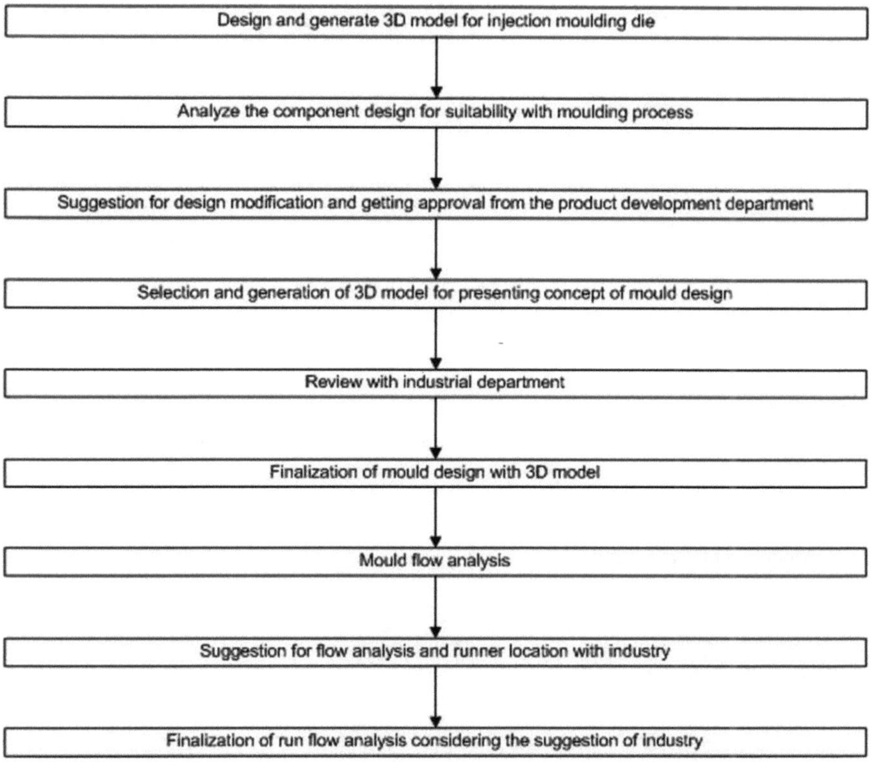

Fig. 1 Detailed methodology for design and simulation of seat handle

3.2 Melt Front Time

Melt front time is another important aspect to achieve the balanced gate contribution of each gate and all flow path should reach the cavity wall at the same time in order to reduce the hesitation. It indicates the melt front position of skin material with respect to time during filling stage. It is essential to select the proper parameters such as mold temperature, change gate location, injection flow rate, and melt flow index in order to reduce the hesitation during the injection process. Figures 4 and 5 represent the melt front time for the different locations of gate. Optimum location of gate is achieved by performing number of flow analysis on product that is indicated in Fig. 5. Area in the red color shows the minimum time to fill the pattern geometry between 0 and 0.7404 s, and area in the blue color shows the maximum time to fill the pattern geometry between 4.812 and 5.553 s. If the part is having problem such as hesitation, it appears with no color in the geometry.

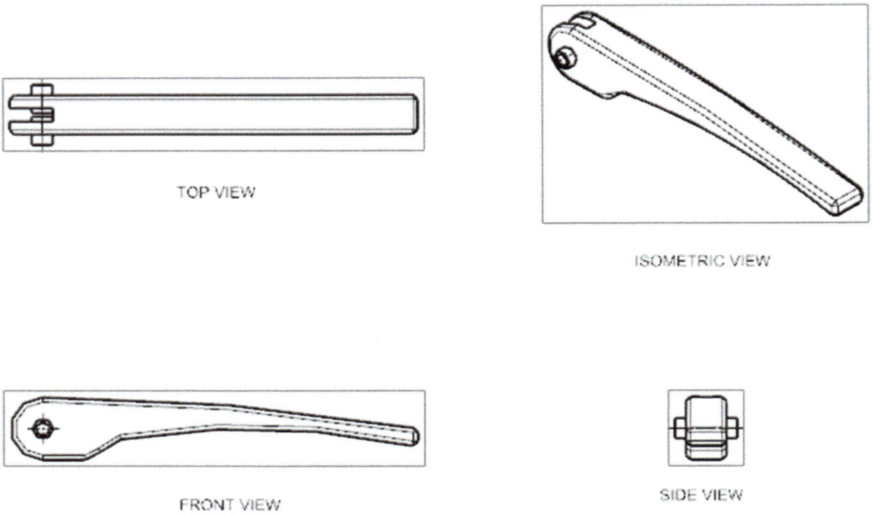

Fig. 2 Modeling of seat handle

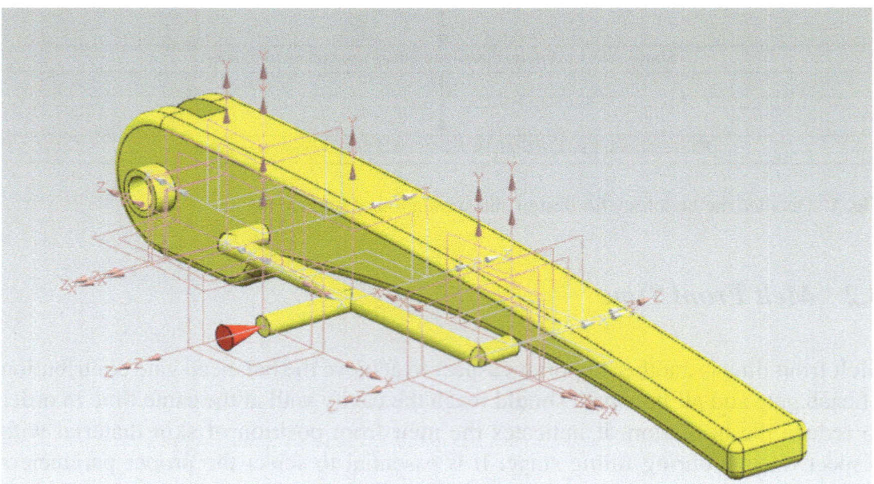

Fig. 3 Location of gate

3.3 Melt Front Temperature

In order to avoid the problems like weld lines, flow marks, hesitation, (and) material degradation during the injection molding process, it is essential to study the melt front temperature.

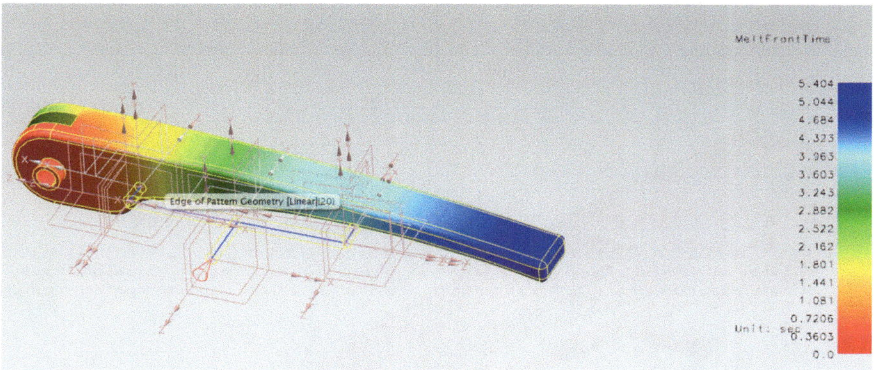

Fig. 4 Melt front time for first location of gate

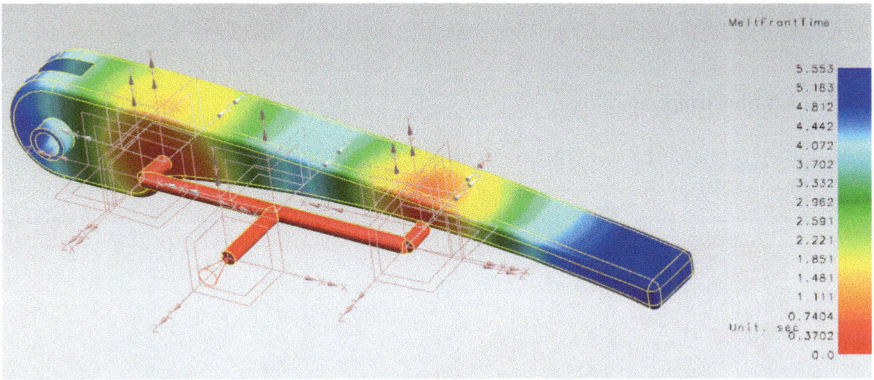

Fig. 5 Melt front time for second location of gate (optimum)

As shown in figure during Moldflow process, ripples are generated near the gate location. The melt front temperature during the analysis is distributed uniformly. The temperature during the process is up to 210 °C, which is uniform throughout the component but in the circular section of handle, there is minor temperature drop between 203.1 and 207.3 °C. As the temperature during process is high, problems such as material degradation are not observed.

3.4 Pressure Drop

It is necessary to analyze the pressure drop as it reveals the proper distribution of pressure throughout the component in order to avoid the problems like material degradation and overpacking which caused due to non-uniform pressure distribution.

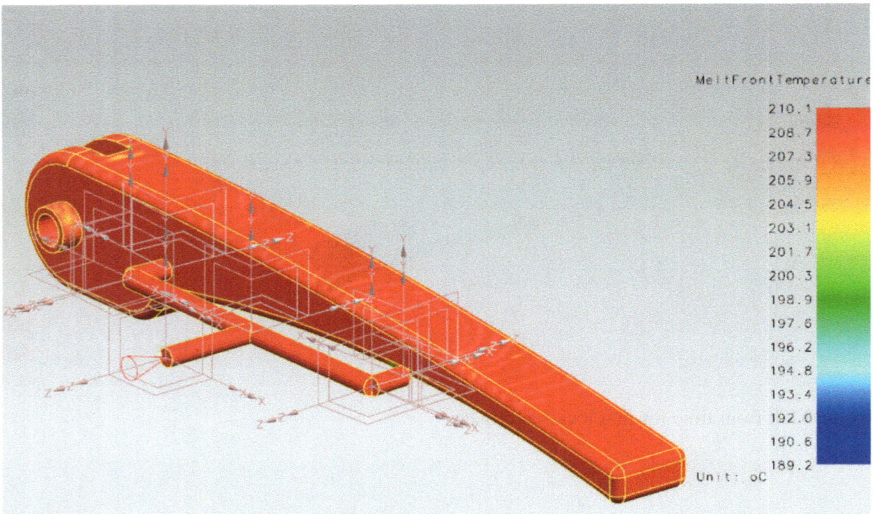

Fig. 6 Melt front temperature

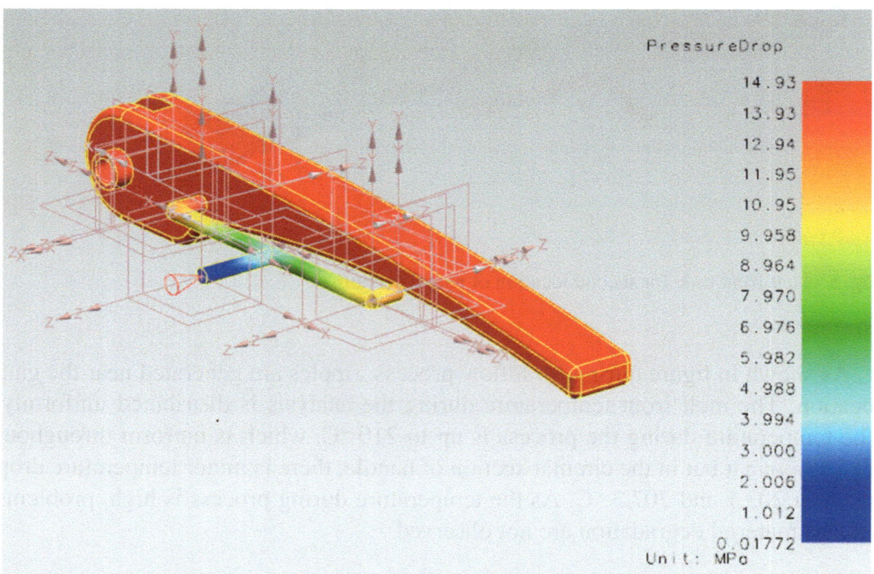

Fig. 7 Pressure drop

As shown in Fig. 6, there is sudden increase in pressure at runner section and at the part section, the pressure remains uniform that is represented by red color. By selecting the proper gate size and runner location, it can reduce the non-uniform distribution in the pressure. The values of pressure obtained in this analysis are minimum between 0.01772 and 1.012 MPa that appears in blue color and maximum between 13.93 and 14.93 MPa that appears in red color (Fig. 7).

4 Conclusions

In order to avoid the problems in the injection molding process such as improper surface finish, flow marks, hesitation, short shots, improper location of runner, and over packing during the manufacturing of components in automobile industry. It is essential to analyze the flow parameter such as melt front time, filling time, pressure drop, melt front temperature, location of runner, air traps, and weld lines which affects the quality of the product. Therefore, this study analyzes the part accurately for obtaining optimum results; number of trials are done by varying the different parameters using Moldflow analysis. It has been observed that all the parameters such as the melt front temperature, pressure drop, and maximum temperature remain uniform. Therefore, designer concludes that the simulation shows the minimum defects. It also helps the mold designer to design the perfect mold with minimum defects.

Acknowledgements Authors are grateful for Ashapura mold tech. for their support.

References

1. Longzhi, Z., Binghui, C., Jianyun, L., Shangbing, Z.: Optimization of plastics injection molding processing parameters based on the minimization of sink marks. In: International Conference on Mechanic Automation and Control Engineering (MACE), pp. 593–595. IEEE (2010)
2. Shakkarwal, P., Yadav, L.: 'Design and mould flow analysis of injection mould for connecting link. IJET **4**(1), 182–185 (2013)
3. Anil Kumar, K., Suresh Babu Y.: Design and analysis of industrial helmet. Int. J. Comput. Eng. Res. **3**, 48–53 (2013)
4. Stenek, M., Manas, D., Manas, M., Javorik, J.: Simulation of injection molding process by cadmould rubber. Int. J. Math. Comput. Simul. **5**(5), 422–429 (2011)
5. Martinsen, K., Gellein, L.T., Boivie, K.M.: Sensors embedded in surface coatings in injection moulding dies. Proc. CIRP **62**, 386–390 (2017)
6. Oktem, H., Erzurumlu, T., Uzman, I.: Application of Taguchi optimization technique in determining plastic injection molding process parameters for a thin-shell part. Mater. Des. **28**(4), 1271–127 (2007)
7. www.dc.engr.scu.edu

Wear Behaviour of Boron Carbide Added Friction Surfaced Cladded Layer

Kedar Badheka and Vishvesh J. Badheka

Abstract Friction surfacing is a solid-state cladding process based on the linear friction concept, process conceptualized from friction welding. Friction surfacing creates the metallic bonding between two similar or dissimilar materials with the minimum dilution. Deposition of aluminium on mild steel could be the interesting and viable option in the field of solid-state cladding because both materials joining with the fusion-based process are incompatible due to the formation of iron aluminide and Fe and Al are immiscible. Friction surfacing starts a new era of repair and reclamation application in the marine and naval application field. AA 6351 T6 as a consumable rod with 22 mm diameter and SA 516 Gr 70 as a steel substrate material with 6 mm plate thickness. During friction surfacing process, rotating rod is rubbed against the substrate material under the applied axial load under frictional pressure and temperature conditions resulting in a visco-plastic layer at the end of rod tip. Experimentally optimized process parameters were used for the trials. Here, friction surfacing process is carried out in two different variants: normal friction surfacing and friction surface cladding with adding metal matrix boron carbide particles. Pin on disc wear testing is carried out on two different variant loading conditions: 20 and 40 N. Mechanical interlocking is observed between coating layer and substrate material with the fine grain particles. Deposited layer results in superior wear resistance properties due to hot forging action.

Keywords Friction surfacing · B_4C metal matrix particles · Fine grain microstructure · Wear resistance

---

K. Badheka (✉)
Department of Mechanical Engineering, MEFGI, Rajkot 360003, India
e-mail: kedar.badheka@gmail.com

V. J. Badheka
School of Technology, Pandit Deendayal Petroleum University, Gandhinagar 382010, India
e-mail: vishvesh79@gmail.com

D. Deb et al. (eds.), *Innovations in Infrastructure*, Advances in Intelligent Systems and Computing 757, https://doi.org/10.1007/978-981-13-1966-2_35

1 Introduction

Friction surfacing is a solid-state surface modification process based on machine tool technology [1]. It is a variant process of linear friction welding concept. Friction surfacing process was first patent by Klopstock and Neelands in 1941 [2].

Friction surfacing is an advanced technology highly used to alter the surface or subsurface properties of substrate material with producing a fine grain cladding [3]. Friction surfacing is majorly differed from friction stir welding as that, here the consumable rod is used instead of non-consumable tool used in FSW. Temperature generated through frictional heat has a great influence on developing visco-plastic layer [4]. Schematic diagram of the friction surfacing process is shown in Fig. 1. Friction surfacing process has a feature which attracts the new application field in surface modification like joining of advanced material, higher energy efficiency and hazard-free processing (green processing technology). Friction surfacing process is most suitable for the dissimilar material which was difficult to join with the fusion-based process due to metallurgical immiscibility; here, friction surfacing has fewer chances of metallurgical defect because the process is carried out with minimum dilution [4]. Temperature and pressure (force) both are the key elements of friction surfacing process, uniform width and uniform thickness that are only possible in friction surfacing process by controlling process parameters: spindle speed, traverse speed, a rate of feeding (feed control) [5], a diameter of rod and clamping of a substrate. The process is done under direct mechanical processing through the thermo-physical cycle which is responsible for dynamic recrystallization and consequent grain refinement [1, 2]. The higher thermal conductivity of substrate material either consumable material is the main cause for disbonding in a case of dissimilar material like brass and aluminium [6, 7]. Metallurgical incompatibility is a probable reason for Al–steel joining [8]. Iron base material is commonly used as a structural material, but they have some problems regarding the formation of scaling and poor resistance against coating, which are overcome by depositing Al layer [9]. Wear

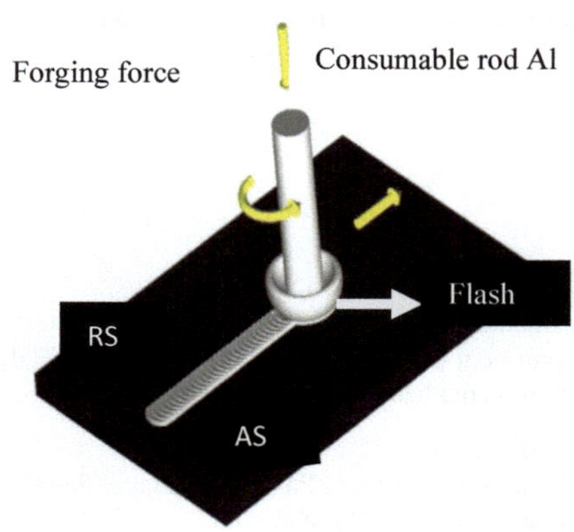

Fig. 1 Friction surfacing process

characteristics are analysed with the different patterns for feeding the metal matrix composite particles and its effect are analysed [10]. High strength-to-weight ratio of non-ferrous alloys like aluminium, magnesium and titanium alloys is considered to be the possible replacement of widely accepted steel of transportation and automobile industries. Boron carbide particle mixed with coated layer enhances the ballistic behaviour of the friction surfaced layer. Metal matrix friction surfaced coated layer exhibits strong metallurgical bond with the substrate, and some kind of interdiffusion of element is also observed at the deposited layer interface [11–13].

Present study is focussed on the wear behaviour of normal friction surfacing layer and layer with adding boron carbide particles at 20 and 40 N loading conditions.

2 Material Specification

Friction surfacing of steel is carried out with Al cladding as shown in Fig. 1. The process is carried out on customized friction surfacing machine available at Pandit Deendayal Petroleum University developed under BRFST project as shown in Fig. 2, with the advanced features like control penal and load cell. AA 6351 T6 is selected as

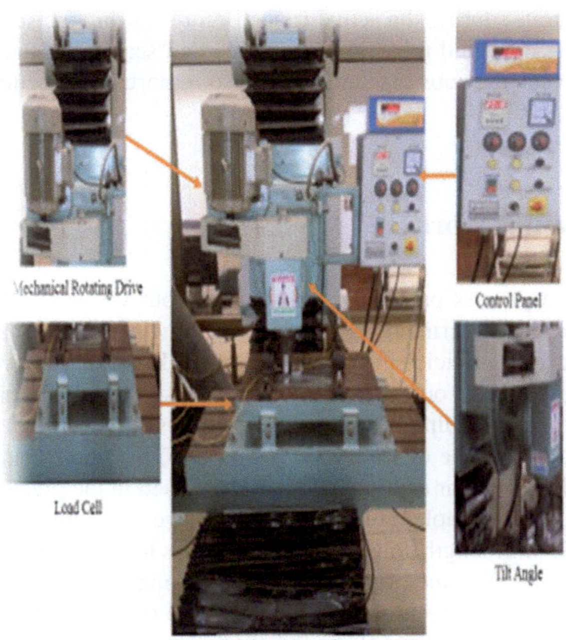

Fig. 2 Actual set of friction surfacing at PDPU

Table 1 Chemical composition of substrate material

Chemical composition of base metal					
	Carbon (%)	Sulphur (%)	Phosphorous (%)	Manganese (%)	Silicon (%)
SA 516 Gr 70	0.196	0.008	0.018	1.160	0.366

Table 2 Chemical composition of consumable material

Chemical composition of consumable material					
	Magnesium (%)	Silicon (%)	Iron (%)	Manganese (%)	Aluminium (%)
AA 6351 T6	0.620	0.870	0.178	0.660	97.600

a consumable material with the size of 100 mm length and 22 mm diameter, while SA 516 Gr 70 is selected as a substrate material with the size of 300 mm × 100 mm × 6 mm. The chemical composition of both materials is shown in Tables 1 and 2. Customized friction surfacing machine has 3000 rpm max. rotational speed, while 140 mm/min max. travel speed from which optimum process parameter is selected for the experimental investigation. All the experimental investigations are carried out at downward feed control mode (Consumable rod is pushed along its axial direction against the substrate) during the actual trials. Before starting the surfacing process, substrate plate was ground to remove the oxide and scaling on the surface, while flat contact between the substrate and rod is also important for better coating layer generation.

3 Experimental Work

With keeping the process parameter constant for both variants of friction surfacing, the following are experimentally investigated: single layer and single layer with adding boron carbide particles indicated in Table 3. First, single-layer friction surfacing cladding is carried out after pre-cleaning of substrate top surface, and then friction surfacing with adding boron carbide particles is performed. Boron carbide particle selected in the size of 1500 mesh (8–12 μm) is to be fed in an aluminium consumable rod. Triangular drilling hole pattern is used in the present investigation shown in Fig. 3. All three holes are drilled with the size of 2.4 mm in diameter and at 120° angle. In 80-mm-length consumable rod, hole is to be drilled up to 50–55 mm depth. In single hole, volume of powder fed is 0.33 mg. During surfacing process, feeding side of powder is fixed into the collet, while opposite side is initially contacted with the substrate after initial friction rod will consume and afterwards powder with the visco-plastic material mixed with cladded layer was deposited on a substrate material.

Table 3 Process parameter used in actual investigation work

S. No.	Process parameter	Single-layer FS	Double-layer FS	Triple-layer FS
1	Rotational speed RPM	2400	2400	2400
2	Traverse speed in mm/min	50	50	50
3	Rate of feeding in mm/min	10–50	10–50	10–50

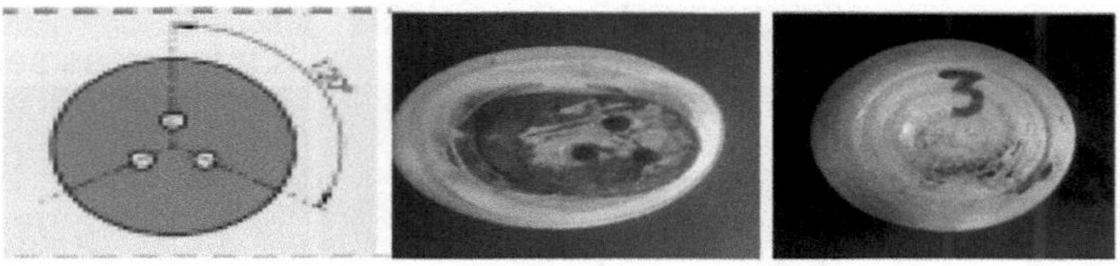

Fig. 3 Conceptual pattern of adding MMC particles in consumable rod

4 Results and Discussion

4.1 *Visual Examination*

The visual appearance of friction surfaced cladding is quite positive in both single and single layer with adding metal matrix particles friction surfacing. Uniform ripples are formed on top surface of the cladding, and uniform width and thickness are generated in both variants of friction surfacing. Table 4 shows uniform ripple on the top side of the cladding.

4.2 *Macro-examination*

1. Macro-observation gives clear understanding about the interfacing bonding between Al and steel due to mechanical interlocking.
2. Uniform thickness and width at advancing and retreating side are observed in Table 4.
3. Width at the deposited layer on the top and at the interface is different, and at the interface slightly lower bond width is observed compared to actual width.

Table 4 Friction surfacing sample after cutting from cross section

Sr.no	Specimen detail	Specimen Macro examination	photograph
1	Single layer FS	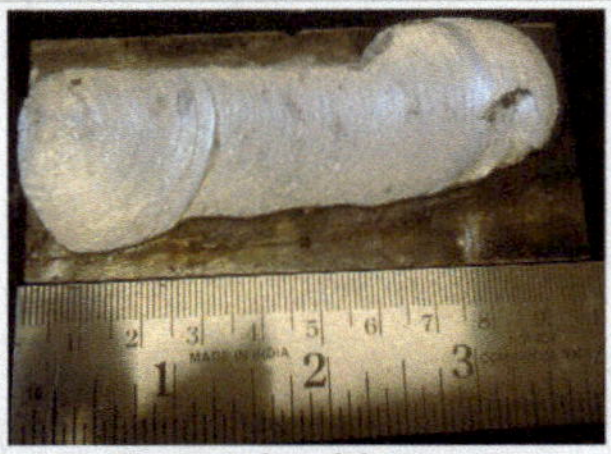 Deposited layer	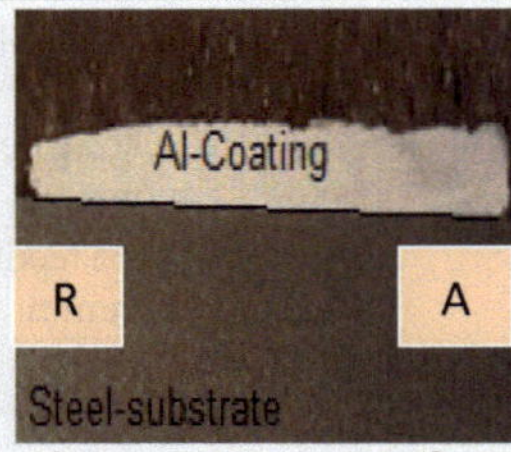 Layer cross section
2	Boron carbide added FS layer	 Deposited B_4C layer FS	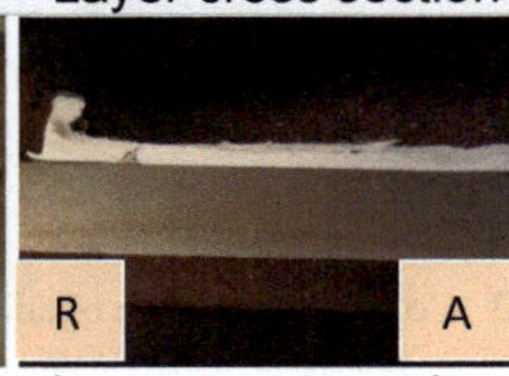 Layer cross section

4.3 Micro-examination

Specimen preparation for the micro-examination plays a very crucial role for dissimilar Al–steel joining system. Specimen from the deposited layer was cut from the centre (perpendicular direction to cladding) subjected for the metallographic examination with different etchant reagents. Here, for the dissimilar cladding layer, different etchant reagents are used for both metals with the swabbing method. For Al, Keller's etchant is used, while for steel, Nital is used for microstructure development. The image of the microstructure was captured at 400X magnification. Extremely fine microstructure of the Al was observed during the examination since the hot forging action was observed during surfacing operation. Figure 4 indicates the microstructure of the single layer and single layer with adding boron carbide particles, which clearly brings out that the bonding at the Al–steel interface is because of mechanical interlocking of asperities [14, 15]. Figure 5 clearly gives indication with optical and SEM image about the mechanical interlocking of the asperities. The reason behind the interlocking of Al–steel asperities was consumable rod rotating and travelling simultaneously in the cladding direction as well as from the steel plate surface. Al is soft material compared to steel, so the broken asperities ratio is higher of AL. Both the materials of broken asperities were observed at the interface region that also increases the strength of the joints. While from second variant of examination friction surfacing with adding the metal matrix particles, effective presence of boron carbide particles resulted on the surface or subsurface region shown in Fig. 4.

Fig. 4 Microstructure of Al–steel friction surfacing cladding

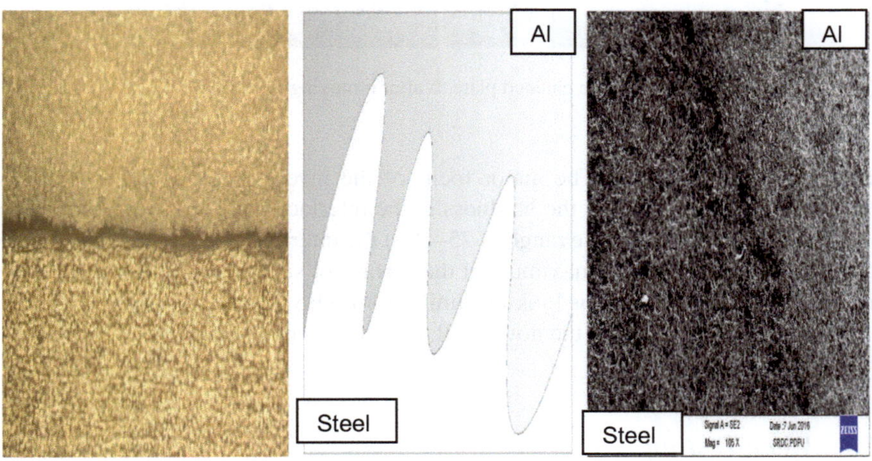

Fig. 5 Mechanical interlocking of Al–steel asperities

4.4 Microhardness

The Vickers microhardness was measured in the through direction from the centre of the cross section of the surfaced plate using WILSON WALPERT 401 MVD test specimen. The indentation was made at 300 mg load and 10 s dwell time set for hardness reading. It was observed that the hardness of the Al layer for single-layer sample was greater than 50HV. Hardness was improved with adding the boron carbide particles, 60HV is close to interface, while hardness at the interface in both the cases is observed same. Figure 6 presents the microhardness profile of both variants of

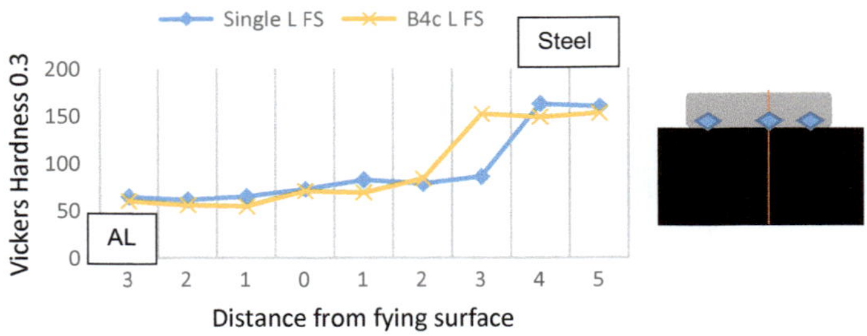

Fig. 6 Hardness profile of friction surfacing variant

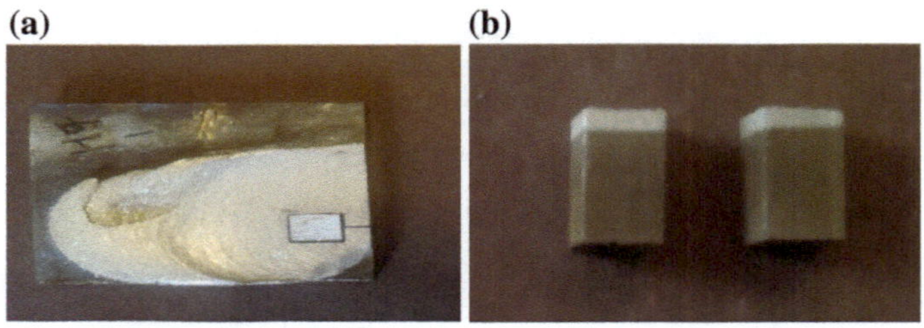

Fig. 7 **a** Sample cut out from the cladded plate, **b** after removing from plate

friction surfacing sample. The major focus of the investigation of microhardness values for surfacing trials is the hardness at the interface and top of AL layer. The VHN was found to vary in the range of 75–88 at the interface for both variants, while the presence of particles is maximum at the surface or subsurface level which is the reason that hardness profile is looking similar for both variants. Improvement in the hardness totally depends on the how the B_4C particles are spread on cladded layer.

4.5 Wear Testing

Wear testing was carried out according to ASTM:G 99-05(2010) [10], using a pin on disc tribometer (Ducom Model: Rotary Tribometer TR-20LE-PHM 400-CHM 400) comprising a 6 mm × 6 mm × 6 mm cubical pin, tested at 20 and 40 N load and total track length of 300 m. A sliding velocity of 0.1 m/s was fixed for the testing. The test was performed in dry atmospheric condition. Sample was weighed using analytical balance (least count of 0.1 mg) weighing machine before and after the wear test to determine the mass loss. Wear testing was done on two different samples at two different loading conditions 20 and 40 N on single layer and single layer with adding B_4C particles. The volume loss and wear rate were calculated for all the sample using the method suggested in previous literature [16, 17]. The image of the sample before testing and location from which it was expected is shown in Fig. 7.

Table 5 Results of weight loss before and after wear testing at 20 N

S. No.	Sample Id	Weight before PoD (gms)	Weight after PoD (gms)	Weight loss (gms)
1	Single layer	1.55898	1.55819	0.00300
2	Single layer with B_4C	1.33090	1.32840	0.00250

Results of weight loss before and after wear testing at 40 N

S. No.	Sample Id	Weight before PoD (gms)	Weight after PoD (gms)	Weight loss (gms)
1	Single layer	1.55819	1.55524	0.00295
2	Single layer with B_4C	1.28109	1.27872	0.00237

Table 6 Results of wear rate calculation in $mm^3/mm/N$

S. No.	Detail	20 N(μ)	40 N(μ)
1	Single L. FS	0.00044	0.00041
2	Single L. with B_4C	0.00038	0.00031

The wear test results are presented in Table 5. Figures 8 and 9 which indicate the wear rates provide some interesting results. The results indicate that with adding of B_4C particles, wear resistance is improved. In both loading conditions 20 and 40 N in the case of single layer, friction surfacing 74 and 138 μ wear of material are observed, respectively. While in cladding with the addition of B_4C particles, the wear resistance was improved and 64 and 105 μ materials were worn out, respectively, for 20 and 40 N loads. In both the cases initially some kind of fluctuation is observed in initial period because in starting the contact between pin and disc was point type, so material is loss from some specific area after some time the surface contact generated between the pin and disc which gives more effective results regarding material loss. The results in the form of weight loss as indicated in Table 5 further clarify that adding B_4C particles reduces material loss. This clearly brings out that the benefits of adding B_4C particles in surface layer are beneficial for weight loss against time. The results in Table 6 indicate the specific wear rate ($mm^3/mm/N$) of samples from normal friction surfacing and friction surfacing with the addition of B_4C particles for 20 and 40 N loading conditions.

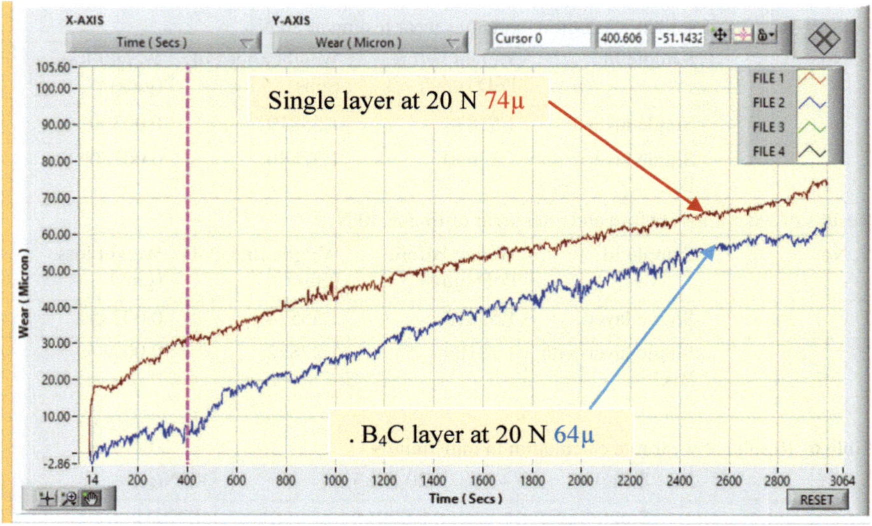

Fig. 8 Compare single-layer FS versus single-layer FS with B_4C particles at 20 N

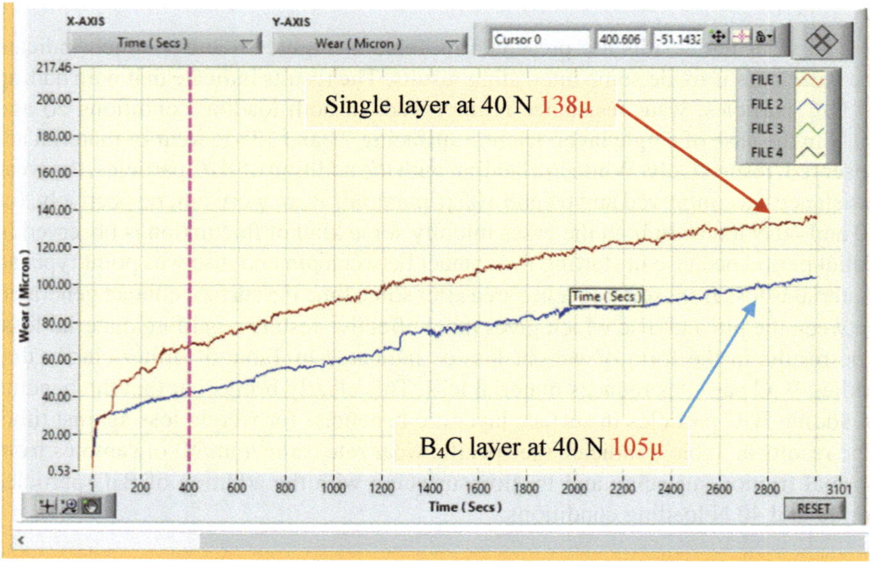

Fig. 9 Compare single-layer FS versus single-layer FS with B_4C particles at 40 N

5 Conclusions

From the following study and the results of macro- and micro-examination, it confirms that mechanical interlocking of the surfaced asperities between Al and steel materials is observed at coating interface. From the micro-examination results, it is clear that effective distribution of B_4C particle is observed at surface and subsurface of cladding.

The wear resistance increases by 13 and 24% with adding the B_4C particles, 20 and 40 N, respectively. The main reason of increasing wear resistance is ballistic behaviour of boron carbide particles on deposited coated layer. Fine grain microstructure which could the major reason for the superior wear resistance of boron carbide added friction surfaced cladded layer due to hot forging action.

Acknowledgements The authors wish to thank Pandit Deendayal Petroleum University for providing technical and experimental facility.

References

1. Gandra, J., Krohn, H., Miranda, R.M., Vilaca, P., Quintino, L., Santos dos, J.F.: Friction surfacing—A review. J. Mater. Process. Technol. **214**(2014), 1062–1093
2. Klopstock, H., Neelands, A.R.: An improved method of joining or welding metals. Patent application no 572789 (1941)
3. Miranda, R.M., Gandra, J., Vilaca, P.: Surface Modification by Solid State Cladding Process. Woodhead Publication Ltd., Cambridge (2014). ISBN 978-0-85709-468-1
4. Janakiraman, S., Reddy, J., Kailash, S.V., Udaya, K.: Bhat, Surface modification of steel using Friction stir surfacing. Mater. Sci. Forum **710**, 258–263 (2012)
5. Badheka, K., Badheka, V.: Friction surfacing of aluminium on steel: an experimental approach. Mater. Todays: Proc. **4**(9), 9937–9941 (2017)
6. Gandra, J., Miranda, R.M., Vilaca, P.: Performance analysis of friction surfacing. J. Mater. Process. Technol. **212**, 1676–1686
7. Vijaya Kumar, B., Madhusudhan Reddy, G., Mohandas, T.: Influence of process parameters on physical dimensions of AA6063 aluminium alloy coating on mild in friction surfacing. Defense Technol. 1–7 (2015)
8. de Macedo, M.L.K., Pinheiro, G.A., dos Santos, J.F., Strohaecker, T.: Deposition by friction surfacing and its application. Welding Int. **24**(6), 422–431. https://doi.org/10.1080/09507110902844535
9. Batchelor, A.W., Jana, S., Koh, C.P., Tan, C.S.: The effect of metal type and multi-layering on friction surfacing. J. Mater. Process. Technol. **57**, 172–181 (1996)
10. ASTM G 99-05 (2010) pin on disc Wear testing standard
11. Madhusudhan Reddy, G., Srinivasa Rao, K., Mohandas, T.: Friction surfacing: novel technique for metal matrix composite coating on aluminium-silicon alloy. Surf. Eng. **25**, 25–30 (2009)
12. Madhusudhan Reddy, G., Satya Prasad, K., Rao, K.S., Mohandas, T.: Friction surfacing of Titanium alloy with aluminium metal matrix composite. Surf. Eng. **27**(2), 92–98 (2009)
13. Sudhakar, I., Madhusudhan Reddy, G., Srinivasa Rao, K.: Ballistic behaviour of boron carbide reinforced AA 7075 aluminium alloy using friction stir processing—An experimental study and analytical approach. Defense Technol. **12**, 25–31 (2016)
14. Janakiraman, S., Udaya Bhat, K.: Formation of composite surface during friction surfacing of steel with aluminium. Adv. Tribol. **2012**, Article ID 614278

15. Margam, C., William, B.A., Sukumar, J.: Friction surfacing of metal coatings on steel and aluminium substrate. J. Mater. Process. Technol. **72**, 449–452 (1997)
16. Gandra, J., Vigarinho, P., Pereira, D., Miranda, R.M., Velhinho, A., Vilaca, P.: Wear characterization of functionally graded Al-Sic composite coating produced by friction surfacing. Mater. Des. **52**, 373–383 (2013)
17. Pereira, D., Gandra, J., Pamies-Teixeira, J., Miranda, R.M., Vilaca, P.: Wear behaviour of steel coatings produced by friction surfacing. J. Mater. Process. Technol. **214**(2014), 2858–2868

Effect of Speed on Various Machinability Criteria in Dry Turning of Nickel–Iron–Chromium-Based Superalloy

Anadh Gandhi, Kumar Abhishek and Soni Kumari

Abstract The present experimental work investigates the effect of cutting speed on various machinability criteria such as surface roughness, resultant machining force, material removal rate and chip thickness ratio in straight turning of nickel–iron–chromium-based superalloy (Inconel 825) using C-type and T-type inserts on manually operated lathe. The experimental results conclude that at 557 RPM, C-shape inserts provide better surface finish. Machining force decreases with increase in spindle speed. Chip thickness ratio decreases with increase in spindle speed.

Keywords INCONEL 825 · Turning · Chip thickness ratio
Resultant machining force

1 Introduction

Superalloys find numerous applications in aerospace, defence, marine and nuclear sectors due to their unique combination of properties but they have poor machinability because of the following properties: (1) They maintain their strength during machining. (2) They have poor thermal conductivity. (3) They undergo work hardening during machining. (4) Presence of carbide and abrasive particles. (5) During machining, high temperature is generated at cutting zone that accelerates diffusion wear rate.

Choudhary and Baradie [1] concluded that nickel-based alloys can be machined at the speed of order 50 m/min with carbide tools. Uncoated carbide tools provide better surface finish compared to coated carbide tools; hence, coating has no significant effect on machining performance while machining Inconel. Jindal et al. [2]

A. Gandhi · K. Abhishek (✉)
Mechanical Engineering, IITRAM, Ahmedabad, India
e-mail: krabhishek1987@gmail.com

S. Kumari
Mechanical Engineering, GLA University, Mathura, India

D. Deb et al. (eds.), *Innovations in Infrastructure*, Advances in Intelligent Systems and Computing 757, https://doi.org/10.1007/978-981-13-1966-2_36

investigated the effect of various types of coating on tool life. TiAlN-, TiCN- and TiN-coated carbide tools were taken for machining of Inconel 718. TiAlN coating provides better tool life. Arunachalam and Mannan [3] concluded that positive rake angle provides better machinability while machining high-temperature alloys. Flank wear and notch wear are dominating tool failure criteria during machining of nickel-based high-temperature alloys. Arunachalam et al. [4] studied the surface integrity for machining of Inconel 718 using coated carbide tool in dry and wet environmental conditions. In dry condition, negative rake type insert is preferable whereas for wet machining, positive rake type insert provides better results. Li et al. [5] focused on tool wear and cutting forces in dry end milling operations of Inconel 718 using coated carbide tool. Flank wear is dominating tool failure criteria in dry machining of Inconel 718. Blunt cutting edges increase the cutting force and cutting temperature. Thakur et al. [6] have experimentally investigated the effect of cutting speed under the dry machining of Incoloy 825. Surface roughness and tool wear increase with increase in cutting speed. Uncoated carbide tool provides better surface finish compared to coated carbide tool. Tool wear rate of coated carbide insert is lower than uncoated insert. Parida and Maity [7] have investigated the effect of nose radius on cutting forces using finite element analysis for Inconel 718. The investigation revealed that the forces increase with increase in the nose radius. With increase in the nose radius, shear plan angle decreases and so chip thickness increases. Satyanarayana et al. [8] have optimized the cutting parameters such as speed, feed and depth of cut for flank wear and interface temperature in dry high-speed turning of Inconel 718 using generic algorithm. The ANOVA analysis concludes that cutting speed has major influence on decision parameters. Munde and Pansare [9] have experimentally evaluated the effect of feed and spindle speed on surface roughness and tool wear during dry turning of Inconel 825 using uncoated carbide tool. Surface finish deteriorates with increase in spindle speed. Tool wear increases with increase in feed and spindle speed.

Literature highlights mainly the influence of machining parameters on the machining evaluation characteristics mainly surface roughness and material removal rate but less attempt has been made on to study the influence on cutting forces and chip thickness ratio and specific cutting energy. Therefore, present experimental study revealed the effect of cutting speed on various machinability criteria such as machined surface quality, resultant machining force, material removal rate, specific cutting energy and chip thickness ratio during dry plain turning of Inconel 825 using coated carbide tool.

2 Experimental Procedure

Inconel 825 with dimensions of $\phi 45 \times 500$ (cutting length of 20 mm) has been used as working material. The chemical composition of the Inconel 825 has been illustrated in Table 1.

Table 1 Chemical composition of work material Inconel 825

Element	C	Si	Mn	S	P	Cr	Ni	Fe
Content (%)	0.036	0.362	0.118	0.017	0.012	22.695	38.246	31.083

(continued)

Table 1 (continued)

Element	Mo	Cu	Al	Ti	V	Co	W
Content (%)	2.774	2.779	0.045	0.653	0.055	0.036	0.355

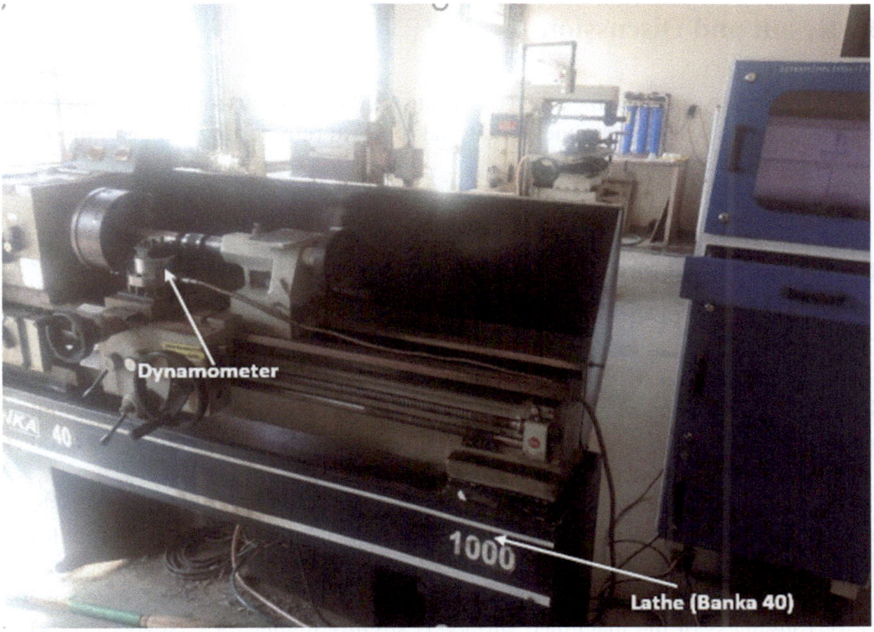

Fig. 1 Experimental setup

Figure 1 shows the experimental setup for straight turning. For experimentation, manually operated lathe BANKA 40 is used.

For straight turning, C-shape and T-shape inserts with 0.8 mm corner radius were used. MCLNR2020K12 and WTJNR2020K16 tool holders were used to hold the C-shape and T-shape inserts, respectively. In the dry turning, depth of cut and feed are set as 0.8 mm and 0.333 mm/rev, respectively.

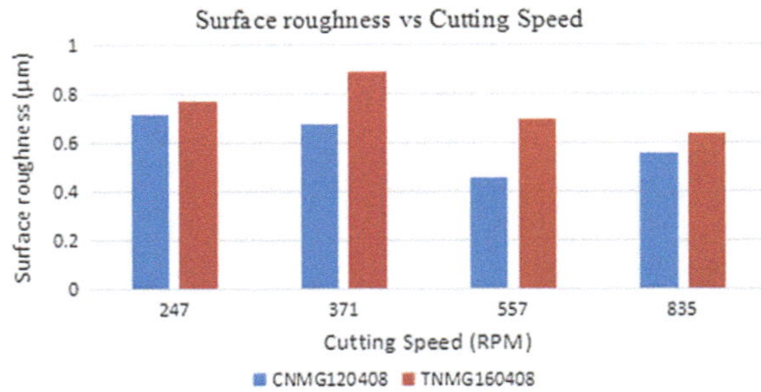

Fig. 2 Variation of average surface roughness with spindle speed

3 Result and Discussion

The present study revealed the effect of cutting speed on various machinability criteria such as average surface roughness, material removal rate, resultant machining force and chip thickness ratio.

3.1 *Average Surface Roughness (Ra)*

The quality of machined surface can be examined by these roughness criteria. Here, surface roughness is measured using Taylor Hobson surface roughness tester Surtronic S128 having standard stylus type 112-1502 DCN 001. Sampling length is taken as 5 mm up to the range of 400 μm. For a single run, 10 readings were taken and their average is considered as final output.

Figure 2 shows the variation of Ra with spindle speed for C-shape and T-shape inserts at different speed levels. Figure 2 clearly states that average roughness decreases with increase in spindle speed as chip detaches from the workpiece easily. Tool wear rate increases with increase in cutting speed. At high spindle speed of 835 RPM, roughness of the machined surface is slightly higher for C-shape insert. For T-shape insert, surface finish improves with increase in spindle speed from 371 RPM to 835 RPM. At every level of speed, C-shape insert performs better than T-shape insert. CNMG120408 insert has point angle of 80° and TNMG160408 insert has of 60°; hence, the end cutting edge angle of C-shape insert is smaller than that of T-shape. C-shape insert created a shallower slope of feed marks than the T-shape insert.

3.2 *Resultant Machining Force* (**R**)

During the experiments, the forces are measured using dynamometer (Kistler made, Type 9272). Dynamometer separately measures the component forces in X, Y and Z

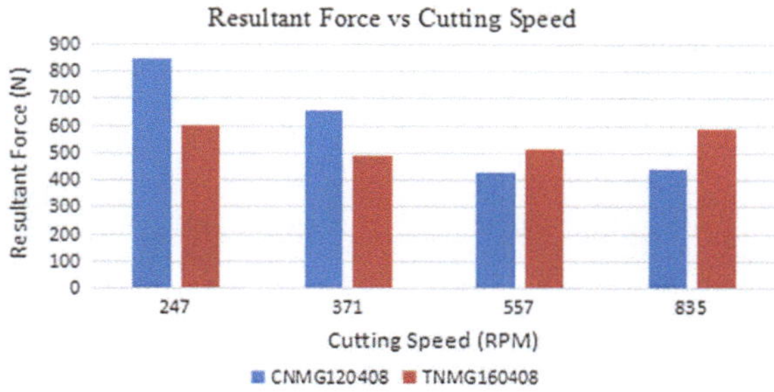

Fig. 3 Variation of resultant force with spindle speed

directions, respectively. The resultant machining force has been assessed using the below relation:

$$R = \sqrt{F_c^2 + F_t^2 + F_r^2}.$$

Here,

F_c is the cutting force in Z-direction,
F_t is the feed force in Y-direction and
F_r is the radial force in X-direction.

Figure 3 shows the variation of resultant force for C-shape and T-shape inserts with spindle speed. For CNMG120408, resultant machining force decreases with increase in spindle speed due to reduction in friction coefficient up to 557 RPM. At 835 RPM, resultant force is higher than that of 557 due to tool wear. For T-shape insert, resultant force increases with increase in spindle speed from 371 to 835 RPM. For spindle speed of 247 and 371 RPM, T-shape inserts perform better than C-shape insert. At high speed, CNMG120408 is more suitable.

Cutting zone temperature increases with increase in spindle speed that may soften the work material and decreases the coefficient of friction but at high speed, wear rate is high that increases the tool—workpiece contact region and coefficient of friction—so resultant force is higher.

3.3 Material Removal Rate (MRR)

Material removal rate is defined as quantity (volume) of material machined per unit time. MRR is calculated using the following relation:

$$\text{MRR} = \frac{W_i - W_f}{\rho \cdot t_m} (\text{mm}^3/\text{s})$$

Here,

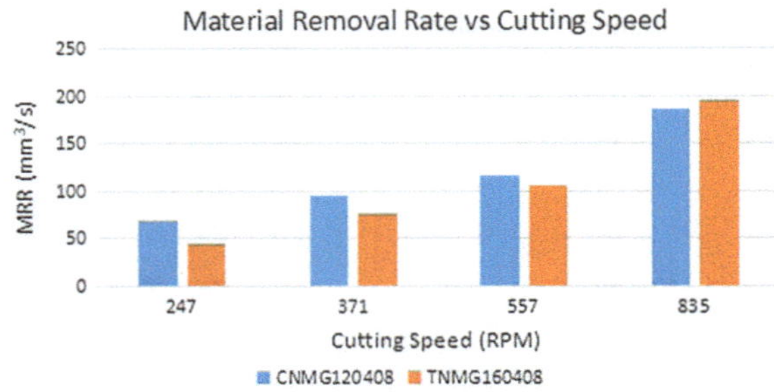

Fig. 4 Variation of MRR with cutting speed

W_i is the initial weight of the workpiece in gm,
W_f is the final weight of the workpiece in gm,
ρ is the density of the work material and
t_m is the machining time in second.

To calculate weight before machining and after machining, high precision digital balance meter SAB E10 (made by SCALE-TEC) was used.

Figure 4 shows the effect of cutting speed on material removal rate in bar chart form. It is clear from the diagram that MRR increases with increase in cutting speed. At high speed of 835 RPM, T-shape insert machines large amount of material compared to C-shape insert.

3.4 Chip Thickness Ratio

Chip behaviour plays an important role to understand the machining condition. Apart from machining parameters, it depends on various parameters such as material behaviour, system vibration, machining environment, tool geometry, etc. Chip thickness ratio is defined as ratio of chip thickness after machining to uncut chip thickness. For oblique cutting, the uncut chip thickness is calculated by below formula:

Uncut chip thickness $(T1) = f * \sin\alpha$
Here, f is the feed in mm/rev and
α is the approach angle.

During the plain turning, feed and approach angle of insert remain constant so uncut chip thickness remains constant.

Figure 5 shows the graphical representation of variation of chip thickness ratio with spindle speed in dry turning of alloy 825. From Fig. 5, it is very clear that chip thickness ratio decreases with increase in spindle speed. As the spindle speed increases, chip will be easily moved away from the work material. With the increase in spindle speed, cutting temperature increases but due to poor thermal conductivity and rapid detachment from the workpiece, less amount of heat can be conducted by the chip from the cutting zone.

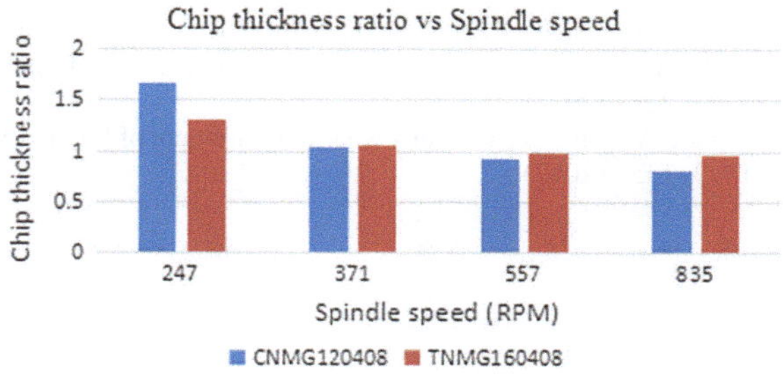

Fig. 5 Variation of chip thickness ratio with spindle speed

4 Conclusions

The present experiments show the effect of spindle speed on various machinability criteria such as material removal rate, cutting force, average surface roughness and chip thickness ratio in dry straight turning of Inconel 825 using C-shape and T-shape insert by keeping feed and depth of cut as constant parameters. From the above discussion, the following conclusions can be derived:

1. Material removal rate increases with increase in spindle speed.
2. Resultant machining force decreases with increase in spindle speed.
3. Surface finish improves with increase in spindle speed. At 557 RPM, C-shape insert provides best result among the selected inputs.
4. Chip thickness ratio decreases with increase in spindle speed.

Acknowledgements Authors are thankful to B K ENTERPISE, Ahmedabad for providing tooling and facilities for the experimentation works.

References

1. Choudhury, I.A., El-baradie, M.A.: Machinability of nickel base super alloys: a general review. Int. J. Mater. Process. Technol. **77**, 278–284 (1998)
2. Jindal, P., Santhanam, A.T., Schlienkofer, U., Shuster, A.F.: Performance of PVD TiN, TiCN and TiAlN coated cemented carbide tools. Int. J. Refract Metal Hard Mater. **17**, 163–170 (1999)
3. Arunachalam, R., Mannan, M.: Machinability of nickel based high temperature alloy. Mach. Sci. Technology: Int. J. **4**(1), 127–168 (2000)
4. Arunachalam, R.M., Mannan, M.A., Spowage, A.C.: Surface integrity when machining age hardened Inconel 718 with coated carbide cutting tool. Int. J. Mach. Tools Manuf. **44**, 1481–1491 (2004)
5. Li, H.Z., Zeng, H., Chen, X.Q.: An experimental study of tool wear and cutting force variation in end milling of Inconel 718 with coated carbide inserts. J. Mater. Process. Technol. **180**, 296–304 (2006)
6. Thakur, A., Mohanty, A., Gangopadhyay, S.: Comparative study of surface integrity aspects of incoloy 825 during machining with uncoated and CVD multilater coated inserts. Appl. Surf. Sci. **320**, 829–837 (2014)
7. Parida, A., Maity, K.: Effect of nose radius on forces, and process parameters in hot machining of Inconel 718 using finite element analysis. Eng. Sci. Technol.: Int. J. **20**, 687–693 (2017)

8. Satyanarayana, B., Reddy, M.D., Nitin, P.R.: Optimization of controllable turning parameters for high speed dry machining of super alloy: FEA and Experimentation. Mater. Today: Proc. **4**, 2203–2212 (2017)
9. Munde, V., Pansare, V.: Flank wear measurement of Inconel 825. Int. J. Res. Eng. Technol. **4**(8), 1343–1351 (2017)

A Review on Dissimilar Friction Stir Welding of Aluminum Alloys to Titanium Alloys

Surya Jain, Krishna Bhuva, Palak Patel and Vishvesh J. Badheka

Abstract The joining of titanium alloy with aluminum alloy is the current area of interest owing to the increasing demand to reduce weight and simultaneously achieve high strength for the aviation, automotive, and ground transportation industries in order to reduce fuel consumption. However, the dissimilar metal joining of Al–Ti poses several challenges due to the difference in metallurgical and mechanical properties. It is difficult to achieve commercially acceptable sound joints with appropriate strength using fusion welding technique due to the problem of segregation, porosity, cracks, slag inclusion, aluminum-rich Ti_xAl_y intermetallic phases, and inferior mechanical properties of the weld. To overcome all these defects of fusion welding, solid-state welding is employed, which is the coalescence of metals under the application of pressure and at a temperature lower than the melting point of the parent metals. This leads to lower heat input and the formation of less intermetallic compounds. This technique is opening doors for new research possibilities for joining dissimilar metals like titanium and aluminum. This paper focuses on presenting the progress in the field of joining Al–Ti using friction stir welding along with a detailed analysis regarding the effect of different parameters on the quality of the weld.

Keywords Friction stir welding · Aluminum alloy · Titanium alloy · Dissimilar

1 Introduction

The joining of dissimilar metals is becoming increasingly popular due to its wide application in the industry, such as cryogenics, nuclear, microelectronics, and thermal power stations, because of both economic and technical reasons. This enables engineers to create flexible product designs as well as efficiently use materials using materials for specific functions that match with their properties. With these advantages of joining dissimilar metals, there are many problems faced due to thermal and

---

S. Jain · K. Bhuva · P. Patel (✉) · V. J. Badheka
Pandit Deendayal Petroleum University, Gandhinagar 382007, Gujarat, India
e-mail: palak98patel@gmail.com

D. Deb et al. (eds.), *Innovations in Infrastructure*, Advances in Intelligent Systems and Computing 757, https://doi.org/10.1007/978-981-13-1966-2_37

metallurgical properties incompatibility between the parent metals which leads to the failure of their welds due to the formation of intermetallic compounds. These intermetallic compounds may consist of low and high melting phases formed due to mismatch in chemical properties and brittle phases, which later lead to failure of the weld. Another challenge faced is the mismatch in coefficient of thermal expansion which leads to thermal distortion, residual stresses, and quench cracking.

Properties of aluminum like low density, high conductivity, high corrosion resistance, high ductility, and high strength-to-weight ratio make it the most extensively used non-ferrous material in the world. However, there are various weldability issues while dealing with aluminum alloys which are difficult to avoid like porosity, cracking, and precipitation hardening leading to lower joint efficiency. On the other hand, titanium apart from having good corrosion resistance and low weight also has high stiffness, toughness, and strength. Titanium alloys are in general weldable under controlled temperatures. The weld between these two dissimilar metals opens doors for various high-strength high-temperature applications. Currently, Al–Ti bond is being researched upon, so that it can be applied in the frames of aero bodies and missiles.

Talking about welding of aluminum and titanium, it is difficult to achieve commercially acceptable sound joints with appropriate strength using fusion welding technique due to the problem of segregation, porosity, cracks, slag inclusion, aluminum-rich Ti_xAl_y intermetallic phases, and inferior mechanical properties of the weld. To

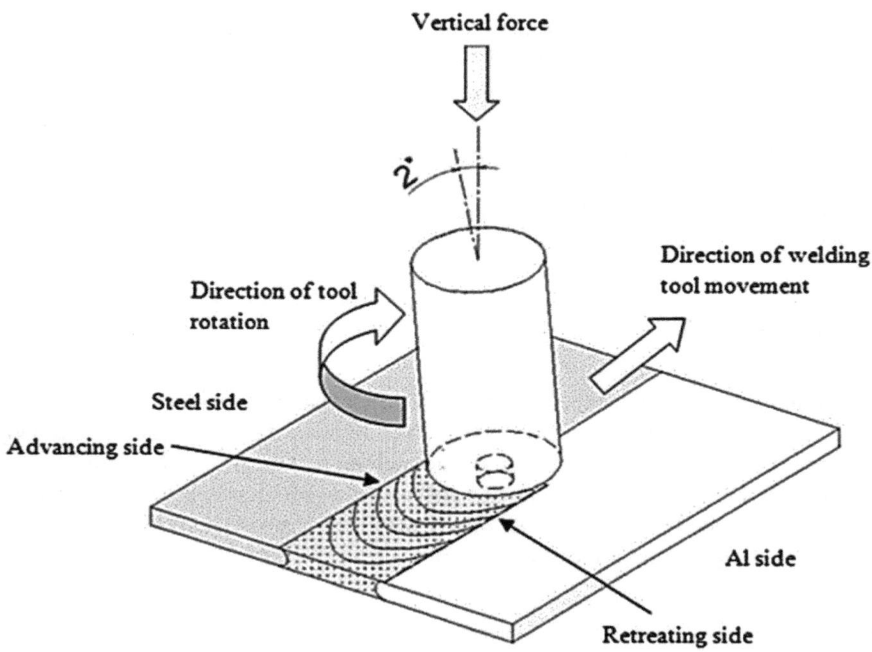

Fig. 1 Friction stir welding schematic diagram

overcome all these defects of fusion welding, solid-state welding is employed. Friction stir welding (FSW), as shown in Fig. 1, is a solid-state joining process which involves the coalescence of metals under the application of pressure and at a temperature lower than the melting point of the base metals. There is no need of any filler material, and low amount of heat input is required. With the industrialization, there is a growing need to reduce cost and weight, and improve mechanical and metallurgical properties, structural rigidity, resistance toward harsh environmental conditions, etc.; friction stir welding (solid state) is finding applications in a wide variety of areas such as in shipbuilding industry, railway carriage, aerospace industry, etc. Recently, NASA's Orion spacecraft also used FSW technique to join lightweight aluminum alloy and Apple used FSW to join the aluminum surfaces of its iMac. This technique is the opening door for new research possibilities for joining dissimilar metals, e.g., Ti and its alloys, aluminum and steel, copper and nickel, etc. The parameters that affect the quality of the weld are shown in Table 1. Each of these parameters is discussed and reviewed in detail with respect to the Al–Ti weld using FSW.

2 Friction Stir Welding Tool

2.1 Tool Material

Tool material is more of the mandatory consideration parameter rather than optimizing one. Effect of tool material on the quality of the weld may be deeper rooted than it is thought of. Using a superior tool material might not improve the weld quality but will definitely increase the reliability of the welding process by decreasing tool wear and withstanding the high temperatures and shear forces accompanying the FSW process. The requirements for an ideal tool material should be high melting point, strength at ambient and elevated temperatures, corrosion resistance, hardness greater than that of the workpiece materials, abrasion resistance, fatigue strength, shear strength, chemical stability concerning the workpiece metals, creep resistance, machinability, and low coefficient of thermal expansion [1].

When WC-13% Co is used as tool material for welding Ti–6Al–4V and Al–6Mg under modified butt joint (L shape) configuration with tapered tool, no visible defects were observed on the tool although it was used at speeds higher than 1000 rpm. However, there was no mention about the microstructure changes in the tool before or after the welding process [2]. Similar results were obtained when Dressler et al. [3] welded Ti–6Al–4V and AA2024–T3 using standard tool steel under the travel speed of 30 mm/min and tool rotation speed of 800 rpm. Bang et al. [4] chose to use SKD61 as the tool material while investigating the weldability of titanium and duralumin at various rotational speeds ranging from 100 to 300 rpm keeping the other parameters constant.

Table 1 Summary of recommended parameters and properties of Al–Ti welds

S. No.	Workpiece materials	Thickness (mm)	Tool design	Process parameters	Remarkable properties of joint	References	Type of weld
1	Ti-6Al-4V and Al-6Mg	2-mm-thick Ti sheet and 4-mm-thick Al plate with 2.2-mm-depth groove	TM: WC-13% Co PSp: Tapered SD: 15 mm RD: 6 mm TD: 4 mm PL: 2 mm	WS: 60 mm/s RS: 1200, 1800, 2400 rpm Offset: 0.5, 1.0, 1.5, 2 mm TTA: 0° PlD: 0.2 mm Shielding gas: 99.9% pure argon Offset: pin radius(r) mm	IMCs: TiAl, $Ti(Al)_3$, Ti_3Al UTS_{max}: 92% Al	[2]	Modified butt joint (L shape)
2	Ti-6Al-4V and AA2024-T3	2 mm	TM: standard tool steel PSp: threaded and tapered SSp: concave SD: 18 mm PD: 6 mm	WS: 30 mm/min RS: 800 rpm Offset: pin radius(r) mm	UTS: 348 MPa HtL: 0.3 kg Hardness: 135–140 HV	[3]	Butt joint
3	Ti-6Al-4V and AA6061-T6	5 mm	PSp: circular truncated cone probe RD: 6 mm TD: 4 mm SD: 15 mm PL: 4.5 mm	WS: 2.5 mm/s RS: 1000 rpm TTA: 3° PlD: 0.2 mm Offset: pin radius(r) mm	UTS: 134 MPa HtL: 490 mN	[4]	Butt joint
4	7075-T651/2024-T3 and Ti-6Al-4V	2.0 mm	TM: SKD61 SD: 15 mm PD: 6 mm PL: 1.9 mm	WS: 100, 200, 300 mm/s Offset: 0.5 mm RS: 850 rpm TTA: 3° APL: 7.8 kN	IMCs: TiAl3 UTS: 311 MPa	[6]	Square butt joint
5	TC1 Ti alloy and LF6 Al alloy	2 mm	PSp: Threaded cone SSp: Concave shoulder SD: 15 mm RD: 5 mm PL: 1.85 mm	WS: 95–180 mm/min RS: 600–1180 rpm Offset: 0 mm TTA: 2°	IMCs: TiAl, $Ti(Al)_3$, Ti_3Al ShS_{max}: 48 MPa	[8]	Lap joint

(continued)

Table 1 (continued)

S. No.	Workpiece materials	Thickness (mm)	Tool design	Process parameters	Remarkable properties of joint	References	Type of weld
6	TC1 Ti alloy and LF6 Al alloy	2 mm	PSp: Threaded cone SSp: Concave shoulder SD: 15 mm RD: 4 mm PL: 2.1 mm	WS: 95–180 mm/min RS: 600–1180 rpm Offset: 0.5 mm TTA: 2°	IMCs: TiAl, $Ti(Al)_3$, Ti_3Al UTS_{max}: 131 MPa	[8]	Butt joint
7	Ti-6Al-4V and Aluminum 1060	3 mm	TM: Tungsten carbide PSp: Cutting stir pin SSp: Concave PD: 6 mm SD: 25 mm	WS: 150, 235, 300, 375, 475 mm/min RS: 950 rpm TTA: 0° PID: 3.2 mm Offset: 0 mm	IMC: $TiAl_3$ TSS_{max}: 1910 N 100% Al1060	[10]	Lap joint
8	TC1 Ti alloy and LF6 Al alloy	2 mm	TM: SSp: Concave PSp: Threaded cone SD: 15 mm RD: 4 mm PL: 2.1 mm	WS: 60, 95, 118, 150 mm/min RS: 600, 950, 1500 rpm Offset: 0 mm TTA: 2°	IMCs: Ti–Al compounds TSS_{max}: 2800 N	[9]	Lap joint
9	ADC12 cast AL and Ti	4 mm Al and 2 mm Ti	TM: WC-Co SD: 15 mm PD: 5 mm PL: 3.9 mm	WS: 60, 90, 120 mm/min RS: 1500 rpm Offset: 0 mm TTA: 3° APL: 5.39 kN	IMC: Al, Si, Ti, $TiAl_3$ phases FL: 16 kN	[12]	Lap joint
10	Al-5083 and commercially pure Ti	3 mm	PSP: conical	RS: 1120 rpm WS: 50 mm/min.	Hardness: (16% more of Al and 60% more of Ti)	[11]	Butt joint
11	Ti6Al4V and AA2024	TiAl4V: 1.6 mm AA2024: 1.2 mm	SD: 13 mm MD: 4.5 mm SCA: 30° PL: 2.4 mm	RT: 20° HEC: 1.9 N/(s/mm/C). SG: Argon. WS: 0.5 mm/rev. RS: 900, 1200, 1500 r/min FR: 200, 50, 100 mm/min	WP: waived Hardness max: 370 HV	[7]	Lap joint

(continued)

Table 1 (continued)

S. No.	Workpiece materials	Thickness (mm)	Tool design	Process parameters	Remarkable properties of joint	References	Type of weld
12	Al alloy 1100 and Titanium aluminum alloy	5 mm	TM: HSS PD: 4 mm SD: 20 mm	RS: 1100 rpm WS: 5 mm/min	Hardness: 225 KHN	[13]	–
13	Ti6Al4V and A6061-T6	3 mm	PSP: cylindrical SP: conical	RS: 500, 750, 1250 Offset: (0–1.2 mm)	UTS MAX: 215 Mpa (68% Al at 750 rpm). IMC: $TiAl_3$	[5]	Butt joint

TD tool diameter, *TM* tool material, *PD* Pin diameter, *SD* shoulder diameter, *SSp* shoulder shape, *PL* Pin length, *Pld* plunge depth, *WP* weld profile, *PSp* Pin shape, *RT* room temperature, *HEC* heat exchange coefficient, *SG* shielding gas, *FR* feed rate, *ShS* shear strength, *TSS* tensile shear strength, *FL* failure load, *RS* rotation speed, *WS* working speed, *TTA* tool tilt angle, *APL* applied load, *Htl* hardness test load

2.2 Tool Design and Geometry

Superior welds are produced by different profiles while dealing with different materials. Tool profile can be chosen based on experience, requirement, or based on experimental results. The D/d ratio affects the effective intermixing of the materials being joined, and hence the tensile strength of the weld. Different material combinations have different optimum ratios which can be determined experimentally. On decreasing the shoulder diameter, contact area between the tool and workpiece is reduced which reduces the heat generated due to friction, and hence joint becomes weaker. Longer pins reduce forging and intermixing because of the reduced thermo-mechanical effect of the shoulder

3 Process Parameters

3.1 Tool Pin Offset

Tool offset parameter of FSW leads to changes in the direction and the amount of intermixing, surface bonding, etc. The pin offset toward the softer and metal with low melting point controls the amount of heat generation in the FSW weld by reducing the degree of deformation of harder and higher melting side, and hence lowering the frictional heat input [2, 5]. Since the welding parameters of the dissimilar metals are different due to their different mechanical and metallurgical properties, tool offset is required to provide different rotating speeds and welding speeds [3]. It was observed that when the welding tool was more toward the Al alloy, the stirred zone is on the Al side. In the stir zone, it was observed that there were fine recrystallized grains of Al alloy and titanium was pushed away from titanium surface due to the stirring action of the tool. In research done by Li et al. [2], the probe was placed such that it hardly made contact with the titanium sheet. They mentioned that such arrangement increases the efficiency of the process reducing the heat input (which reduces the Al sheet melting) and tool wear, and hence decreasing the cost. Wu et al. [5] claimed that when the offset is 0, no joint was formed, offset between (0.9–1.2 mm) required an array of rotating rates (500–1000 rpm) and welding speed (120–280 mm/min) to form sound joint, whereas with small tool offset (0.3 mm and 0.6 mm) higher rotating rate of 1250 rpm was needed [5]. In this study, probe was inserted more on the Al side and the edge of the probe was offset in the titanium alloy.

3.2 Rotational Speed

Increasing the speed of rotation gives better distribution of particles in the heat-affected zone but it simultaneously also increases the heat input, and hence optimum speed must be chosen depending on the materials being joined. Rotational speed in general affects the metallurgical and mechanical properties of the weld to a great extent. Li et al. [2] claimed that as tool rotation speed is increased while keeping other parameters like tool offset and travel speed constant, it can be seen that the average tensile strength and average percent elongation both decrease considerably. This phenomenon was seen when the tool rotational speed was increased from 1200 to 1800 while joining Ti–6Al–4V and Al–6Mg under modified butt joint configuration. Likewise, the results of research done by Aonuma et al. [6] showed that a laminar structure was formed due to the stirring action of the high rotation tool when 2-mm-thick plates of 7075-T651/2024-T3 and Ti–6Al–4V were welded under square butt joint configuration.

According to Wu et al. [5], the highest strength of the joint was observed at 750 rpm. At 1250 rpm, the joint failed due to the formation of thicker intermetallic layer, and at 500 rpm the joint failed in the stir region. It was seen by Buffa et al. [7] that tool rotation has a major impact on the joints strength and temperature than feed rate.

Chen et al. [8] concluded that an easy weld craze is obtained at 1180 rpm of tool rotation rate according to their experiment. It was seen that high rotation speeds form intermetallic compounds in the weld due to an increase in frictional heat between the workpiece and stir head, and this hence increases the possibility of weld crazes. With low tool rotation speeds like 600 rpm, flash formation and grooves are formed in the weld, and hence its surface is rough. So, an optimum speed was concluded to be 750/950 rpm. According to Hua et al. [9], as there is increase in tool rotation rate, there is increase in the Ti alloy particles that are being stirred by the force of the pin in the stir zone, and hence the failure of the lap joints decreased with the higher rotation rate of 1500 rpm.

3.3 Travel Speed

Tool travel speed or the welding speed is the speed with which the tool is moved forward along the joint line. The travel speed basically affects the intermixing of the materials, heat input, and consequently the quality of the weld. A very high travel speed leads to non-uniform particle distribution, while low travel speeds lead to the formation of undesirable IMCs apart from unnecessarily increasing the process time. While welding 2-mm-thick plates of 7075-T651 and Ti–6Al–4V, Aonuma et al. [6] observed that defects appeared near the groove when working at the travel speed of 300 mm/min. The joint, hence, formed was effortlessly broken near the weld interface. He also declared that the width of the stir zone was nearly the same for Ti–6Al–4V/2024 and Ti–6Al–4V/7075 joints although the width increased on the Al side on decreasing the travel speed. Buffa et al. [7], in their research, claimed that the strength of the weld increases with the increase in feed rate.

With an increase in welding speed when described by Chen et al. [8] to 95 and 118 mm/min, the particles of Ti in the stir zone decrease due to less welding heat input that reduces the plastic flowability of Ti alloy. With a further increase in weld speed to 150 mm/min, a crack appears at the Al and Ti alloy interface. Wei et al. [10] discover that there is a decrease in the Ti cut-off layer when the welding speed increases due to the relationship of welding speed and energy input per length being inversely proportional. Voids were discovered at lower weld speeds. Hence, it was concluded that a proper weld was formed at intermediate speeds such as at 300 mm/min.

4 Metallurgical Properties of Dissimilar Al–Ti FSW

4.1 Macrostructure

According to Li et al. [2], on paralleling the two sides of the butt joint, the macrostructure analysis of SNZ showed that the macrostructure of aluminum was immensely deformed near the joint interface. On the other hand, only a "hook"-like structure owing to the plastic deformation was formed near the joint on the titanium side.

While investigating the weld properties, Aonuma et al. [6] discovered that the stir zone was mainly composed of finely recrystallized aluminum alloy grains strengthened by the coarse titanium fragments which might be bought into the zone due to the stirring action of the probe. According to Buffa et al. [7], the Al side near the joint region included stir zone, TMAZ, and HAZ, whereas the Ti side included stir zone and HAZ. The highest hardness (480 VHN) was observed in the stir zone. Three zones, namely, Al-based metal, Ti-based metal, and intermetallic compound zone of Ti and Al.

Ghogeri et al. [11] saw that three zones were formed in the stir zone, namely, aluminum-based metal, titanium-based metal, and the area composed of intermetallic compounds of aluminum and titanium. It is also observed that the joint region in the aluminum side consists of stir zone, TMAZ, and HAZ which were formed, while the joint region on the titanium side includes stir zone and HAZ. The microstructure of stir zone includes coaxial micro-grains that are recrystallized. Deformed and elongated grains were observed in the thermo-mechanically affected zone. In the HAZ, the microstructure and mechanical properties of the material have been altered because of the entered thermal cycle, without mechanical deformation.

4.2 Microstructure

According to Li et al. [2] on examining the weld on microlevel, it was seen that no microcracks existed at the interface. The weld was furthermore investigated with the use of SEM/EDS for analysis purpose at different process conditions.

Research done by Bang et al. [4] showed a swirl-like structure with darker and lighter regions when viewed in the SEM micrograph of the weld interface area. One material is clasped into the other demonstrating proper intermixing.

5 Weld Defects

Inappropriate heat input or incorrect process parameters like tool tilt angle, rotational speeds, travel speed, or offset majorly contribute toward the formation of defects such as surface grooves, macro- and microcracks, tunnels, voids, surface galling, excessive flash, and nugget collapse. Apart from tool and process parameters, the basics of workpiece placement problem like wide gaps or mismatch between the two mating plates also lead to the welding defects.

Inappropriate depth of penetration may cause defects such as the common roof flaw. According to the results obtained by Li et al. [2], since the defect is caused due to lack of penetration, a little higher plunge of the pin can effectively avoid the defect.

Bang et al. [4], while finding defect in the 2-mm-thick plates for welding 7075-T651/2024-T3 and Ti-6Al-4V, it was found that compared to the surrounding aluminum particles, the titanium particles elongate to a smaller extent which leads to the formation of cavities at the edge of titanium units.

It was observed by Wu et al. [5] that lower higher speed and low rotating rate produces wormhole defect in stir zone. High value of tool offset along with lower tool rotating rate and higher welding speed leads to defect in the Ti alloy. At low tool offset, low tool rotating rate, and high welding speed, no bonded defect were observed at the bottom interface of the weld.

6 Conclusion

The effect of various process parameters on the mechanical and metallurgical properties on the weld formed by dissimilar metal joining of aluminum alloys with titanium alloys has been discussed in the paper. The paper reviews the literature with respect to the important process parameters like tool rotational speed, travel sheet, tool pin offset, and tool material. It was observed that although the welds in general seemed visually acceptable, analysis on microscopic level showed several defects like root flaw, cavities, wormholes, voids, and cracks. The aluminum-rich Ti_xAl_y intermetallic phases formed due to excessive heat input reduces the strength of the weld by increasing the hardness, and hence encouraging the brittle behavior of the joint. This is the area of concern mainly whenever either the process parameters or the working temperatures are not controlled. However, there are no such formulae or relations to determine the optimum working parameters and they must be experimentally determined and optimized according to the welder needs depending on the strength requirements, economic constraints, acceptable process time, and availability of required equipment. The industrial application of titanium–aluminum bond is held back due to reasons like higher economic cost and lack of availability of effective stir tools made up of high-temperature wear proof materials and lower service life [2]. The review showed that there was a lack of appropriate statistics regarding the effect of various parameters like tool tilt angle, tool design, axial force, shoulder diameter, and tool pin plunge depth on the micro- and macrostructural properties of the weld. It can be concluded that there is a scope of research in the field of determining the character of the effect of previously mentioned parameters and other on the weld properties.

Acknowledgements This work is supported by the Student Research Program of Office of Research & Sponsored Program, Pandit Deendayal Petroleum University, Gandhinagar, India. No. ORSP/R&D/SRP/2017/PPVB.

References

1. Chiteka, K.: Friction stir welding/processing tool materials and selection. In: International Journal of Engineering Research & Technology (IJERT), vol. 2, issue 11. Delhi Technological University, India (2013)
2. Li, B., Zhang, Z., Shen, Y., Hu, W., Luo, L.: Dissimilar friction stir welding of Ti-6Al-4V alloy and aluminium alloy employing a modified butt joint configuration: influences of process variables on the weld interfaces and tensile properties. Mater. Des. **53**, 838–848 (2014)
3. Dressler, U., Biallas, G., Mercado, U.: Friction stir welding of titanium alloy TiAl6V4 to aluminum alloy AA2024-T3. Mater. Sci. Eng. A **526**, 113–117 (2009)
4. Bang, K., Lee, K., Bang, H., Bang, H.: Interfacial microstructure and mechanical properties of dissimilar friction stir welds between 6061-T6 Aluminum and Ti-6%Al-4%V alloys. Mater. Trans. **52**(5), 974–978 (2011)

5. Wu, A., Song, Z., Nakata, K., Liao, J., Zhou, L.: Interface and properties of the friction stir welded joints of Titanium alloy Ti6Al4V with aluminum alloy 6061. Mater. Des. **71**, 85–92 (2015)
6. Aonuma, M., Nakata, K.: Dissimilar Metal joining of 2024 and 7075 aluminum alloys to titanium alloys by friction stir welding. Mater. Trans. **52**(5), 948–952 (2011)
7. Buffa, G., Lisi, M., Sciortino, E., Fratini, L.: Dissimilar titanium/aluminum friction stir welding lap joints by experiments and numerical simulation. Adv. Manuf. **4**, 287 (2016)
8. Chen, Y., Liu, C., Liu, G.: Study on the joining of titanium and aluminum dissimilar alloys by friction stir welding. Open Mater. Sci. J. **5**, 256–261 (2011)
9. Hua, C., Quan, N., Ming, K.: Interface characteristic of friction stir welding lap joints of Ti/Al dissimilar alloys. Trans. Nonferrous Met. Soc. China **22**, 299–304 (2012)
10. Wei, Y., Li, J., Xiong, J., Huang, F., Zhang, F., Raza, F.: Joining aluminum to titanium alloy by friction stir lap welding with cutting pin. Mater. Charact. **71**, 1–5 (2012)
11. Ghogheri, M., Asgarani, M., Amini, K.: Friction stir welding of dissimilar joint of aluminum alloy 5083 and commercially pure titanium. Kovove Mater. **54**, 71–75 (2016)
12. Chen, Y.C., Nakata, K.: Microstructural characterization and mechanical properties in friction stir welding of aluminum and titanium dissimilar alloys. Mater. Des. **30**, 465–475 (2009)
13. Dhindaw, B., De, P., Jayashree, P.: Friction stir welding of aluminum alloy 1100 and Titanium-Al alloy. In: Proceedings of the ASME 2016 International Manufacturing Science and Engineering Conference, MSEC (2016), Blacksburg, Virginia, USA (2016)

Vibration-Assisted Multipurpose Cutter

Ajay Kale, PL. Ramkumar and Dipankar Deb

Abstract The article presents an innovative approach to develop a multipurpose cutting tool for effective hole-making and milling operations. The present invention aims in reducing the cutting forces by inducing vibrations in various axes depending upon the direction of cutting. The present setup also facilitates automatic tool changing and tool setup. The assembly includes automatic tool changing assembly, robotic arm, multipurpose cutting tool and guide rails. Two possible configurations of the system are presented. The basic idea of this invention is to develop a setup which can perform three operations, i.e. drilling, milling and reaming using different configurations of the same setup.

Keywords Annular cutter · Multipurpose cutter · Automatic tool changer

1 Introduction

Manufacturing industry is considered one of the most important contributors in the development of any country. The aim of any industry in manufacturing sector is to increase the productivity and achieve maximum output for given inputs. Increase in productivity can be achieved by reducing the machining time.

A significant percentage of time is consumed in changing a tool and setting up another tool for machining the parts which require different types of operations on various locations on a workpiece. This invention aims to provide solution for such problems.

The effect of inducing vibration while performing any kind of machining operation is to provide an impact force. This force helps in reducing the overall cutting force as the impact of tool on workpiece helps in material removal process.

A. Kale · PL. Ramkumar (✉) · D. Deb
Institute of Infrastructure Technology Research and Management, Khokara, Ahmedabad 380026, India
e-mail: pl.ramkumar@iitram.ac.in

D. Deb et al. (eds.), *Innovations in Infrastructure*, Advances in Intelligent Systems and Computing 757, https://doi.org/10.1007/978-981-13-1966-2_38

1.1 Recent Developments in Field of Multipurpose Cutting

Many inventions have been made to reduce the cutting force on a tool by inducing vibration while performing cutting operation on a workpiece. For example, US7687975B2 has given a vibration-assisted machining (VAM) system [1]. This system uses piezoelectric material stacks which are adapted to generate a substantially elliptical tool path in a vibration plane of the system. This system also includes a third PZT actuator coupled between the frame and either the vibration element or the workpiece holder to provide additional vibration as per requirement. US 20120125171A1 has presented a borer for forming holes in a workpiece which has a fastening portion for mounting on an oscillating tool and a cutter which is connected with the fastening portion [2]. The cutter is driven in a vibrating manner by the oscillating tool such that a feeding direction of the cutter is substantially parallel to a longitudinal axis about which the borer is driven in a vibrating manner. Stephen et al. have claimed in US 20170096862A1, a vibration-assisted rotary drilling (VARD) tool that provides axial compliance and low amplitude axial displacements at the drill bit while transmitting the full rotary speed and torque of the drill string to increase drilling penetration rate [3].

Saeid Amini et al. have presented their investigations on performance of longitudinal–torsional (L-T) vibration in UAD of Al 7075-T6 with HSS tool. The results revealed that this kind of vibration significantly reduces cutting forces compared with the conventional drilling (CD). Furthermore, this reduction was clarified by investigation of generated chips and tool rake angle. Moreover, effect of ultrasonic vibration on drill skidding and surface quality has been analysed [4]. Xue Hui Shen & Gue Feng Xu analysed the milling force variation in ultrasonic vibration-assisted end milling process through a series of slot-milling experiments. Experimental results revealed that ultrasonic vibration can change traditional milling conditions and realize separate-type milling, and the peak value of the feed-direction cutting force can also be greatly decreased by adopting reasonable vibration amplitude [5]. Jianhua Zhang et al. have carried out comparative study on ultrasonic vibration-assisted milling (UVAM) and conventional milling. The pressing model showed that the tool minor cutting edge left clear traces with certain width because of the tool minor cutting edge angle. Scratches by tool minor cutting edge were intermittent and regularly varied when feed-direction vibration was introduced [6]. M. A. Moghaddas et al. developed a robust UAD module, fully compatible with computer numerical control (CNC) systems, and able to function in a production environment [7].

US3609056A presents an annular hole cutter having a relatively thin, cylindrical sidewall with a plurality of teeth formed in the thickened lower edge thereof [8]. The teeth have cutting edges extending across the full width of the thickened edge. This annular hole cutter provides reduction in cutting forces, heat generation and power consumption. D. E. Brehl et al. have presented the review work on vibration-assisted drilling on various machining operations like drilling, milling, turning and grinding [9].

Annular cutters can be used for making through holes with larger diameters by cutting annular material from the workpiece [10]. Yung Tseng has presented an innovative approach to assemble various annular cutters having different diameters [11]. But even after intensive research, annular cutters have been limited to hole cutting only.

To overcome this limitation, the present invented system is used to perform multiple operations using a single vibrating annular cutter.

2 Setup for Experimentation

One of the possible configurations of the presented invention is shown in Fig. 1. This configuration can be used for drilling and reaming. The multipurpose cutter remains perpendicular to workpiece. The guide rails help the cutter to move in *X*–*Y* plane. The tool changing mechanism is shown in Fig. 1 consisting of robotic arm and automatic tool changer (ATC). Robotic arm pulls the multipurpose tool and places it in ATC, and then takes a new tool from ATC and attaches it back in the apron on guide rails.

Another possible configuration of the system is shown in Fig. 2. This can be used for milling purpose. The tool axis remains parallel to workpiece to carry out slab milling operation. For effective cutting, vibration is induced in the tool using piezoelectrics. Guide rails are designed in a way to provide smooth transition from configuration shown in Figs. 1 and 2. The automatic tool changing assembly remains the same as shown in Fig. 1. The tool cannot be directly changed, and it has to be brought in configuration in Fig. 1 before changing.

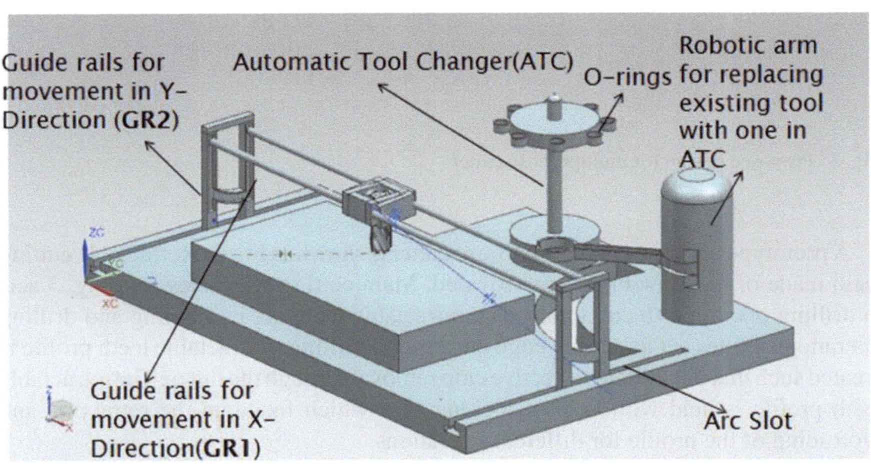

Fig. 1 Proposed configuration for drilling and reaming using multipurpose cutting assembly

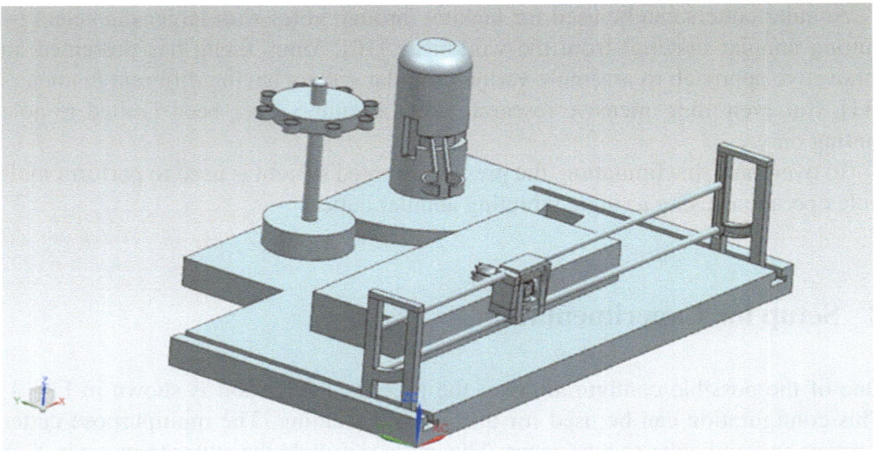

Fig. 2 Proposed configuration for milling using multipurpose cutting assembly

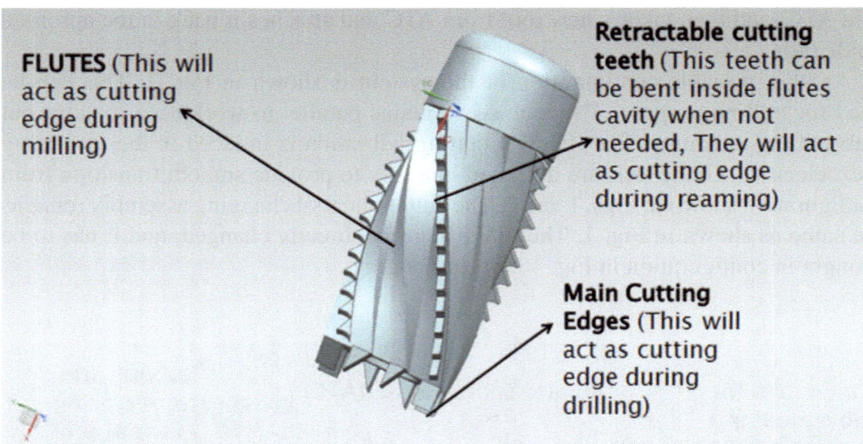

Fig. 3 Proposed design for multipurpose cutter

A prototype of proposed multipurpose cutter is shown in Fig. 3. Retractable cutting teeth made of hacksaw blade are provided. Main cutting edge shown in Fig. 3 acts in drilling operation to cut a hole. The retractable teeth act in reaming and drilling operations. Flutes act as cutting edge during slab milling. Retractable teeth profile is created such that it facilitates effective chip removal through the flutes. The retractable teeth profile is held with help of locking pins which facilitate the retracting and protruding of the profile for different operations.

One of the possible assemblies of cutter, holder and apron is shown in Fig. 4. The piezoelectric which vibrates the tool in X and Y direction is shown. The piezo-

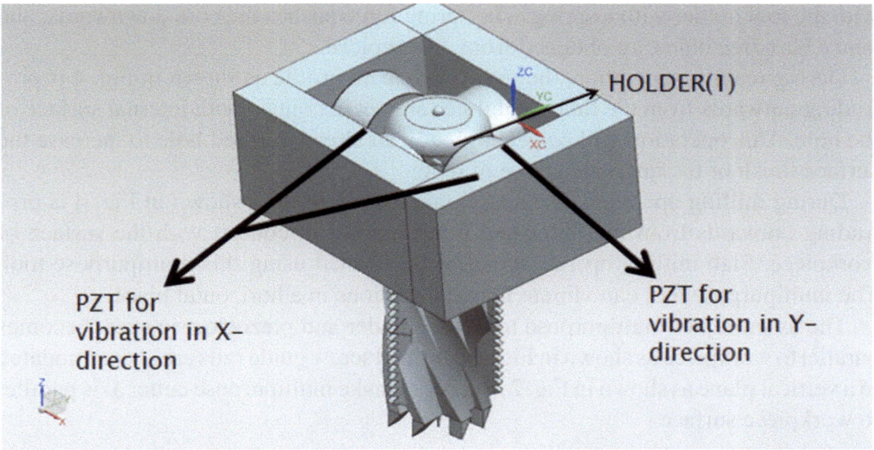

Fig. 4 Proposed design for multipurpose cutter and piezoelectric stack assembly

electric stack which vibrates the tool in *Z* direction is between the tool holder and multipurpose tool. The box-like structure accommodating this stacks and holder is apron. This apron will be mounted on the guide rails as shown in Fig. 1.

3 Detailed Design of System

The whole setup basically consists of subsystem as described below:

1. Multipurpose cutter,
2. Automatic tool changer assembly and
3. Tool holder and piezoelectric assembly.

1. **Multipurpose Cutter**

This subsystem consists of three retractable teeth profile placed on the flutes at an angle of 120° to each other on its periphery as shown in Fig. 3. The retractable teeth profile can be engaged or disengaged with help of locking pins according to a predefined input programme. The retractable teeth profile is a series of V-shaped teeth similar to that in a hacksaw blade but with increased thickness. Height of this retractable teeth profile is less than the height of flutes provided, such that when it is retracted, it is accommodated within the flutes and does not interact with surface of workpiece. During drilling operation, teeth will be in retracted position and will not make contact with internal surface of hole. The peripheral teeth at end of tool come in contact with workpiece and material is cut and chips flow through the flutes provided. The core that is not being cut, i.e. the internal material which is not cut by the peripheral teeth, is pushed down with the help of centre locater pin attached with the tool holder with a spring. The spring force pushes the core downwards, and hence burr-free holes are obtained from a workpiece.

During reaming operation, the retractable teeth profile as shown in Fig. 4 is protruding outwards from the flutes and flutes are not in contact with internal surface of the hole. This operation will be performed in an already drilled hole to increase the surface finish of the internal surface of hole.

During milling operation, the retractable teeth profile as shown in Fig. 4 is protruding outwards from the flutes and flutes are not in contact with the surface of workpiece. Slab milling operation can be performed using this multipurpose tool. The multipurpose tool can vibrate in two directions in a horizontal plane.

The assembly of multipurpose tool, tool holder and piezoelectric stack becomes parallel to workpiece as shown in Fig. 2. The lead screw guide rails are accommodated in a vertical plane as shown in Fig. 2 in order to make multipurpose cutter axis parallel to workpiece surface.

2. **Automatic Tool Changer Assembly**

Tools made of different geometry and materials can be mounted on the automatic tool changer. Eight such tools can be mounted. The tools can be attached in the O-ring provided on automatic tool changer. A tool in operation can be exchanged with a spare tool on automatic tool changer with help of a robotic arm. The original position of robotic arm is in middle of arc slot as shown in Fig. 1. The lead screw guide rails and T-slot guide rails are brought to the arc slot provided. The servo-based robotic arm pulls only the multipurpose tool from the tool holder assembly; piezoelectric stacks and other assembly remain there as it is. Meanwhile, the servo mounted on automatic tool changer rotates it such that the empty O-ring is positioned just above the arc slot. The robotic arm now moves radially in the arc slot with help of the other servo and arrives exactly below the empty O-ring on automatic tool changer. The multipurpose cutter is pushed to fit inside the O-ring and then the robotic arm retreats. After that, the servo on automatic tool changer rotates it such that the tool to be used is present in the O-ring just above the arc slot. The robotic arm pulls this tool from automatic tool changer and moves radially again into the arc slot and then attaches the tool to the tool holder assembly and the robotic arm retreats to its original position.

3. **Tool Holder and Piezoelectric Stack Assembly**

As shown in Fig. 4, piezoelectric stack is fitted between tool holder and multipurpose tool. During both drilling and reaming operation the multipurpose tool, tool holder and piezoelectric stack assembly remain perpendicular to the workpiece as shown in Fig. 1. Only piezoelectric stack which is fitted inside tool holder is actuated. The multipurpose cutter vibrates in axis parallel to its central axis.

During milling operation, piezoelectric stacks are actuated. The piezoelectric stack is attached to the apron as shown in Fig. 4 and they can be actuated individually depending on direction of tool travel. The multipurpose cutter has its axis parallel to the workpiece surface as shown in Fig. 2. The multipurpose cutter vibrates in an axis perpendicular to the cutter axis.

The present invention can be used in manufacturing of internal combustion engines which has pathways for the lubrication oil to flow made with slab milling operation and has several holes on it for various applications made by drilling operation. Hence, this system finds several applications in manufacturing industries which perform hole-making and slab milling operations on the same workpiece.

4 Conclusion

The present invention aims at creating a multipurpose tool which can perform drilling, milling and reaming operations. This system aims at reducing the overall cost of production by reducing the tools cost, tool setup cost and tool setup time. The cutter has induced vibrations which will provide reduction in cutting force and enhanced surface roughness. The present invention consists of multipurpose cutter with retractable teeth profile, piezoelectric stacks, robotic arm, automatic tool changer, lead screw guide rail, T-slot guide rail and workpiece. Robotic arm has two degrees of freedom and consists of two servo motors. The assembly of automatic tool changer and robotic arm can be detached if single type of tool is used. Piezoelectric stacks provide vibration in transverse as well as axial direction of the multipurpose cutter. The present invention is capable of performing vibration-assisted drilling for making through holes in fully automated mode in CNC machine and semi-automatic mode on a magnetic portable drill. The setup is capable of performing semi/fully automated vibration-assisted reaming operation with help of retractable teeth profile provided on flutes of the multipurpose cutter whole setup can be retrofitted in standard machine tools.

References

1. Lee, T., Ethington, J.: Vibration assisted machining system with stacked actuators. US patents, US7687975B2
2. Chen, L., et al.: Borer for an oscillating tool. US patents, US20120125171A1
3. Butt, S., et al.: Vibration assisted rotary drilling tool. US patents, US 20170096862A1
4. Amini, S., et al.: Effect of longitudinal−torsional vibration in ultrasonic-assisted drilling. Mater. Manuf. Processes **32**(6), 616–622 (2017)
5. Shen, X.-H., Guo-Feng, X.: Study of milling force variation in ultrasonic vibration-assisted end milling. Mater. Manuf. Processes **33**(6), 644–650 (2018)
6. Zhang, J., et al.: Feed-direction ultrasonic vibration–assisted milling surface texture formation. Mater. Manuf. Processes **32**(7), 193–198 (2018)
7. Moghaddas, M.A., et al.: Performance of an ultrasonic assisted drilling module. Int. J. Adv. Manuf. Technol. **94**(12), 3019–3028 (2018)
8. Hougen, E.D.: Annular hole cutter. US patents US3609056A
9. Brehl, D.E., et al.: Review of vibration assisted machining. Precis. Eng. **32**(3), 153–172 (2008)
10. www.metaller-machines.com. Metaller Machines. Retrieved 26 May 2014
11. Tseng, Y.: Cutter assembly. US patents, US9144847B2

Effect of Sheet Thickness and Grain Size on Forming Limit Diagrams of Thin Brass Sheets

Dhruv Anand and D. Ravi Kumar

Abstract Various metals and alloys have huge applications in the form of very thin sheets to manufacture parts such as micro metallic components for electronics and MEMS. In manufacturing of metallic components using very thin sheets, the material behaviour changes significantly with miniaturization due to the size effects—the 'grain size effect' and the 'feature/specimen size effect'. Due to this, formability of very thin sheets depends on sheet thickness to grain size ratio (*t*/*d*). But, forming limit diagrams use in FE software meant for prediction of failure are plotted without incorporating the grain size effect and the plane strain condition of Forming Limit Diagram (FLD_0) is calculated based on strain hardening exponent (*n*) and sheet thickness (*t*). In view of this, the present work is aimed at an investigation of grain size effect (*t*/*d*) on FLD_0 of very thin brass sheets to achieve an improvement in accuracy of formability prediction in FE Analysis (FEA). In the present work, forming limit diagrams of very thin sheet specimens of CuZn36 brass (50–200 μm) have been experimentally determined.

Keywords Formability · Grain size effect · Forming limit diagrams

1 Introduction

In the current years, there has been a huge demand for micro technical products in various fields of engineering and science such as telecommunication, microelectromechanical systems, medicine, biotechnology, automotive industry, etc. [1]. Due to the gradually growing field of applications for miniature, there is a great demand for micro metallic parts which need to be produced in big numbers. For example,

---

D. Anand (✉)
Vishwakarma Government Engineering College, Chandkheda, Ahmedabad, India
e-mail: dhruv_30@rediffmail.com

D. Ravi Kumar
Indian Institute of Technology, Delhi, India

D. Deb et al. (eds.), *Innovations in Infrastructure*, Advances in Intelligent Systems and Computing 757, https://doi.org/10.1007/978-981-13-1966-2_39

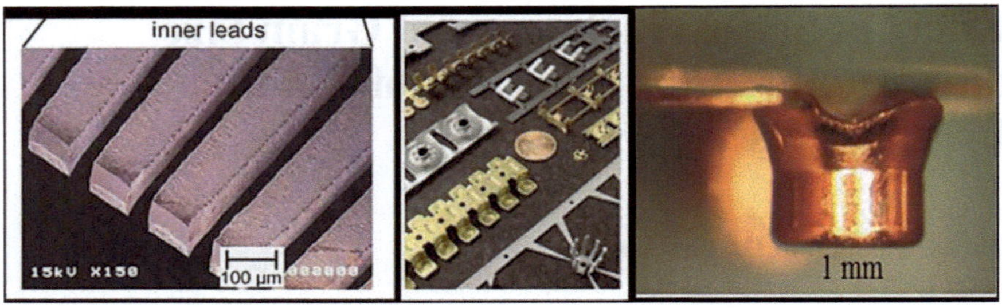

(a) Lead frame for microprocessor (b) Micro stamping (c) Micro deep drawn cup

Fig. 1 Applications of very thin sheet forming in micro manufacturing [6]

cellular phones became smaller with every new generation; additional functions are being added like integrated organizers.

Due to this continuously increasing trend towards the use of miniaturized products in various fields of application, manufacturing of very small components using very thin sheet metals or metal foils has also become increasingly important. It clearly shows the need to produce micro metallic parts in huge quantity have drawn the attention by researchers in the area of metal forming of miniaturized parts known as micro forming [2–5].

Very thin sheet metals have large applications in manufacturing of both micro and macro products for various fields. Various metals and alloys such as brass, copper, stainless steel and aluminium have applications in the form of thin sheet metals to manufacture micro parts such as micro-resonators, micro-surgical tools, lead frames, IC sockets, circuit boards, electronic connectors, micro cups and micro gears [6]. Some typical micro parts manufactured from very thin sheet metals for various applications are shown in Fig. 1.

In the case of very thin sheet forming, the thickness of sheet metals is usually less than a millimetre, going down to less than 100 μ in certain cases. When a forming process or a specimen size is scaled down from macro scale to the submillimetre range, some aspects of the workpiece not changed such as the microstructure and the surface topology. Because the ratio between the dimensions of the part and parameters of the microstructure or surface to change, and is commonly known as the 'size effect'. Because of the size effects, the material and process behaviour changes significantly with miniaturization. Therefore, understanding the miniaturization effects on formability of very thin sheet metals is essential.

Brass can easily be cast to shape or fabricated by extrusion, rolling, drawing, hot stamping and cold forming. Brass has excellent ductile behaviour when cold worked, high strength and corrosion resistance and excellent formability characteristics for sheet metal operations. Brass has excellent mechanical and electrical properties which are essential for electronic devices. Hence, it is used for components in microelectromechanical systems (MEMS) applications such as IC sockets, circuit boards, micro cups and electronic connectors.

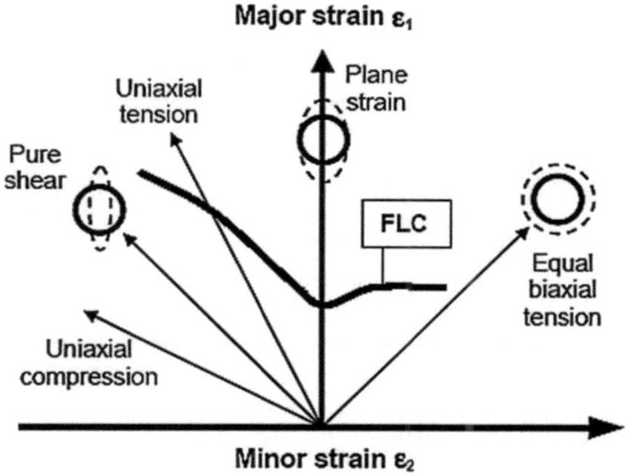

Fig. 2 Forming limit curve of a sheet metal [8]

Forming Limit Diagram (FLD) or Forming Limit Curve (FLC) provides comprehensive information on the overall formability of a sheet metal. It is a pictorial presentation of maximum principal surface strains that the sheet material can withstand without necking or failure under all the possible conditions of deformation, i.e. biaxial stretching, plane strain stretching, uniaxial tension and combined tensile and compressive stresses [7]. It is plotted between two principal strains measured or calculated on the surface of the sample (usually referred to as major strain and minor strain). Typical forming limit diagram of a sheet metal is shown in Fig. 2. In the below diagram, strain paths corresponding to different modes of deformation are also shown. The combinations of major strain and minor strain lying below the FLC represent safe region of working. If the strains exceed the limit strains predicted by FLC (the region lying above the FLC), it represents failure. The strains measured from the necked area or the critical strains lie close to the FLC. Right side of the diagram represents strain data from biaxial stretching mode and the strain data obtained from uniaxial condition and combination of tension and compression lie on the left-hand side of the diagram. The major strain value when the minor strain value is zero represents the limiting strain in plain strain stretching and it is also referred to as plain strain intercept (FLD_0). It has been observed that forming limit diagram is sensitive to material parameters such as sheet thickness, strain hardening coefficient, anisotropy, strain rate sensitivity index, etc. It was shown that limit strains increase with increase in sheet thickness [9].

2 Size Effects

The size effect is characterized by the relation of dimensions of the billet or the initial blank to material or process-related parameters like grain size. When a forming process is reduced in size, the relative ratio between these two changes significantly

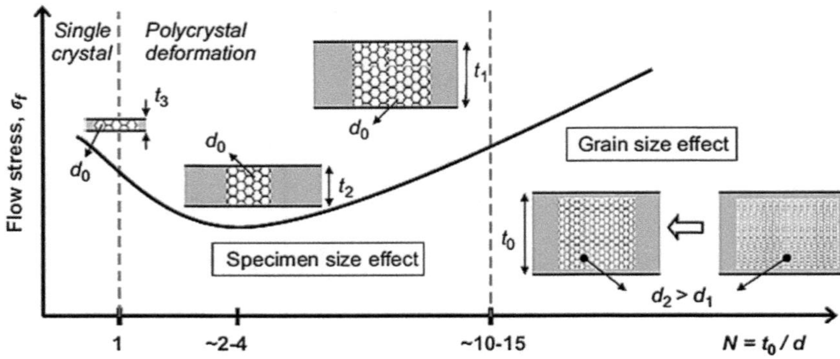

Fig. 3 Effect of *t*/*d* ratio on flow stress [14]

which leads to change in metal flow behaviour compared to macro scale processes [10].

In manufacturing of metallic components using very thin sheets, the size effects can be investigated in two categories. One is the 'grain size effect' which occurs when the grain size increases from d to D while the feature/specimen size is kept constant. The other is the 'feature/specimen size effect', when the specimen size decreases while the grain size remains constant [2, 10–13]. The experimental work conducted towards identifying the size effects was carried out by scaling down the workpiece/blank dimensions and test apparatus applied at the conventional macroscale level. The materials that were tested presented a wide range of industrial materials that are being used in current microfabrication practices, mainly electronics and biomedical industries.

A summary of the literature on effect of *t*/*d* on the flow stress is presented in Fig. 3. The flow stress decreases with the increasing miniaturization is related to *t*/*d* which is an important parameter for flow behaviour of very thin sheets. For a given grain size, as *t*/*d* decreases, flow stress decreases. Because the grains located at/near the free surface are less restricted than the grains located inside. With decreasing specimen size of the equal microstructure (size invariant), they contribute to the surface grains increases. The surface grains deform easier than the inside grains leading to decrease in resistance to deformation and a reduction of hardening so, decrease in flow stress [14].

3 Materials and Chemical Composition

In the present work, very thin sheet materials of brass (CuZn36) in the range of 50–200 μm have been used for investigation of size effects on formability. These materials were chosen because of their wide application in the form of very thin sheets and foils.

Table 1 Composition of the materials (by weight %) Element

Element	Cu	Zn	Sn	Pb	Fe	Ni
Brass (50–200 μm)	63.2 (±1.3)	36.64 (±0.9)	0.04	0.02	0.03	0.01

The brass thin sheets were as received have been used. A large range of grain size (*d*) is required to study the effect of sheet thickness to grain size ratio on formability. In view of this, the samples of very thin brass sheets were annealed with combinations of annealing temperature and time in a furnace in normal atmosphere. The samples were annealed at 600 °C and the holding time was 2 h.

The chemical composition of materials (by weight %) used in the present study has been analysed by spectroscopy as per ASTM E1507 and it is given in Table 1.

4 Microstructure/Grain Size

The microstructure of sheet specimens of all the materials used in this study were examined for determination of the grain size. All the specimens of very thin sheets were prepared to study the microstructure and grain size by standard metallographic techniques.

The microstructures were then examined using an optical microscope, with different magnifications (25X–1000X). From the optical micrographs, the average grain size was determined using standard linear intercept method as per ASTM E112. (Fig. 4).

The value of grain size varied is of 9–50 μm for sheets of thickness 50–200 μm. This resulted in specimens with a large range of ratios of sheet thickness to grain size ratio (*t*/*d*).

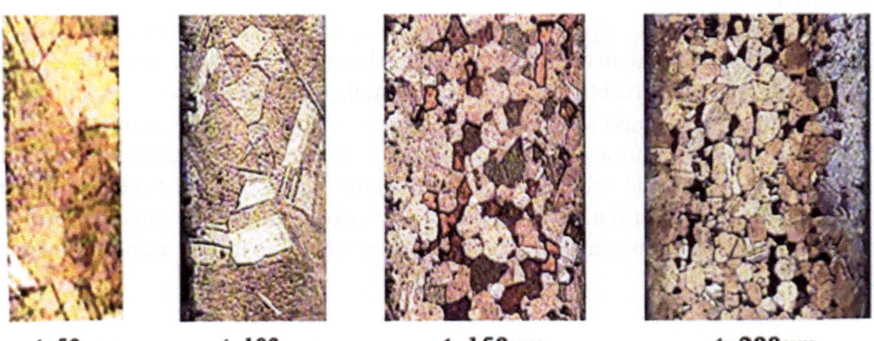

Fig. 4 Microstructure of annealed (6000 °C, 2 h) brass sheets at 200X in thickness direction

5 Characterization of Formability

Several laboratory tests have been developed for evaluation of formability which predict sheet metal failure in different modes deformation such as Erichsen cup test, Olsen cup test, Fukui conical cup test, Swift flat bottom cup test and Limiting dome height (LDH) test [15, 16]. A modified cup test has been developed of the penetrator size to 101.6 mm and having a draw bead and this test is known as the limiting dome height test [17, 18]. The height of the dome at the onset of necking is used as a measure of stretchability. Strain distribution in deformed samples can also be measured and analysed in this test.

To investigate grain size effect on formability, limiting dome height tests were carried out. FLDs were also experimentally determined Experimental procedure developed by Hecker [7] was followed to determine forming limit diagrams. The experiments were conducted on a double action 100-ton hydraulic press with 50 mm diameter hemispherical bottom punch. To determine FLDs, the specimen width was varied in the range of 20–90 mm in steps of 10 mm for getting limit strains in all the possible modes of deformation (biaxial stretching, plane strain stretching and draw-in forming modes). Three to four samples were tested for each width to account for scatter of results. The thin sheet metal samples of all widths were laser marked with a 2 mm diameter circular grid. The sheet specimen was clamped firmly between the die and the blank holder before it was stretched over the punch. The amount of drawing in while punch stretching varies with the width of the blank. Since the draw bead diameter is 70 mm, draw-in occurs in specimens with width less than 70 mm (as there is no complete clamping). As the width increases, the drawing component decreases and stretching component increases. All the FLD tests were conducted on the 100 tonne hydraulic press in dry condition (without lubrication) at a punch velocity of 10 mm/min. As in the case of LDH tests, some trial experiments were conducted to determine an optimum blank-holding force which has been found to be in the range of 20–70 kN. The experiments were stopped when a visible neck or initiation of fracture was observed on the specimens. The deformed samples in FLD tests are shown in Fig. 5.

The major and minor strains on the tested samples (in FLD tests) were measured using the deformed grid on the tested samples. The circles deformed into ellipses due to the deformation of the blanks. A digital toolmakers travelling microscope having a least count of 1 μm was used to measure major axis and minor axis diameters of the ellipses. An approximation was done to measure 3D length on the deformed surface by adjusting the sample orientation under the microscope such that the grid under observation was almost flat. This was repeated to obtain major and minor diameters of all the required ellipses and the corresponding major and minor strains have been calculated.

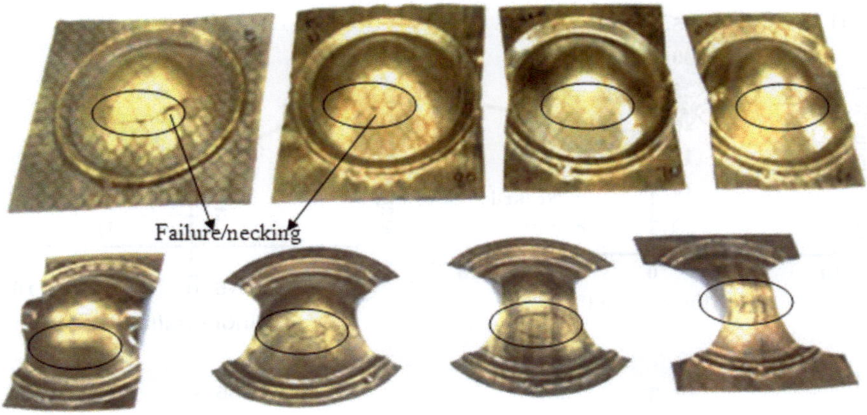

Fig. 5 Specimens of different width tested for the determination of FLD

6 Forming Limit Diagrams

Corresponding major and minor strains measured from FLD test were plotted. The strains measured from safe, necked and failed regions are shown with different symbols. The curves represent the limit strains up to which these sheet metals can be formed in all modes of deformation (stretching, plane strain stretching and draw-in forming modes).

The forming limit curves of thin brass sheets of different thickness are shown in Fig. 6a–d. From the Fig. 6 the FLD_0 for very thin brass sheets are found to be 18, 15, 17 and 12%, respectively, for 200, 150, 100 and 50 μm. Due to very low thickness and *R* values below one, the limit strains of these sheets are also low indicating inferior formability when compared to brass sheets of thickness in the conventional range.

The forming limit curves of the very thin brass sheets of all four different thickness are compared in Fig. 7. As the thickness decreased, the forming limits decreased in all three regions–tension–compression, plane strain and biaxial stretching. However, the forming limits of 100 μm thick sheet have been found to be higher than those of 150 μm thick sheet. It could be attributed to much higher ductility and '*n*' value of 100 μm sheet than the 150 μm sheet apart from the effect of *t*/*d* ratio on formability.

The experimentally determined FLD has been used to predict necking/failure in FE simulation as input that can be helpful in identifying the best process configuration before engaging in expensive experimental trials. In contrast to conventional metal forming, where numerical process simulation is already successfully applied, size effects have to be considered in process simulation in very thin sheet metal forming.

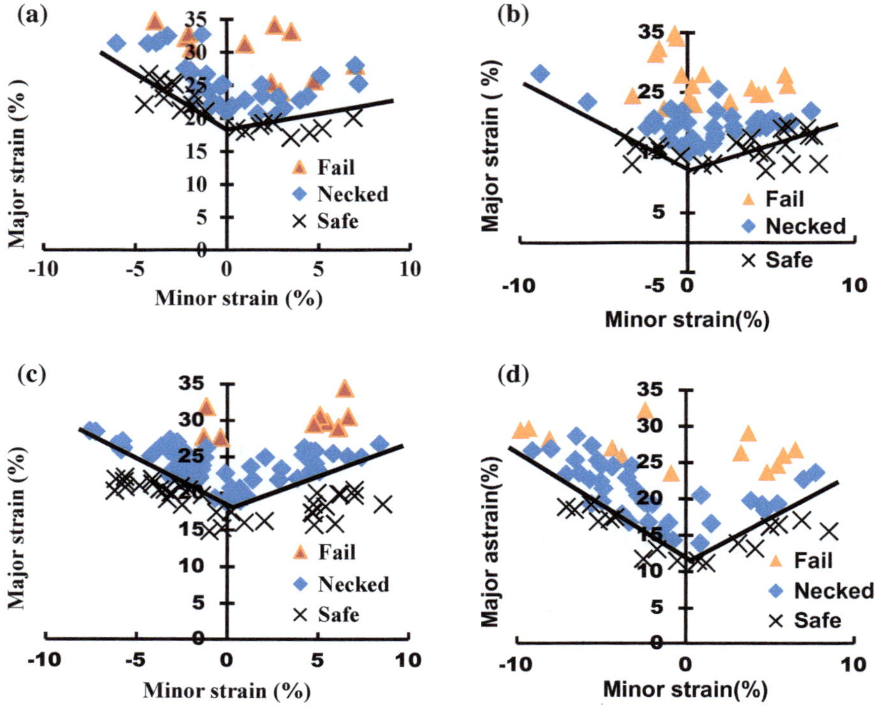

Fig. 6 Experimentally determined FLD of brass sheet of thickness **a** 200 μm, **b** 150 μm, **c** 100 μm and **d** 50 μm

Fig. 7 Comparison of FLDs of brass sheets with different thickness

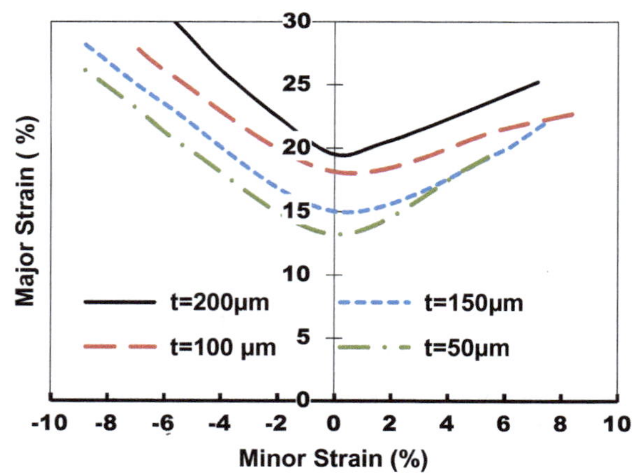

7 Conclusion

The following conclusions or main findings are drawn:

From the formability tests to study the grain size effect, it has been found that FLDs of very thin sheets as the thickness decreased, the forming limits decreased in general in all modes of deformation. Formability decreases as thickness of sheet decreases from 200 to 50 μm.

For considerable increase in accuracy of formability predictions need to integrate the grain size effect for FE simulation and to identify the failure in simulations, the experimentally determined FLDs of brass sheets of each thickness have been used in simulations.

The theory of plasticity and the knowledge of conventional metal forming technology developed for macro scale cannot be directly applied to thin sheet forming.

References

1. Vollertsen, F., Schulze Niehoff, H., Hu, Z.: State of the art in micro forming. Int. J. Mach. Tools Manuf. **46**(11), 1172–1179 (2006)
2. Geiger, M., Vollertsen, F., Kals, R.: Fundamentals on the manufacturing of sheet metal microparts. CIRP Ann.—Manuf. Technol. **45**(1), 277–282 (1996)
3. Geißdörfer, S., Engel, U., Geiger, M.: FE-simulation of micro forming processes applying a mesoscopic model. Int. J. Mach. Tools Manuf. **46**(11), 1222–1226 (2006)
4. Vollertsen, F., Hu, Z., Niehoff, H.S., Theiler, C.: State of the art in micro forming and investigations into micro deep drawing. J. Mater. Process. Technol. **151**(1–3), 70–79 (2004)
5. Cao, J., Krishnan, N., Wang, Z., Lu, H., Liu, W.K., Swanson, A.: Micro forming: experimental investigation of the extrusion process for micropins and its numerical simulation using RKEM. Trans. ASME **126**, 642–652 (2004)
6. Geiger, M., Kleiner, M., Eckstein, R., Tiesler, N., Engel, U.: Micro forming. CIRP Ann.—Manuf. Technol. **50**(2), 445–462 (2001)
7. Ghosh, A., Hecker, S.: Failure in thin sheets stretched over rigid punches. Metall. Mater. Trans. A **6**(4), 1065–1074 (1975)
8. Holmberg, S., Enquist, B., Thilderkvist, P.: Evaluation of sheet metal formability by tensile tests. J. Mater. Process. Technol. **145**(1), 72–83 (2004)
9. Hiam, J., Lee, A.: Factors influencing the forming limit curves of sheet steel. Sheet Metal Ind. **55**, 631–643 (1978)
10. Messner, A., Engel, U., Kals, R., Vollertsen, F.: Size effect in the FE-simulation of micro-forming processes. J. Mater. Process. Technol. **45**(1–4), 371–376 (1994)
11. Vollertsen, F., Biermann, D., Hansen, H.N., Jawahir, I.S., Kuzman, K.: Size effects in manufacturing of metallic components. CIRP Ann.—Manuf. Technol. **58**(2), 566–587 (2009)
12. Al-Qureshi, H.A., Klein, A.N., Fredel, M.C.: Grain size and surface roughness effect on the instability strains in sheet metal stretching. J. Mater. Process. Technol. **170**(1–2), 204–210 (2005)
13. Yeh, F.-H., Li, C.-L., Lu, Y.-H.: Study of thickness and grain size effects on material behavior in micro-forming. J. Mater. Process. Technol. **201**(1–3), 237–241 (2008)
14. Mahabunphachai, S., Koç, M.: Investigation of size effects on material behavior of thin sheet metals using hydraulic bulge testing at micro/meso-scales. Int. J. Mach. Tools Manuf. **48**(9), 1014–1029 (2008)

15. Hosford, W., Caddel, R.M.: Metal Forming Mechanics and Metallurgy. Cambridge University Press, Cambridge, MA (2007)
16. Mellor, P.B.: Sheet metal forming. Int. Metall. Rev. **1**, 1–20 (1981)
17. Ghosh, A.K.: The effect of lateral drawing-in on stretch formability. Metal. Eng. Quart. **15**, 53–64 (1975)
18. Hecker, S.S.: A cup test for assessing stretchability. Met. Eng. Quart. **14**, 30–36 (1974)

Electrochemical Deburring of Al6082 Using NaCl Electrolyte: An Exploratory Study

Satisha Prabhu, Abhishek Kumar and Vishvesh J. Badheka

Abstract Often burrs are termed as perennial thorns, real productivity killers and production engineer's headache. These burrs are the thin ridges, projections, loose metal pieces that develop during production operation like turning, milling, drilling, sheet metal operations, etc. Workman injury, short circuits, and assembly problems were reported while handling parts with burrs during machining. In hydraulic and pneumatic systems burrs and metal pieces can alter the flow conditions, which may result into catastrophic failure. In the absence of standardized procedure and adequate standards for removal of burrs of various shapes, dimensions, and properties, electrochemical deburring (ECD) method is considered as an alternate method. In particular, deburring of internal geometry is more difficult, ECD has proved to be beneficial and inexpensive. Therefore, in automation and high-efficiency processes of deburring cross holes, development of ECD is considered to be of prime importance. The present research uses Sodium chloride (NaCl) as a neutral electrolyte for the deburring Al6082 specimen. The performance is evaluated by varying input parameters such as Current applied, concentration, current density, time, etc. Machining characteristics are analyzed and reported.

S. Prabhu (✉) · V. J. Badheka
Department of Mechanical Engineering, Pandit Deendayal Petroleum University, Gandhinagar 382007, India
e-mail: satishaprabhu@gmail.com; satisha.pphd14@sot.pdpu.ac.in; satishaprabhu679@gmail.com; satisha.prabhu@manipal.edu

V. J. Badheka
e-mail: vishvesh.badheka@spt.pdpu.ac.in

S. Prabhu
Department of Mechanical Engineering, Manipal Institute of Technology, Manipal 576102, India

A. Kumar
Department of Industrial Engineering, Pandit Deendayal Petroleum University, Gandhinagar 382007, India
e-mail: abhishek.k@sot.pdpu.ac.in

D. Deb et al. (eds.), *Innovations in Infrastructure*, Advances in Intelligent Systems and Computing 757, https://doi.org/10.1007/978-981-13-1966-2_40

Keywords Burrs · Recast metal · Deburring · Tooling · Fixtures · Cross holes
Pseudo alloy

1 Introduction

Precision components are often produced by grinding, EDMing (electro discharge machining), laser machining, conventional machining, polishing, etc., may have unwanted effects like rough surfaces, microcracks, recast metal, burrs, metallurgical phase changes, visible scratches or damaging residual stresses, and EDMed surfaces may have recast metal subsurface damage. These can cause rejection of the parts and service damage if used.

Ordinance factory needs stringent quality requirements since the components have to work only once and there are no second chances [1]. They must perform under some of the highest forces that mechanical systems can withstand. The forces from the incredible velocities, accelerations, and RPMs that cause many of the components to function, may prove to be disastrous if sliding, indexing, or rotating surfaces and edges are not properly deburred.

Fuel injection nozzles must have well radiused internal and external edges. The ports edge condition depend upon nozzle configuration as with other orifices. The orifices affect the pressures and velocities downstream and spray patterns. The orifices must have well-defined edge geometries. Edge radii, shape, hardness, resistance to erosion, and roughness are all important aspects for heavily pressurized fuel injection systems. Diesel injection nozzles use holes of diameter ranging from 100 to 200 μ having 0.1–0.4 'Ra' (CLA) surface roughness. Entry and exit holes require fine finishes without edge breaks or burrs. Any loose burr condition can influence fuel efficiency and emissions into the atmosphere. Pressure drops through the hole and reduced size of the fluid stream may adversely affect the engine performance. It has been a challenge for diesel engine manufacturers and high-pressure fuel systems to reduce the stress inside the components, where the pressure can be of the order of 2000 bar [2]. Repetitive pressure changes and pulsating fuel spray can cause fatigue failures of the components. To be free from these drawbacks, all surface cracks need to be removed and uniform radius and smooth edges must be provided at intersecting cross holes.

Complex geometry and limited accessibility of the burrs at intersections may become a bottleneck in the manufacturing systems. In case parts having several intersecting boreholes, cross holes at the intersection areas, the presence of burr can adversely influence the correct operation of the hydraulic systems. The blocking and detachment of burr can cause catastrophic failures. The burrs are inaccessible and removing them by the unsuitable method can damage the functionality of the components. Removing these burrs are more challenging when three or more holes intersect and their diameter ratios are close to unity. Cross holes burrs are termed as potato chip burrs and are considered as nagging problems in the manufacturing industry. These burrs are located inside the components with blind holes and intersections. Adding to the complexity burrs at angles and off-centers can create partial breakout and irregular edges. It is always been a challenge for manufacturing engineers to device a suitable method to deburr these areas. Typical examples are hydraulic manifolds, fittings, elbows, camshafts, crankshafts, connecting rods, large oilfield parts and aerospace, medical, semiconductor components for industrial control applications.

2 Process Description

Electrochemical deburring process (ECD) is a localized material removal process uses electrolytic dissolution mechanism to remove burrs at the specific areas like intersecting and cross holes at inaccessible areas [1]. The components having very fine burrs are placed on locator at close proximity of cathode and workpiece (anode) maintaining a small gap (IEG) of the order of few microns. These locators or fixtures are the nonmetallic type (nylon, polypropylene) designed and tested to withstand forces during electrolyte circulation, corrosion, and possible ill-usage. The workpiece serves as an anode, the tool is cathode and electrolyte is continuously pumped during machining at predetermined pressure and velocity. The flow of electrolyte establishes the electrical contact and allows the loose deburred metal pieces to move away from the machining zone. These loose metal pieces otherwise cause electrical short circuit and damage to the system. The burrs are detached from the workpiece due to electrochemical action and a smooth radius is generated at the intersecting areas which are often desirable.

Fixture design and testing is an important aspect of ECD [3]. A fixture may contain locator to hold the part, insulation for the non-machining area and electric contacts for anode and cathode. The electrode (cathode tool) confirms to the part configuration, deburring zone, and remaining area masked with polypropylene coatings. The flow of electrolyte is through the passage created by the fixture and the gap between the tool and the workpiece (IEG). Two types of flow are often used in practice among which reverse flow of electrolyte is preferred due to metal piece flushing away from machining zone preventing short circuits and damage. The manufacturing engineer controls the following variables to obtain a desirable radius at the workpiece intersecting area. The variables include interelectrode gap, current density, electrolyte concentration, voltage, and time of deburring. ECD will change the dimension of the part only to the extent of burr removal and controlled radius generation. This dimensional change is desirable and generally required to make a part "to print" [1]. Electrolytic action will take place in the vicinity of exposed areas of cathode if designed and insulated properly. The locator or fixture is properly designed to protect critical areas and surfaces including threads although ECD is effectively used in removing burrs from thread form. The burrtech deburring machine used for current research as shown in Fig. 1.

ECD is exclusively used in inaccessible areas and regarded as high technology process and provides very high-quality deburring. Several benefits of ECD are men-

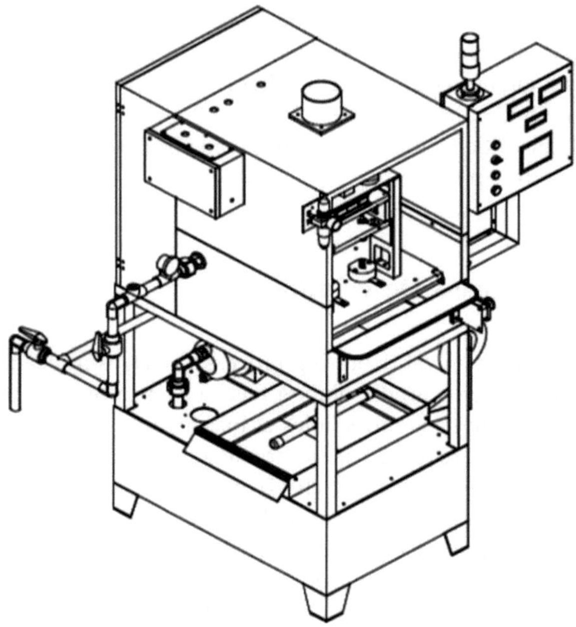

Fig. 1 ECD machine configuration (*Source* Mystec Pvt. Ltd)

tioned, which includes fixed manufacturing costs without variances, elimination of time-consuming and labor-intensive process, consistent burr removal, reduction of burr removal costs, increase of quality levels, and reliability and controlled radius generation.

3 Samples of Al6082 and Measurement of 'g' Value

Many nozzle edge configurations have fuel injection ports and orifices of very fine diameter. Burrs on the edges of these orifices affect the pressures and velocities down streams as well as spray patterns. Edge radii, shape, micro-indentation hardness, resistance to erosion, corrosion and roughness are all issues for heavily used injection systems. This research uses the specimen made of Aluminum 6082, known as a structural alloy. As a relatively new alloy, Al6082 replaces Al6010 for many applications. The addition of a large amount of Manganese controls the grain structure which in turn results in a stronger alloy. Referring to the Fig. 2, the specimen has a large central hole with two diametrically opposite (180°) intersecting holes and a groove on the upper region. The ECD is performed with tooling (−ve), which can produce a radius at the corners of the deburred area [4]. Experimental investigations are carried out with Tungsten Copper tool.

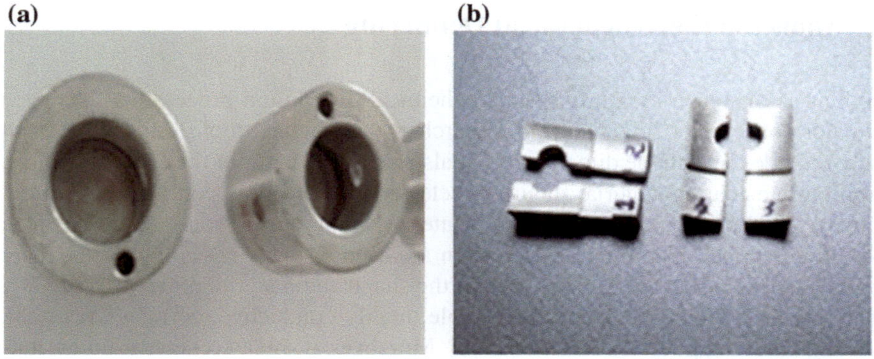

Fig. 2 **a** Aluminum 6082 specimen before sample preparation. **b** Specimen for investigation of burr value (WEDM)

3.1 Burr Value 'g' and Its Relation with Input Variables

Prior to deburring the burr height, thickness, root radius, and geometry have to be measured and quantified. If long burr exists between the electrodes, then the frequent short circuit will arise resulting in machine stoppages. Hence, the burr size must be reduced by using hand files so that it should be below 50 μ (Inter electrode gap).

This will help in the smooth deburring operation. On the other hand, gap should not be more than required IEG. More gap may cause no machining in that region. Samples prepared from destructive method, i.e., by WEDM (Wire cut EDM) is shown in Fig. 2. The burr geometry is inspected by RAPID-I (3D) vision measuring system and tabulated as below. From the empirical relation by Schafer [5] for calculating 'g' value, it is clear that the burr root dimension has more impact. Higher the 'g' value, greater will be the time required for deburring to remove the burr completely (Table 1).

Table 1 Geometrical values obtained from burr measurement and calculated 'g' value [5]

Specimen Id	Average Burr width (Br)	Burr width (Bg) mm		Burr height Max (ho) mm		Burr root radius (rf) mm		'g' avg (mm)
	Max (mm)			1	2	r_1	r_2	
ECD/SP/29	0.170325	0.1646	0.1604	0.1226	0.1936	0.0992	0.3239	0.1792969
ECD/SP/58	0.364175	0.1691	0.4811	0.2317	0.2877	0.2064	0.3816	0.3238719
ECD/SP/27	0.213825	0.1500	0.1813	0.1946	0.1811	0.1151	0.3697	0.2090781

4 Tools for Electrochemical Deburring

Tooling is one such area in the electrochemical dissolution process that has been a major concern for scientists and researchers. A practical tool design should not only provide the cathode dimensions but also a suitable path for the electrolyte flow. Appropriate coating material should be selected to avoid stray cutting, over current damage, and dimensional instability. Materials like copper, brass, stainless steel, beryllium copper, platinum, and tungsten are used in many cases. For tools, there are no such standard available to predict the shape and material removal. The alloy material used should be chemically stable, highly conductive, corrosive resistant, and high melting point characteristics [6]. Moreover, it should retain its dimensions and have prolonged usage life without much wear.

The present research focuses on the use of a pseudo-alloy of W: Cu composition having good electrical characteristics as the tool material. The alloy must withstand machining forces, i.e., electrolyte pressure within the small working gap of about 50 μ (Approx.) and should be chemically stable against corrosive electrolyte flow. In addition, it should withstand high localized current density and provide the same degree of accurate machining throughout its useful life.

5 Results and Discussions

Electrochemical deburring uses DC current for a short period of time to dissolve the burr electrolytically [7]. Other factors such as electrolyte concentration, current density which greatly influence the accuracy of the machining. The current research aims for the investigation of performance using uncoated W: Cu electrodes. The metal removal rates are analyzed. The tool behavior is investigated for a given set of conditions. The said tool is consistent and indicated proportional output variations as shown in the graphs. Time of deburring greatly influence the radius generated and the linear region can be used for practical calculations. Smut and pittings [4] cannot be ruled out under electrolytic dissolution conditions. Since sodium chloride is an aggressive electrolyte the stray machining and striations could not be readily controlled. Smut and over-etching is readily observed in the micrographs. The higher radius of deburring is also achieved during machining over a longer period of time (Table 2).

5.1 Machining Characteristics of the Electrolytic Composition

Electrochemical dissolution starts at the certain composition of the electrolyte [7]. Electric current flow is established between the electrodes resulted in metal erosion

Table 2 Measurement details of Al6082 specimen before and after electrochemical deburring

Predeburring			Postdeburring		
Sample Id	**Micrographs before deburring**	**Measured details**	**Sample Id**	**Micrographs after deburring**	**Variables measured**
ECD058	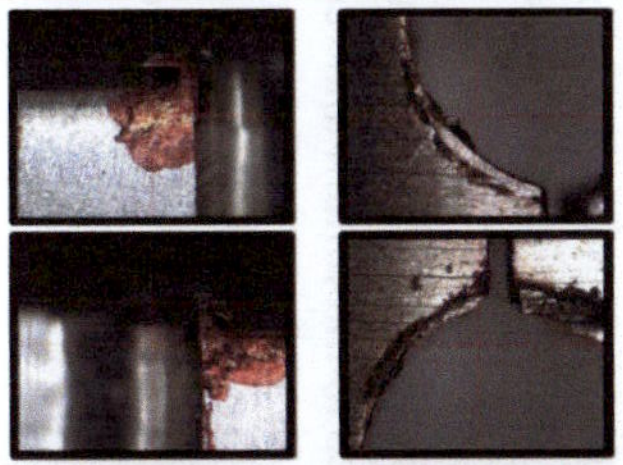	Burr width max 0.36175mm. Burr width 0.1691. Burr height 0.2317. Root radius 0.2064. Micro hardness at burr interface 180Hv.	**ECD073**	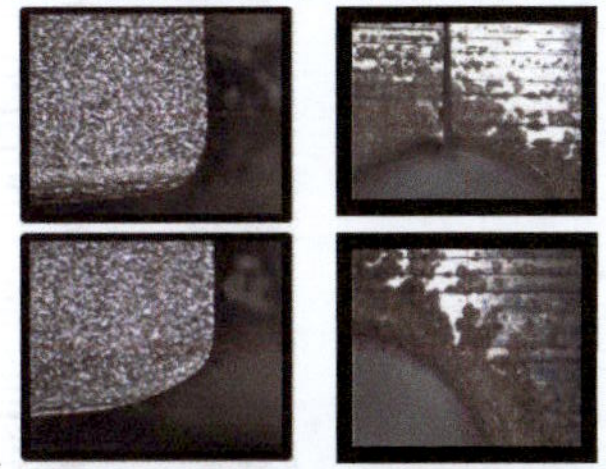	Electrolyte -NaCl pH-6.66 Concentration-15% Time debur-30Sec Voltage-12.45V Current-103.05A Tool- Uncoated Material removal rate-0.2354gm/min Current Density-126.23A/cm^2 Current Efficiency-90.54% Microhardness-128.3Hv Radius generated after deburring-0.8660mm
ECD027	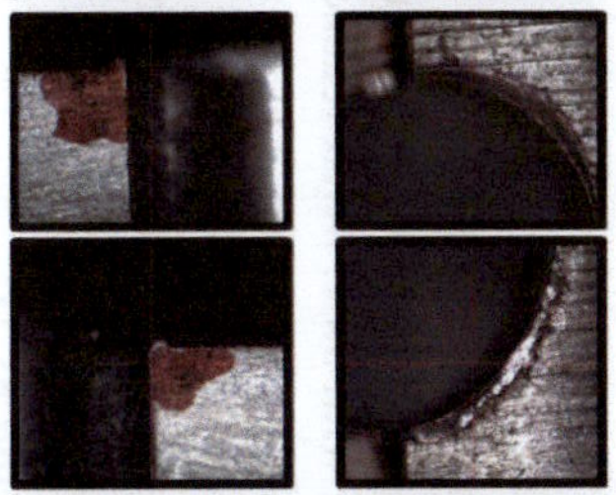	Burr width max 0.213825mm. Burr width 0.15. Burr height 0.1946. Root radius 0.1151. Micro hardness at burr interface 277.2Hv.	**ECD052**	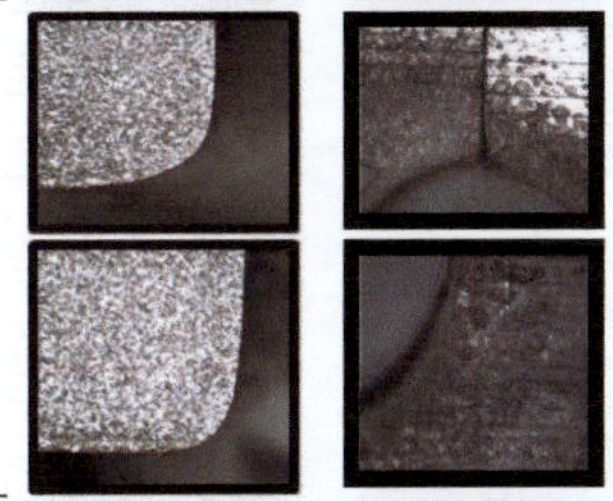	Electrolyte used-NaCl PH-6.53 Concentration-20% Time of deburring-30Sec Voltage-12.45V Current-104.8A Tool- Uncoated Material removal rate-0.232gm/min Current Density-128.38A/cm^2 Current Efficiency-85.93 Microhardness-116.6Hv Radius generated after deburring-0.4367mm
ECD029		Burr width max 0.170325mm. Burr width 0.1646. Burr height 0.1226. Root radius 0.3249. Micro hardness at burr interface 93.9Hv.	**ECD017**	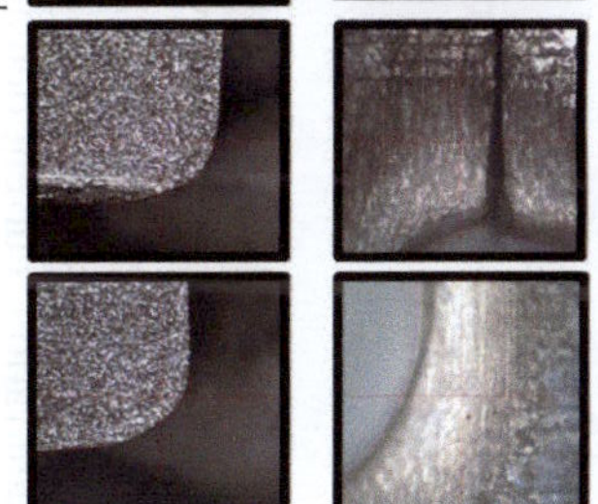	Electrolyte used-NaCl PH-6.43 Concentration-25% Time of deburring-30Sec Voltage-12.45V Current-117.35A Tool- Uncoated Material removal rate-0.2402gm/min Current Density-143.75A/cm^2 Current Efficiency-80.07% Microhardness-119.6Hv Radius generated after deburring-0.4384mm

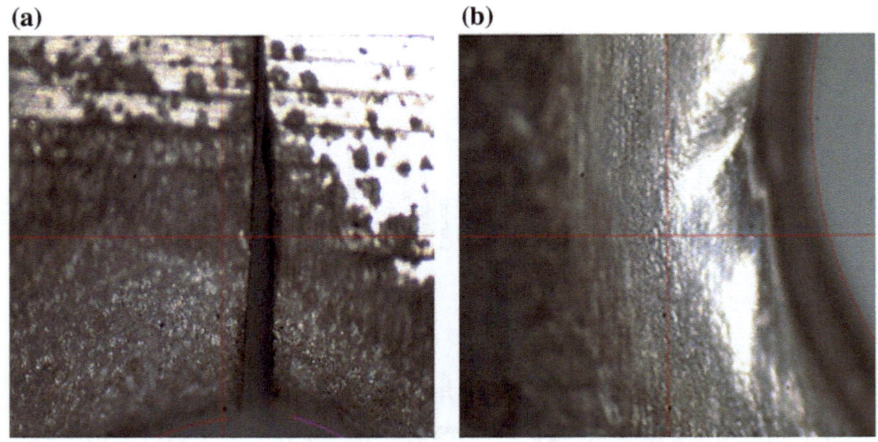

Fig. 3 Micrographs of the deburred areas with NaCl electrolytes. **a** Smuts and pittings **b** Striated edge condition

and removal. The present research uses Sodium chloride (NaCl) as the neutral electrolyte. The composition is varied from 15, 20, and 25% and trials are carried out at 10, 20, 30, and 40 s intervals.

The performance is evaluated by varying input parameters such as current applied, concentration, current density, time, etc., and the results are tabulated. Machining characteristics are analyzed and are as below. Also, micrographs of the deburred area show striations with the use of NaCl electrolytes (Fig. 3).

Sodium chloride uses higher deburring current producing striations, more smut, and pitting.

(i) **Material removal rate (MRR)**

Graphs are plotted by varying the deburring time and concentration of electrolyte. For a given concentration, MRR increases up to a certain value and become proportionately equal. Higher MRR is observed at higher deburring time and concentrations. But for higher concentrations electrolytic material removal rate becomes constant after reaching the maximum value (Fig. 4).

Subsequent graphs show the variations of current density and material removal rate. Variations are identical with time. MRR reaches to maximum and reduces due to variations in experimental conditions such as inter-electrode gap, precipitation of salts, increase in temperature, etc. (Fig 5).

Similarly, variations of metal removal with reference to applied current are as shown in Fig. 6.

(ii) **Radius generated during ECD**

Since the radius is generated and controlled during the electrochemical deburring process the variations need to be analyzed graphically. From the graphs, it is clear

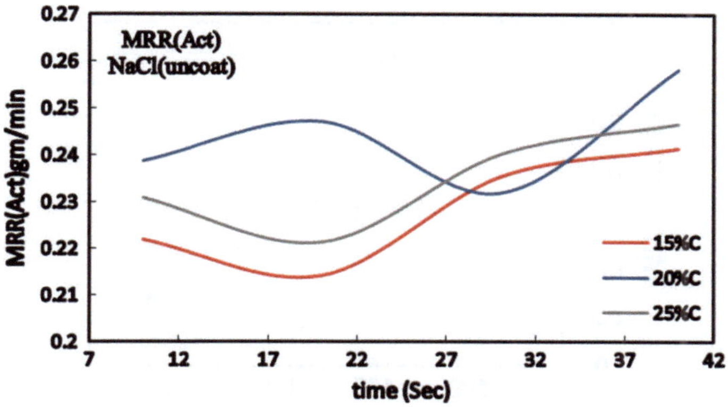

Fig. 4 Variation of MRR with concentrations (machining with NaCl electrolyte)

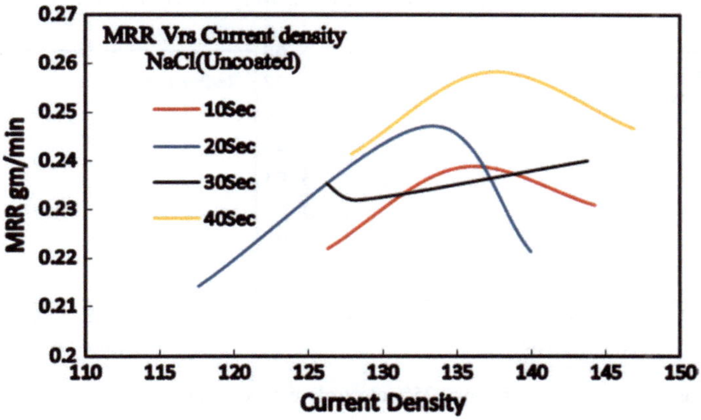

Fig. 5 Variation of MRR with current densities while using NaCl electrolytes

that the edge condition is varied with concentration and time. 25% of concentration has generated higher radius compared to other levels. The control over the radius can be achieved by adjusting the variables stated above (Fig. 7).

(iii) **Current density variations**

Current density can be directly related to applied current and variation in the concentrations. As the concentration increases current density increases proportionately. The figure below shows the variations discussed. Since current density plays an important role in metal removal and performance of ECD its adjustment with other parameters have to be evaluated graphically (Fig. 8).

(iv) **Current efficiency effects**

All trials are conducted and results are reported assuming the fact that the current efficiency is 100%. But practically this is not true and its variations are reported

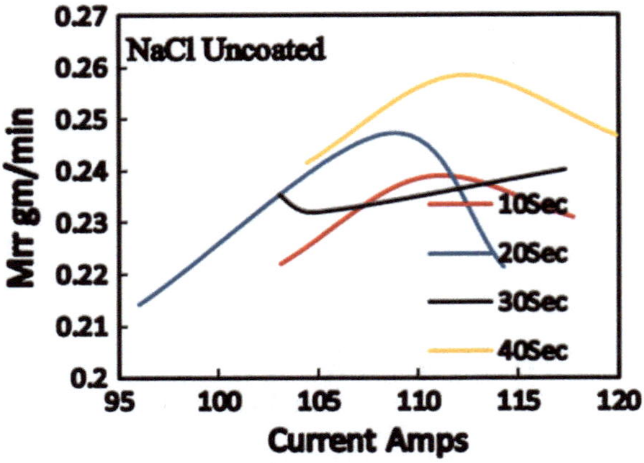

Fig. 6 Variation of MRR with the applied current while using $NaNO_3$ and NaCl electrolytes

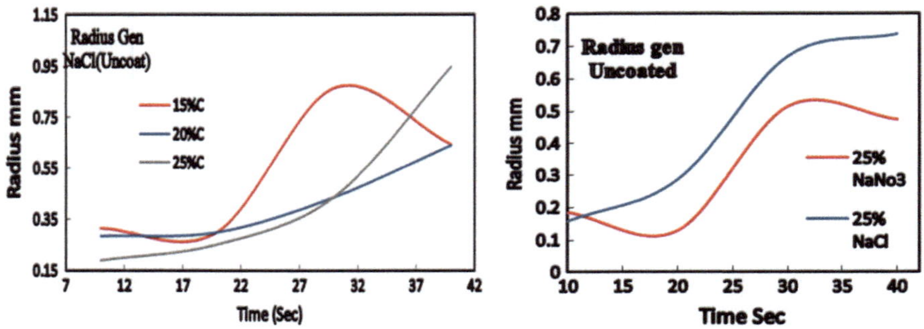

Fig. 7 Radius-generated comparison at 25% electrolyte concentration

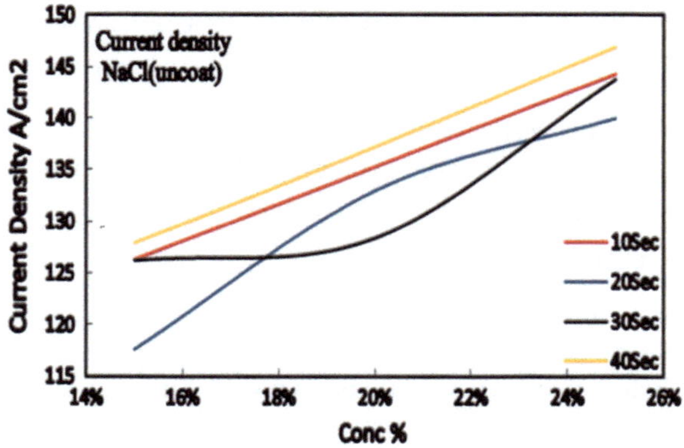

Fig. 8 Variations of current density with time and concentrations

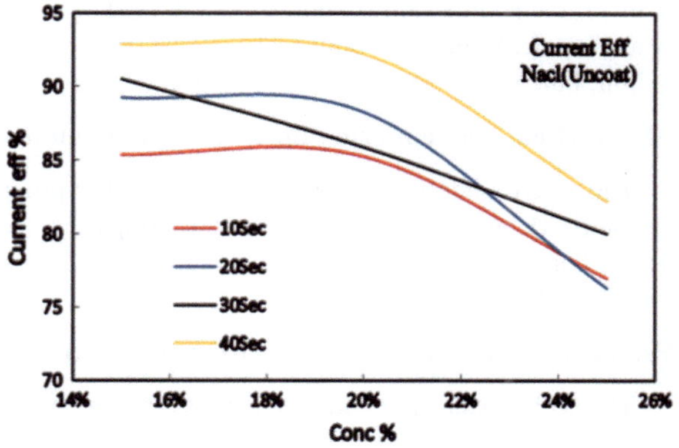

Fig. 9 Variations of current efficiency with time and concentrations

graphically. But in practice other side reaction occurrence, precipitation, high valence dissolution decreases the efficiency to 80–90%. Thus, the amount of metal dissolved is smaller than the theoretical value due to these reasons. Electrical conductivity plays an important role in current efficiency estimation. Heat transfer, electrolyte flow, concentration, electrode structural geometry, inter-electrode gap (IEG), affect the electrical conductivity. Heat generation, gas bubble generation, sludge precipitation all have contributed to variations of current efficiency.

(v) **Micro-indentation Hardness**

Electrolytic action at the burr area has reduced the hardness by few Vickers numbers (Hv). Its relationship with concentration, time, and current density is insignificant. This variation has not given any contributing output on experimentation (Fig. 9).

6 Conclusions

The research investigation described above deals with electrochemical deburring of cross holes. The experimental verification carried out with Al6082 alloy by varying the process parameters. The conclusions drawn from the investigation are as follows.

Intersecting and cross-hole deburring

In case parts having several intersecting boreholes, cross holes at the intersection areas, the presence of burr can adversely influence the correct operation of the systems. The electrochemical method of removing these burrs can be the solution to overcome the problems.

Measurement of burr geometry and 'g' value

Since burrs are inconsistent and it is not easily controlled for the same machining conditions. Hence, it is important to measure the burr accurately and predict the results. The assessment of burr based on one feature in accordance with the standard is insufficient. The same can be characterized by the burr value 'g'. This burr value is a numerical number obtained by measuring many geometrical details, viz., burr height, burr width, root width, and root radius.

Electrochemical deburring for intersecting and inaccessible burrs

The electrochemical deburring method is the process that can be is successfully applied to deburr inaccessible and intersecting areas. A controlled radius is generated as the burr is electrolytically dissolved. This ECD process is extremely consistent and cost-effective.

Performance evaluation by variations of electrolyte and composition

The performance is evaluated by varying input parameters such as current applied, concentration, current density, time, etc., and results are tabulated. Machining characteristics are analyzed and reported. Also, micrographs of the deburred area show striations with the use of NaCl electrolytes.

Controlled radius generation

Since radius is generated and controlled during the electrochemical deburring process the variations are analyzed graphically. And it is clear that the edge condition is varied with concentration and time. Higher concentration has generated larger radius compared to other compositions. The control over the radius can be achieved by adjusting the variables.

References

1. Halladay, J.: Practical applications of thermal deburring and electrochemical deburring. In: National Technical conference (2011)
2. Solutions for Deburring Applications—Electrochemical Deburring (1999). Retrieved from http://www.sermatec.com
3. Hakim, M.A.E., Abdel Mohsen Mahdy, M., Abdulla Sayed, M.: The effect of some electrode design factors on electrochemical deburring (ECD). Int. J. Mater. Prod. Technol. **4**(3) (1989)
4. Laroux, K.G.: Hand Book of Deburring and Edge Finishing. Copyright © 1999. Society of Manufacturing Engineers, ISBN 0-87263-501-5 (1999)
5. Pres, P., Skoczynki, W., Jaskiewicz, K.: Research Modeling work piece edge formation process during orthogonal cutting. Archieves Civil Mech. Eng. **14**, 622–635 (2014)
6. Swain, A.K., Sundaram, M.M., Rajurkar, K.P.: Use of coated micro tools in advanced manufacturing: an exploratory study in electrochemical machining (ECM) context. J. Manuf. Processes (2011). https://doi.org/10.1016/j.jmapro.2011.11.005
7. Lee, E.-S., Won, J.-K., Shin, T.-H., Kim, S.-H.: Investigation of machining characteristics for electrochemical micro-deburring of the AZ31 lightweight magnesium alloy. Int. J. Precis. Eng. Manuf. **13**(3), 339–345 (2012)
8. Aurich, J.C., Dornfeld, D., Arrazola, P.J., Franke, V., Leitz, L., Min, S.: Burrs-analysis, control and removal. CIRP Ann. Manuf. Technol. **58**, 519–542 (2009)
9. Hakim, M.A.E., Abdel Mohsen Mahdy M., Abdulla Sayed, A.: ECD tooling design. Int. J. Mater. Prod. Technol. **139**, 3–6 (1998)
10. Bolunduţ, I.-L.: Electrochemical deburring of mining hydraulic elements. Ann. Univ. Petroşani, Mech. Eng. **12**, 23–32 (2010)

11. LaRoux, K.G.: Process control for burrs and deburring. Allied Sign. Aerosp. KCP-613-5526 (1994)
12. Tang, L., Zhao, G.: Technique of ECM deburring for intersecting holes of pump. Adv. Mater. Res. **411**, 323–326 (2012)
13. Mitra, S., Sarkar, S.: Experimental Investigation on electrochemical deburring (ECD) process. J. Mater. Process. Technol. **75**, 198–203 (1998)
14. Rajurkar, K.P., Sundaram, M.M., Malshe, A.P.: Review of electrochemical and electro discharge machining. Procedia CIRP **6**, 13–26 (2013)
15. Niknam, S.A., Zedan, Y., Songmene, V.: Machining Burrs Formation & Deburring of Aluminium Alloys, Chap. 5, pp. 99–122 (2014). https://doi.org/10.5772/58361
16. Zheng, X., Wei, Z., Yu, Z.: Electrochemical deburring of miniature holes. Adv. Mater. Res. **690–695**, 3226–3230 (2013)
17. Jain, V.K.: Electrochemical Allied Processes. ASM Handbook, vol. 13A, Corrosion Fundamentals, Testing and Protectio, pp. 153–156 (2003). https://doi.org/10.1361/asmhba0003597
18. Rajurkar, K.P., Jerzy, K., Chatterjee, A.: Nonabrasive Finishing Methods. ASM Handbook, vol. 05, Surface Engineering, pp. 110–117 (1994). https://doi.org/10.1361/asmhba0001233

Harmonic XFEM Simulation of 3-D Cracks

Saurabh Kumar Yadav, Nathi Ram and I. V. Singh

Abstract Fracture mechanics is widely used nowadays to study and analyze the failure of components/structures. Several failures in structures are related to fracture mechanics. When failure due to fracture occurs, they are mostly unpredicted, quick, and disastrous in nature. Therefore, it becomes necessary for us to pay attention to minimize the failure when designing and analyzing modern-day structures. The stress intensity factors are extracted from the HXFEM solution by volume-based interaction integral approach using the curvilinear coordinate system and are compared against the standard XFEM and the analytical result available in the literature. In this volume-based interaction integral approach, the volume is virtually extended and integrated by applying Gauss divergence theorem. The numerical results are obtained for edge and penny crack problems.

Keywords FEM · XFEM · Enrichment functions · Fracture mechanics

1 Introduction

The conventional solid mechanics approach assumes a component to be flawless. But it is impossible that crack is not present in a real structure/component. The cracks are found during maintenance, and inspection in aircraft fuselages, nuclear

S. K. Yadav (✉)
Department of Mechanical Engineering, Institute of Infrastructure Technology Research and Management, Ahmedabad, India
e-mail: saurabhme.iitr@gmail.com

N. Ram
Department of Mechanical & Automation Engineering, Indira Gandhi Delhi Technical University for Women, New Delhi, India
e-mail: n.r.jaan10@gmail.com

I. V. Singh
Department of Mechanical Engineering, Indian Institute of Technology Roorkee, Roorkee, Uttarakhand, India

D. Deb et al. (eds.), *Innovations in Infrastructure*, Advances in Intelligent Systems and Computing 757, https://doi.org/10.1007/978-981-13-1966-2_41

component, pressure vessels, automobile components, nautical, and castings industry. These cracks must be shown to be inoffensive from the safety point of view; if not then the component is replaced, repaired, or discarded. The failure in these components appears due to the growth of initial small crack on the surface of the component, which becomes elliptical or semi-elliptical shape after some time due to the application of the repeated load.

By using the theory of elasticity, the failure due to crack cannot be predicted because it does not have any consideration of asymptotic crack field near the crack front in three dimensional and crack tip in two dimensional. Theory of elasticity is only able to predict failure due yield criterion [1–3]. To predict the stress field near the crack front the use of principles of fracture mechanics is necessary. Nowadays, fracture mechanics is widely used to predict the failure in structure. The SIF (Stress Intensity Factor) is an important parameter to characterize the cracks. Analytical methods in the literature are only able to calculate SIF of some regular and analytical crack geometry. Therefore, XFEM is a suitable choice to calculate the behavior of complex three-dimensional cracks.

In order to solve the problem of fracture mechanics, various methods have been developed. People used the finite difference method to solve differential equations. It works well with regular shapes of crack but unable to provide the good result for complex crack geometries. Finite Element widely used to perform a simulation of complex geometry and crack growth. It has flexibility, effectiveness, and accuracy to predict the failure of geometry. But despite these advantages, it has some disadvantage like that it is very difficult to treat crack in FEM simulation. We require singular and conformal mesh near the crack front which is an expensive and time-consuming task. Therefore, to alleviate these burdens in the last few decades XFEM emerged as a viable option to simulate the fracture mechanics problem without conformal mesh at the crack tip. It uses the enrichment function near the crack front instead of a conformal mesh.

In 2000, Sukumar et al. [4] presented three-dimensional XFEM formulations for the modeling of the static crack. They also discussed the issues in computational geometry, associated with the representation of crack, and enrichment of the finite element approximations. But, it was not a general formulation of the XFEM.

Between 2000 and 2002, various types of the enrichment function for holes inclusion, bimetallic joint and cracks have been purposed by many researchers in two-dimensional XFEM [4]. Recently, Sukumar et al. [5] developed a Harmonic XFEM which is a powerful technique to simulate multiple cracks in XFEM. Sukumar et al. [5, 6] proposed that this enrichment scheme is effective and is much better than the traditional enrichment scheme. In this method, they used numerical enrichment function instead of analytical enrichment function.

3-D model crack problems such as edge and penny cracks in the finite domain are solved by harmonic XFEM. Shifted enrichment is used in Harmonic XFEM to capture crack tip singularity [4]. In this enrichment approach, only one enrichment function is used instead of using two types of enrichment functions. Numerical harmonic enrichment function is derived from the solution of the Laplace domain.

Few model problems are solved by a developed algorithm, and the results obtained from this algorithm are compared with those available in the literature.

2 Harmonic Enrichment Functions

Crack field enrichment function for crack is obtained by using the solution of the Laplace equation [5]:

$$\nabla^2 F = 0 \text{ in } \Omega^{l} \tag{1}$$

F is the numerical **Harmonic** enrichment function of the crack front and crack tip. Ω^{l} is the Laplace domain in which a subgrid mesh is generated as shown in Fig. 1. The Laplace domain Ω^{l} contains all enriched elements for crack c. Solving a Laplace equation separates the problem for enrichment as it is done by discretizing the Laplace domain and a different type of meshing is used for the solving elastostatics problems. Laplace equation and domain is solved by FEMs.

The harmonic-extended finite element approximation for crack modeling with harmonic enrichment functions is [5]:

$$u^h(X) = \sum_{j=1}^{n} N_j(x)\left[u_j + \left(F(X) - F\left(X_j\right)\right)a_j\right] \tag{2}$$

Here, $F(X)$ is the numerical harmonic enrichment function of crack front and crack tip.

3 Solution Procedure

The enriched element is found out in which crack is passing. These elements make Laplace domain Ω^{l} of the problem as shown in Fig. 2. Ω^{l} is discretized using quadrilateral finite element as shown in Fig. 2. A background mesh is generated in each element. The element that shares the enriched node is also counted in the Laplace domain.

Numerical harmonic enrichment function F which is numerically computed on the subgrid mesh.

The complete solution procedure as shown in Fig. 2 has been implemented in the following steps.

1. By combining enriched elements and blending elements Laplace domain Ω^{l} is generated. This domain has cracked.
2. Discretize Ω^{l} in each element which is divided into $\mathrm{m} \times \mathrm{m} \times \mathrm{m}$ subelement mesh elements near crack front domain.

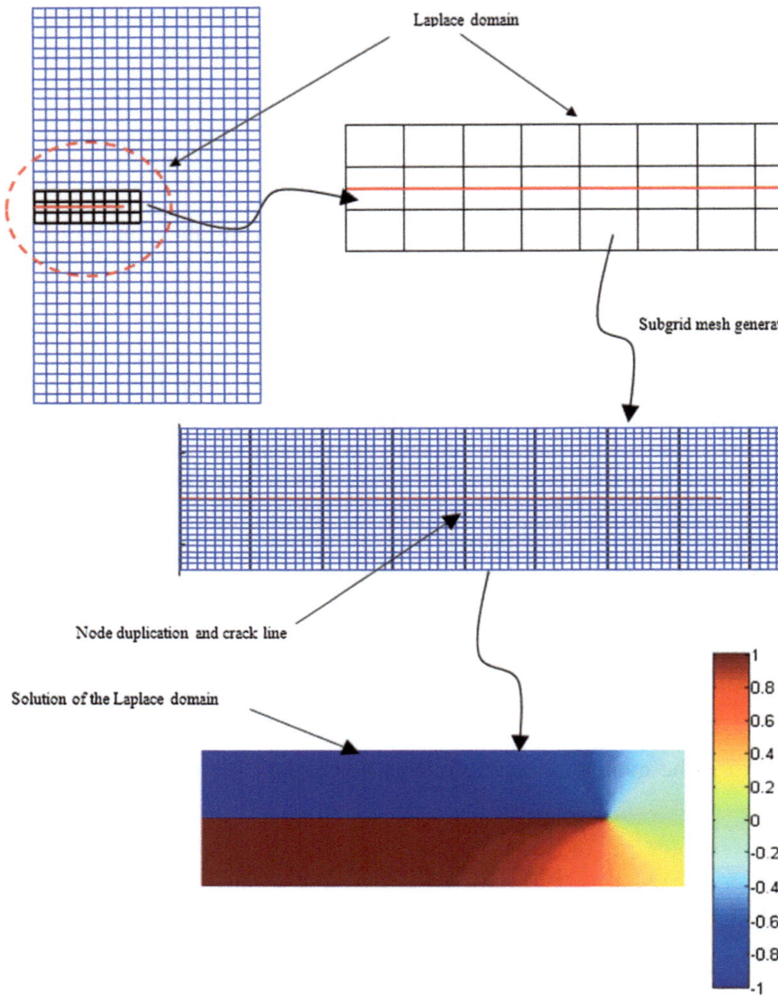

Fig. 1 Harmonic enrichment procedure

3. Find out the rasterized approximated path Γ_l^r of crack in the discretized domain.
4. Regenerate all the nodes that are coincident on the rasterized path Γ_l^r and change the connectivity of the sub-element so that the sub-elements above and below Γ_l^r are connected to different copies of nodes.
5. Now apply the boundary condition to generate the discontinuity on crack $F_c = 1$ for the nodes connected to the subelement above Γ_l^r and below $F_c = -1$ for the below Γ_l^r. Assign $F_c = 0$. If the crack c has a tip.
6. Repeat all the steps from 1 to 5 for all other cracks.

Fig. 2 Solution procedure of harmonic XFEM

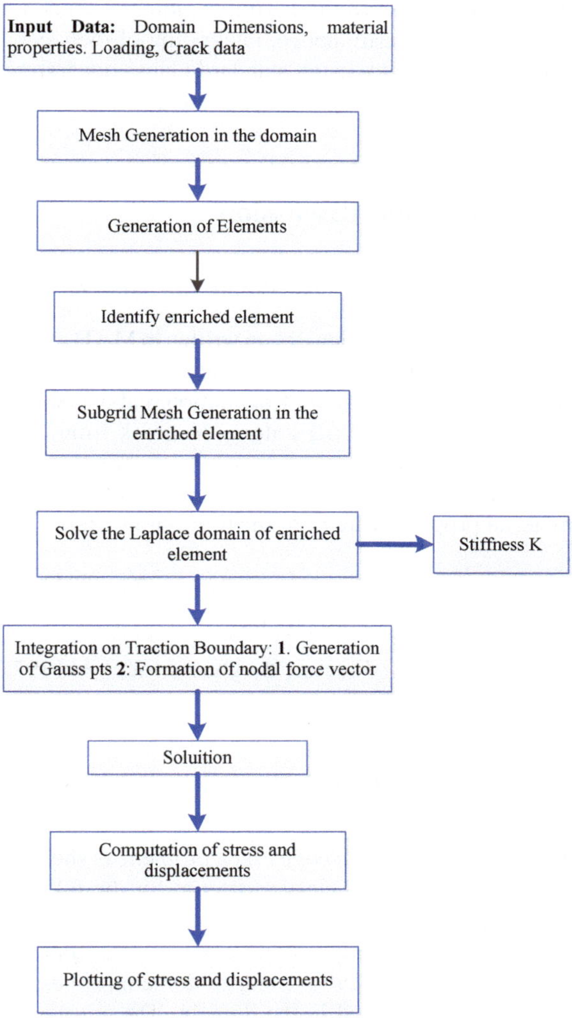

4 Numerical Integration

Standard Gauss quadrature requires smoothness in the integrand polynomial. In XFEM standard displacement function has discontinuity/singularity in derivative above and below the crack surface; therefore, the computation of stiffness matrix entries in enriched elements requires additional care. So integration in standard XFEM is done by the partition of finite element. However, numerical harmonic enrichment function is piecewise linear in the subgrid mesh. Therefore, Harmonic XFEM does not require partition of the finite element for integration. Instead of dividing into conforming partitions, integration is done on subgrid of the elements to resolve singularity hence, integration of the weak form in the Harmonic XFEM can be done exactly using standard Gauss quadrature over the subgrid and there is no need to introduce special quadrature rules over the enriched elements.

5 Results and Discussion

Static Crack by using Harmonic Enrichment

The simulation codes have been written in MATLAB. Numerical results are presented to demonstrate the accuracy of the algorithm. Average time taken by simulation single step crack growth with SIF Calculation is about 10 min. A significant time is invested on the calculation of SIFs along the crack front. Two well-recognized benchmark problems such as Edge crack and Penny crack are simulated by Harmonic XFEM. A subgrid of $6 \times 6 \times 6$ is used for the simulation of both the cracks. The following material properties are used in all the simulations:

$$\begin{aligned} &\text{Elastic modulus}: && 74,000\,\text{N/mm}^2 \\ &\text{Poisson ratio}: && \mu = 0.3 \\ &\text{Fracture toughness}: && K_{IC} = 1897.36\,\text{N/mm}^{\frac{3}{2}} \end{aligned}$$

5.1 3D Edge Crack Growth

The simulations for an edge crack problem as shown in Fig. 3, have been performed using isotropic linear elastic material for the following data: No. of nodes $= 16 \times 16 \times 32$, size of domain, $W = 100$ mm, $L = 100$ mm, $H = 200$ mm, Far-field stress, $\sigma_0 = 100\,\text{N/mm}^2$, Edge crack length, $a = 20$ mm, virtual extension, $c = 10$ mm. In this study, a hollow cuboid shaped domain [7] has been used to calculate SIF. 2000 Gauss points are used in the domain. The exact value of Mode-I SIF for an edge crack for plane strain condition with above dimension is $K_{\text{exact}} = 34.08$ MPa m$^{1/2}$. Figure 3 presents the effect of hollow shaped domain size on Mode-I SIF along crack front. Crack front and crack surface's numerical harmonic enrichment functions are shown in Fig. 3. From the results presented in Fig. 3, it has been noticed that the value of mode-I SIF is increased toward center due to plane strain condition in the middle of the cuboid.

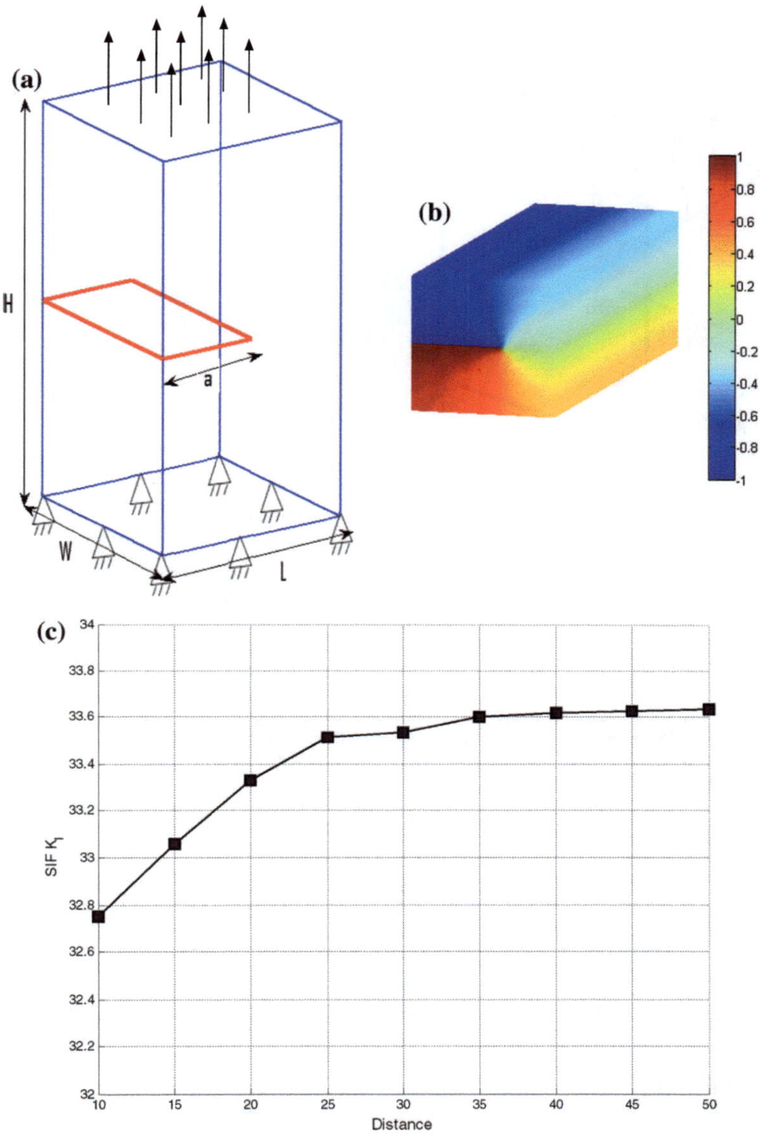

Fig. 3 **a** Edge crack under tensile loading **b** Enrichment function for edge crack **c** Mode-I SIF variation along the crack front

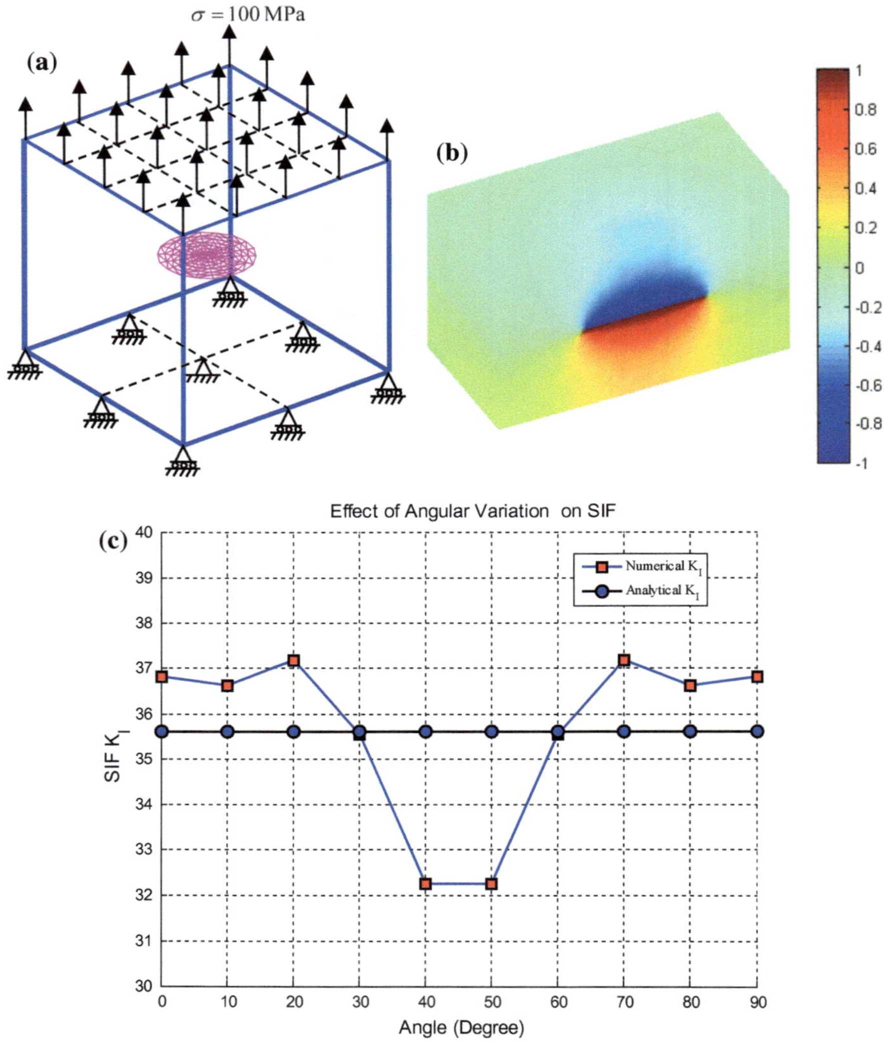

Fig. 4 **a** Penny shape crack in infinite domain **b** Sectional view of enrichment function for the penny shape crack **c** Variation of mode-I SIF along the crack front

5.2 Penny Shape Crack in Infinite Domain

Let r be the radius of the penny in a 2 m × 2 m × 2 m cube as shown in Fig. 4. The analytical solution of a penny in the infinite domain under axial tensile loading is given by [7].

$$K_I = 2\sigma\sqrt{\frac{r}{\pi}},\ K_{II} = 0,\ K_{III} = 0 \tag{4}$$

The specimen is subjected to a uniform axial load in the Z-direction. Uniform hexahedral mesh refinement has been done near penny crack. Far-field stress is applied in the Z-directions. The radius of penny is taken as $r = 0.1$ m so that the effect of finite size can be minimized. The results are presented here for a mesh size of $20 \times 20 \times 20$.

This problem is solved for hollow cuboid that is shown in Fig. 4. The dimensions of virtually extended hollow cuboid contour are taken $V_e = 0.03$ m and $b = 0.09$ m.

A sectional plot along X–X plane of numerical harmonic enrichment function is shown in Fig. 4. The values of mode-*II* and mode-*III* stress intensity factors are almost negligible as compared to mode-*I* stress intensity factor, therefore, the results are presented only for mode-I in Fig. 4. Figure 4 shows a good comparison of Mode-I SIF with analytical results for a hollow cuboid domain.

6 Conclusion

On the basis of the simulations, it was found that the harmonic enriched XFEM is a viable option to analyze 3D cracks. Due to its ease in simulating multiple three-dimensional cracks, it should be used further for the simulation of multiple cracks in infrastructure structures. The efficiency of harmonic XFEM is better than the conventional XFEM. Because conventional XFEM uses multiple enrichments while conventional harmonic XFEM uses simple enrichment function.

References

1. Henshell, R.D., Shaw, K.G.: Crack tip finite elements are unnecessary. Int. J. Numer. Meth. Eng. **9**, 495–507 (1975)
2. Barsoum, R.S.: Triangular quarter-point elements as elastic and perfectly-plastic crack tip elements. Int. J. Numer. Meth. Eng. **11**, 85–98 (1977)
3. Akin, J.E.: The generation of elements with singularities. Int. J. Numer. Meth. Eng. **10**, 1249–1259 (1976)
4. Sukumar, N., Moes, N., Moran, B., Belytschko, T.: Extended finite element method for three-dimensional crack modeling. Int. J. Numer. Meth. Eng. **48**, 1549–1570 (2000)
5. Daux, C., Moes, N., Dolbow, J., Sukumar, N., Belytschko, T.: Arbitrary branched and intersecting cracks with the extended finite element method. Int. J. Numer. Meth. Eng. **48**, 1741–1760 (2000)
6. Mousavi, S.E., Grinspun, E., Sukumar, N.: Higher-order extended finite elements with harmonic enrichment functions for complex crack problems. Int. J. Numer. Meth. Eng. **86**, 560–574 (2011)
7. Pathak, H., Singh, A., Singh, I.V., Yadav, S.K.: A simple and efficient XFEM approach for 3-D cracks simulations. Int. J. Fract. **181**(2), 189–208 (2013)

Effect of Microbial Growth on Internal Resistances in MFC: A Case Study

Ambika Arkatkar, Arvind Kumar Mungray and Preeti Sharma

Abstract Exoelectrogens are widely used in Microbial Fuel Cell (MFC) for renewable energy generation. In the present study, we used four such microorganisms, named as BR, MS, NiR and SW for generation of bioelectricity. The growth cycle of all microbes had a distinct pattern, which had a parallel relation with their current production trend. Through this study, we concluded that the charge transfer resistance is directly co-related with growth phases of the microbial culture. Different microbial cultures affect the internal resistances at different time intervals. The resistances are also affected by the biofilm forming and/or mediator secreting capacity of individual microbial culture. When subjected for a longer fed-batch condition, microbial culture BR produced 670.09 folds more power density than MS, 170.84 folds more power density than NiR and 110.86 folds more power density than SW.

Keywords Impedance · Charge transfer resistance · Growth cycle · Microbial fuel cell

1 Introduction

Microbial cultures which can donate electron outside their cell wall govern the power production process in Microbial Fuel Cell (MFC) system. These microbes are termed as 'Exoelectrogens'. Exoelectrogens like *Geobacter metallireducens* and *Geobacter sulfurreducens* are well known for their higher power production capacity than mixed culture (current density of 4.56 Am^{-2} and power density of 1.88 Wm^{-2}) [1]. *Shewanella oneidensis*, also an exoelectrogen is known for its ability to utilise a wide range of substrates while producing power [2]. The pure microbial cultures of

A. Arkatkar · A. K. Mungray
Department of Chemical Engineering, Sardar Vallabhbhai National Institute of Technology, Surat 395007, India

A. Arkatkar · P. Sharma (✉)
Department of Biotechnology, Veer Narmad South Gujarat University, Surat 395007, India
e-mail: preetisharma26@yahoo.com

D. Deb et al. (eds.), *Innovations in Infrastructure*, Advances in Intelligent Systems and Computing 757, https://doi.org/10.1007/978-981-13-1966-2_42

exoelectrogenes are primarily studied in a MFC system to understand the electron transport mechanism followed by them for donation of the electrons [3]. Inside MFC system, these microbial cultures form a biofilm on the anode surface. The energy generated after degradation of the given substrate is utilised for (i) growth of microbe and (ii) electron donation or electrochemical reactions [4].

The growth of microbial culture in any environment follows a distinct cycle comprising of four phases [5]. Many growth cycles may be followed by microbial culture during formation of biofilm on an anode surface under a fed-batch condition. In a fed-batch condition, these phases can have an impact on the transfer of electrons from bacterial cell wall towards the electrode surface. The electron when given out of the microbial cell travels towards nearest electron acceptor passing through different resistances. These resistances are termed as 'Internal resistances'. The growth rate and electrochemical behaviour of a microbial culture affect the internal resistance of a MFC system [6]. Internal resistance can be examined through Electrochemical Impedance Spectroscopy (EIS) technique. The non-destructive nature of this technique helps in the in situ examination of internal resistance [7].

The details of the growth cycle of microbial culture (in fed-batch mode) and its electrochemical behaviour has not been documented in earlier studies. Ramasamy et al. [8] reported the effect of initial biofilm growth on the impedance of anode. The study followed an analysis period of first day, fifth day, and 3 weeks in continuous feed mode. The sampling period in this study might not give exact information regarding the growth phases of the microbe. Thus, a study reporting the growth cycle of a pure microbial culture correlated with its electrochemical behaviour is lagging.

In our study of initial biofilm growth, microbes isolated from chemically treated municipal sludge sample were used. The growth cycle was analysed in fed-batch mode to have a complete understanding of growth phases of that microorganism. Simple non-destructive impedance analysis was done to understand the electrochemical behaviour of bacterial culture during its growth phases. It is an effort to understand the current generating phase of microbial life cycle. The knowledge of growth phase during which the microbial culture produces better current can be used to harvest stable power form MFC.

2 Materials and Methods

2.1 Construction of MFC

Basic H-shaped MFC setup was used for this study. The capacity of each chamber was 150 mL (working volume 100 mL). Carbon paper (2.5 cm × 4.5 cm) was used as an anode and carbon fibre (15 cm length) was used as cathode [2]. Synthetic water having acetate as an electron donor [9] was used as anolyte and saline phosphate buffer having potassium ferricyanide as an electron acceptor was used as catholyte [10]. The anode and cathode chamber were internally separated by Nafion 117 membrane

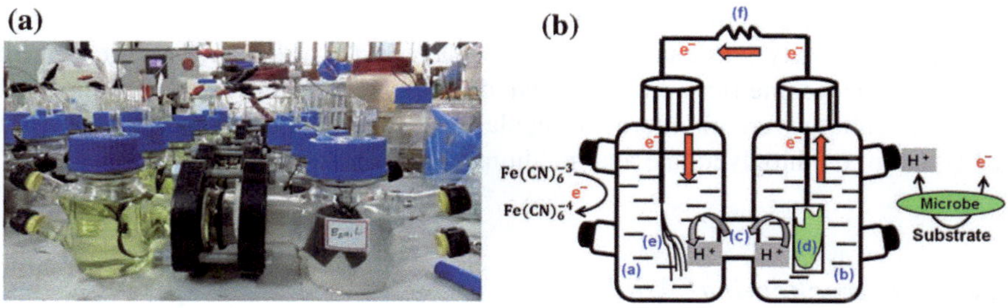

Fig. 1 **a** Photograph of Microbial Fuel Cell (MFC) set used in the laboratory and **b** Schematic representation of the MFC set (showing the electron, proton formation and consumption in respective chambers). The components of MFC in the figure are (a) Cathode chamber, (b) Anode chamber, (c) Proton permeable (Nafion 117), (d) Microbial biofilm on anode, (e) Cathode, (f) External resistance (500 Ω)

and externally the electrodes were connected by stainless steel (SS) wire. Throughout the study 500 Ω external resistance was used during current production [11] (Fig. 1).

2.2 *Working of MFC*

Anaerobic sludge sample, (municipal sewage treatment plant) was inoculated in the MFC setup after different chemical electron acceptor treatment [12]. Morphologically different microorganisms were isolated from anode and anolyte of these MFC's on thioglycollate agar, namely MS, NiR, BR and SW (the nomenclature was based on their morphological appearance on a thioglycollate agar plate). The isolated microbial cultures were inoculated in new MFC set-ups (1% of the working volume of anode chamber). The anolyte was replenished at a regular time interval in accordance with the growth and current generation values. The sampling and addition of fresh anolyte were done under anaerobic condition. All the experiments were conducted in triplicate for reproducibility.

2.3 *Current Generation and Growth of Microbial Culture*

Close circuit potential (CCP) across 500 Ω was measured at regular time interval with a digital multimeter (Risabh-16S) for each MFC reactor. The current was calculated from the following formula:

$$I = E/R \tag{1}$$

where I is current; E is the voltage across external resistance; R is external resistance [2]. Current density was calculated by normalising the value of the current with respect to the geometrically exposed anode area [13]. The growth and purity of microbial culture were monitored at regular time interval by Colony Forming Unit (CFU) count on thioglycollate agar medium at different dilutions [14].

2.4 Electrochemical Analysis

Electrochemical impedance spectroscopy (EIS) technique was used to study internal resistance at periodic intervals. CH instruments electrochemical workstation (Model CHI 760, bi-potentiostat) with the three-electrode arrangement was used during EIS analysis. The Ag/AgCl electrode was used as reference electrode. Impedance was recorded in the frequency range of 1–1,000,000 Hz. EIS data fitting was done with the available software in the CHI electrochemical workstation [15, 16] using Randles electrical circuit having constant phase element.

2.5 Power Generation

To evaluate the power generation capacity of microbial culture an open circuit potential (OCP) and later a close circuit potential (CCP) across 500 Ω was measured at the regular time interval for individual bacterial culture. Power was calculated from the following formula:

$$P = I^2 \times R \tag{2}$$

where I is current, R is external resistance and P is power [2].

3 Result and Discussion

3.1 Growth of Microbial Culture and Current Production

All four organisms exhibited a growth curve with four distinct phases as—lag phase, log phase, stationary phase followed by death phase. The lag phase was marked with low current generation and the cell growth was also less. The log phase had exponential nature of microbial growth and current production also increased simultaneously. In stationary phase, the growth and current both were stabilised. The death phase was indicated by a drop in both CFU/mL and current generation (Fig. 2) [5]. The

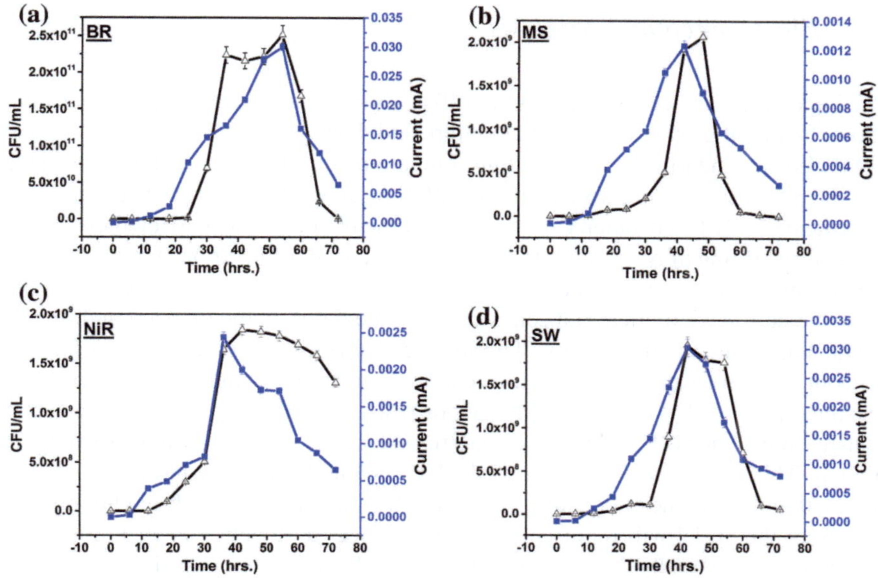

Fig. 2 Comparative graphical representation of growth cycle (CFU/mL) and current production for individual cultures, **a** BR; **b** MS; **c** NiR; **d** SW (In Figure CFU/mL is represented by —△— and current (mA) is represented by —■—)

maximum current generation time for each culture was correlated with different time span in a stationary growth phase.

Comparatively, among the microbial culture studied in this study, BR achieved a higher current production (highest current 0.03 mA). The current production was higher in the late stationary phase of its growth. It was observed that the culture remarkably changes its colour from white to pale yellow which can be attributed to the secretion of extracellular metabolites in the late stationary phase. This metabolite secretion had a positive effect on current production (Fig. 2a). The lowest current producer was MS culture (highest current 0.001 mA). The growth of MS was not having a stable stationary phase and a short lag phase, indicating that the rate of growth and thus substrate consumption was faster (Fig. 2b).

The growth of culture named NiR had a short lag phase of less than 20 h, short log phase or exponential growth phase and a stable stationary phase for a longer duration. The stationary phase was not accompanied by a stable current generation (highest current 0.002 mA) (Fig. 2c). The second current generator was SW culture (highest current 0.003 mA). The stationary growth phase of this culture showed a simultaneous stable current generation, followed by death growth phase and decrement in current (Fig. 2d). Thus, the growth and current production were having a parallel relation, as observed in Fig. 2.

3.2 Electrochemical Analysis

The impedance data generated after the EIS analysis and curve fitting gave an insight about the initial growth of microbial culture and development of internal resistance in MFC system. During the curve fitting simulation, R1 represented ohmic resistance or system resistance (due to electrolyte solution, membrane, etc.); charge transfer impedance 1 (CT1) (considered as resistance R2 with a constant phase element (CPE) in electrical fitting of Nyquist plot). CT1 is experienced by electrons during different conditions. If the anolyte has any mediator then CT1 will be applicable during the flow of electron from microbial cell wall to the mediator. In the second condition, if the mediator is absent then CT1 will be applied when electron travels within a biofilm or from biofilm to anode. The third scenario is when no organism is present; then this impedance (CT1) is considered as anode resistance (Table 1). The second charge transfer impedance is CT2 [considered as resistance R3 with constant phase element (CPE)]. This impedance is experienced when the anode chamber has a mediator or shuttle then CT2 comes in picture. Whereas when the mediator is absent but biofilm is present then CT2 will represent anode resistance. In the absence of biofilm and mediator CT2 generally does not exist. In circuit, W as warburg impedance and OFD as open finite length diffusion was used to analyse diffusion resistance in the anolyte (Table 1).

In case of BR the change in the anolyte colour, increase in current and appearance of CT2 supports the theory that the secreted secondary metabolite is acting like a mediator. The ohmic resistance (R1) and diffusion resistance were reduced due to the growth of BR in the anolyte. The death phase period showed a slight increase in the value of R1 (Table 1). In the case of MS, the growth of culture immediately shows the appearance of CT2 along with CT1. As we are not able to support the mediator secretion phenomenon in this culture from the existing experiments we consider the formation of biofilm as a factor affecting charge transfer for MS culture. The growth of culture positively affects R1, CT1, and CT2. After 48 h of growth CT1and CT2 both have increased in MS-inoculated MFC and the ohmic resistance does not show any noticeable change indicating the hindrance on electron flow. The growth and current production also decline in MS set after 48 h. The biofilm if formed in this MFC was able to grow faster on anode surface but was not able to transfer electrons for a longer duration.

In the MFC set inoculated with NiR and SW the growth of biofilm was assured with the appearance of CT2 after approximately 48 h. In NiR culture exhibits a very long stationary phase (from CFU/mL data), reduction in R1 and CT2, but the decline in current generation indicates the possibility of microbial culture entering in a dormant stage. The death phase of SW culture was indicated by an increase in R1, CT1 and CT2 values.

The capacitance exhibited by a biofilm of MS, NiR and SW increases with their growth on the anode surface. The capacitance (value of CPE in Table 1) included in CT1 does not change significantly for BR culture. The diffusion resistance (W)

Table 1 Periodical changes in different components of internal impedance developed in MFC system (evaluated through EIS analysis)

Microbial inoculation	Time (Hrs.)	R1 (ohms)	R2 (ohms)	R3 (ohms)	CPE (for R2) $\times 10^{-7}$ (Simens sec^{n})	CPE (for R3) $\times 10^{-7}$ (Simens sec^{n})	W $\times 10^{-5}$ (Simens sec$^{1/2}$)	Electrical circuits used for curve fitting simulation	Electron transfer assumed for circuit simulation
BR	0	32.97	349.10	*	3.37	*	297.90	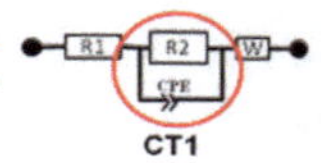	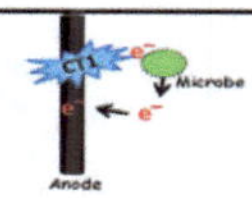
	12	6.200	138.60	*	19.30	*	573.00		
	24	8.041	78.09	*	32.07	*	852.70		
	36	5.654	113.50	*	53.44	*	92.00		
	48	4.826	58.68	*	57.10	*	1300.00		
	60	4.649	64.32	239.20	68.56	4752.00	571.00	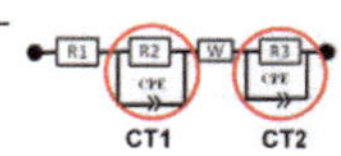	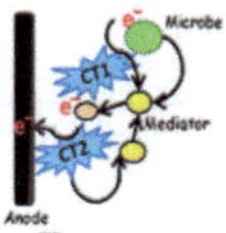
	72	9.641	75.97	76.53	21.01	364.40	699.70		
MS	0	32.97	349.10	*	3.37	*	297.90	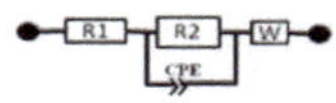	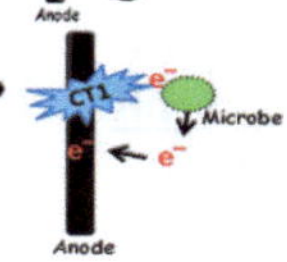
	12	7.475	23.79	53.76	105.90	7920.00	936.10	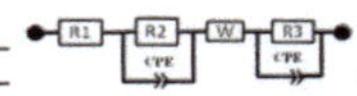	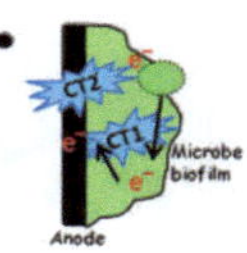
	24	8.748	53.63	117.2	30.13	316.50	356.90		
	36	3.330	92.88	87.27	312.30	27.30	367.20		
	48	4.300	46.31	25.16	288.70	69.62	520.10		
	60	2.610	121.61	95.95	528.90	29.40	428.80		
	72	3.385	186.40	127.30	87.80	1136.00	163.40 (OFD)		
NiR	0	32.970	349.10	*	3.37	*	297.90	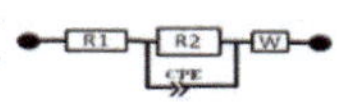	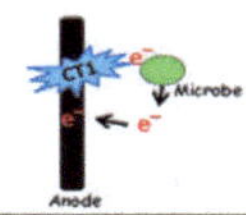
	12	12.290	102.00	*	10.93	*	13.84		
	24	7.075	113.10	*	24.61	*	388.30		
	36	8.330	69.18	*	47.12	*	782.40		
	48	11.610	81.64	66.49	96.51	11.56	342.40	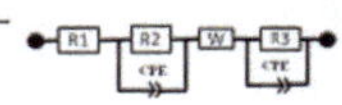	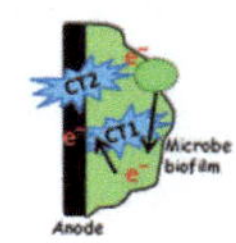
	60	11.130	62.39	60.59	10.69	57.77	781.80		
	72	6.220	38.65	70.07	80.16	361.9	285.60		
SW	0	32.970	349.10	*	336.90	*	297.90	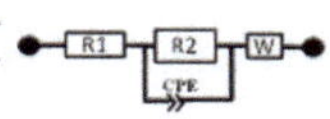	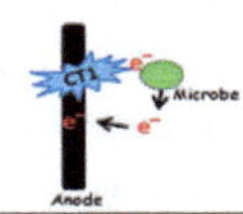
	12	10.870	85.32	*	22.14	*	1017.00		
	24	3.900	48.59	*	188.60	*	678.30		
	36	5.092	36.61	*	34.92	*	1168.00		
	48	3.837	81.60	*	490.50	*	502.90		
	60	2.830	18.59	40.91	58.35	215.20	1043.00	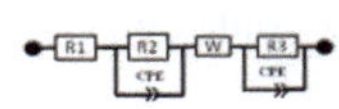	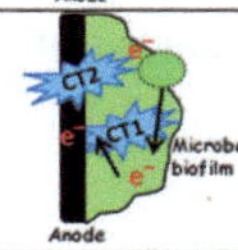
	72	3.984	15.26	66.94	136.80	952.70	217.10		

*Not used in curve fitting circuit; *R*1 ohmic resistance, *CPE* contrast phase element; (resistance *R*2 with contrast phase element)-Charge transfer impedance 1(CT1), (resistance *R*3 with Contrast phase element)—Charge transfer impedance 2(CT2); *W* Warburg impedance; *OFD* open finite length diffusion; OFD is used in a circuit to support the theory that this particular system has a capacity of finite diffusion of ions and electrons

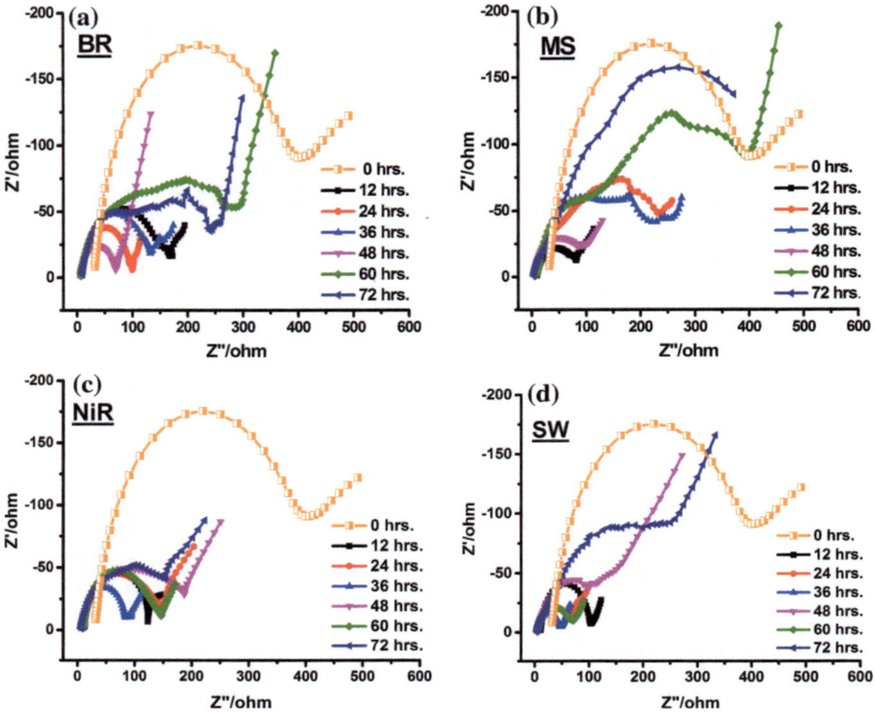

Fig. 3 Comparative Nyquist plots of impedance analysis depicting different internal resistance **a** BR; **b** MS; **c** NiR; **d** SW

irrespective of the microbial culture reduced after the growth started but was less affected by the duration of the growth cycle.

It can be clearly seen in Fig. 3 how charge transfer impedance decreases in an impedance data. The figure is given as a support to the tabular data (Table 1). The comparative Nyquist plots show the reduction in the magnitude of charge transfer resistance during consecutive growth hours of different microbial cultures

3.3 Power Generation

The MFC sets were monitored in a fed-batch mode. The microorganisms were allowed to grow in an open circuit condition for successive six feeding cycles (Fig. 4a). This allowed the microbial culture to attain a stable cell growth. The decrease in current after the addition of fresh media can be attributed to the dilution of anolyte.

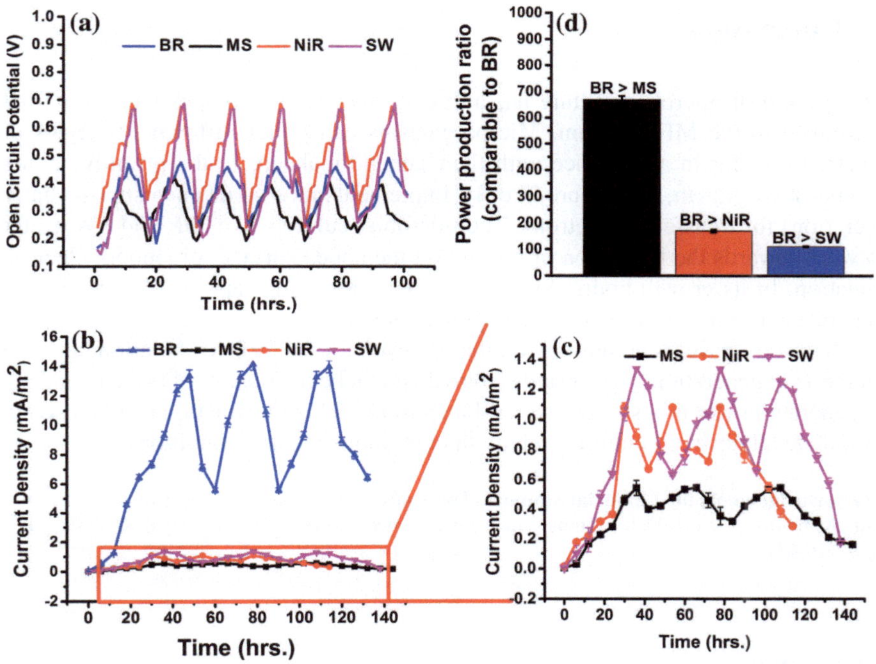

Fig. 4 Voltage and Current recorded for different microbial cultures **a** Open Circuit Potential (OCP) monitored for six consecutive cycles; **b** Current density calculated from the current recorded and geometrically exposed surface area of anode electrode for all microorganisms; **c** Inset for small magnitude current density values. **d** Power production ratio calculated for comparative analysis of power production among different microbes. (Power ratio was calculated by comparing the maximum power density values of all microbial cultures throughout the study period)

The MFC inoculated with BR attained an OCP of 0.489 V and produced a maximum current density of 14.151 mA/m^2. It was the highest power producer among the studied cultures. Whereas the lowest power generator was MS culture which attained a maximum OCP of 0.417 V and produced a maximum current density of 0.546 mA/m^2. The MFC inoculated with NiR attained an OCP of 0.687 V but produced a maximum current density of 1.082 mA/m^2. Similarly, SW-inoculated MFC also achieved an OCP of 0.686 V but the maximum value of current density was 1.344 mA/m^2 (Fig. 4a–c). Thus, the production of current in the closed circuit mode across 500 Ω was not dependent on the open circuit voltage achieved. A system achieving less OCP can produce higher current as in case of BR-inoculated MFC, a system which has a higher OCP value may produce less current depicted in the case of NiR and SW culture MFC and MFC with low OCP can further produce low current as shown by MFC with MS microbial culture.

As compared to other cultures, BR produced 670.09 folds more power density than MS, 170.84 folds more power density than NiR and 110.86 folds more power density than SW (Fig. 4d).

4 Conclusion

The growth of microbial culture has a direct effect on the components of internal resistance in the MFC system. Microorganisms can prefer different strategies of electron transfer in accordance with their genetic makeup. In the primitive hours of microbial growth, the culture like BR implements its own mechanism (mediator secretion) for transfer of electrons. The microbial cultures like NiR and SW prefer to work towards the formation of biofilm on the anode surface for smooth electron donation. In its growth hours, MS culture forms a biofilm rapidly but struggles to generate and transfer electrons at an appreciable rate.

This study gave us an outlook on the life cycle of our isolates, relation between their power generation, growth and electrochemical behaviour and also their capacity to generate power in a successive fed-batch mode. This information can be used to formulate a known co-culture study with continuous feeding conditions.

Acknowledgements for Financial Support The study was funded by Department of Science and Technology, New Delhi, India, under the project code DST-WOS-A (SR/WOS-A/ET-1054/2014(G)).

References

1. Coates, J.D., Phillips, J.P.E., Lonergan, D.J., Jenter, H., Lovley, D.R.: Isolation of *Geobacter species* from diverse sedimentary environments. Appl. Environ. Microbiol. 1531–1536 (1996)
2. Logan, B.E.: Microbial Fuel Cells. Wiley, Hoboken (2008)
3. Nimje, V.R., Chen, C.Y., Chen, C.C., Jean, J.S., Reddy, A.S., Fan, C.W., Pan, K.Y., Liu, H.T., Chen, J.L.: Stable and high energy generation by a strain of *Bacillus subtilis* in a microbial fuel cell. J. Power Sources **190**, 258–263 (2009)
4. Beyena, H., Babauta, J.T.: Biofilms in Bioelectrochemical Systems. Wiley, Hoboken (2015)
5. Maier, R.M.: Bacterial growth. In: Maier, R.M., Pepper, I.L., Gerba, C.P., (eds.) Environmental Microbiology. Academic Press, San Diego, pp. 92101–4495 (2009)
6. Babauta, J., Renslow, R., Lewandowski, Z., Beyenal, H.: Electrochemically active biofilms: facts and fiction. A review. Biofouling 1–8 (2012)
7. Sekar, N., Ramasamy, R.P.: Electrochemical impedance spectroscopy for microbial fuel cell characterization. J. Microb. Biochem. Technol. **S6**, 1–14 (2013)
8. Ramasamy, R.P., Ren, Z., Mench, M.M., Regan, J.M.: Impact of initial biofilm growth on the anode impedance of microbial fuel cells. Biotechnol. Bioeng. **101**, 101–108 (2008)
9. Pardeshi, P., Mungray, A.: High flux layer by layer polyelectrolyte FO membrane toward enhanced performance for osmotic microbial fuel cell. Int. J. Polym. Mater. Polym. Biomater. **63**, 595–601 (2014)
10. Torres, C.I., Krajmalnik-Brown, R., Parameswaran, P., Marcus, A.K., Wanger, G., Gorby, Y.A., Rittmann, B.E.: Selecting anode-respiring bacteria based on anode potential: phylogenetic, electrochemical, and microscopic characterization. Environ. Sci. Technol. **43**, 9519–9524 (2009)
11. Liu, Y., Harnisch, F., Fricke, K., Sietmann, R., Schröder, U.: Improvement of the anodic bioelectrocatalytic activity of mixed culture biofilms by a simple consecutive electrochemical selection procedure. Biosens. Bioelectron. **24**, 1006–1011 (2008)
12. Yang, T.H., Coppi, M.V., Lovley, D.R., Sun, J.: Metabolic response of *Geobacter sulfurreducens* towards electron donor/acceptor variation. Microb. Cell Fact. **9**, 90 (2010)

13. Modestra, J.A., Mohan, S.V.: Bio-electrocatalyzed electron efflux in Gram positive and Gram negative bacteria: an insight into disparity in electron transfer kinetics. RSC Adv. **4**, 34045–340559 (2014)
14. Harley, J.P., Prescott, L.M.: Laboratory Exercises in Microbiology, 6th edn. The McGraw-Hill Companies, Boston, MA (2005)
15. Guo, X., Zhan, Y., Chen, C., Zhao, L., Guo, S.: The influence of microbial synergistic and antagonistic effects on the performance of refinery wastewater microbial fuel cells. J. Power Sources **251**, 229–236 (2014)
16. Lepage, G., Perrier, G., Merlin, G., Aryal, N., Dominguez-Benetton, X.: Multifactorial evaluation of the electrochemical response of a microbial fuel cell. RSC Adv. **4**, 23815–2 (2014)

Effect of Geothermal and Other Renewables on AGC of an Interconnected Thermal System

Washima Tasnin, Lalit Chandra Saikia, Debdeep Saha, Rumi Rajbongshi and Arindita Saha

Abstract An effort of incorporating geothermal plant (GTP) in automatic generation control and analysing its impact along with other renewables on conventional thermal system is focused in this work. Suitable generation rate constraint is considered for thermal units. Fractional-order (FO) Proportional Integral—FO Proportional Derivative (FOPI-FOPD) is used as a secondary controller implementing the innovative concept of FO cascade controller. Dynamic performance of FOPI-FOPD is compared with the normally used traditional controllers and its parameters are optimized using the Sine Cosine Algorithm. The superiority of FOPI-FOPD is clearly revealed in the analysis over others. Effect of integrating GTP and other renewables is also highlighted on the thermal system. The system is studied considering 1% step load perturbation in area1.

Keywords Load frequency control · Fractional-order cascade controller
Geothermal power plant · Sine cosine algorithm

W. Tasnin (✉) · L. C. Saikia · D. Saha · R. Rajbongshi · A. Saha
Electrical Engineering Department, National Institute of Technology Silchar, Silchar, Assam, India
e-mail: washima.nits@gmail.com

L. C. Saikia
e-mail: lcsaikia@yahoo.com

D. Saha
e-mail: saha_debdeep_rs@yahoo.com

R. Rajbongshi
e-mail: rumi.nits90@gmail.com

A. Saha
e-mail: sahaarindita.91@gmail.com

D. Deb et al. (eds.), *Innovations in Infrastructure*, Advances in Intelligent Systems and Computing 757, https://doi.org/10.1007/978-981-13-1966-2_43

1 Introduction

Several individual utilities are connected altogether forming large dynamic systems basically known as the electrical power system. Tie-lines are responsible for the exchange of power among these utilities. For the need of attaining reliable interconnected operation of the power system, it is necessary that the real and reactive powers are controlled properly thereby maintaining the frequency and voltage profile at their scheduled level along with the load flow configuration [1–3]. This role of maintaining the frequency and tie line power at their nominal value both during normal circumstances as well as during slight disturbances is served by automatic generation control (AGC). Some investigations on AGC reveal interesting facts in this field that has been done in the past decades. Nanda et al. [3] studied the unequal three-area thermal system in AGC with an integral (I) controller. Saikia et al. [4] presented several multiarea systems precisely the two, three, and five areas but their studies were confined to the thermal system only. Sahu et al. [5] at first investigated the two-area thermal system and later extended it up to multi-source one incorporating, gas, and hydro to the thermal units. Studies [3–5] mainly focused on the conventional generation and very least or in fact, no attention was paid to the nonconventional resources. Due to the problem of harmful carbon emissions and the rapid depletion of the conventional sources, that is occurring at an alarming rate, it was very much necessary to discover alternative nonconventional sources for fulfilling the future energy demands. Further, nonconventional sources are also preferred due to its abundance. Solar and wind energy are the most dominant ones and their applications are prevalent in [6]. Harnessing of solar energy in AGC using the parabolic trough is initiated by Sharma et al. [7] by incorporating the solar–thermal (ST) system. Das et al. [8] dealt with the dynamic simulation of wind generator on an isolated system. Another effective and emerging renewable source for electricity generation used throughout the world is geothermal power. It can be visualized as a type of thermal energy which is automatically stored in the core of the earth and derived from the earth crust directly when required. Based on the size of the units and the dissolved particulate matter, geothermal plant (GTP) becomes different from the conventional thermal plant. Using geothermal energy for frequency control strategies are discussed by Heimisson [9]. Although geothermal energy is utilized to harness electricity, its impact on frequency regulation studies has several more prospects yet to be analysed. Literature [6–10] give an overview of the different renewable energy sources in a detailed manner along with their modelling and integration with other sources. Hence, GTP with earlier existing renewable sources comes out as a good combination of sources to be studied for AGC.

Robust secondary controllers, that can ensure the insignificant amount of area control error, are required to maintain the control of frequency smoothly. Their importance is well highlighted in literature [3–5, 11, 12]. Nanda et al. in [3] used the classical I controller and Saikia et al. [4] developed an integral plus double derivative (IDD) controller to solve the AGC problem. The authors have also adopted in AGC the idea of fuzzy, degree-of-freedom (DOF) [5], cascade [11] and fractional-order (FO) [12] controllers. Sahu et al. [5] and Dash et al. [11] explored 2DOF proportional–integral–derivative (PID) controller and proportional derivative (PD)-PID cascade controllers. Works based on both FO and cascade controllers are being analysed separately in AGC and performance evaluation regarding both the concepts in combination is also in the process of exploration. Thus, integrating both the concepts a fractional-order cascade controller (FOCC), FO proportional—Integral (PI)—FOPD (FOPI-FOPD) was chosen for this particular work.

It is very much essential for setting the gains of the controllers at their optimum values which is attained using suitable optimisation techniques. Numerous techniques like genetic algorithm [3], bacterial foraging [4], teaching–learning-based optimization [5], grey wolf optimizer [7], bat algorithm [11] and flower pollination algorithm [12] are nowadays adopted instead of the traditional technique because it is a trial and error process and thus is very much time consuming for providing the sub-optimal results. A recent stochastic algorithm by Mirjalili [13], Sine–Cosine Algorithm (SCA) is also emerging in the field of AGC like in other engineering fields of electronics, mechanics, biomedical and feature selection.

Considering the facts cited above, the objectives for this work can be formulated as

(a) To apply SCA technique for optimizing I, PI, PID and FOPI-FOPD gains in a geothermal, ST and wind incorporated thermal system with generation rate constraint (GRC) for comparing dynamic responses for finding the best controller.
(b) To analyse the impact of renewables like ST and wind, incorporated to the two unequal area thermal system in the occurrence of controller chosen in (a).
(c) To study the combined effect on system dynamics due to incorporating GTP into the system in (b) in the occurrence of controller chosen in (a).

2 System Investigated

An unequal two-area system possessing capacity ratio area1: area2 = 1:4 is being considered. Based on the generating units, altogether three systems evolved namely (i) only thermal; where 100% load is shared by the thermal units (ii) thermal along with ST and wind; where load sharing among the thermal and the renewable units were in the ratio thermal: renewable (ST or wind) = 0.6: 0.4 (iii) GTP incorporated to system in (ii) changing the load sharing ratio to thermal: renewable (ST or wind): GTP = 0.3:0.4:0.3. Thermal units with single reheat turbine having GRC of 3%/min are considered. Nominal values for parameters of thermal, solar, and wind are drawn from [4, 7] and [8], respectively, as shown in Appendix. The controllers I, PI, PID and FOPI-FOPD; which is the considered FOCC are considered one by one as the secondary controller and the performance of the system is being analysed. 1% step load perturbation (SLP) at area1 is considered to analyse the system dynamics. Coding and modelling are done in MATLAB environment. The system model is shown

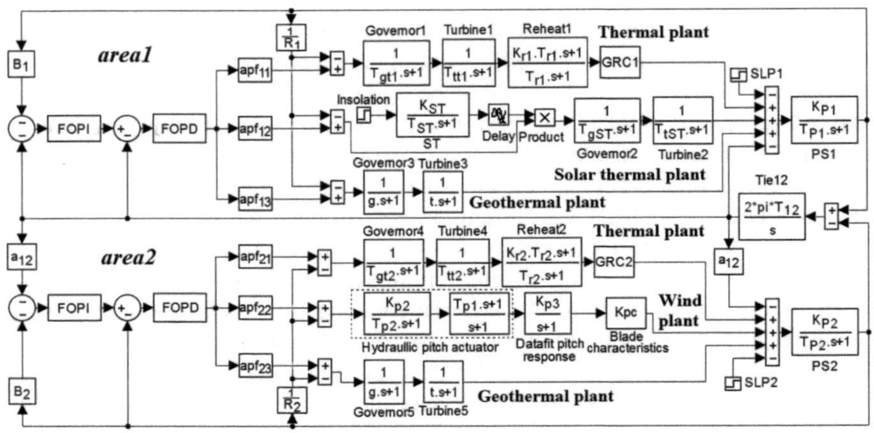

Fig. 1 Model of unequal two-area system incorporating GTP, ST, wind and thermal units with FOPI-FOPD controller

in Fig. 1. SCA is used for optimizing the controller parameters. Integral squared error (ISE) performance index (J) is considered and represented by (1).

$$J = \int_0^t \left\{ (\Delta f_i)^2 + \left(\Delta P_{\text{tie } i-j}\right)^2 \right\} \mathrm{d}t. \tag{1}$$

where i and j are area number. $i = 1, 2$; $j = 1, 2$ and ($i \neq j$).

3 Components Description of the Entire System

3.1 Geothermal Plant (GTP)

Geothermal is a potential candidate in terms of generation of electricity and space heating. The global geothermal industry is expected to reach about 18.4 GW by 2021. The United States has the largest installed capacities, followed by the Philippines. The operation cycles in GTP follow the basic laws of thermodynamics and its behaviour is quite identical to the non-reheat thermal power plant which means the absence of boiler for the purpose of reheating the steam. Thus, its modelling is similar to the thermal system [9]. The modelling of GTP is referred from [1] to [2].

3.2 Fractional-Order Cascade Controller (FOCC)

FOCC is the result of incorporating both the concepts of fractional calculus and cascade control. FOPI and FOPD are cascaded. FOPI is the outer one as $C_1(s)$ and FOPD forms inner controller $C_2(s)$ given by (2) and (3).

$$C_1(s) = K_p + \frac{K_i}{s^{\lambda}}. \tag{2}$$

$$C_2(s) = K_p + K_d s^{\mu}\left(\frac{N}{s^{\mu} + N}\right). \tag{3}$$

Closed-loop transfer function analysing the entire cascaded system performance is represented by (4).

$$\begin{aligned} Y(s) = & \left[\frac{G_1(s)G_2(s)C_1(s)C_2(s)}{1 + G_2(s)C_2(s) + G_1(s)G_2(s)C_1(s)C_2(s)}\right]R(s) \\ & + \left(\frac{G_1(s)}{1 + G_2(s)C_2(s) + G_1(s)G_2(s)C_1(s)C_2(s)}\right)d_1(s). \end{aligned} \tag{4}$$

3.3 Sine–Cosine Algorithm (SCA)

SCA generates several solutions randomly and allows them to oscillate in the direction of the optimal solution by means of a mathematical model depending on sine and cosine functions. This model is based on the updating function of position for any search agent X_i [13] and is given as follows:

$$X_i^{t+1} = \left\{ \begin{array}{l} X_i^t + r_1 \times \sin(r_2) \times \left|r_3 P_i^t - X_i^t\right| \\ X_i^t + r_1 \times \sin(r_2) \times \left|r_3 P_i^t - X_{ii}^t\right| \end{array} \right\} \begin{array}{l} \text{if } r_4 < 0.5 \\ \text{if } r_4 \geq 0.5 \end{array}. \tag{5}$$

where P_i is location of target point in ith dimension and X_i^t is present solution position in ith dimension at tth iteration. r_1, r_2, r_3, r_4 are four major parameters, where r_1 depicts movement direction. r_2 denotes till what extent it should move towards or outwards the target. r_3 allots a arbitrary weight to the target. r_4 switches between sine and cosine function to balance exploitation and exploration and finally converge to optimal point. Sine cosine range in (5) is altered by (6).

$$r_1 = a - t\frac{a}{T}. \tag{6}$$

where a is a constant, t = present iteration and T = maximum number of iterations. The tuned parameter values for this optimization are a = 2, iterations = 100 and search agent = 10.

4 Results and Discussion

4.1 Performances Comparison of I, PI, PID and FOPI-FOPD

The two-area thermal system is investigated separately using I, PI, PID and FOPI-FOPD and every time, the controller gains are optimized by means of SCA. The optimal controller values are given in Table 1 and the consequent responses are compared in Fig. 2. Critical observation highlights the superiority of FOPI-FOPD over the other three classical ones from settling time, fluctuations and peak deviations viewpoint. Hence, the remaining studies are performed by means of FOPI-FOPD controller.

4.2 Impact of Renewables

The impact of renewables on the dynamics of the system is explored by sharing the load on thermal units with the renewables. Thus including renewable sources like solar and wind in area1 and area 2, respectively, the load on the thermal units was reduced to a great extent keeping same area capacity. Table 2 shows the FOPI-FOPD optimal controller gains, optimized by means of SCA. Responses are acquired and compared with the ones of the thermal system as in Fig. 3. Observation of the peak deviations and oscillations, speaks that responses for system with renewables are superior to the system without renewables.

Table 1 Optimal gains for I, PI, PID and FOPI-FOPD

I	$K_{i1}* = 0.5698$	$K_{i2}* = 0.2876$		
PI	$K_{p1}* = 0.1266$	$K_{i1}* = 0.7522$	$K_{p2}* = 0.1000$	$K_{i2}* = 0.4268$
PID	$K_{p1}* = 0.1080$	$K_{i1}* = 0.8442$	$K_{d1}* = 0.1662$	$K_{n1}* = 100$
	$K_{p2}* = 0.1000$	$K_{i2}* = 0.3934$	$K_{d2}* = 0.1128$	$K_{n2}* = 17.0625$
FOPI-FOPD	$K_{p1}* = 0.1221$	$K_{i1}* = 0.6121$	$\lambda_1* = 0.9999$	$K_{p11}* = 0.3831$
	$K_{d11}* = 0.8689$	$\mu_{11}* = 0.8384$	$K_{p2}* = 0.2694$	$K_{i2}* = 0.5661$
	$\lambda_2* = 0.9999$	$K_{p22}* = 0.8502$	$K_{d22}* = 0.5119$	$\mu_{22}* = 0.1000$

Table 2 Optimal gains of FOPI-FOPD for system with renewables

$K_{p1}* =$ 0.9000	$K_{i1}* =$ 0.8836	$\lambda_1* = 0.9999$	$K_{p11}* =$ 0.7555	$K_{d11}* =$ 0.9000	$\mu_{11}* =$ 0.1000
$K_{p2}* =$ 0.8469	$K_{i2}* =$ 0.1000	$\lambda_2* = 0.5788$	$K_{p22}* =$ 0.1056	$K_{d22}* =$ 0.7301	$\mu_{22}* =$ 0.2091

4.3 Effect of GTP

To analyse the effect of GTP, GTP is integrated into the Sect. 4.2 system keeping same area capacity and the controller parameters are optimized by means of SCA (Table 3). Dynamic responses are then compared to a system without GTP in Fig. 4

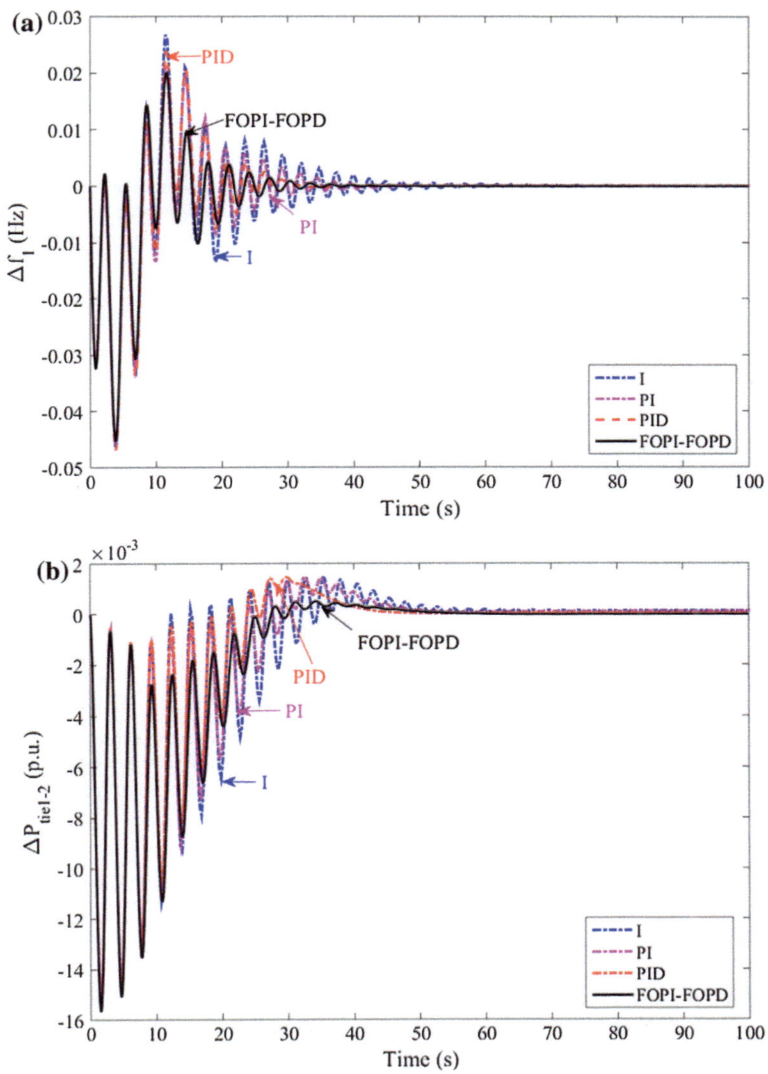

Fig. 2 Dynamic responses comparison of FOPI-FOPD with PID, PI and I for two-area thermal system. **a** $\Delta f_1 = f(t)$ **b** $\Delta P_{\text{tie1}-2} = f(t)$

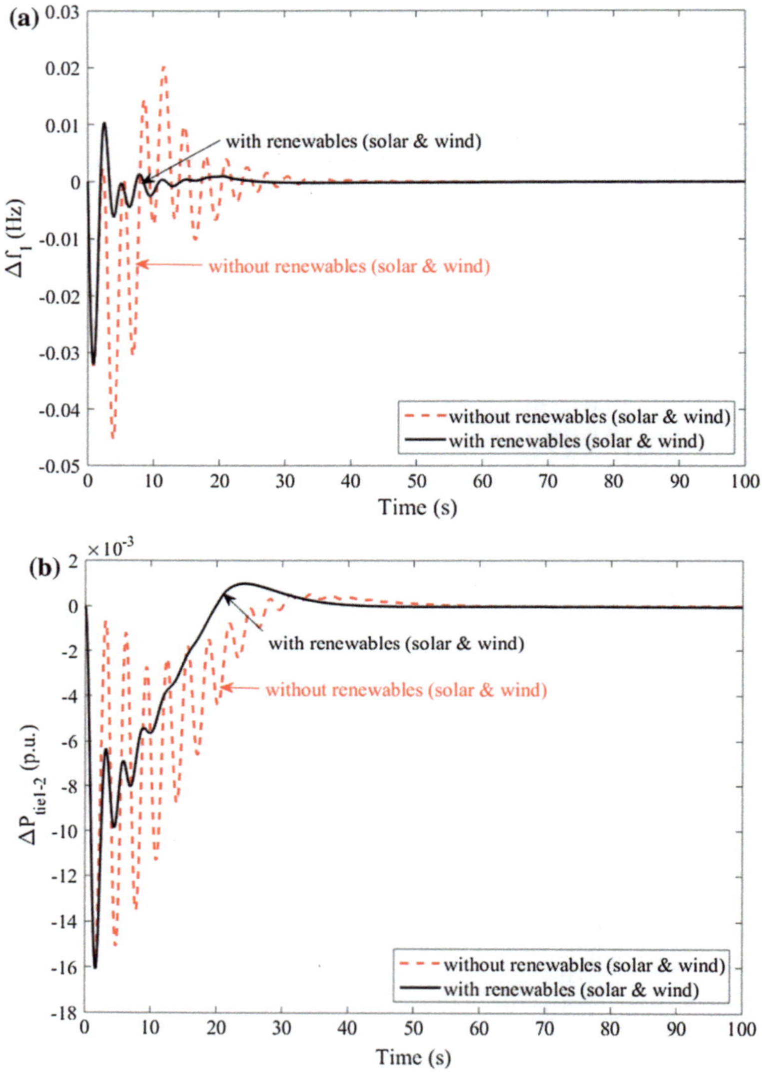

Fig. 3 Dynamic response comparison of the system with and without renewables using FOPI-FOPD. **a** $\Delta f_1 = f(t)$ **b** $P_{\mathrm{tie}1-2} = f(t)$

from which it shows that dynamic responses for a system having GTP are far better to the system without GTP in terms of settling time and fluctuations magnitude. This implies GTP has a significant effect on dynamics of the system.

Table 3 Optimal gains of FOPI-FOPD for a system with both renewable and GTP

$K_{p1}* = 0.1$	$K_{i1}* = 0.9$	$\lambda_1* = 0.9999$	$K_{p11}* = 0.9$	$K_{d11}* = 0.9$
$\mu_{11}* = 0.1$	$K_{p2}* = 0.1$	$K_{i2}* = 0.1$	$\lambda_2* = 0.1214$	$K_{p22}* = 0.8116$
$K_{d22}* = 0.1669$	$\mu_{22}* = 0.1603$	$g = 0.1095$	$t = 0.0731$	

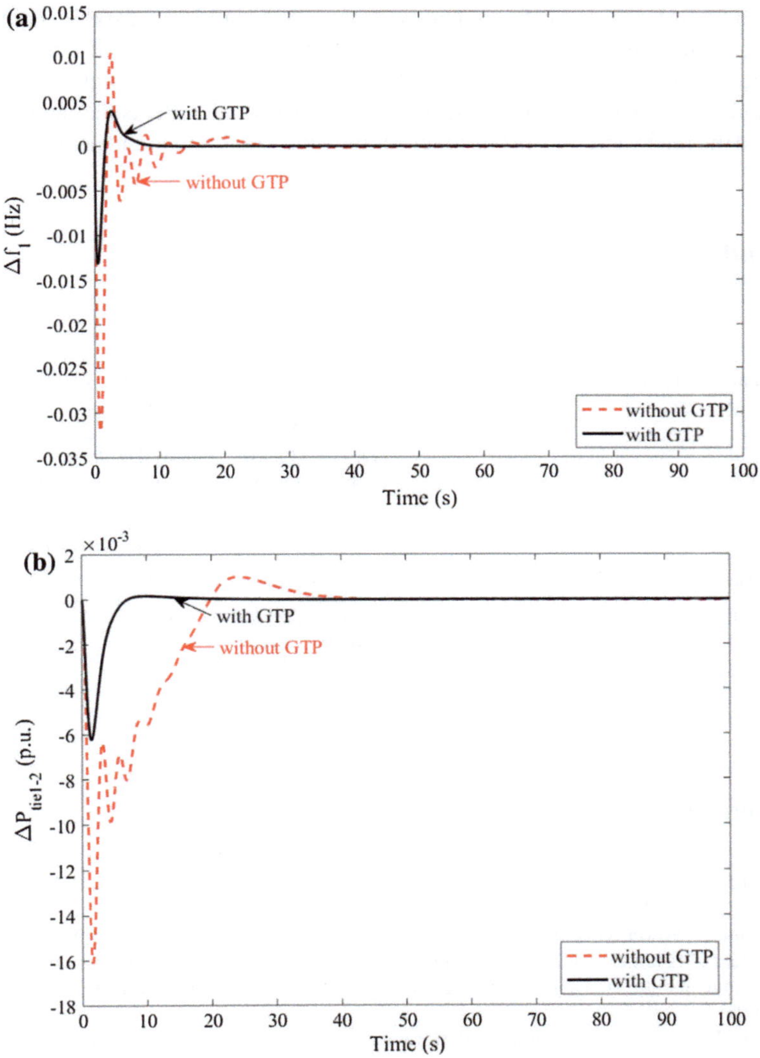

Fig. 4 Response comparison for system with and without GTP using FOPI-FOPD. **a** $\Delta f_1 = f(t)$ **b** $\Delta P_{\text{tie } 1-2} = f(t)$

5 Conclusions

An effort has been done to incorporate GTP along with other renewables like solar and wind in AGC of a conventional thermal system. FOCC concept is used forming the FOPI-FOPD controller. Controller parameters are optimized simultaneously by means of SCA. Load on the thermal units was reduced from 100% to 60% and further to 30%. Performance comparison of FOPI-FOPD with PID, PI and I reflects the enhanced performance of FOPI-FOPD. GTP impact and other renewables like solar and wind are being examined on the thermal system which reflects that inclusion of GTP and the renewables results in better setting time, oscillations and peak deviation.

Appendix

System nominal parameters	$f = 60$ Hz; loading = 50%, $K_{pi} = 120$ Hz/pu MW; $H_i = 5$ s; $T_{pi} = 20$ s; $T_{ij} =$ 0.086 puMW/rad; $Bi = \beta i = 0.425$ pu MW/Hz; $D_i = 8.33 \times 10^{-3}$ puMW/Hz; $R_i =$ 2.4 puHz/MW; $apf_{11} = 0.3$; $apf_{12} = 0.4$; $apf_{13} = 0.3$; $apf_{21} = 0.3$; $apf_{22} = 0.4$; $apf_{23} = 0.3$
Thermal system	$T_{gi} = 0.08$ s; $T_{ti} = 0.3$ s; $K_{ri} = 0.5$, $T_{ri} = 10$ s
Solar system	$K_{ST} = 1.8$; $T_{ST} = 1.8$ s; $T_{gST} = 1$ s; $T_{tST} = 3$ s
Wind system	$K_{P2} = 1.25$; $K_{PC} = 0.8$; $T_{P1} = 0.6$ s; $K_{P3} =$ 1.4; $T_{P2} = 0.041$ s

References

1. Elgerd, O.I.: Electric Energy Systems Theory: An Introduction, 2nd edn. Tata McGraw-Hill, New Delhi (1983)
2. Kundur, P.: Power System Stability and Control. McGraw-Hill, New York (1994)
3. Nanda, J., Mishra, S., Saikia, L.C.: Maiden application of bacterial foraging-based optimization technique in multi area automatic generation control. IEEE Trans. Power Syst. **24**(2), 602–609 (2009)
4. Saikia, L.C., Nanda, J., Mishra, S.: Performance comparison of several classical controllers in AGC for multi-area interconnected thermal system. Int. J. Electr. Power Energy Syst. **33**(3), 394–401 (2011)
5. Sahu, R.K., Panda, S., Rout, U.K., et al.: Teaching learning based optimization algorithm for automatic generation control of power system using 2-DOF PID controller. Int. J. Electr. Power Energy Syst. **77**, 287–301 (2016)
6. Bevrani, H., Ghosh, A., Ledwich, G.: Renewable energy sources and frequency regulation: survey and new perspectives. IET Renew. Power Gener. **4**(5), 438–457 (2010)

7. Sharma, Y., Saikia, L.C.: Automatic generation control of a multi-area ST—thermal power system using Grey Wolf optimizer algorithm based classical controllers. Elec. Power Energy Syst. **73**, 853–862 (2015)
8. Das, D., Aditya, S.K., Kothari, D.P.: Dynamics of diesel and wind turbine generators on an isolated power system. Int. J. Electr. Power Energy Syst. **21**, 183–189 (1999)
9. Heimisson, B.: Improved frequency control strategies for geothermal power plants. Master of Science Thesis, Chalmers University of Technology Sweden (2014)
10. Hossain, M.S., et al.: Role of smart grid in renewable energy: An overview. Renew. Sustain. Energy Rev. **60**, 1168–1184 (2016)
11. Dash, P., Saikia, L.C., Sinha, N.: Automatic generation control of multi area thermal system using Bat algorithm optimized PD–PID cascade controller. Int. J. Electr. Power Energy Syst. **68**, 364–372 (2015)
12. Debbarma, S., Dutta, A.: Utilizing electric vehicles for LFC in restructured power systems using fractional order controller. IEEE Trans. Smart Grid **8**(6), 2554–2564 (2017)
13. Mirjalili, S.: SCA: a sine cosine algorithm for solving optimization problems. Knowl.-Based Syst. **96**, 120–133 (2016)

Numerical Comparison of Defect-Induced Performance Degradation in CZTS and CZTSSe Solar Cells

Jaykumar Patel, Dharmendar Kumar and Kshitij Bhargava

Abstract This report highlights the importance of reducing the density of defect states in the absorber layer of the CZTSSe solar cells in order to reap the benefits of making a technological transition from CZTS to CZTSSe solar cells. Although, the highest power conversion efficiency (PCE) value of the simulated CZTSSe and CZTS solar cells under no defect condition is 12.3 and 7.49%, respectively, however, with increasing defect density the PCE value deteriorates to 4.12% in CZTSSe solar cell which is 1% lower than that of CZTS solar cell (5%) for the highest level of defect density.

Keywords CZTS · CZTSSe · Solar cell · SCAPS-1D · Defect density

1 Introduction

The commercialization of high-efficiency (>20%) $Cu(In,Ga)Se_2$ (CIGS) solar cells is currently challenged by the limited reserves of Indium (In) which makes it very costly in the photovoltaic market. In this context, the Cu_2ZnSnS_4 (CZTS) and $Cu_2ZnSn(S_x, Se_{1-x})$ (CZTSSe) solar cells are widely investigated as the cheaper alternative of CIGS solar cells [1]. Although the efficiency of these cells have yet not been able to achieve the level of CIGS cells, researchers worldwide are continuously dedicated towards improving their efficiency by tuning the electronic properties of various layers constituting the structure of the solar cell. Till date, the achieved efficiency of CZTSSe solar cells is only 12.6% and still, some serious technological investments are needed to reach the performance comparable to that of CIGS solar cell. As it is well known that the efficiency of a solar cell is critically dependent on the optical and electronic properties of the absorber layer, and therefore, it becomes necessary to

J. Patel · D. Kumar · K. Bhargava (✉)
Department of Electrical Engineering, Institute of Infrastructure Technology Research and Management (IITRAM), Ahmedabad 380026, Gujarat, India
e-mail: kshitijbhargava@iitram.ac.in

D. Deb et al. (eds.), *Innovations in Infrastructure*, Advances in Intelligent Systems and Computing 757, https://doi.org/10.1007/978-981-13-1966-2_44

both understand and improvise the thin film properties of absorber layer constituting the solar cell structure in order to achieve the desired photovoltaic performance.

Due to the excellent bandgap and absorption coefficient of both CZTS and CZTSSe materials, these have been deeply investigated in the recent past to reach the power conversion efficiency (PCE) values of 9.2 and 12.6%, respectively. Just like some other thin film photovoltaic devices, the presence of defect states in the absorber layer of the cell too has been observed to deteriorate the performance of CZTS and CZTSSe solar cells [2]. In order to optimize and improve the performance of these solar cells, the numerical simulations can provide deep insight into the physics governing the performance of cells in the presence of defects in the absorber layer [3–7]. Although it has been reported in past that CZTSSe solar cell shows better performance over its CZTS counterpart it would be interesting to study the relative influence of varying defect densities in the absorber layer of the simulated structures.

In this paper, we simulate and compare the performances of CZTS and CZTSSe solar cells. The results were quantitatively compared on the basis of their performance parameters such as the fill-factor (FF), power conversion efficiency (PCE), open circuit voltage (V_{OC}), and short circuit current density (J_{SC}) under different values of defect densities present in the absorber layer.

2 Simulation Details

Both CZTS and CZTSSe solar cells were simulated using the SCAPS-1D (version 3.3.05) device simulator. The schematics of the simulated structures are shown in Fig. 1. The defined structures in the SCAPS simulation domain were the conventional structures with Mo back contact, CZTS or CZTSSe absorber layer, CdS buffer layer, and ZnO as transparent conducting oxide (TCO). All the device layers were assumed to be the same except the absorber layer. The layer thicknesses and the simulation parameters were also kept the same for the sake of comparison. The list of simulation parameters was adopted from the literature and are tabulated in Table 1. The motive of the study is to elucidate the significance of the presence of defect states at different levels, in the most important layer of a solar cell, i.e., the absorber layer, over the performance of solar cells. As it has already been demonstrated in the literature that the CZTSSe layer acts as the better absorber material in solar cells than the CZTS layer, and therefore, it is important to understand the influence of defect states over their respective performances as defects are bound to be generated during the deposition of thin films and has been the subject of intense investigation in the field of nanoelectronics. In this context, numerical simulations can provide a wealth of insight about the deteriorating performance of solar cells in terms of the values of its performance parameters. The similar study if done experimentally would be quite inconvenient as it will consume both time and resources.

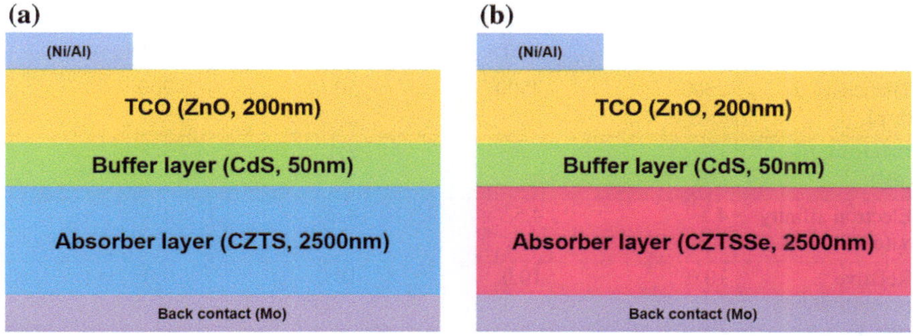

Fig. 1 Schematic of simulated **a** CZTS **b** CZTSSe solar cell structures

The SCAPS-1D software solves basic steady-state 1-D semiconductor equations. The electrons and holes transport are described by drift-diffusion Poisson equations along with necessary boundary conditions at interfaces.

$$\varepsilon_r \varepsilon_0 \frac{\partial^2}{\partial x^2}\phi(x) = q\left(p(x) - n(x) + N_D^+ - N_A^- + \rho_{\text{def}}\right) \tag{1}$$

$$-\frac{\partial}{\partial x}J_p(x) + G(x) - R(x) = \frac{\partial p}{\partial t} \tag{2}$$

$$-\frac{\partial}{\partial x}J_n(x) + G(x) - R(x) = \frac{\partial n}{\partial t} \tag{3}$$

3 Results and Discussion

The CZTSSe material is expected to be a better replacement of CZTS material as an absorber material in thin film solar cells owing to its superior light absorbing and charge transport properties. However, it becomes necessary to understand the effect of the presence of defect states in these absorber layers, which are bound to be generated in some extent or the other during the fabrication process, over the performance of thin film solar cells. Therefore, the performance of the simulated CZTS and CZTSSe solar cells were evaluated and compared on the basis of their performance parameter values. The typical performance parameters of solar cell viz. FF, V_{OC}, $J_{SC,}$ and η were estimated under the presence of different levels of defect densities in the absorber layer. For this purpose, the current density versus voltage (*J–V*) characteristics and quantum efficiency versus wavelength (*QE*-λ) characteristics were utilized. A detailed discussion pertaining to the effect of defects on the performance of both the simulated solar cells is described as follows.

Table 1 List of parameters defined in the SCAPS simulation domain [8]

Parameters	CZTSSe	CZTS	CdS	ZnO
Thickness, d (nm)	2500	2500	50	200
Bandgap, E_g (eV)	1.2	1.5	2.4	3.3
Electron affinity χ (eV)	4.1	4.5	4.2	4.4
Relative permittivity, $(\varepsilon/\varepsilon_r)$	10.0	10.0	10.0	9.0
Effective density of state in conduction band, N_C (cm^{-3})	2.2×10^{18}	2.2×10^{18}	2.2×10^{18}	2.2×10^{18}
Effective density of state in valence band, N_V (cm^{-3})	1.8×10^{19}	1.8×10^{19}	1.8×10^{19}	1.8×10^{19}
Electron mobility, μ_n (cm^2/V)	100	100	100	100
Hole mobility, μ_p (cm^2/V)	25	25	25	25
Donor concentration, N_D (10^{17} cm^{-3})	0	1.0×10^{0}	1.0×10^{17}	1.0×10^{17}
Acceptor concentration, N_A (10^{14} cm^{-3})	2.0×10^{14}	1.0×10^{18}	0	0
Electron thermal velocity, v_n (10^7 cm/s)	1.0×10^{7}	1.0×10^{7}	1.0×10^{7}	1.0×10^{7}
Hole thermal velocity, v_p (cm/s)	1.0×10^{7}	1.0×10^{7}	1.0×10^{7}	1.0×10^{7}
Defect type	d	D	–	–

3.1 Comparison on the Basis of '*QE*-λ' Characteristics

Figure 2 shows the comparison of '*QE*-λ' curves of the simulated structures under several values of defect density. It is important to notice that both magnitude as well the range of '*QE*' curve was superior in CZTSSe cell. In case of simulated CZTS solar cell, the majority of the '*QE*' curve is restricted within the visible region (400–800 nm) of the solar spectrum, while in the case of CZTSSe solar cell it is also

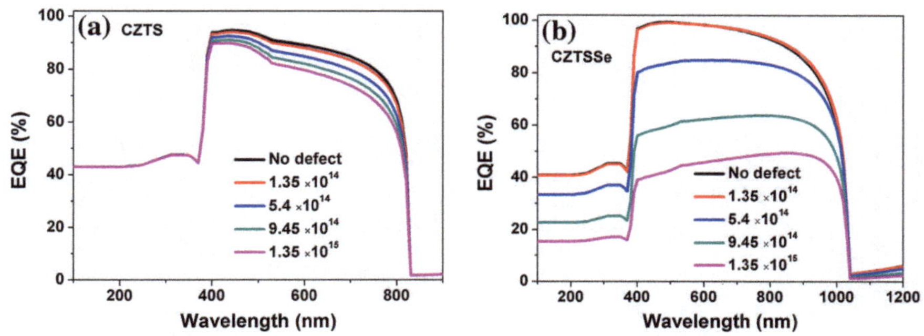

Fig. 2 Comparison of '*QE*-λ' curves for simulated **a** CZTS **b** CZTSSe solar cells at different values of the absorber layer defect densities

covering some part of the infrared region. Further, in both the cases the '*QE*' values decreased with the increasing defect density which is attributed to the reduced extent of photo-generated charge carrier population reaching the respective electrodes. Also, for the same range of defect density values the level of degradation of '*QE*' in CZTSSe solar cell was quite dominant as compared to that observed in CZTS solar cell thus elucidating the role of absorber layer defect states toward the performance degradation in CZTSSe solar cells even with superior '*QE*' characteristics over CZTS solar cells.

3.2 Comparison on the Basis of J–V Characteristics

The comparison of '*J–V*' characteristics (Fig. 3) of the simulated structures represents that although the area under the curve was greater for CZTSSe cell but it tends to reduce more severely in comparison to CZTS cell. This particular trend again points toward the same fact of the greater extent of performance degradation in CZTSSe solar cell as compared to CZTS solar cell even though the properties of the CZTSSe layer is deemed to be more appropriate for solar cell applications.

3.3 Comparison on the Basis of Energy Band Alignment

The observations of '*J–V*' and '*QE*-λ' characteristics were also to some extent evident through the energy band alignment diagrams for the two cases. These band diagrams were obtained for extreme values of defect densities, i.e., 0 (no defect) and 1.35×10^{15} cm^{-3} (max. defect) conditions for the sake of simplicity. Although there was hardly any variation in case of CZTS solar cell (Fig. 4a, b), there was a significant difference in the energy band levels throughout the absorber layer (from 0 to 2.5 μm

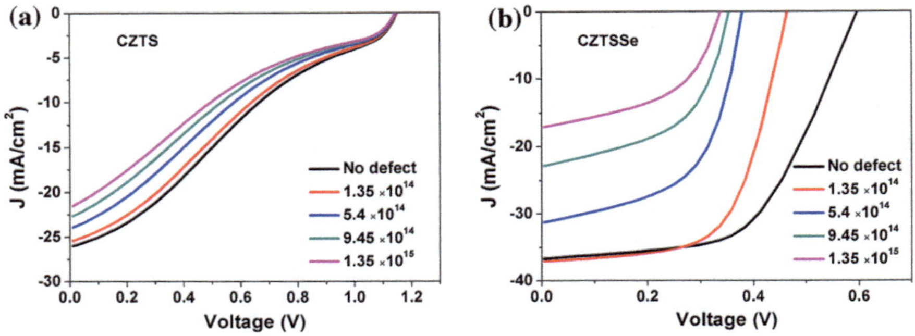

Fig. 3 Comparison of 'J–V' characteristics for simulated **a** CZTS **b** CZTSSe solar cells at different values of the absorber layer defect densities

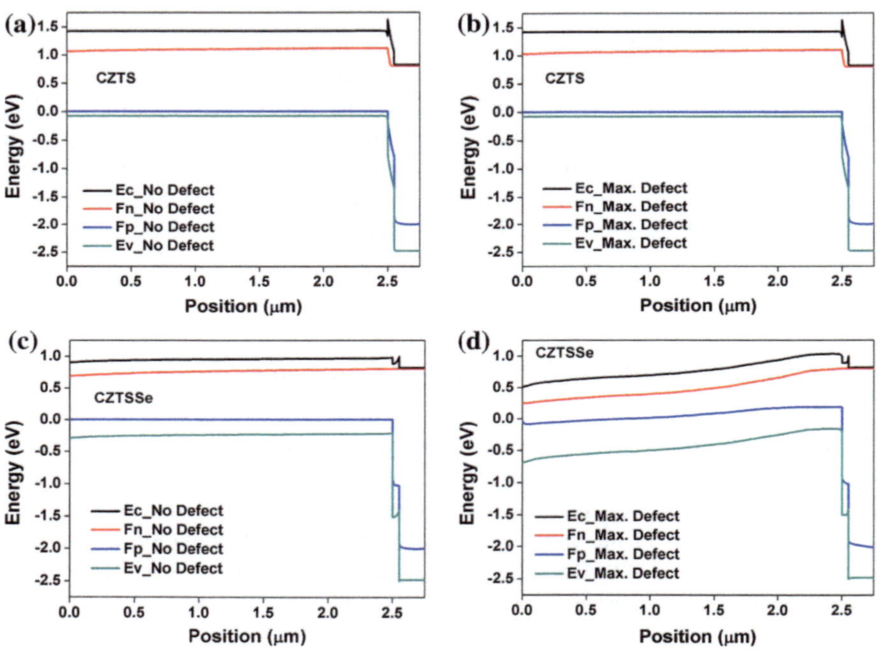

Fig. 4 Comparison of energy band alignment for simulated CZTS and CZTSSe solar cells under extreme conditions of defect densities, i.e., under no defect and maximum defect condition (1.35×10^{15} cm^{-3}) in the absorber layer

of position) for CZTSSe solar cells (Fig. 4c, d) thus highlighting the dominance of defect states on CZTSSe solar cell over CZTS solar cell.

Table 2 Comparison of estimated solar cell performance parameters for the simulated structures

Solar cell	Defect density	Performance parameters			
		J_{SC} (mA/cm^2)	V_{OC} (V)	FF (%)	η (%)
CZTS	0 (No defect)	26.06	1.15	25	7.49
	1.35×10^{14}	25.54	1.15	24.19	7.10
	5.4×10^{14}	24.16	1.15	22.55	6.25
	9.45×10^{14}	22.99	1.15	21.52	5.67
	1.35×10^{15}	21.99	1.15	20.84	5.25
CZTSSe	0 (No defect)	36.66	0.6	56.10	12.30
	1.35×10^{14}	37.12	0.56	52.78	11.06
	5.4×10^{14}	31.65	0.51	47.63	7.79
	9.45×10^{14}	23.84	0.49	47.06	5.57
	1.35×10^{15}	18.30	0.48	46.73	4.12

3.4 Comparison on the Basis of Performance Parameters

Table 2 shows the comparison between the parameter values for each defined solar cell structure at different defect density values. The tabulated data signifies the deterioration in the performance of both types of simulated solar cells with respect to the increasing defect density in the absorber layer. Although the parameter values of CZTS solar cell were comparatively quite inferior to those of CZTSSe cell, it was interesting to observe that for the highest simulated defect density value the performance parameters of CZTSSe solar cell were even more inferior to those of CZTS solar cell. These results highlight the importance of reducing the presence of defect states in the absorber layer of the CZTSSe solar cell so as to achieve better performance over CZTS solar cell. For reducing the density of defect states in the absorber layer of the CZTSSe solar cells more advanced techniques of thin film growth would be required in order to further achieve its efficiency value similar to that achieved in CIGS solar cells.

4 Conclusion

The simulation-based comparison of CZTS and CZTSSe solar cells were performed at several absorber layer defect density values. The results reveal the importance of reducing the density of defects in the absorber layer of the CZTSSe solar cells in order to achieve the power conversion efficiency value higher than CZTS solar cell as is also claimed by several reports in the literature. This report will be helpful to researchers working in this area to concentrate their efforts towards the development of thin film growth techniques critical for increasing the crystallinity of CZTSSe absorber thin film.

Acknowledgements The author Kshitij Bhargava is grateful to Prof. Marc Burgelman, Department of Electronics and Information Systems (ELIS), University of Gent, Belgium, for providing the SCAPS-1D simulation software.

References

1. Yang, K.-J., et al.: A band-gap-graded CZTSSe solar cell with 12.3% efficiency. J. Mater. Chem. A **4**, 10151–10158 (2016)
2. Hironiwa, D., et al.: Simulation of optimum band-gap grading profile of $Cu_2ZnSn(S,Se)_4$ solar cells with different optical and defect properties. Jpn. J. Appl. Phys. **53**, 071201(1–9) (2014)
3. Bhargava, K., et al.: Electrical characterization and parameter extraction of organic thin film transistors using two dimensional numerical simulations. J. Comput. Electron. **13**, 585–592 (2014)
4. Khoshsirat, N., et al.: Analysis of absorber layer properties effect on CIGS solar cell performance using SCAPS. Optik **126**, 681–686 (2015)
5. Asaduzzaman, M., et al.: An investigation into the effects o band gap and doping concentration on $Cu(In,Ga)Se_2$ solar cell efficiency. SpringerPlus **5**, 578(1–) (2016)
6. Ouédraogo, S., et al.: Numerical analysis of copper-indium-gallium-diselenide- based solar cells by SCAPS-1D. Int. J. Photoenergy **2013**, 1–9 (2013)
7. Chihi, A., et al.: Investigation on the performance of CIGS/TiO_2 heterojunction using SCAPS software for highly efficient solar cells. J. Electron. Mater. **46**, 5270–5277 (2017)
8. Sanghyun, L., et al.: Variation of quantum efficiency in CZTSSe solar cells with temperature and bias dependence by SCAPS simulation. J. Energy Power Eng. **11**, 69–77 (2017)

Wavelet Transform and Variants of SVR with Application in Wind Forecasting

Harsh S. Dhiman, Pritam Anand and Dipankar Deb

Abstract Accurate wind prediction plays an important role in grid integration. In this paper, we analyze the performance of a hybrid forecasting method comprising of wavelet transform and different variants of Support Vector Regression (SVR) like ε-SVR, Least Square Support Vector Regression (LS-SVR), Twin Support Vector Regression (TSVR) and ε-Twin Support Vector Regression (ε-TSVR). Each of these methods is trained and tested for a wind farm Sotavento, Galicia, Spain. Wavelet transform is used to filter the raw wind speed data from any kind of stochastic volatility. Among the different variants of SVR, the forecasting results of ε-TSVR and TSVR are compared with ε-SVR and LS-SVR to evaluate various quantitative measures like RMSE, MAE, SSR/SST and SSE/SST.

Keywords Wind forecasting · Support Vector Regression (SVR) Least Square Support Vector Regression (LS-SVR) · Twin support vector regression (TSVR) · ε-Twin Support Vector Regression (ε-TVSR)

1 Introduction

Growing electricity demands have led to increased usage of renewable energy resources like solar energy, wind energy, and biomass. Among various sources of energy, wind energy has a lot of potential in terms of clean and efficient power generation. However, random nature of wind resource causes many technical challenges like that of power system security, grid interconnection, and market clearance. Wind

H. S. Dhiman (✉) · D. Deb
Institute of Infrastructure Technology Research and Management,
Ahmedabad 380026, Gujarat, India
e-mail: harsh.dhiman.17pe@iitram.ac.in

D. Deb
e-mail: dipankardeb@iitram.ac.in

P. Anand
Department of Computer Science, South Asian University, New Delhi 110021, India
e-mail: ltpritamanand@gmail.com

D. Deb et al. (eds.), *Innovations in Infrastructure*, Advances in Intelligent Systems and Computing 757, https://doi.org/10.1007/978-981-13-1966-2_45

forecasting thus plays an important role to ensure power system reliability and security. Time-based wind speed forecasting is essentially carried out in order to cater for different purposes [8]. Wind forecasting methods can further be classified as numerical-weather prediction models and statistical models. Statistical models use past data to forecast wind speed ahead in time, also called as data-driven models. Time series models like ARMA and ARIMA models provide good prediction accuracy, but fail to deliver in case of nonlinearities in wind speed. Studies have shown that hybrid forecasting methods that exploit the advantages of individual methods perform better than the conventional techniques.

Yuan et al. have proposed a hybrid forecasting model employing Autoregressive Fractionally Integrated Moving Average (ARIFMA) and least square support vector regression to predicts short-term wind power [14]. Various machine-learning-based forecasting models have proven their superiority over time series models. Among machine-learning methods, support vector regression estimates a regression function in the spirit of classical support vector machines developed by Vapnik [11]. Liu et al. have proposed a hybrid short-term wind speed forecasting method using Wavelet transform and support vector machines optimized by genetic algorithm [4]. Another hybrid method that studies wind speed forecasting in Mexico uses ARIMA and ANN models that is presented in [1]. Liu and Tian have used wavelet and wavelet packet along with ANN to predict wind speed and have compared their findings with standard models like ANFIS, wavelet-RBF, and persistence method [5]. Xiao et al. have demonstrated combined prediction models employing weight calculation for individual models according to their superiority among various other models [13]. In [3], the authors have used ANN with statistical weight preprocessing technique to predict he coefficient of performance of ground source heat pump (GCHP) system. Further studies have shown that using optimization methods for selecting model parameters yields better results than individual methods. A commonly used optimization technique is genetic algorithm owing to its fast convergence. Su et al. have used Particle Swarm Optimization to selected optimal model parameters for ARIMA model combined with Kalman filter for wind speed forecasting [9].

Through this paper, we aim to analyze the short-term wind speed forecasting using Wavelet transform and different variants of SVR. Further, we compare the performance metrics like Root Mean Squared Error (RMSE), Sum of Squared Error (SSE), Sum of Squared errors in Testing (SST), and computational time for different hybrid models.

This paper is divided as follows: Sect. 1 discusses the introduction. In Sect. 2, wavelet transform and its different types are discussed. Section 3 discusses variants of support vector regression (SVR), that is, ε-SVR, LS-SVR, TSVR, and ε-TSVR which is followed by Results and Conclusions.

2 Wavelet Transform: A Decomposition Technique

Wavelet transform (WT) is a multi-resolution decomposition technique used in various fields of engineering applications like image processing, feature extraction for image segmentation, and noise reduction. Wavelet transform is of two types; that is, Continuous Wavelet Transform (CWT) and Discrete Wavelet Transform (DWT). Owing to its computational advantage DWT is more often used than CWT. WT decomposes a given input signal into an approximation signal and detail signal with former containing low-frequency components of the main signal. The DWT for an input signal $x(t)$ can be mathematically expressed as

$$W(m, n) = 2^{-m/2} \sum_{t=0}^{T-1} x(t)\phi\left(\frac{t - n.2^m}{2^m}\right), \tag{1}$$

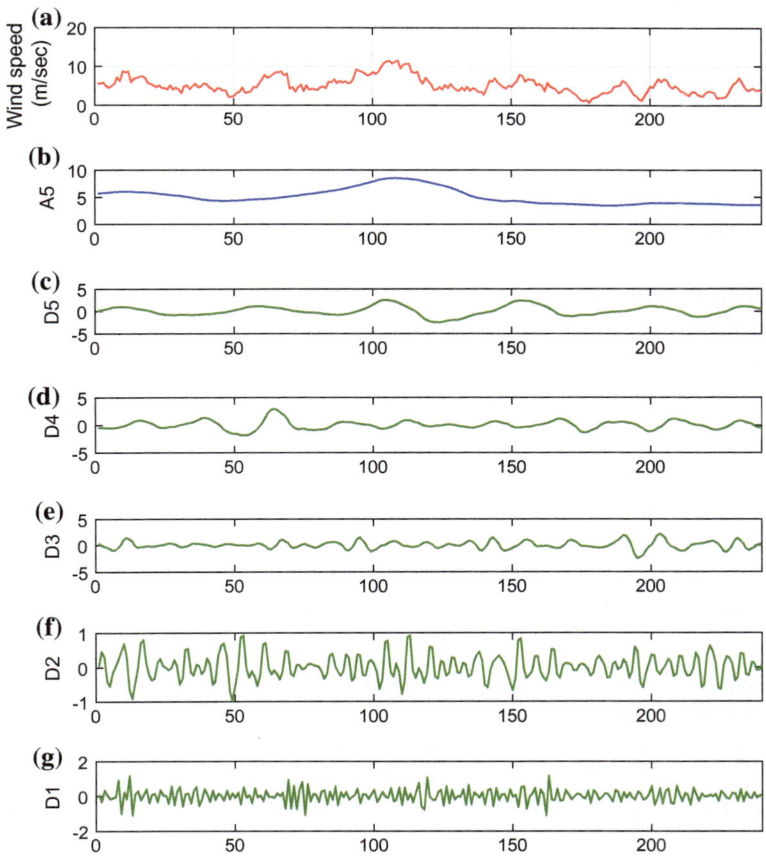

Fig. 1 Five-level wavelet decomposition for wind speed in Sotavento, Galicia, Spain

where T represents the length of the signal $x(t)$ where scaling and translation parameters are the functions of integer variables m and n as stated by Catalão et al. [2]. WT is carried out with respect to a mother wavelet $\phi(t)$, which is of several types like Haar wavelet, Daubechies wavelet, Coiflet wavelet, and Symmelet wavelet. In our study, we apply Daubechies (db4) wavelet transform to the original wind series. A five-level signal decomposition is carried out and its results are shown in Fig. 1.

3 Support Vector Regression

Support vector regression (SVR) is a machine-intelligent algorithm derived from the concept of support vector machines (SVM) developed by Vapnik [11]. The algorithm aims to find a regressor $f(x)$ having most deviation ε from the training targets and is as flat as possible. The above-mentioned objective is achieved by optimizing the following risk function:

$$\begin{aligned} &\min \frac{1}{2}w^T w + C(e^T\xi + e^T\xi^*), \\ &\textbf{s.t. } Y - (Aw + b) \le e\varepsilon + \xi \quad , \xi \ge 0 \\ &(Aw + b) - Y \le e\varepsilon + \xi^* \quad , \xi^* \ge 0 \end{aligned} \tag{2}$$

where $C \ge 0$ is a regularization factor that weighs trade-off between training errors and flatness of the linear regressor $f(x)$, ξ and ξ^* are the slack variables introduced to relax the ε-insensitive condition. In case of nonlinear regression problem, a positive semi-definite kernel matrix with elements as $K_{i,j} = K(x_i, x_j)$ is built to transform the input data to a higher dimension feature space. Commonly used kernel function is a radial basis function (RBF), that is, $K(x_i, x_j) = \exp\left(-\frac{\|x_i - x_j\|^2}{2\theta^2}\right)$ with parameter $\theta \ge 0$ that defines the bandwidth of the RBF. However, due to computational complexities and large size of the training data, a least square support vector machine algorithm which considers square of the error tolerances is proposed by Suykens and Vandewalle [10].

3.1 Least Square Support Vector Regression (LS-SVR)

As stated earlier due to computation complexity, ε-SVR model cannot given accurate predictions and may consume large computation time, and so to overcome this a LS-SVR model that solves linear equation for training instead of quadratic equation in case of classical SVR was proposed [10]. Given the training dataset $(A_1, Y_1), (A_2, Y_2), \ldots (A_k, Y_k)$, where $A_i \in \Re^n$ is the ith input from the input space and $Y_i \in \Re$ is the corresponding output and k is the total number of training samples. The objective function corresponding to LS-SVR model is

$$\min \frac{1}{2}(w^T w + C\xi^T \xi^*),$$
$$\textbf{s.t. } Y = \phi(A)w + eb + \xi, \tag{3}$$

where $\phi(.)$ is the mapping function used to transform input space into higher dimension space. An application of LS-SVR along with ARFIMA was applied on wind power forecasting where the hybrid method proved out to be superior to ε-SVR [14].

3.2 Twin Support Vector Regression (TSVR)

Twin Support Vector Regression (TSVR) was invented in the spirit of Twin Support Vector Machines (TSVM) [6]. TSVR aims to derive two nonparallel hyperplanes each being close to one of the classes and as far as possible from each other. Further instead of solving one large Quadratic Programming Problem (QPP), TSVR solves two QPPs thus reducing the computation time and complexity. The mathematical formulation of the two QPPs is given as

$$\min \frac{1}{2}(Y - e\varepsilon_1 - (Aw_1 + eb_1))^T(Y - e\varepsilon_1 - (Aw_1 + eb_1)) + C_1 e^T \xi$$
$$\textbf{s.t. } Y - (Aw_1 + eb_1) \geq e\varepsilon_1 - \xi, \xi \geq 0 \tag{4}$$
$$\min \frac{1}{2}(Y - e\varepsilon_2 - (Aw_2 + eb_2))^T(Y - e\varepsilon_2 - (Aw_2 + eb_2)) + C_2 e^T \eta$$
$$\textbf{s.t. } (Aw_2 + eb_2) - Y \geq e\varepsilon_2 - \eta, \eta \geq 0 \tag{5}$$

where $C_1, C_2 > 0$, $\varepsilon_1, \varepsilon_2 \geq 0$ are parameters and ξ, η are the slack variables. A detailed explanation of the dual QPPs along with their KKT conditions can be found in [6]. In case of nonlinear regression, kernel technique can be used to solve the QPP. The kernel- based TSVR aims to find out the mean of the two regressors $f_1(x) = K(x^T, A^T)w_1 + b_1$ and $f_2(x) = K(x^T, A^T)w_2 + b_2$.

3.3 ε-Twin Support Vector Regression

Similar to TSVR, Shao et al. [7] have proposed a novel TSVR technique that minimizes structural risk function rather than empirical risk function as in the case of TSVR. The ε-TSVR formulation is given as

$$\min \frac{1}{2}c_3(w_1^T w_1 + b_1^2) + \frac{1}{2}\xi^{*T}\xi + c_1 e^T \xi,$$
$$\textbf{s.t. } Y - (Aw_1 + eb_1) = \xi^*, \quad Y - (Aw_1 + eb_1) \geq \varepsilon_1 e - \xi, \xi \geq 0 \tag{6}$$
$$\min \frac{1}{2}c_4(w_2^T w_2 + b_2^2) + \frac{1}{2}\eta^{*T}\eta + c_2 e^T \eta,$$
$$\textbf{s.t. } (Aw_2 + eb_2) - Y = \eta^*, \quad (Aw_2 + eb_2) - Y \geq \varepsilon_2 e - \eta, \eta \geq 0 \tag{7}$$

where c_1, c_2 and ε_1, ε_2 are the positive parameters. Major difference between TSVR and ε-TSVR is the extra regularization term $\frac{1}{2}c_3(w_1^T w_1 + b_1^2)$, which is used to minimize the structural risk function. The performance of the ε-TSVR can be further improved by adjusting the values of c_3 and c_4 by using grid search algorithm and various other optimization techniques. The computational ability of TSVR and ε-TSVR can be improved by using successive over-relaxation technique (SOR).

4 Hybrid Wavelet-SVR Model for Wind Speed Prediction

A hybrid wavelet-SVR model for predicting wind speed ahead of time is studied for different variants of SVR, that is, ε-SVR, LS-SVR, TSVR, and ε-TSVR. Hybrid methods prove out to be better candidates as they incorporate the advantages of individual models. SVR is based on kernel regression model that captures the nonlinearity in the features of the model. DWT is a decomposition technique used to remove any stochastic variation present in a nonstationary signal. A five-level *db*4 DWT is carried out on the wind speed series. Any DWT decomposition leads to two signals, that is, approximation signal and detail signal. The approximation signal contains the low-frequency information regarding the signal while detail signal consists of high-frequency data.

The inputs to the SVR model selected are approximation signal (A5), detail signals (D1-D5), and the direction associated with the corresponding wind speed. The testing set comprises of 80 setpoints for all the datasets stated. The wind speed data of 10 days (for dataset D1 and D3) and 20 days (datasets D2 and D4) is collected for a wind farm site Sotavento located in Galicia, Spain. The model performance was assessed based on metrics like Sum of Squared Regression (SSR), Root Mean Squared Error (RMSE), Sum of Squared Error of testing (SSE), Sum of squared deviation of testing (SST), and CPU time which are listed in Table 1. CPU time (in seconds) is time taken

Table 1 Performance metrics for short-term wind forecasting

Parameter	Mathematical expression
SSR	$\sum_{i=1}^{n}(\hat{y}_i - \bar{y})^2$
SSE	$\sum_{i=1}^{n}(y_i - \hat{y}_i)^2$
SST	$\sum_{i=1}^{n}(y_i - \bar{y}_i)^2$
SSE/SST	$\sum_{i=1}^{n}(y_i - \hat{y}_i)^2/(y_i - \bar{y}_i)^2$
SSR/SST	$\sum_{i=1}^{n}(\hat{y}_i - \bar{y})^2/(y_i - \bar{y}_i)^2$

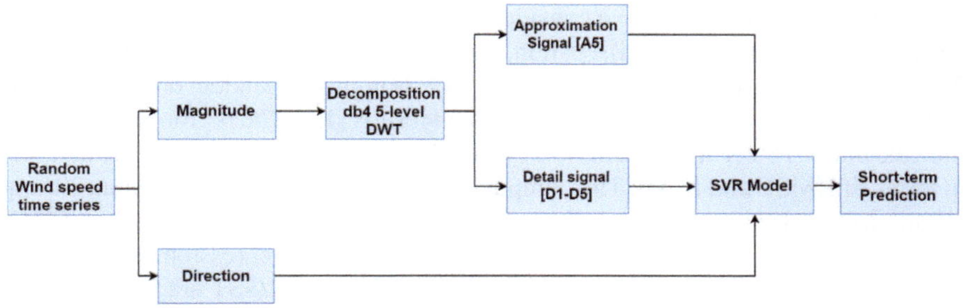

Fig. 2 Block diagram for hybrid wind speed forecasting using wavelet and SVR

by the processor for training and testing the SVR model. CPU time plays an important role in real-time wind forecasting where large training data is involved depending on the prediction horizon. Here, y_i is the ith testing data, $\hat{y}_i$ represents the predicted value of ith test data and $\bar{y}$ is the mean of all testing samples.

Figure 2 shows the flow of the forecasting process and various inputs selected for SVR model. The wind speed time series is fragmented into wind speed magnitude (in m/sec) and direction (in degrees). The original signal is decomposed by DWT into approximate and detail signals. A five-level decomposition is carried out thereby resulting in five detail and approximate signals. The inputs to the SVR include fifth approximate signal (A5), detail signals (D1-D5), and direction time series.

5 Results and Discussion

The present study on wavelet-SVR based forecasting model is applied on four datasets taken from wind farm site in Sotavento, Galicia, Spain. Table 2 shows various datasets taken with their training and testing sample sizes.

Table 2 Different datasets for short-term wind speed forecasting

Dataset	Description	Total size	Training/Testing
D1	Feb. 1, 2018–Feb. 10, 2018	240	160/80
D2	Feb. 1, 2018–Feb. 20, 2018	2880	2800/80
D3	Dec. 1, 2017–Dec. 10, 2017	240	160/80
D4	Dec. 1, 2017–Dec. 20, 2017	2880	2800/80

For our simulations, we have considered a RBF kernel and the values of parameters like C (regularization term), epsilon, and sigma are selected from the set of values $2^i | i = -9, -8, \ldots 10$. The performance of SVR is highly dependent on hyperparameters and proper tuning yields better regression results [12]. The performance indices indicate that ε-TSVR outperforms ε-SVR, LS-SVR and TSVR for respective datasets in terms of RMSE and MAE. The computation time of LS-SVR is minimum among all due to smaller sized optimization problem. Further, the CPU time for ε-TSVR is less than classical ε-SVR, indicating a better choice for regression especially on larger datasets like D2 and D4. The ratios SSE/SST and SSR/SST are found to be in accordance with the results of [7] suggesting ε-TSVR is a better candidate among all. The wavelet-based hybrid wind speed forecasting can be further improved by selecting appropriate decomposition level by using minimum entropy principle of wavelet transform. Figures 3 and 4 shows the short-term wind forecasting results for the datasets D1, D2, D3, and D4 using hybrid wavelet-SVR model (Table 3).

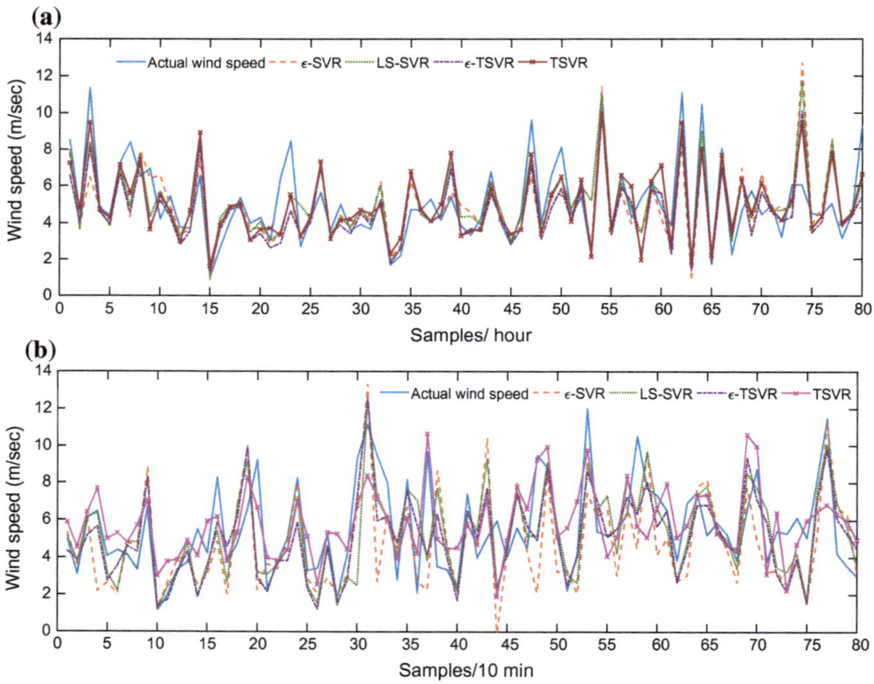

Fig. 3 Short-term wind power forecasting using hybrid wavelet-SVR model **a** Dataset 1 **b** Dataset 2

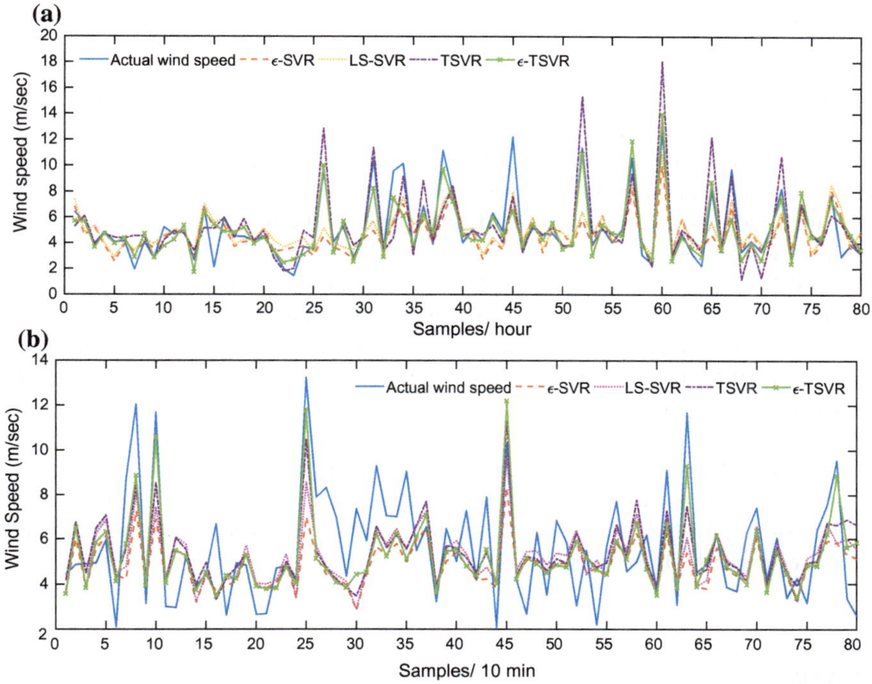

Fig. 4 Short-term wind power forecasting using hybrid wavelet-SVR model **a** Dataset 3 **b** Dataset 4

Table 3 Performance metrics of wavelet- ε-SVR, LS-SVR, TSVR and ε-TSVR on different datasets of Sotavento, Galicia, Spain

Dataset	Model	RMSE (m/sec)	MAE	SSE/SST	SSR/SST	CPU time (s)
D1	ε-SVR	1.7438	100.09	0.6142	0.1568	0.3440
	LS-SVR	1.5461	99.62	0.4828	0.3802	0.0598
	TSVR	1.4271	79.21	0.4128	0.4892	0.3316
	ε-TSVR	1.4200	79.12	0.4073	0.5053	0.1242
D2	ε-SVR	2.5090	150.24	0.9310	0.0225	16.692
	LS-SVR	2.3965	147.97	0.8494	0.1126	1.15004
	TSVR	2.3491	145.69	0.8161	0.3171	16.510
	ε-TSVR	2.2389	137.00	0.7404	0.3491	5.8531
D3	ε-SVR	1.8322	96.53	0.5097	0.3692	0.4377
	LS-SVR	1.5708	87.07	0.3746	0.3788	0.0500
	TSVR	1.3633	75.85	0.2822	0.7541	0.3487
	ε-TSVR	1.1157	57.25	0.1890	0.7629	0.1325
D4	ε-SVR	2.1936	134.74	0.7998	0.2635	21.238
	LS-SVR	2.0402	129.91	0.6919	0.2942	1.4632
	TSVR	1.8835	120.42	0.5897	0.3678	21.005
	ε-TSVR	1.7083	111.48	0.4851	0.5026	7.4469

6 Conclusion

In this paper, we study the short-term wind speed forecasting based on hybrid method consisting wavelet transform and different variants of SVR. The evaluation of the hybrid forecasting method is done on four datasets. A 5-level db4 DWT is carried out on the original wind speed time series and approximation and detail signals along with the direction time series are used as an input to the SVR model. Based on our results, it was observed that wavelet-based ε-TSVR outperformed ε-SVR, LS-SVR, and TSVR for all the datasets as listed in the paper. Further, the computational time for all the models was evaluated and it was found that LS-SVR took least CPU time than all the other models owing to its smaller sized constrained optimization. The ratio SSR/SST gives an estimation of goodness of fit of the prediction model and extracts maximum information from the dataset thereby ε-TSVR is found to be the best regressor among all. Further, a low ratio SSE/SST implies good estimation between actual values and predicted values. The hybrid model can be extended by using wavelet packet transform that considers additional decomposition of the original signal, further SVR hyperparameters can be optimized based on grid search and pattern search algorithms.

References

1. Cadenas, E., Rivera, W.: Wind speed forecasting in three different regions of mexico, using a hybrid ARIMA–ANN model. Renew. Energy **35**(12), 2732–2738 (2010)
2. Catalão, J., Pousinho, H., Mendes, V.: Short-term wind power forecasting in portugal by neural networks and wavelet transform. Renew. Energy **36**(4), 1245–1251 (2011)
3. Esen, H., Inalli, M., Sengur, A., Esen, M.: Forecasting of a ground-coupled heat pump performance using neural networks with statistical data weighting pre-processing. Int. J. Therm. Sci. **47**(4), 431–441 (2008)
4. Liu, D., Niu, D., Wang, H., Fan, L.: Short-term wind speed forecasting using wavelet transform and support vector machines optimized by genetic algorithm. Renew. Energy **62**, 592–597 (2014)
5. Liu, H., qi Tian, H., fu Pan, D., fei Li, Y.: Forecasting models for wind speed using wavelet, wavelet packet, time series and artificial neural networks. Appl. Energy **107**, 191–208 (2013)
6. Peng, X.: TSVR: an efficient twin support vector machine for regression. Neural Netw. **23**(3), 365–372 (2010)
7. Shao, Y.H., Zhang, C.H., Yang, Z.M., Jing, L., Deng, N.Y.: An ε-twin support vector machine for regression. Neural Computi. Appl. **23**(1), 175–185 (2012)
8. Soman, S.S., Zareipour, H., Malik, O., Mandal, P.: A review of wind power and wind speed forecasting methods with different time horizons. In: North American Power Symposium 2010. IEEE, Piscataway (2010)
9. Su, Z., Wang, J., Lu, H., Zhao, G.: A new hybrid model optimized by an intelligent optimization algorithm for wind speed forecasting. Energy Convers. Manage. **85**, 443–452 (2014)
10. Suykens, J., Vandewalle, J.: Least squares support vector machine classifiers. Neural Process. Lett. **9**(3), 293–300 (1999)
11. Vapnik, V.N.: The Nature of Statistical Learning Theory. Springer, New York (2000)

12. Wang, W., Xu, Z., Lu, W., Zhang, X.: Determination of the spread parameter in the gaussian kernel for classification and regression. Neurocomputing **55**(3–4), 643–663 (2003)
13. Xiao, L., Wang, J., Dong, Y., Wu, J.: Combined forecasting models for wind energy forecasting: a case study in china. Renew. Sustain. Energy Rev. **44**, 271–288 (2015)
14. Yuan, X., Tan, Q., Lei, X., Yuan, Y., Wu, X.: Wind power prediction using hybrid autoregressive fractionally integrated moving average and least square support vector machine. Energy **129**, 122–137 (2017)

Exploring Low Power Design Through Performance Analysis of FinFET for Fin Shape Variations

Sangeeta Mangesh, P. K. Chopra, K. K. Saini and Amit Saini

Abstract With concern of global warming, low power design is an important research domain for scientist and engineers. Focusing upon the energy saving trend, this paper compares the performance analysis of all possible fin shapes in a 16 nm Bulk FinFET device from low power design perspective. Performance metrics include transconductance, transconductance generation factor (TGF), on/off current ratio, Subthreshold swing (SS), Drain Induced Barrier Lowering (DIBL) and power consumption. Low power system design feasibility with the optimized round fin shape resulting from the comparative analysis is justified by implementing N and P FinFET devices with perfectly matched VI characteristics. Prospective usage to meet recent developments in system design and control engineering with power optimization and scalability success in round FinFET device is also reviewed through this work.

Keywords SS-Subthreshold swing · DIBL-Drain induced barrier lowering
FOM-Figure of merit

S. Mangesh (✉)
A.P.J. Abdul Kalam Technical University, Lucknow, India
e-mail: sangeetamangesh@gmail.com

P. K. Chopra
Department of Electronics & Communication Engineering,
Ajay Kumar Garg Engineering College, Ghaziabad, India

K. K. Saini
National Physical Laboratories, New Delhi, India

A. Saini
Cadre Design Systems, Delhi, India

D. Deb et al. (eds.), *Innovations in Infrastructure*, Advances in Intelligent Systems and Computing 757, https://doi.org/10.1007/978-981-13-1966-2_46

1 Introduction

Today the semiconductor industry has revolutionized the way of life through innovative applications related to communication, travel, entertainment, energy harnessing, agriculture, infrastructure development bio-medical and bio-technology. The goal is to be able to scale the technology in overall performance [1]. With the scaling of technology, device design innovations are either focusing on new device architectures or use of new materials. We have considered FinFETs for analysis review as they are the promising and most explored device structures that meet the scaling challenges [2]. Unlike independent analysis of fin shapes [3–9], through this review work a comparison of five possible fin shape variations is carried out for exploring low power circuit design and implementation.

Bulk FinFET device is implemented using Cogenda 3D Visual TCAD simulator tool having same physical structure specifications but varying fin shapes. The performance metrics are analyzed from low power design perspective that includes parameters such as transconductance, transconductance generation factor, on/off current ratio, thermal reliability, subthreshold swing etc. that are essential for implementing a low power design [10].

2 Device and Process Specifications

Adhering to the design guidelines, [11–15] a 16 nm FinFET mask layout is implemented using Klayout open source tool. Cogenda GDS2MESH device design tool is used to implement 3D device structure. The device is simulated using Cogenda GENIUS 3D device simulator. We have implemented the device with five fin structures—rectangular, round, triangular and trapezoidal with two different fin angles as indicated in Fig. 1, for performance evaluation.

For the device design, both n and *p*-well doping is maintained at $10^{18}/\text{cm}^3$, source/drain doping at $10^{20}/\text{cm}^3$. The length of spacer between gate and source/drain pad is 5 nm. Though the height of fin is 20 nm, for numerical calculations it is considered to be 10 nm due to diffusion profile within the fin. Similarly, effective channel length is 16 nm even though it is 20 nm according to the mask layout. Both top and side gate oxide thickness is assumed to be the same with effective oxide thickness of less than 2 nm [16].

With drain voltage maintained constant at 0.05 V for linear and 0.5 V for saturation, the device is simulated for varying gate to source voltage in the range of 0–1 V to obtain the drain characteristics. The device and process specifications are tabulated in Table 1.

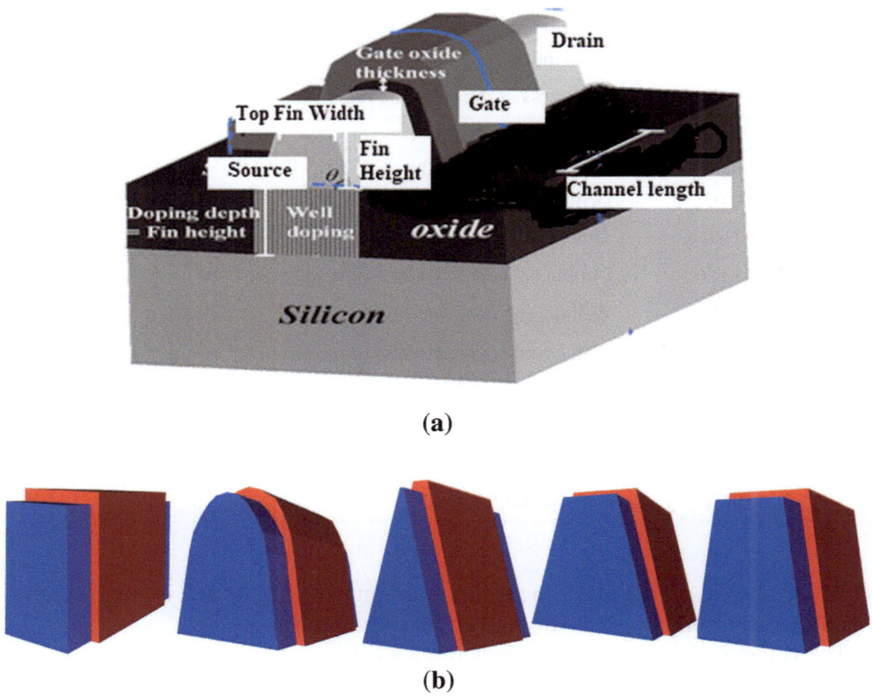

Fig. 1 **a** Bulk Si FinFET device structure. **b** Fin Shapes—Rectangular, Round, Triangular, Trapezoidal-1 and Trapezoidal-2

3 Simulation Details

Quantum corrected or density gradient corrected drift-diffusion (DG-DDM) equation model from Cogenda GENIUS 3D device simulation tool is used for the drain current estimation. The tool estimates current base on moments of the Wigner function equation of motion having carrier equation expressed as [17].

$$\nabla \cdot \epsilon \nabla \psi = -q\left(p - n + N_D^+ - N_A^-\right) \tag{1}$$

where ψ is the electrostatic potential at the vacuum, n and p are the electron and hole concentration N_D^+ and N_A^- are the ionized impurity concentrations. q is the magnitude of the charge of an electron.

The continuity equation for electrons and holes is determined by equations

$$\frac{\delta n}{\delta t} = \frac{1}{q} \cdot \nabla \cdot \overrightarrow{J_n} - (U - G) \text{ and } \frac{\delta n}{\delta t} = -\frac{1}{q} \cdot \nabla \cdot \overrightarrow{J_p} - (U - G) \tag{2}$$

Table 1 FINFET device and process specifications

Parameter	Dimension	Description
Tsub	0.1	Thickness of the substrate region (um)
Hfin	0.02	Height of the fin (um)
Wfin	0.016	Width of the fin (um)
TSTI	0.03	Depth of the STI trench (um)
Tox	1.7e-3	Thickness of the gate oxide (Physical) (um)
Nsub	1e17	Doping concentration in *p*-type substrate (cm^{-3})
Nwel_n	1e18	Well doping concentration (acceptor) for nMOS (cm^{-3})
Nwel_p	1e18	Well doping concentration (donor) for pMOS (cm^{-3})
Nsd_n	1e20	S/D doping concentration (donor) for nMOS (cm^{-3})
Nsd_p	1e20	S/D doping concentration (acceptor) for pMOS (cm^{-3})
ΦM	4.5, 5.1 eV	N-FinFET, P-FinFET gate material work function
K	9	HfO_2 permittivity

where $\overrightarrow{J_n}$ and $\overrightarrow{J_p}$ are the electron and hole current densities, U and G are the recombination and generation rates for both electrons and holes.

The value of electron and hole densities are estimated by expressions

$$\overrightarrow{J_n} = q\mu_n n\nabla(\Psi - \Lambda) + qD_n\Delta_n \tag{3}$$

$$\overrightarrow{J_P} = -q\mu_P p\nabla(\Psi - \Lambda) - qD_p\Delta_p \tag{4}$$

where Λ is quantum corrected potential calculated by

$$\Lambda = -\frac{\gamma\hbar^2\nabla^2\sqrt{n}}{6m\sqrt{n}} \tag{5}$$

The model is sensitive to carrier concentration and can predict more accurate drain current values in the nanoscale device [17].

4 Discussion on Simulation Results

The drain current values for gate voltage variations (shown in Fig. 2) are analyzed to compute the value of subthreshold slope by formula [2] $SS = \frac{\partial V_G}{\log 10(\partial I_D)}$.

The values of transconductance and transconductance generation factor (TGF) is estimated from equation $TGF = \frac{I_D}{g_m}$ [18] (Fig. 3). The new figure of merit (FOM) defined to investigate leakage current is estimated as $\frac{\Delta V_{DIBLSS}}{(I_{dsat}/I_{sdleak})}$ [19]. The values

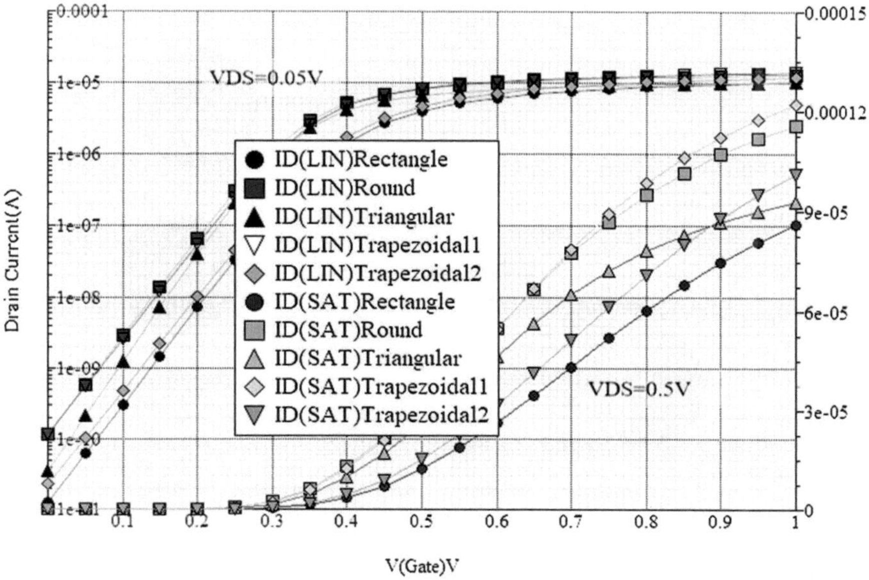

Fig. 2 Drain current variations against gate voltage for VDS = 0.05 V (on log scale) and VDS = 0.5 V (on linear scale)

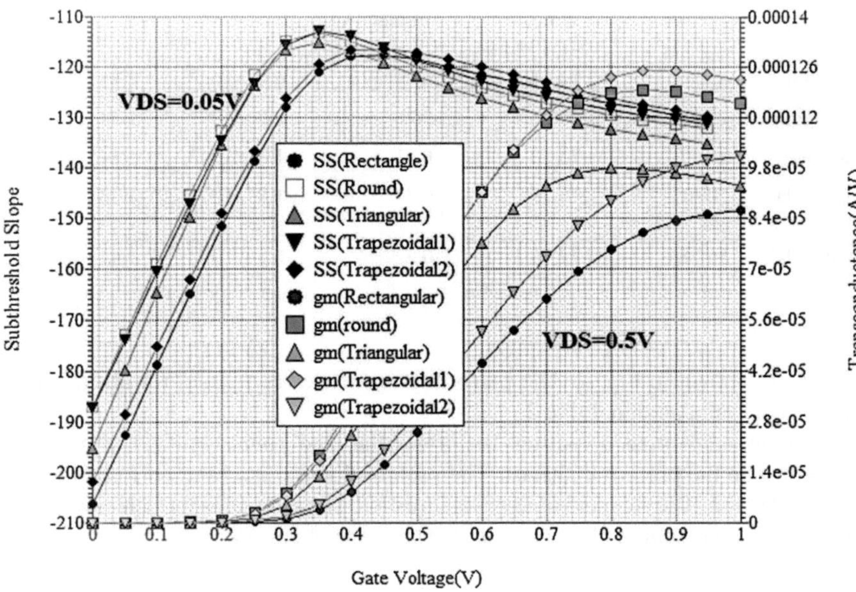

Fig. 3 Subthreshold slope and transconductance against gate voltage

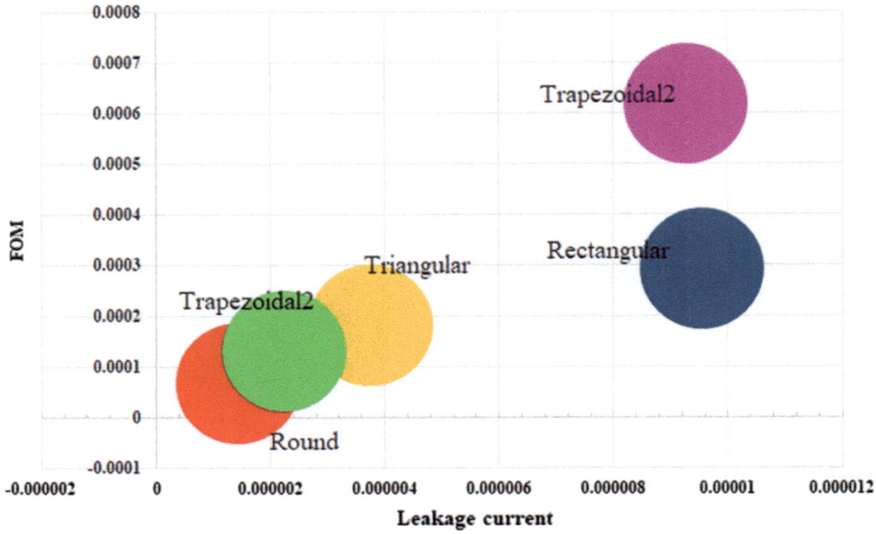

Fig. 4 Figure of merit for devices implemented having all fin shape variations

Table 2 Performance metrics for all the fin shape variations

Fin shape	DIBL mV/V	Subthreshold swing mV/decade	Threshold voltage (V)	FOM	TGF_{AVG}
Round Fin	32.89	73.11	0.35	6.9×10^{-5}	1.55
Rectangular	29.77	75.26	0.28	2.9×10^{-4}	1.51
Triangular	28.77	73.50	0.31	1.3×10^{-4}	1.84
Trapezoidal Fin-1	26.61	75.28	0.34	1.8×10^{-4}	1.43
Trapezoidal Fin-2	36.88	72.16	0.32	6.1×10^{-4}	1.44

of FOM are shown in Fig. 4. The performance metrics computed for all fin shape variations are tabulated in Table 2.

Based on performance metrics the triangular and round FinFET shapes are observed to be the preferred shapes. The preferred shape is observed to be the one with rounded top. The steeper value of subthreshold slope justifies its faster switching ability that is an essential characteristic for high speed design.

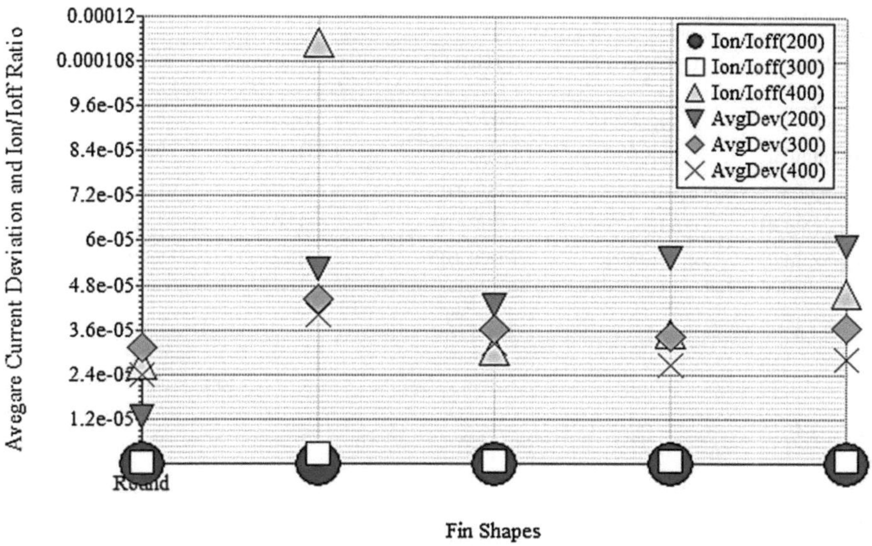

Fig. 5 Impact of temperature on ratio (I_{on}/I_{off}) and Average drain current deviation

4.1 Positive Bias Temperature Impact in N-FinFET(PBTI)

As discussed in [20] the devices are simulated to validate the impact of temperature at 200, 300 and 400 K. The values of on/off current ratio and average current deviation so as to validate their employability in a low power design with reduced voltage swing approach [10, 21]. The round FinFET is observed to have minimum drain current deviation for all the working temperatures as well as it has exhibited second maximum value of current ratio that expresses its leakage behavior (indicated in Fig. 5).

4.2 Internal Capacitances

To estimate the values of internal capacitive effects the device is simulated with signal frequency set at ±0.5 V ($1V_{pp}$). The maximum value obtained for all the internal capacitances gate-substrate, gate-gate, gate-source and gate-drain are tabulated in the Table 3. Gate-substrate capacitance value of rectangular fin shape is minimum value and for gate, gate to drain and gate to source capacitances rectangular, trapezoidal and round all fin shape devices have capacitive effect of the same order. Based on internal capacitances round fin shape has better performance than triangular one. This also justifies employability for high frequency analog applications.

Table 3 Internal capacitances for devices

Fin shape	C (Gate-sub) F	C (Gate) F	C (Gate-source) F	C (Gate-drain) F
Round	8.39E-020	1.085E-017	7.18E-019	7.26E-019
Rectangular	4.19E-022	1.10E-017	7.31E-019	7.38E-019
Triangular	1.87E-019	9.00E-018	6.45E-018	1.45E-017
Trapezoidal1	4.03E-021	1.10E-017	7.25E-019	7.36E-019
Trapezoidal2	4.22E-021	1.09E-017	7.24E-019	7.32E-019

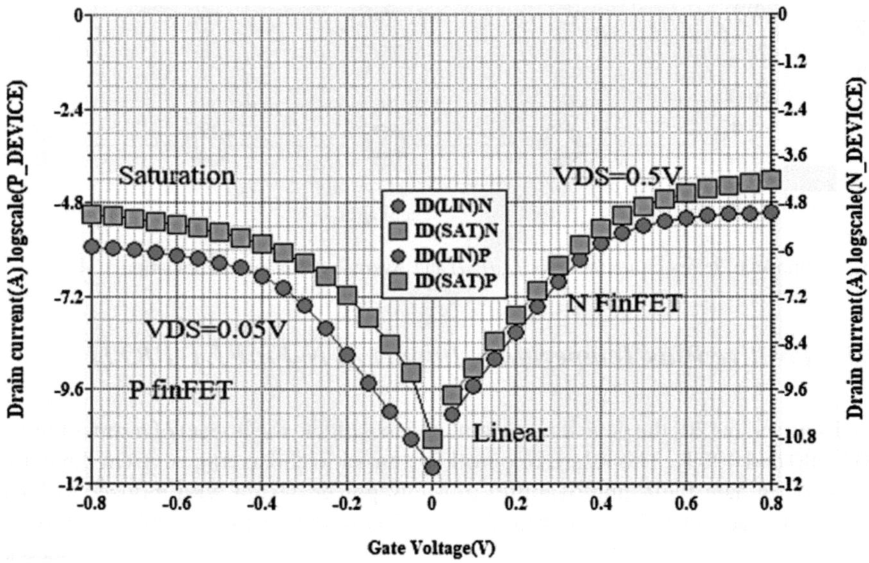

Fig. 6 Matched drain current versus Gate voltage for 16 nm round *n* and *p* FinFETs

4.3 CMOS Implementation and Possibility of Scaling

To explore the low power design feasibility of the round fin device its matched p-FinFET is implemented. The device is simulated for linear as well as saturation region operation and the drain characteristics obtained are plotted in Fig. 6.

To explore performance in the scaled devices the power analysis of round FinFET devices having channel length of 10 and 12 nm is carried out. The ratio $\left(\frac{I_{\text{on}}}{I_{\text{off}}}\right)$ that exhibits leakage current behavior has value in the range of 10^{-5} for 12 nm device and 10^{-4} for 10 nm device indicating reliable device design. The plot of power versus gate voltage in 10 and 12 nm device is shown in Fig. 7.

The device is also simulated without considering quantum effect by using simple energy balanced drift and diffusion modeling of drain current. The value of drain current using classical model is higher than the quantum corrected model justifying the operational reliability and optimized power consumption. Also, a plot of conduction band electron energy Ec(eV) along the channel for both drain current models indicates peak value approximately 1 nm apart indicating reduced inversion layer capacitance due to quantum effect improving resultant conductivity through the device near oxide interface (as indicated in Fig. 8a, b).

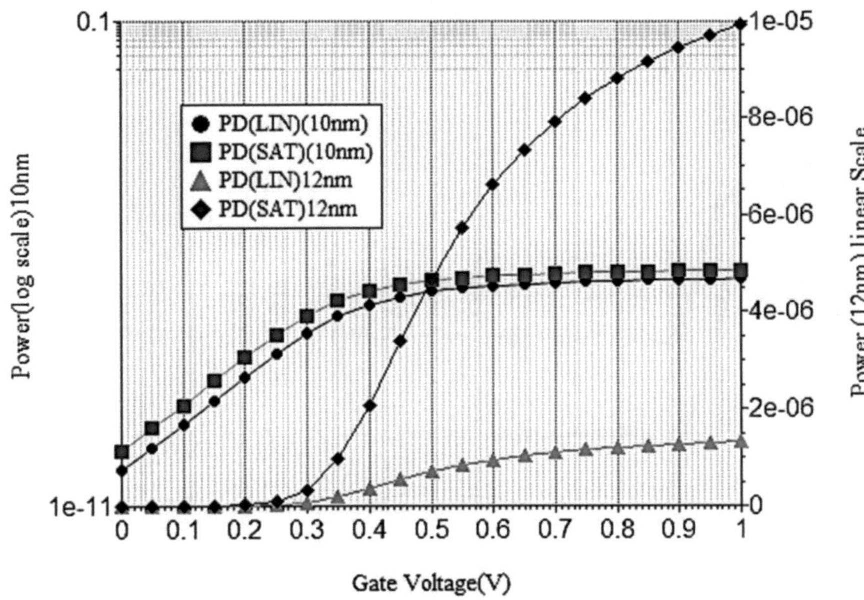

Fig. 7 Round FinFET—Power versus Gate voltage for 10 and 12 nm channel length

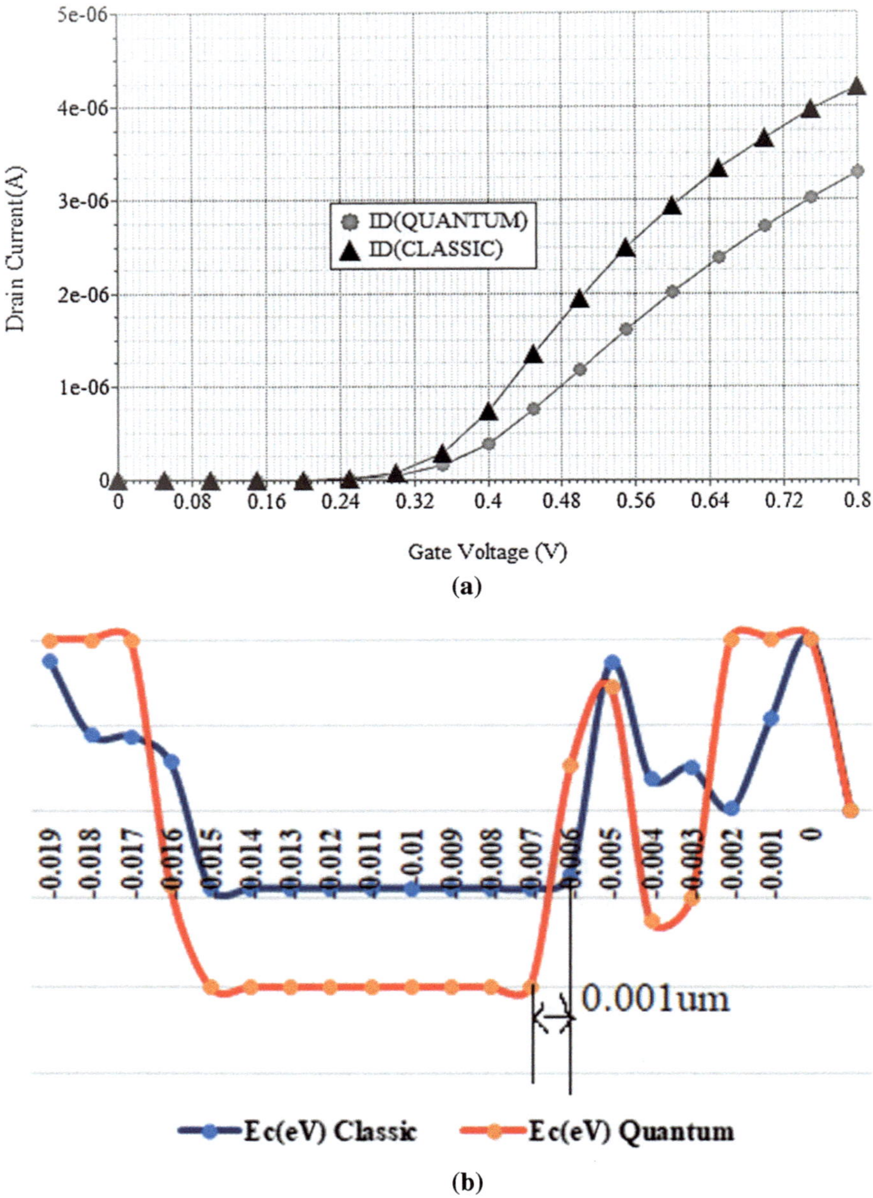

Fig. 8 **a** Drain Current models plotted against Gate Voltage. **b** Electron Energy Ec(eV) for two different drain current models showing difference in the peak values

5 Conclusion

With minimum figure of merit value, maximum threshold voltage and subthreshold swing and drain induced barrier lowering values within moderate range the round FinFET design can be concluded as the best choice. The implementation of low power design of round fin shape device is justified from implementation of matched p-FinFET. The capacitive effects, subthreshold slope and average drain current deviation prove the employability of the device for faster switching applications. Thermally stable and scaled device performance with optimum power demand and drain current sensitivity makes the device suitable for biomedical applications.

In future, for performance improvement, material variations in device design, gate stacking, source-drain resistance improvement by doping variations can also be explored.

References

1. Moore, M.: International Technology Roadmap (2015)
2. Ferain, I., Colinge, C.A., Colinge, J.-P.: Multigate transistors as the future of classical metal–oxide–semiconductor field-effect transistors. Nature **479**, 310–316 (2011)
3. Yi-Bo, Z., Lei, S., Hao, X., Jing-Wen, H., Yi, W., Sheng-Dong, Z.: Comparative study of silicon nanowire transistors with triangular-shaped cross sections. Jpn. J. Appl. Phys. **54**, 04DN01 (2015)
4. Fasarakis, N., Tassis, D.H., Tsormpatzoglou, A., Papathanasiou, K., Dimitriadis, C.A.: Compact modeling of Nano-Scale Trapezoidal Cross-Sectional FinFETs, pp. 13–16. IEEE, Piscataway (2013)
5. Gaynor, B.D., Hassoun, S.: Fin shape impact on FinFET leakage with application to multithreshold and ultralow-leakage FinFET design. IEEE Trans. Electron Devices **61**, 2738–2744 (2014)
6. Nam, H., Shin, C.: Impact of current flow shape in tapered (versus rectangular) FinFET on threshold voltage variation induced by work-function variation. IEEE Trans. Electron Devices **61**, 2007–2011 (2014)
7. Xu, W., Yin, H., Ma, X., Hong, P., Xu, M., Meng, L.: Novel 14-nm Scallop-Shaped FinFETs (S-FinFETs) on Bulk-Si Substrate. Nanoscale Res. Lett. **10**, 249 (2015)
8. Li, Y., Hwang, C.H.: Effect of fin angle on electrical characteristics of nanoscale round-top-gate bulk FinFETs. IEEE Trans. Electron Devices **54**, 3426–3429 (2007)
9. Shukla, S., Gill, S.S., Kaur, N., Jatana, H.S., Nehru, V.: Comparative Simulation Analysis of Process Parameter Variations in 20 nm Triangular FinFET. Active and Passive Electronic Components (2017)
10. Yeap, G.K.: Practical Low Power Digital VLSI Design (1998)
11. Cao, Y.: Predictive Technology Model for Robust Nanoelectronic Design (2011)
12. Boukortt, N., Hadri, B., Patanè, S., Caddemi, A., Crupi, G.: Electrical Characteristics of 8-nm SOI n-FinFETs. Silicon **8**, 497–503 (2016)
13. Bhattacharya, D., Jha, N.K.: FinFETs: From Devices to Architectures. Adv. Electron. **2014**, 1–21 (2014)

14. Mangesh, S., Chopra, P.K., Saini, K.K.: Quantum effect in Nanoscale SOI FINFET device structure: a simulation study. In: Proceedings of 2nd International Conference on 2017 Devices for Integrated Circuit, DevIC 2017 (2017)
15. Vidya, V., Sciences, C.: Thin-Body Silicon FET Devices and Technology. Spring (2007)
16. Chauhan, Y.S., Lu, D.D., Venugopalan, S., Karim, M.A., Niknejad, A., Hu, C.: Compact Models for sub-22 nm MOSFETs 2, pp. 720–725 (2011)
17. Genius Simulator User's Guide from Cogenda
18. Mohapatra, S.K., Pradhan, K.P., Sahu, P.K., Kumar, M.R.: The performance measure of GS-DG MOSFET: An impact of metal gate work function. Adv. Nat. Sci. Nanosci. Nanotechnol. **5** (2014)
19. Eng, Y.C., Hu, L., Chang, T.F., Hsu, S., Chiou, C.M., Wang, T., Yang, C.W., Lin, C.T., Wang, I.C., Chen, M.C., Lai, A., Wang, P.W., Hsu, C.J., Pang, W.Y., Kuo, C.H., Cheng, O., Wang, C.Y.: A new figure of merit, Δ VDIBLSS/(Id, sat/Isdleak), to characterize short-channel performance of a bulk-Si n-channel FinFET device. IEEE J. Electron Devices Soc. **5**, 18–22 (2017)
20. Yeh, W.K., Zhang, W., Yang, Y.L., Dai, A.N., Wu, K., Chou, T.H., Lin, C.L., Gan, K.J., Shih, C.H., Chen, P.Y.: The observation of width quantization impact on device performance and reliability for High-k/metal tri-gate FinFET. IEEE Trans. Device Mater. Reliab. **16**, 610–616 (2016)
21. Rabaey, J.M., Pedram, M.: Low Power Design Methodology (1996)

Evaluation of Response Reduction Factor for Un-reinforced Masonry-Infilled RC Buildings

Nirav Patel and Sandip A. Vasanwala

Abstract The design of earthquake-resistant structures has been a challenge as well as a motivation for all designers. The design philosophy which includes seismic resistant idea by the nonlinear response of the structure via appropriate Response reduction factor (*R*). *R* is a most imperative number which illustrates the potential of the structure to dissipate energy through inelastic behavior. It grants the designer permission to apply a linear elastic force-based design for consideration of nonlinear (NL) behaviour and deformation limits. This *R* factor is unique and different on behalf of the type of structures and design. However, Indian Standards IS 1893:2002 Agarwal and Shrikhande (Earthquake-resistant design of structures [1], 2010) gives the sole value of *R* based on the particular category of detailing and type of structures and design. The NL Static Analysis was used to evaluate *R* for G + 5-story buildings with various infill wall conditions. These findings are noteworthy for Indian seismic design code which, at hand, does not capture into account the deviation in *R*.

Keywords Earthquake-resistant structure · Response reduction factor
Pushover · Infill walls

1 Introduction

Earthquakes, naturally occurring are considered as one of the critical hazards those outcomes in rigorous not only economical but social impact too. A simple impact of a seismic activity has a drastic effect on the not only manmade structures but have major cost on lives too. About, 15,000 individuals dribble every year because of quakes,

N. Patel (✉) · S. A. Vasanwala
SVNIT, Surat, Gujarat, India
e-mail: nppatelnirav@gmail.com

S. A. Vasanwala
e-mail: vsandip18@gmail.com

N. Patel
Navrachana University, Vadodara, India

D. Deb et al. (eds.), *Innovations in Infrastructure*, Advances in Intelligent Systems and Computing 757, https://doi.org/10.1007/978-981-13-1966-2_47

while yearly monetary misfortunes keep running in billions of rupees, taking a real toll on country's economy. Looking at a glance, though natural hazards are inevitable, it is the duty of structural designer to craft earthquake-resistant structures.

Expression used to obtain seismic design forces in codes of almost all countries shows that there is one parameter to serve as calibration factor to incorporate the observed seismic performance of a particular structure-type under actual earthquake shaking. The accuracy of this calibration depends on information gathered from post-earthquake damage survey. ATC-19 [2] and FEMA [3] accept that the calibration factor, specifically named as response reduction factor in current use has a little technical basis. Response reduction factor link elastic level and design level lateral strength of the system. Response reduction factor provides the link between linear behavior and nonlinear behavior. Response reduction factor also provides the basis for comparing analysis and design methodologies based on linear behavior and nonlinear behavior.

I.S 1893 [1] specifies the earthquake design lateral force for accounting a reduction of ($2R$) intended for elastic level forces. This decrease is by virtue of the over quality emerging out of inelastic material conduct enumerating and malleability emerging from the extra hold quality in structure far beyond plan quality. Yet, no formal direction is accessible to assess these commitments. It can be seen that the term ($Z \times I \times \mathrm{Sa}/g$) is based on seismic hazard estimation and not on the structural aspect of the building to be designed. Therefore, the choice of reduction factor has the greatest effect on the desired performance of the structure.

2 Response Reduction Factor

2.1 Concept of Response Reduction Factor

Gravity loads (dead and imposed) are the base for design criteria, but unlike them horizontal loads, as seismic and winds form a different trend of behavioral pattern. Owing to the frequency of loading scenario, design for wind loads is a key requisite. However, in areas of high seismicity, the structures are intended to resist seismic actions also. It is actually uncertain to predict the occurrence of the seismic, so it is a tough call for a structural to decide the performance level so as to withstand the forces. Structures as buildings are usually analyzed for horizontally acting wind loads in a variation of 2–4% of their weight while seismic load might attain 30% to even 40% of the weight of the structure. It may result in enormously intense and costly structures if the same concepts adapted for primary loads considering elastic design are been employed for seismic loads as well. Consequently, the earthquake design philosophy utilizes the concepts of restricted damage and prevention to collapse or failure.

In seismic engineering, it is sought to comprise a control on the location, type, and level of the damage which is shown in Fig. 1, where the linear elastic and nonlinear

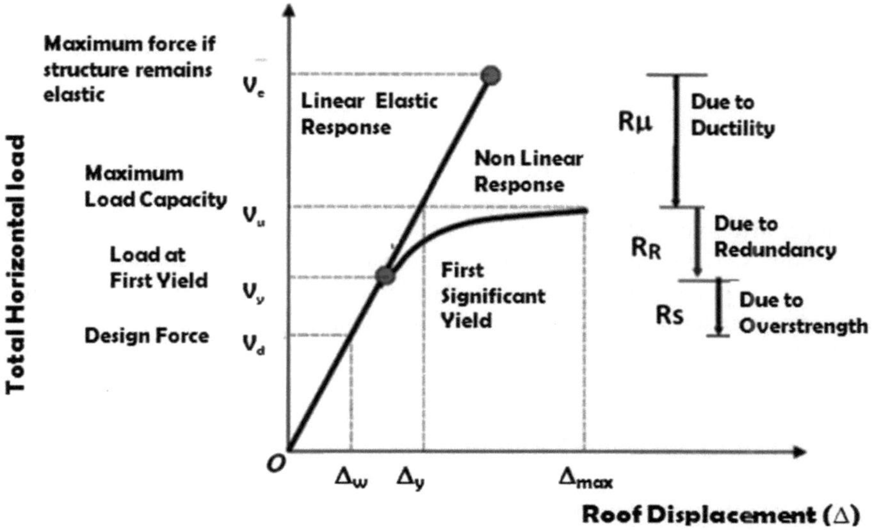

Fig. 1 Concept: response reduction factor [4]

responses along with the concept of equal energy is considered to reduce the force from elastic V_e to design force levels V_d.

As discussed previously for seismic analysis, *R* factor is a crucial design parameter, which incorporates the inelasticity levels expected in structural systems during an earthquake. The explanation throws light on the perceptiveness for the seismic response and the predictable performance of a building in the design earthquake as per prescribed code.

Response Reduction factor incorporates the ability of the structure to dissipate energy through inelastic behavior, which trims down the forces considered for design as well as accounts for factors, viz., ductility, overstrength.

The conventional seismic design procedures incorporate force-based design criteria contrasting to displacement-based in which the structure is designed for a prescribed target displacement rather than a strength level. Therefore, for structures subjected to earthquakes, deformation, which eventually is the chief reason of damage can be controlled during the design. Nonetheless, still implemented is the conventional practice of plunging the seismic forces using *R* factor, to achieve a design force level. So as to rationalize this reduction, seismic codes rely on reserve strength and ductility, which improves the capability of the structure to absorb and dissipate energy [4]. The values assigned to the response modification factor (*R*) of the US codes FEMA [3] are intended to account for both reserve strength and ductility. Various studies mention the redundancy as an individual parameter as well. But here, redundancy is measured as a factor contributing to overstrength, opposing to the suggestion of ATC-19 [2], bifurcating response reduction factor into three parts as overstrength, redundancy, and ductility.

Usually, R is articulated as a gathering of various parameters.

$R = R_s R_\mu R_R R_\xi$ where R_s is strength factor, R_μ ductility factor, R_R Redundancy factor, and R_ξ damping factor.

2.1.1 Overstrength Factor (R_s)

An imperative role of structural overstrength is played in the collapse prevention of the buildings. The strength factor (R_s) is a measure of overstrength achieved by the ratio of the maximum base shear (V_u) related to actual performance of the structure to the design base shear (V_d) as per codal provisions [5].

So, $R_s = V_u / V_d$

2.1.2 Ductility Reduction Factor (R_μ)

R_μ evaluates the overall nonlinear response of a structural system with respect to its plastic deformation capacity. During the past three–four decades, noteworthy work for establishing R_μ due to various types of ground motions based on Single Degree of Freedom (SDOF) systems has been going on.

$R\mu$ incorporates energy dissipating capacity of a well detailed as well as properly designed structures and so, primarily depends on the global ductility demand, μ of the structure denoting μ as the ratio between the maximum roof displacement and yield roof displacement. Large structural ductility permits the structure to move as a mechanism using the maximum potential strength, ensuing dissipating a large amount of energy. The level of inelastic deformation experienced by the structural system subjected to a given ground motion or a lateral loading is given by the displacement ductility ratio.

The equation for R_μ was introduced by Miranda and Bertero [6] derived using detailed study of 124 ground motions recorded on a spectrum of ample soil condition range. The expressions for R_μ are calculated by

$$R_\mu = \frac{\mu - 1}{\phi} + 1$$

where

$$\phi = 1 + \frac{1}{10T - \mu T} - \frac{1}{2}\mathrm{e}^{-1.5(\ln(T)-0.6)^2} \quad \text{for rock sites}$$

$$\phi = 1 + \frac{1}{12T - \mu T} - \frac{1}{5T}\mathrm{e}^{-2(\ln(T)-0.2)^2} \quad \text{for medium soil}$$

$$\phi = 1 + \frac{T_1}{3T} - \frac{3T_1}{4T}\mathrm{e}^{-3\left(\ln\left(\frac{T}{T1}\right)-0.25\right)^2} \quad \text{for soil site}$$

Table 1 Redundancy factors [11]

Lines of vertical seismic framing	Draft redundancy factor
2	0.76
3	0.86
4	1.00

where T_1 is the predominant period of the ground motion corresponding to the period at which the relative velocity of a linear system with damping as 5%, maximum within the entire period ranges [7].

2.1.3 Redundancy Factor (R_R)

A redundant seismic framing system ought to be compiled of several vertical lines of frame, every designed to convey seismic-induced forces to the base of the structure. In these systems, the horizontal load is distributed to different frames based on characteristics of its lateral stiffness and strength for an individual frame. ATC-34 [8] recognized redundancy as one of the three elements of the 'R' factor and introduced redundancy effects into building design. In their study, the number of vertical lines of moment frames was used to measure the redundancy of a building. There are a variety of other codes too for adopting an R_R. The values shown in Table 1 are projected for design professionals and researchers to encourage them.

2.1.4 Damping Factor (R_ξ)

R_ξ considers the effect of supplementary viscous damping for structures with additional energy dissipating devices. If no such devices are provided R_ξ is usually considered as 1.0 and is excluded from the determination of the R factor.

3 Description and Modeling of the Structural Systems Considered

In this current study, SAP2000, FEM-based software as a tool has been used to model RC ductile frames to evaluate response reduction factor. The structure frames are analyzed and designed as per provision of I.S codes following both I.S:1893:2000 [1] and I.S 13920 [9]. Considering 5 models: one symmetric, one having infill wall at the outer core, another at both inner at outer walls and two having soft stories at different heights. The general information regarding the buildings and preliminary design consideration is tabulated in tables as below. The structure is having 3 bays in both directions having span 5 m and stories of story height 3.2, adapting M25

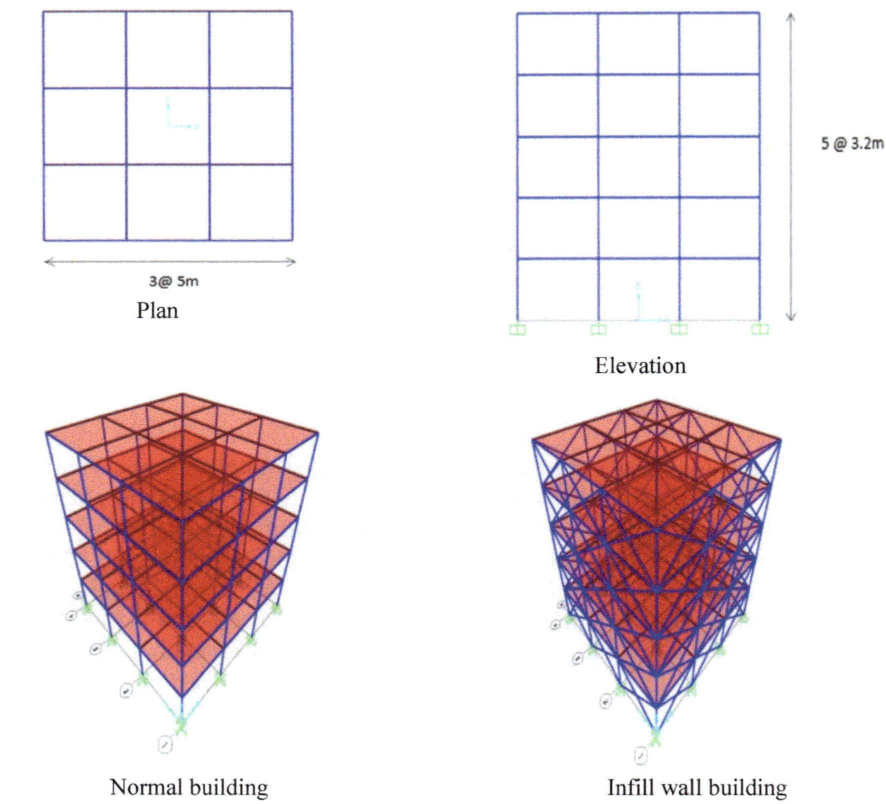

Fig. 2 View of 5-story normal and infilled wall building

as concrete grade and for reinforcement Fe 415. The sizes of Column elements as 0.5 m × 0.5 m and beams as 0.3 m × 0.45 m, considering medium soil, importance factor as 1 and seismic zone *V*. The outer wall thickness considered is 0.23 m while internals are 0.115 m (Fig. 2).

SAP 2000 was used as a software tool for nonlinear (NL) static pushover (PO) analysis of the buildings. Discrete plastic hinges at the two ends of each element were assigned to the modeled frame elements for columns and beams beyond the elastic limit. As a result, M3 hinges to beams and P-M2-M3 hinges to columns have been assigned. SAP 2000 determines plastic hinge properties as provisions given in FEMA 356 [3]. Apart from it, default hinge properties can be provided based on ATC-40 [10].

For masonry walls, considered as diagonal struts, user-defined hinges were assigned. In the study, axial hinges at the center of struts were used.

3.1 Estimation of Performance Point

Safety level, which means sizeable damage occurring to the structure intending of reducing a sizeable amount of its original stiffness. Conversely, prior collapse, a considerable margin for supplementary horizontal deformation is also residual. The NL Static PO analysis is performed to examine the performance point of the building frame considering displacement and base shear via various load combinations. Subsequent to PO analysis, the capacity, and demand curves are derived to obtain the performance point of the structure as per ATC 40 [10] capacity spectrum method.

3.2 Response Reduction Factor (R) Computation

The R value is controlled by the performance limit which is not provided in IS 1893 directly but is defined in PBSD guidelines differently, like ATC-40 [10] and FEMA-356 [3]. These codes have defined the performance limits with a trivial deviation. The computation for R factor is as per ATC-19 [2] for 5-story Reinforced Concrete normal building. The values of the various components of R factor and its component factor for different story buildings are calculated as shown below and tabulated including a sample calculation for the same.

$$\begin{aligned} R &= R_s \times R_\mu \times R_R \times R_D \\ &= 1.658 \times 2.983 \times 1 \times 1 \\ &= 4.952 \end{aligned}$$

4 Analysis of RC Buildings Considered

The building was analyzed and the following factors are drawn out using data tabulated in Table 2:

1. For over strength factors (R_s), it can be seen that for infill wall buildings it is more than that of bare-framed buildings. The value for bare-framed building was found to be 1.658, whereas that of the infill walls at the inner and outer faces was 7.395, which is 4.46 times more than that of simple bare-framed building. The building with infill walls at the outer faces had R_s of 6.47, which is 3.9 times more than that of simple bare-framed building.
2. It was found that the ductility factor (R_μ) of bare-framed building was the highest. R_μ of bare-framed building was 2.983, whereas, the infill wall at the inner and outer faces was only 1.014, which is less than a third of a simple bare-framed building. The building with infill wall at the outer faces had R_μ 1.059, which is 2.82 times that of the simple bare-framed building.

Table 2 Evaluation of R value

Type of building	R_s	R_μ	R_R	R
Normal building	1.658	2.983	1	4.952
Outer faces filled with	6.470	1.059	1	6.850
Outer and inner faces filled	7.395	1.014	1	7.500
Weak story at ground floor	3.324	1.540	1	5.120
Soft story at middle floor	4.394	1.379	1	6.060

Table 3 Increase in overstrength factor compared to normal building (%)

Type of building	% increase in R_s
Infill wall at outer faces	290.23
Infill wall at inner and outer faces	346.20
Soft story at ground floor	100.50
Soft story at middle floor	165.01

Table 4 Decrease in ductility factor compared to normal building (%)

Type of building	% decrease in R_μ
Infill wall at outer faces	64.50
Infill wall at inner and outer faces	66.00
Soft story at ground floor	48.37
Soft story at middle floor	53.77

Table 5 Increase in *R* factor compared to normal building (%)

Type of building	% increase in *R*
Infill wall at outer faces	38.33
Infill wall at inner and outer faces	51.45
Soft story at ground floor	03.40
Soft story at middle floor	22.38

3. The response reduction factor (*R*), infill wall buildings are more than the bare-framed building. The *R* value of bare-framed building showed 4.952 whereas the infill walls at inner and outer faces showed 7.50, which is 1.52 times more than simple bare-framed building. The building with infill wall at outer faces provided an R factor of 6.85, which is 1.38 times more than simple bare-framed building.

If we see the values in terms of percentage, the ratios will give a broader view (Tables 3, 4 and 5).

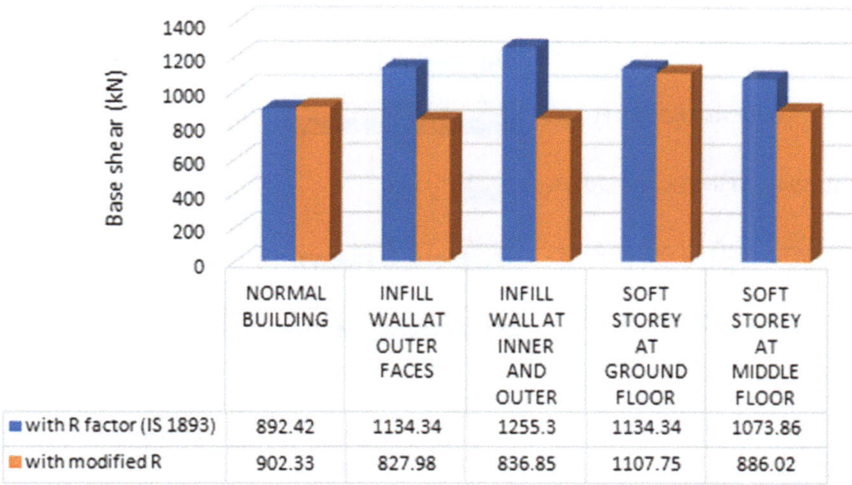

Fig. 3 The effect of *R* factor on base shear

Table 6 Effect of modified *R* value on base shear compared to IS 1893 in (%)

Type of building	% decrease in base shear
Infill wall at outer faces	27.00
Infill wall at inner and outer faces	33.30
Soft story at ground floor	02.35
Soft story at middle floor	53.77

Table 7 Effect of modified *R* value on roof displacement compared to IS 1893 in (%)

Type of building	% decrease in roof displacement
Infill wall at outer faces	35.20
Infill wall at inner and outer faces	99.30
Soft story at ground floor	05.70
Soft story at middle floor	22.26

From Fig. 3, it is clearly understood that in the case of building with infill wall at the inner and outer faces, the base shear got decreased by 33.3% with the usage of modified *R* value, compared the base shear corresponding to the *R* value from IS:1893. Also, in the case of building with infill wall at the outer faces only, the base shear got decreased by 27% with the modified *R* value (Table 6).

Using Fig. 4 and Table 7, it is understood that the maximum roof displacement got reduced by 99.3% in the case of building with infill wall at the inner and outer faces using modified *R* factor, compared the base shear corresponding to the *R* value from IS:1893, whereas, in the case of building with infill wall at the outer faces only the maximum roof displacement got reduced by 35.2%.

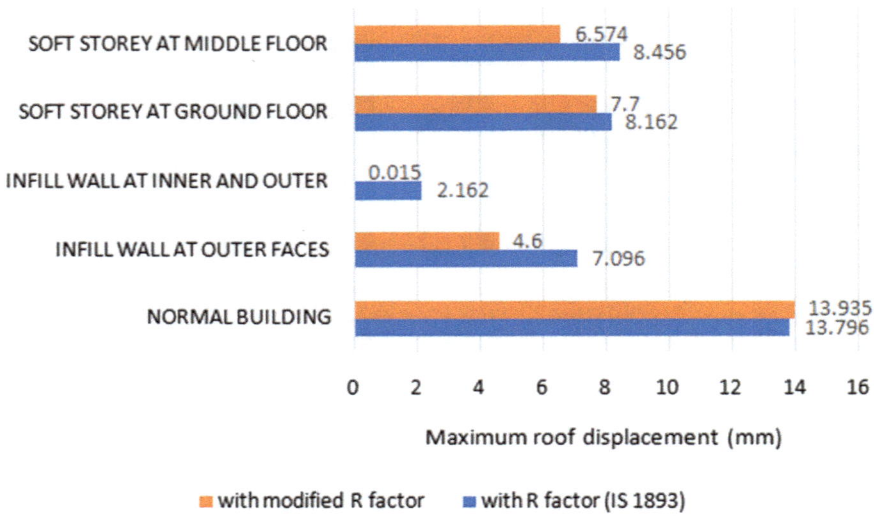

Fig. 4 The effect of *R* factor on maximum roof displacement

5 Conclusions

For the building frames considering different load combinations using pushover analysis and targeting performance by means of point capacity spectrum method following conclusions were drawn:

1. The *R* factor is noticeably having an effect for the un-reinforced masonry infill wall. The stiffness of the building increases by using infills.
2. The dependency of ductility factor is most significant with respect to infill wall. When the infill wall was taken into consideration the ductility factor was found to be reduced.
3. When the infill wall was taken into consideration the overstrength factor was found to be increased.
4. When the infill wall was taken into consideration the maximum roof displacement was found to be decreased.
5. When the infill wall was taken into consideration the IDR was found to be decreased.
6. When the modified R value was used the base shear was found to be decreasing.
7. Estimation of Response reduction factor with exact analysis will help in an economical design. To ensure consistent level of damaged, values of *R* should be according to seismic zone, fundamental period of frame and type of soil. It was observed that consider the normal and infilled 5-story buildings response reduction varies from 4.95 to 7.5 in seismic zone *V*.

8. The *R* factor provided in IS 1893 [1] is a single value, rather should be provided with the equivalent overstrength and ductility factors too, for scrutiny of structural safety.

References

1. I.S 1893 (Part1): Indian Standard Criteria for Earthquake Resistant Design of Structures Part-1 General Provisions and Buildings. Bureau of Indian Standards, New Delhi (2002)
2. ATC-19: Structural Response Modification Factors ATC-19 Report. Applied Technology Council Redwood City California (1995)
3. FEMA Prestandard and commentary for the seismic rehabilitation of building (FEMA-356). Federal Emergency Management Agency, Washington (USA) (2000)
4. Patel, N., Vyas, P.: Evaluation of Response Modification Factor For Moment Resisting Frames. Kalpa Publications in Civil Engineering (2017)
5. Mondal, A., Ghosh, S., Reddy, G.R.: Performance-based evaluation of the response reduction factor for ductile RC frames. Eng. Struct. **56**, 1808–1819 (2015)
6. Miranda, E., Bertero, V.V.: Evaluation of strength reduction factors for earthquake resistant design. Earthq. Spectra **10**(2), 357–359 (1994)
7. Kappos, A.J.: Evaluation of behavior factors on the basis of ductility and over strength studies. Eng. Struct. **21**(1999), 823–835 (1999)
8. ATC-34: A Critical Review of Current Approaches to Earthquake-Resistant Design, Applied Technology Council (1995)
9. IS 13920L: Indian Standard Code of Practice for Ductile Detailing of Reinforced Concrete Structures Subjected to Seismic Forces. Bureau of Indian Standards, New Delhi (1993)
10. ATC 40: Seismic Evaluation and Retrofit of Concrete Buildings, Volume ATC-40 Report, Applied Technology Council (1996)
11. ASCE SEI/ASCE 7-05. Minimum design loads for buildings and other structures. Reston (USA): American Society of Civil Engineers (2005)

Optimal Cycle Time and Payment Option for Retailer

Poonam Mishra, Azharuddin Shaikh and Isha Talati

Abstract In business transaction, it is evident that offering a credit period to the end customers plays an effective role in boosting the demand. In addition, the high risk of default is associated with lengthier credit period. It is also a natural phenomenon that compared to advance payment, the offer of credit period accelerates the sales. Therefore, it is marked that lengthier credit period yields more sales with the risk of defaulters. On the other side, no default risk is associated with the advance payment. We link the above-mentioned scenario in this paper and formulate the mathematical model; the objective is to maximize retailer's profit with payment period (amongst advance period, cash on delivery and credit period) and cycle time as a decision variable. To validate the model numerical examples are presented and sensitivity analysis of parameters is done to examine their effect on retailer's decision.

Keywords Advance payment · Cash on delivery · Permissible delay in payment

1 Introduction

In a supply chain, members of supply chains make use of different payment options to settle their business transactions. Basically they opt for one of the following options: (1) Advance payment in which the customer pays the purchase cost of item well in advance before the delivery of product, (2) Cash on delivery in which, customer pays the purchase cost of item at the time receiving the product, (3) Credit period in which the customer gets a permissible interest free period to pay the purchase cost of item.

A. Shaikh (✉)
Department of Mathematics, Institute of Infrastructure Technology
Research and Management, Ahmedabad, Gujarat 380026, India
e-mail: azharuddin_shaikh7077@yahoo.com

P. Mishra · A. Shaikh · I. Talati
Department of Mathematics, School of Technology, Pandit Deendayal
Petroleum University, Raisan, Gandhinagar 382007, India

D. Deb et al. (eds.), *Innovations in Infrastructure*, Advances in Intelligent Systems and Computing 757, https://doi.org/10.1007/978-981-13-1966-2_48

In case of advance payment, there is no risk of default involved and interest can be earned using the advance payment received. At the same time, advance payment option may lead to a decrease in sales of a product. Several researchers have modelled the phenomenon of advance payment in their research work. It also saves customer time, as to make an advance payment of $60 for 3-month water bills is better than paying $20 monthly. This idea was proposed by Zhang [1]. Mateut and Zanchettin [2] indicated that advance payments are a sign of consumer's trustworthiness. Zhang et al. [3] found that advance payment is beneficial for the seller as to avoid the risk of consumer's order cancellation. Advance payment is also favourable for the customer when the attractive discount is offered while opting for it.

While offering credit period acts a promotional tool to boost the sales. It attracts the customer because it reduces the buying cost, but as a drawback, it involves the default risk. An inventory model with credit period option that charges interest on the unsettled amount at the end of the credit period was proposed by Goyal [4]. Shortages were included by Shah et al. [5] in Goyal's model. Mandal and Phaujdar [6, 7] further extended Goyal's model by adding interest earned on sales revenue. Huang [8] studied that retailer gets benefited when the credit period received from the supplier is passed onto customers. A complete review of trade credit in the inventory model can be found in Shah et al. [9]. Sarkar [10] offered a delay in payments for deteriorating inventory item. Shah et al. [11] presented a model for deteriorating items with a fixed lifetime. Mishra and Shaikh [12] formulated an EOQ model for items with fixed lifetime under two-level trade credit with a fraction of defective items received. An integrated inventory model with two warehouses for stock-dependent demand and trade credit dependent on order size was formulated by Mishra and Shaikh [13]. Mishra and Shaikh [14] also developed EOQ model with deterioration following two-parameter Weibull distribution. Li et al. [15] presented an economic order quantity model to determine optimal payment option.

In the literature, most of the inventory models consider the idea of trade credit as a payment option. On the other hand, inventory models with advance payment are not studied to that extent. Other than this most of the researchers have considered only one of the payment option without considering the other options. So, in this paper we try to formulate the model in which we try to incorporate the following ideas: (1) The demand is hiked when credit period is offered and it reduces when advance payment is required (2) Offering a lengthier credit period increases the sales with a higher default risk (3) Asking for prepayment has no default risk involved and interest is earned from the payment received. Focusing on this, the idea is to find the optimal cycle time and to choose the best payment option which maximizes the retailer's total profit. The remaining part of the paper is as follows: Sect. 2 has the notations and assumption used for mathematical modelling. In Sect. 3 retailer's profit function has been formulated for all three payment options. Section 4 includes the numerical example and sensitivity of inventory parameters followed by a conclusion in Sect. 5.

2 Notation and Assumptions

To formulate the mathematical model, we will use the following notations:

C_r Retailer's unit purchase cost
h_r Holding cost per unit per annum
A_r Ordering cost per order
I_e Interest rate per unit time
S Retailer unit selling price
f_0 Fixed transportation cost per order
f_1 Unit transportation cost
d Coefficient of default risk on revenue; $d > 0$

Decision variables

L Payment period in unit time
T Replenishment cycle time

Functions:

It has been observed that offering credit period to the customer plays an important role in boosting the demand. Hence, more the credit period higher the demand rate. In a similar way, earlier the advance payment lesser the demand. To demonstrate this scenario, we consider the demand rate as a function of payment period L. We define the demand function by $D(L) = \alpha + \beta_i L$; where $\alpha > 0$ is scale demand, $L < 0$ is for advance payment with coefficient $\beta_1 > 0$, $L = 0$ is for cash payment and $L > 0$ is for credit payment with coefficient $\beta_2 > 0$. In this paper, for notational convenience $D(L)$ and D are used interchangeably.

Further, it is also obvious that offering a longer credit period involves more risk of defaulters compared to shorter credit period. That is longer the credit higher the chances that customer would not be able to repay the debt. To demonstrate this we consider $F(L) = \left(1 - \mathrm{e}^{-dL}\right)$; The rate of default risk given credit period L, where d is zero or positive constant. Notice that the risk of defaulter is zero when cash payment is to be done (i.e., $F(0) = 0$), on other hand, the default risk increases if the credit period increases (i.e., $F(\infty) = 1$).

Using this assumptions and notation, we next have the mathematical model in which we formulate retailer's profit function. The profit function is to be maximized by determining the best payment option (i.e., advance payment, cash on delivery or offering credit period), optimal cycle time, and ordering quantity.

3 Mathematical Model

Next, our aim is to mathematical model the scenario with respect to payment options which are advance, cash and permissible delay. First, we discuss the case of advance payment and then the other case will follow.

For the case when retailer opt for advance payment (i.e., when $L < 0$).
The revenue generated from sales per cycle time T is

$$SR_1 = SQ \tag{1}$$

where $Q = D \cdot T$,
Similarly, the purchase cost per cycle time T is given by

$$PC = C_r Q \tag{2}$$

Ordering cost per order is

$$OC = A_r \tag{3}$$

The inventory holding cost per cycle time T is

$$HC = h_r \int_0^T Dt dt = \frac{1}{2} h_r DT^2 \tag{4}$$

In advance payment option, the retailer has sales revenue on which he earns interest; which is given by the interest rate I_e times revenue multiplied by the advance payment period $(-L)$ (because $L \leq 0$). Therefore, the interest earned per cycle T is given by

$$IE = -I_e SLDT \tag{5}$$

The transportation cost per cycle is the sum of fixed and unit transportation cost which is given by

$$TC = f_0 + f_1 Q \tag{6}$$

Hence, the profit per unit time T is given by

$$\pi_1(L, T) = \frac{1}{T}(SR_1 - PC - OC - HC + IE - TC) \tag{7}$$

Therefore,

$$\pi_1(L, T) = (S - C_r)D - \frac{A_r}{T} - \frac{h_r DT}{2} - I_e SLD - \frac{f_0}{T} - f_1 D \tag{8}$$

with $L \leq 0$ and $T \geq 0$.

Next, for the credit payment option (i.e., when $L > 0$), the revenue received after default risk is the revenue from sales multiplied by $(1 - F(L))$ as follows:

$$SR_2 = SQ(1 - F(L)) \tag{9}$$

In this case, the retailer offers a credit period (L) to the customers and bares an interest loss. Therefore, the interest loss is given by the interest rate times revenue multiplied by the credit period (i.e., L). Hence, interest loss is

$$IL = I_e SLDT \tag{10}$$

Hence, the profit after default risk per unit time T is given by

$$\pi_2(L, T) = \frac{1}{T}(SR_2 - PC - OC - HC - IL - TC) \tag{11}$$

Therefore,

$$\pi_2(L, T) = SDe^{-dL} - C_r D - \frac{A_r}{T} - \frac{h_r DT}{2} - I_e SLD - \frac{f_0}{T} - f_1 D \tag{12}$$

with $L \geq 0$ and $T \geq 0$.

Lastly, to discuss the case of cash on delivery we substitute ($L = 0$) in (8) and (12). On simplification of these two equations we have

$$\pi_1(0, T) = \pi_2(0, T) = (S - C_r)\alpha - \frac{A_r}{T} - \frac{h_r \alpha T}{2} - \frac{f_0}{T} - f_1\alpha \tag{13}$$

Therefore the retailer's total profit is given by combining (8), (12) and (13)

So we have

$$\pi(L, T) = \begin{cases} \pi_1(L, T) & L < 0; \\ \pi_1(L, T) = \pi_2(L, T) & L = 0; \\ \pi_2(L, T) & L > 0; \end{cases} \tag{14}$$

Next, our aim is to find the optimal solution (L^*, T^*) of (14) which yields a maximum profit for retailer. That is we need to maximize the following,

Maximize, $\pi_1(L, T)$ Subject to, $L \leq 0$ and $T \geq 0$	Maximize, $\pi_2(L, T)$ Subject to, $L \geq 0$ and $T \geq 0$

Further to find an optimal solution, we follow this procedure:

Step 1: Assign numerical values to all inventory parameters except for decision variables.

Step 2: Set $\frac{\partial \pi_i}{\partial L} = 0$, $\frac{\partial \pi_i}{\partial T} = 0$ for $i = 1, 2$ and solve simultaneously for L and T under the given constraints.

For the obtained solution, the concavity of profit function is shown in Figs. 1 and 2.

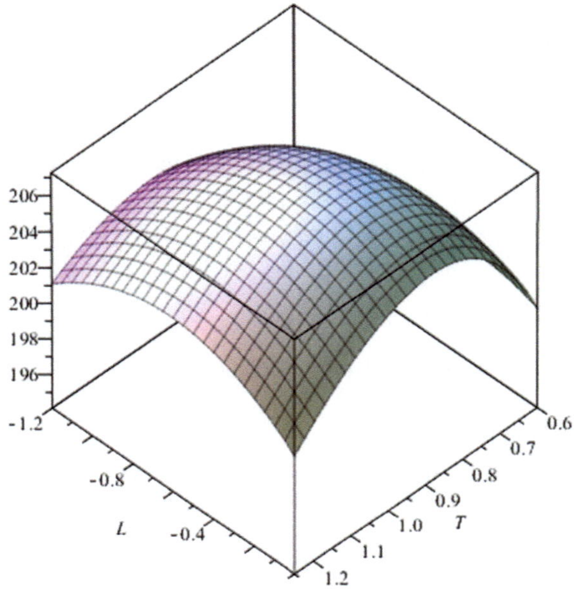

Fig. 1 Concave nature of profit function for advance payment with $\beta_1 = 5$

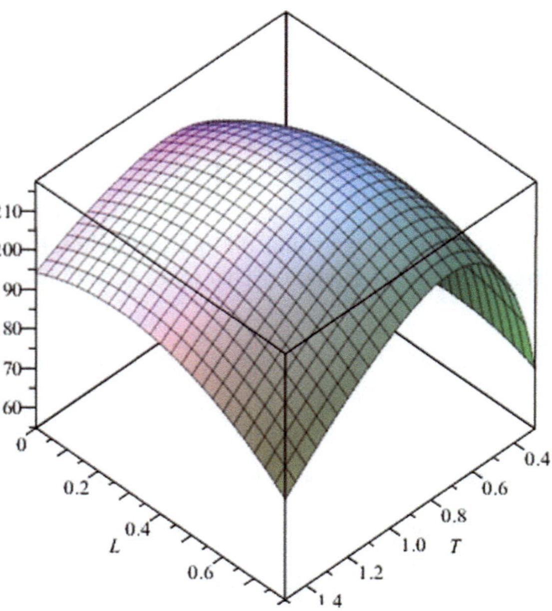

Fig. 2 Concave nature of profit function for the permissible delay with $\beta_2 = 20$

Let (L_1^*, T_1^*) and (L_2^*, T_2^*) be the optimal solution of $\pi_1(L, T)$ and $\pi_2(L, T)$. Then the choice of optimal solution (L^*, T^*) is one which is more profitable to retailer. i.e. the retailer chooses max $\{\pi_1(L_1^*, T_1^*), \pi_2(L_2^*, T_2^*)\}$.

4 Numerical Example

In this section, numerical examples are presented to obtain managerial insights. The interest is to study the effect of business-related parameters on the retailer's decisions and profits. The application of this model can be made by a retailer who has to choose between an advance payment and credit payment. By using historical data, the retailer can approximate the effect of advance payment or credit period on demand, as it is represented by β_1 and β_2. We consider, unit purchasing cost $C_r = 10\$$, the percentage of the default risk $d = 10\%$, the holding cost per unit per unit time $h_r = 3\$$, $\alpha = 30$, the ordering cost per order $A_r = 25\$$, the unit selling price $S = 20\$$, the interest rate $I_e = 8\%$, fixed transportation cost $f_0 = 10\$$ and unit transportation cost $f_1 = 0.5\$$.

Example 1 In this example, we study the effects of β_1 and β_2 on the retailer's decisions and profits where β_1 (or β_2) represents the coefficient of advance payment (or credit payment) on demand. The numerical results are shown in Table 1. Note that the retailer's decision is going to be one which yields more profit. Also, $L^* = 0$ indicates that the retailer's optimal payment option is cash.

The main observations made from Table 1 are; the retailer requires the customer to make advance payment when the value of coefficient of advance payment and credit payment (i.e., β_1 and β_2) is less than 5. When the value of β_1 and β_2 comes in the range of 5–14 retailer opts for cash on delivery option. Lastly, when β_1 and β_2 is more than 14, the retailer offers a credit period to the customer and for this the ordering quantity and profit increases whereas cycle time decreases. As a result, when the value of β_1,β_2 is large the prepayment option reduces the demand and therefore it is gainful for the retailer to offer permissible delay period which in return increases the profit and reduces the cycle time. On the other hand, when the value of β_1, β_2 is small the prepayment option causes a small decrease in demand and therefore it is

Table 1 The effect of β_1 and β_2 on the retailer's decision

β_1	L_1^*	T_1^*	Q_1^*	$\pi_1(L_1^*, T_1^*)$	β_2	L_2^*	T_2^*	Q_2^*	$\pi_2(L_2^*, T_2^*)$
5	−0.46	0.92	25.42	207.27	5	0.00	0.88	26.46	205.63
8	0.00	0.88	26.46	205.63	8	0.00	0.88	26.46	205.63
11	0.00	0.88	26.46	205.63	11	0.00	0.88	26.46	205.63
14	0.00	0.88	26.46	205.63	14	0.07	0.87	26.90	205.86
17	0.00	0.88	26.46	205.63	17	0.28	0.82	28.50	210.01
20	0.00	0.88	26.46	205.63	20	0.43	0.78	30.03	217.62

Table 2 The effect of d on the retailer's decisions when $\beta_1 = 5$ and $\beta_2 = 20$

Parameter		L_1^*	T_1^*	Q_1^*	$\pi_1(L_1^*, T_1^*)$	L_2^*	T_2^*	Q_2^*	$\pi_2(L_2^*, T_2^*)$
d	0.05	−0.46	0.92	25.42	207.27	0.90	0.70	33.47	244.24
	0.1	−0.46	0.92	25.42	207.27	0.43	0.78	30.03	217.62
	0.15	−0.46	0.92	25.42	207.27	0.16	0.84	27.83	207.66
	0.20	−0.46	0.92	25.42	207.27	0.00	0.88	26.46	205.63

Table 3 The effect of interest rate on the retailer's decisions when $\beta_1 = 5$ and $\beta_2 = 20$

Parameter		L_1^*	T_1^*	Q_1^*	$\pi_1(L_1^*, T_1^*)$	L_2^*	T_2^*	Q_2^*	$\pi_2(L_2^*, T_2^*)$
I_e	0.06	0.00	0.88	26.46	205.63	0.60	0.74	31.33	225.93
	0.07	−0.08	0.89	26.27	205.67	0.51	0.76	30.65	221.34
	0.08	−0.46	0.92	25.42	207.27	0.43	0.78	30.03	217.62
	0.09	−0.75	0.94	24.74	210.57	0.36	0.79	29.49	214.60
	0.10	−0.99	0.96	24.18	215.04	0.30	0.80	28.99	212.17

gainful for the retailer to ask the customer for advance payment instead of granting a credit period.

Example 2 In order to study the effect of coefficient of default risk d on retailer's decision and profit. We assume $\beta_1 = 5$ and $\beta_2 = 20$, the reason for this is when $\beta_1 = 5$, $L_1^* < 0$ and when $\beta_2 = 20$, $L_2^* > 0$; The result obtained for different values of d are given in Table 2. It is obvious that default risk d will not affect the prepayment option but, when d increases the cycle time increases; the order quantity, permissible delay period and profit decreases. Moreover, the concavity of profit function can be seen in Figs. 1 and 2.

Example 3 In this case, we study the effect of interest rate on retailer's decision and profit. We assume $\beta_1 = 5$ and $\beta_2 = 20$ and other parameters the same as in example 1. The calculations for different values of interest rate are shown in Table 3. It can be observed that when the rate of interest increases retailers asks for an advance period from buyers for $\beta_1 = 5$ and it yields more profit for him. For $\beta_2 = 20$, retailer bares an interest loss and his profit decreases which is not a favourable deal for him.

In Table 4, sensitivity analysis for other inventory parameters has been done. For this calculation we change one parametric value at a time by −20%, −10%, +10% and +20%. We assume $\beta_1 = 5$, $\beta_2 = 20$, $d = 10\%$ and $I_e = 8\%$ fixed and then calculations are made.

Table 4 Sensitivity of other parameters when $\beta_1 = 5$ and $\beta_2 = 20$

Parameter		L_1^*	T_1^*	Q_1^*	$\pi_1(L_1^*, T_1^*)$	L_2^*	T_2^*	Q_2^*	$\pi_2(L_2^*, T_2^*)$
C_r	8	0.00	0.88	26.46	265.63	0.75	0.72	32.40	301.26
	9	−0.14	0.89	26.15	235.77	0.59	0.75	31.24	257.85
	10	−0.46	0.92	25.42	207.27	0.43	0.78	30.03	217.62
	11	−0.79	0.95	24.66	180.39	0.28	0.81	28.79	180.53
	12	−1.11	0.98	23.87	155.14	0.12	0.85	27.50	146.56
h_r	2.4	−0.41	1.02	28.54	215.34	0.45	0.86	33.75	227.15
	2.7	−0.44	0.97	26.85	211.19	0.44	0.82	31.74	222.25
	3	−0.46	0.92	25.42	207.27	0.43	0.78	30.03	217.62
	3.3	−0.48	0.88	24.19	203.55	0.42	0.74	28.57	213.23
	3.6	−0.50	0.84	23.11	200.00	0.42	0.71	27.29	209.04
α	24	0.00	0.99	23.66	157.01	0.58	0.81	28.85	178.58
	27	−0.17	0.94	24.69	181.43	0.51	0.79	29.45	197.67
	30	−0.46	0.92	25.42	207.27	0.43	0.78	30.03	217.62
	33	−0.75	0.89	26.12	234.60	0.36	0.76	30.60	238.44
	36	−1.04	0.87	26.81	263.43	0.28	0.75	31.16	260.12
A_r	20	−0.43	0.85	23.60	212.93	0.45	0.72	18.16	224.31
	22.5	−0.45	0.88	24.53	210.04	0.44	0.75	28.99	220.90
	25	−0.46	0.92	25.42	207.27	0.43	0.78	30.03	217.62
	27.5	−0.48	0.95	26.27	204.59	0.43	0.81	31.04	214.46
	30	−0.49	0.98	27.10	202.01	0.42	0.83	32.00	211.41
S	16	−1.44	1.01	23.05	98.23	0.00	0.88	26.46	85.63
	18	−0.89	0.96	24.41	151.12	0.23	0.82	28.40	148.64
	20	−0.46	0.92	25.42	207.27	0.43	0.78	30.03	217.62
	22	−0.11	0.89	26.21	265.73	0.60	0.75	31.30	290.92
	24	0.00	0.88	26.46	325.63	0.74	0.72	32.32	367.46
f_0	8	−0.45	0.89	24.71	209.48	0.44	0.75	29.20	220.23
	9	−0.46	0.90	25.07	208.37	0.44	0.77	29.62	218.91
	10	−0.46	0.92	25.42	207.27	0.43	0.78	30.03	217.62
	11	−0.47	0.93	25.77	206.19	0.43	0.79	30.44	216.34
	12	−0.47	0.94	26.11	205.12	0.43	0.80	30.84	215.08
f_1	0.4	−0.43	0.92	25.49	210.05	0.45	0.77	30.16	221.50
	0.45	−0.45	0.92	25.46	208.65	0.44	0.78	30.09	219.55
	0.5	−0.46	0.92	25.42	207.27	0.43	0.78	30.03	217.62
	0.55	−0.48	0.92	25.38	205.89	0.43	0.78	29.97	215.69
	0.6	−0.49	0.92	25.34	204.51	0.42	0.78	29.91	213.77

5 Conclusion

In this paper, we considered the three payment options for the retailer, using this we defined demand as a function of payment period. Offering credit period acts as a promotional tool which involves default risk; while asking for advance payment from customer reduces sale with zero default risk. The objective is to maximize retailer's profit by determining optimal cycle time and payment term amongst the three option. To validate the model numerical examples are presented and sensitivity analysis of parameters is carried out.

The major observations made from this calculations are: When the value of coefficient of advance payment and credit payment is small it is gainful for the retailer to opt for advance payment, on the other hand, when this value is relatively large it is beneficial for retailer to offer a permissible delay period which increases the profit. It is also noted that when the value of the coefficient of default increase retailers profit and credit period offered decreases. Increase in cost price per unit decreases the profit whereas hike in selling price per unit raises the profit.

The proposed inventory model can be extended by incorporating the effect of selling price on demand. One can also consider offering a price discount to the customer when a retailer asks for advance payment option.

References

1. Zhang, A.X.: Optimal advance payment scheme involving fixed pre-payment costs. Omega **24**(5), 577–582 (1996)
2. Mateut, S., Zanchettin, P.: Credit sales and advance payments: substitutes or complements? Econ. Lett. **118**(1), 173–176 (2013)
3. Zhang, Q., Tsao, Y.C., Chen, T.H.: Economic order quantity under advance payment. Appl. Math. Model. **38**(24), 5910–5921 (2014)
4. Goyal, S.K.: Economic order quantity under conditions of permissible delay in payments. J. Oper. Res. Soc. **36**(4), 335–338 (1985)
5. Shah, V.R., Patel, H.C., Shah, Y.K.: Economic ordering quantity when delay in payments of order and shortages are permitted. Gujarat Stat. Rev. **15**(2), 51–56 (1988)
6. Mandal, B.N., Phaujdar, S.: Some EOQ models under permissible delay in payments. Int. J. Manage. Syst. **5**(2), 99–108 (1989)
7. Mandal, B.N., Phaujdar, S.: An inventory model for deteriorating items and stock dependent consumption rate. J. Oper. Res. Soc. **40**(5), 483–488 (1989). https://doi.org/10.1057/jors.1989.75
8. Huang, Y.F.: Optimal retailer's ordering policies in the EOQ model under trade credit financing. J. Oper. Res. Soc. **54**(9), 1011–1015 (2003). https://doi.org/10.1057/palgrave.jors.2601588
9. Shah, N.H., Soni, H., Jaggi, C.K.: Inventory models and trade credit [Review]. Control Cybern. **39**, 867–884 (2010)
10. Sarkar, B.: An EOQ model with delay in payments and time varying deterioration rate. Math. Comput. Model. **55**(3), 367–377 (2012). https://doi.org/10.1016/j.mcm.2011.08.009
11. Shah, N.H., Patel, D.G., Shah, D.B.: Optimal policies for deteriorating items with maximum lifetime and two-level trade credits. Int. J. Math. Math. Sci. **2014**, 1–5 (2014). https://doi.org/10.1155/2014/365929

12. Shaikh, A., Mishra, P.: Optimal ordering policies for retailer with fixed life time defective items under holding cost constraint. J. Basic Appl. Res. Int. **22**(1), 8–16 (2017)
13. Mishra, P., Shaikh, A.: Optimal ordering policy for an integrated inventory model with stock dependent demand and order linked trade credits for twin ware house system. Uncertain Supply Chain Manage. **5**, 169–186 (2017)
14. Shaikh, A.S., Mishra, P.P.: Optimal policies for items with quadratic demand and time-dependent deterioration under two echelon trade credits. In: Shah, N., Mittal, M. (eds.) Handbook of Research on Promoting Business Process Improvement Through Inventory Control Techniques, pp. 32–43. Hershey, PA: IGI Global (2018) https://doi.org/10.4018/978-1-5225-3232-3.ch002
15. Li, R., Skouri, K., Teng, J., Yang, W.: Sellers optimal replenishment policy and payment term among advance, cash, and credit payments. Int. J. Prod. Econ. **197**, 35–42 (2018). https://doi.org/10.1016/j.ijpe.2017.12.015

Attentional Recurrent Neural Networks for Sentence Classification

Ankit Kumar and Reshma Rastogi (nee Khemchandani)

Abstract This paper proposes a novel attention mechanism-based recurrent neural networks for sentence classification. The proposed models are based on attentional implementation of Vanilla Recurrent Neural Network and Bidirectional Recurrent Neural Network architecture over two kinds of recurrent cells: Long Short-Term Memory (LSTM) and Gated Recurrent Unit (GRU). These have been termed as Attentional LSTM (ALSTM), Attentional GRU (AGRU), Attentional Bi-LSTM (ABLSTM), and Attentional Bi-GRU (ABGRU). In order to improve context construction by the network, pretrained word embeddings are used as input for the network. To check the efficacy of proposed framework, we made a comparison of our models with other state-of-the-art methods on six benchmark datasets. The proposed attentional models achieve state-of-the-art result on three datasets and attain performance gain over baseline models on four of the six datasets.

Keywords Sentence classification · Recurrent neural network · LSTM · GRU
Attention mechanism

1 Introduction

Text classification is an emerging research area in Natural Language Processing (NLP). It finds multiple applications in various fields such as medical applications [1], social media analysis [2], financial analysis [3] and many others. Sentence classification is an important domain of text classification that classifies text sentences into different categories depending on their syntactic and semantic structure. Over the past decades, there has been immense growth in the availability of text data which has facilitated as well as complicated the task of sentence classification resulting in

A. Kumar (✉) · R. Rastogi (nee Khemchandani)
Department of Computer Science, South Asian University, New Delhi, India
e-mail: yaduvanshiankit@outlook.com

R. Rastogi (nee Khemchandani)
e-mail: reshma.khemchandani@sau.ac.in

D. Deb et al. (eds.), *Innovations in Infrastructure*, Advances in Intelligent Systems and Computing 757, https://doi.org/10.1007/978-981-13-1966-2_49

the requirement of more structured and scalable ways to deal with this problem. However, the availability of large unstructured social data has resulted into emerging applications related to sentence classification tasks such as opinion mining, review classification, sentence polarity detection, sentiment analysis, question classification among many others.

With the advent of deep learning, NLP has achieved new dimensions. It enables our machines to analyze, understand, and derive meaning out of texts. Most of the text data that we encounter in real life are contextual and hence require taking into account the semantic context in order to be classified. Recently, several important works have been carried out in this direction. Recurrent neural networks are appearing as promising alternative to deal with this challenge.

Recurrent Neural Networks (RNNs) belong to a rich class of dynamic models that have been used in a variety of tasks like language translation [4], image captioning [5], speech recognition [6], intent detection [7] and many others. Theoretically, RNNs are known to preserve long-term dependencies in data. This context or sequential information is preserved in the network's hidden state. However, Vanilla Recurrent Neural Networks are susceptible to vanishing or exploding gradients [8]. To overcome its limitation Long Short-Term Memory units (LSTMs) was introduced by Hoschreiter et al. [9] in 1997 which uses gating mechanism to combat the gradient problem. Since then various variants of LSTMs have been proposed. Each of them is capable of storing long-term temporal dependencies in data and have been used to advance the state of the art for many complex problems including handwriting recognition [10], speech modeling [11], language translation [4] among others. Cho et al. [12] proposed Gated Recurrent Unit (GRU) in 2014 which is a more generalized version of LSTM cell. Both GRUs and LSTMs produce similar results on most tasks as shown by Chung et al. [13] with the only difference being that GRUs are more computationally efficient than LSTMs since GRUs use only two gates (rest and update) while LSTMs use three (input, output, and forget gates). Another variant that improves on RNN is Bidirectional RNN proposed by Schuster et al. [14] in 1997. Recently, Yao et al. [15] showed that Bidirectional LSTMs can learn to extract high-level abstractions of sequential features without requiring any prior knowledge.

Recently, Bahdanau et al. [4] proposed Attention mechanism in 2014 to improve the performance of RNNs. Attention mechanism allows the network to refer back to inputs and move to next state on the basis of it. Another factor that affects the performance of NLP tasks in RNN is the word vector representation. RNNs mainly use one-hot vector as input which suffers from high dimensionality problem with increasing size of vocabulary, along with its inability to produce meaningful results with arithmetic between vectors. Recently Mikolov et al. [16] have delineated an efficient way to revamp the virtue of word vectors in less time. They trained their model on Google News data which consists of 100 billion words and have made it publicly available. We intend to use this word2vec model in our classification task.

In this paper, we propose four novel methods based on vanilla and bidirectional RNN architecture with LSTM and GRU cell units that utilize attention mechanism to classify sentences. These have been termed as Attentional LSTM (ALSTM) and Attentional GRU (AGRU), and Attentional Bi-LSTM and Attentional Bi-GRU

(ABGRU). We are using word2vec model of Mikolov et al. [16] and feeding the resulting feature vectors as input to our model. We are randomly initializing the words that are not present in word2vec model. To check the efficacy of our proposed framework, we have analogized its performance with other state-of-the-art methods on multiple benchmark datasets which confirms our proposed RNN framework's potential to store long-term dependencies in the task of sentence classification.

The rest of the paper is organized as follows. In Sect. 2, we bring in light the other works being carried out in text classification with details of attention mechanism. In Sect. 3 we describe our proposed model. Section 4 incorporates the implementation details, results, and observations made on the basis of the results obtained by our models. We conclude our work with final remarks in Sect. 5.

2 Background

Neural network-based deep learning is gradually improving in the task of text classification with convolution neural networks. Work of Kim [17] on sentence classification, Hughes et al. [1] on medical text classification, Lopez et al. [18] on basics of CNN for text classification are clear shreds of evidence. Recently, RNNs have also been used both in combination with CNN as well as standalone networks for text classification. Work of Lai et al. [19] uses an RNN with CNN while Liu et al. [20] and Kowsari et al. [21] used RNNs alone to classify text.

Our work is inspired by the work of Kim [17] on sentence classification using different variants of CNNs and we use it as our baseline model.

2.1 Attention Mechanism

Bahdanau et al. [4] proposed attention mechanism in 2014. RNN works in a sequential manner and hence takes into account all the outputs of previous cells to produce the output for the current cell. To calculate the output for the current cell it uses the weighted combination of all the input states, not just the last state. Attention weights allow us to decide how much of each previous input should be considered in order to get output for next cell. Thus, it helps in deciding which state's output is more important for the output of the current cell. Mathematically, for a hidden unit h_t, at each time step, attention-based model calculates a context vector c_t as the weighted mean of the state sequence h as follows:

$$c_t = \sum_{j=1}^{T} \alpha_{tj} h_j, \tag{1}$$

where T is the number of time steps, α_{tj} is a computed weight at time step t for hidden state h_j. This context vector is then used to compute a new state sequence s where s_t depends on previous state sequence s_{t-1}, current context c_t and model's output at time $t-1$. The weight α_{tj} is then computed as

$$e_{tj} = a\left(s_{t-1}, h_j\right), \tag{2}$$

$$\alpha_{tj} = \frac{\exp\left(score\left(e_{tj}, h_s\right)\right)}{\sum_{k=1}^{T} \exp\left(e_{tk},\right)}, \tag{3}$$

where a is a learned function that computes a scalar importance value for h_j given the value of h_j and the previous state s_{t-1}.

3 Model

In this section, we discuss our proposed attention mechanism-based implementation of recurrent neural network framework that takes into account the contextual information of the sentence to classify them into different categories. The models in the framework are described below. The key characteristics of our proposed framework can be summarized as follows:

- **Attentional LSTM (ALSTM)**
  The architecture for Attentional LSTM is similar to Vanilla RNN but we have used Graves [22] implementation of LSTM cell in place original tanh function.
- **Attentional GRU (AGRU)**
  Attentional GRU also uses the vanilla RNN architecture with tanh function replaced by GRU cells proposed by Cho et al. [12] in 2014.
- **Attentional Bi-LSTM (ABLSTM)**
  As the name suggests, Attentional Bi-LSTM, uses a bidirectional RNN architecture proposed by Schuster et al. [14] with LSTM cells.
- **Attentional Bi-GRU (ABGRU)**
  Similar to ABLSTM, Attentional Bi-GRU, uses a bidirectional RNN architecture of Schuster et al. [14] with GRU cells instead of LSTM cells.

Let $x_i \in \mathbb{R}^m$ be a m-dimensional word vector corresponding to the ith word in the input sequence. Now, let the length of the input sequence after preprocessing be n, with padding when required. Then, the input sequence fed to our networks can be given as

$$x_{1:n} = x_1 \oplus x_2 \oplus \cdots \oplus x_n, \tag{4}$$

where $\oplus$ is the concatenation operator.

3.1 LSTM Cell

In traditional RNN, the sum of inputs is calculated using tanh whereas an LSTM cell, depicted in Fig. 1a, computes on the sum of inputs with a kth cell memory c_t^k at time t and produces the output as

$$h_t^k = o_t \tanh\left(c_t^k\right), \tag{5}$$

where o_t is the output gate that controls the amount of memory content exhibited in LSTM cell. This output gate is computed as

$$o_t^k = \sigma\left(W_o x_t + U_o h_{t-1} + V_o c_t^k\right)^k, \tag{6}$$

where σ is a sigmoid function, W_o is the input to hidden layer weight, U_o is the hidden layer to hidden layer weight and V_o is the diagonal matrix. At each time step the memory cell c_t^k is updated using following equation:

$$c_t^k = f_t^k c_{t-1}^k + i_t^k \bar{c}_t^k, \tag{7}$$

where $\bar{c}_t^k$ is given as follows

$$\bar{c}_t^k = \tanh\left(W_c x_t + U_c h_{t-1}\right)^k, \tag{8}$$

Forget gate f_t^k controls the amount of existing memory to be forgotten while input gate i_t^k controls the amount of new content to be added. These are evaluated as follows:

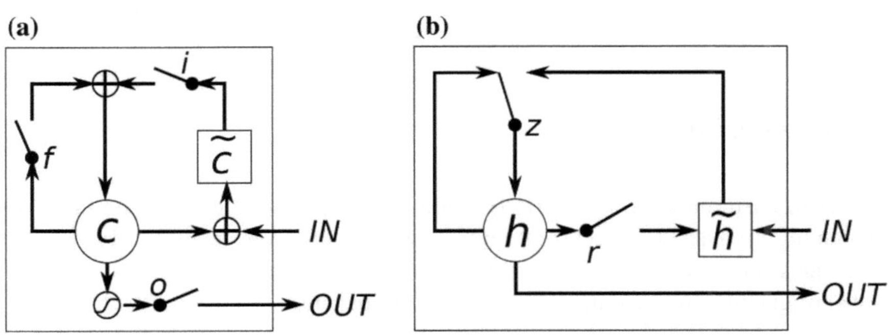

Fig. 1 **a** An LSTM Cell with forget (f), input (i) and output (o) gates, c as memory cell and $\bar{c}$ as new memory cell content (**b**). A GRU cell with reset (r) and update gates (z), and h as the activation and and $\bar{h}$ as the candidate activation

$$f_t^k = \sigma\left(W_f x_t + U_f h_{t-1} + V_f c_{t-1}\right)^k, \tag{9}$$

$$i_t^k = \sigma\left(W_i x_t + U_i h_{t-1} + V_i c_{t-1}\right)^k. \tag{10}$$

Note that V_o, V_f and V_i are diagonal matrices.

3.2 GRU Cell

GRU (Fig. 1b) activation h_t^k for kth recurrent unit at time t is the linear interpolation between previous activation h_{t-1}^k and future candidate activation $\bar{h}_t^k$ and is given as follows:

$$h_t^k = \left(1 - z_t^r\right) h_{t-1}^r + z_t^r \bar{h}_t^k. \tag{11}$$

The update gate z_t^k decides the content to be updated by the unit and can be calculated as

$$z_t^k = \sigma\left(W_z x_t + U_z h_{t-1}\right)^k. \tag{12}$$

Reset gate allows GRU to forget the preceding computations done on input and reads the new input as if it is the first word of the sequence. It is calculated as

$$r_t^k = \sigma\left(W_r x_t + U_r h_{t-1}\right)^k. \tag{13}$$

Here, σ is the sigmoid function, W_r and W_z are the weights from input to hidden layer and, U_r and U_z are the weights from one hidden unit to its next hidden unit in layer.

3.3 Regularization

In order to ensure that proposed frameworks do not produce its result on the basis of some highly over fitted recurrent units, resulting into over generalized system to particular kind of sequence, dropout is applied to the dense recurrent layer d. So, instead of calculating the output o in forward propagation as

$$o = W \cdot d, \tag{14}$$

we calculate it as

$$o = W \cdot (u \odot q), \tag{15}$$

where $q \in \mathbb{R}^m$ is a Bernoulli masking vector and $\odot$ is the element-wise multiplication operator. Note that masked units do not take part in the backpropagation since it acts only on the unmasked units.

4 Experimental Results

4.1 Datasets

We have tested our model on six benchmark datasets namely CR, MPQA, MR, Subj SST-1, and SST-2. A summary of these datasets is given in Table 1.

- **CR**: It consists of customer reviews of various products labeled positive or negative [23].
- **MPQA**: It consists of opinions categorized into negative or positive opinions [24].
- **MR**: MR dataset consists of movie reviews with one sentence per review. Each review is either a positive or a negative review [25].
- **Subj**: A subjective dataset where sentences are labeled as being subjective or objective [26].
- **SST-1**: Stanford Sentiment Treebank dataset. It is an extension of MR but with separate training and testing set. The sentences are labeled very negative, negative, positive, very positive, and neutral. They were re-labeled by Socher et al. [27].
- **SST-2**: SST-2 dataset is similar to SST-1 dataset but with binary levels i.e. positive and negative reviews [27].

4.2 Implementation Details

In the preprocessing of data, we remove the stop words from the input sequence and then obtain a 300-dimensional word vector by using the Mikolov's [16] word2vec

Table 1 Summary of all datasets. Here $Length_{av}$: average length of sentences, $|V|$: Size of vocabulary of dataset, $|V_{pre}|$: Number of words obtained from pretrained model. Test: Size of test data, 10CV: tenfold cross validation (applied whenever test set was not given)

Data	Classes	$Length_{av}$	Dataset-size	$\|V\|$	$\|V_{pre}\|$	Test
CR	2	19	3775	4508	4254	10CV
MPQA	2	3	10,606	6195	5964	10CV
MR	2	20	10,662	19,585	16,303	10CV
Subj	2	23	10,000	22,304	17,799	10CV
SST-1	5	18	18,553	17,790	16,179	2210
SST-2	2	19	9613	16,132	14,737	1821

model available publicly. We randomly initialize the words that are not present in word2vec model. In order for the network to be trained more precisely, we have allowed the word vectors both obtained from word2vec model and randomly initialized to be fine-tuned according to the dataset. For datasets with no explicit test data provided, a random set of test data has been used from the dataset. Also, tenfold cross validation has been used to reduce the effect of randomness. Training is done using Adam [28] optimization algorithm—an extension of stochastic gradient descent, performed over shuffled mini batch.

In addition to these models, we have implemented both the unidirectional as well as bidirectional RNN using LSTM and GRU cells without attention mechanism in order to test the capability of LSTM and GRU cells with both the architectures. We have named them LSTM-Embed and GRU-Embed for unidirectional RNN and, BLSTM and BGRU for bidirectional RNN.

4.3 Hyperparameter Selection

For all the datasets we have used: 70 recurrent units, an embedding size of 300, dense layer with sigmoid activation, categorical cross-entropy as the loss function with the batch size of 35. Apart from this, we have used a dropout rate of 0.03 for regularization. These values were obtained via grid search on all datasets.

4.4 Results

The results for our models against the baseline model of Kim [17] are listed in Table 2. Every model has performed well on all datasets with attentional models outperforming non-attentional models consistently. Also, attentional models, AGRU and ALSTM have performed better than bidirectional attentional models ABGRU and ABLSTM.

Our models have produced better results as compared to the baseline models on CR, MPQA, MR, and Subj datasets while the results on SST-1 and SST-2 datasets are comparable. Three of the results, 82.18% on MR by AGRU, 85.68% on CR by ABGRU and 89.69 on MPQA by AGRU are better than the previous state-of-the-art results achieved by Kim's [17] models.

Table 2 Result of our Attentional RNN models against the baseline CNN models. **CNN-rand**: Baseline model with word vectors initialized randomly and then modified during CNN training. **CNN-static**: Baseline CNN model with pretrained vectors. Both pretrained and randomly initialized words vectors are kept static. **CNN-non-static**: Baseline model, same as CNN-static except that for each task the word vectors are fine-tuned. **CNN-multichannel**: Baseline CNN model with two sets of word vectors. One set is fine-tuned while other is kept static

Model	CR	MPQA	MR	Subj	SST-1	SST-2
CNN-rand	79.8	83.4	76.1	89.6	45.0	82.7
CNN-static	84.7	89.6	81.0	93.0	45.5	86.8
CNN-non-static	84.3	89.5	81.5	93.4	**48.0**	87.2
CNN-multichannel	85.0	89.4	81.1	93.2	47.4	**88.1**
AGRU	84.62	***89.69***	***82.18***	**93.50**	46.74	84.01
ALSTM	84.08	**89.62**	81.16	92.40	46.61	84.07
ABGRU	***85.68***	89.53	81.34	93.30	46.24	83.42
ABLSTM	83.29	89.43	**81.89**	92.90	45.88	84.02
GRU-Embed	82.76	89.23	81.43	93.10	46.54	83.42
LSTM-Embed	84.03	89.34	80.96	91.10	46.65	84.17
BGRU	84.88	89.34	81.24	92.80	45.97	83.91
BLSTM	83.02	89.25	81.19	92.40	46.33	83.80

4.5 Observations

The observations made from above experiments can be summarized as follows:

- Attentional RNNs perform better than the non-attentional RNNs.
- GRU cell-based RNN networks are better at the task of sentence classification than LSTM cells.
- We implemented the word vectors both as static as well as non-static vectors and observed that both non-static, as well as static word vectors, performed well on all datasets with non-static word vectors performing a little better. But they have a drawback too. The training involving non-static word vectors takes more time than static word vectors with the time factor depending on the size of the vocabulary of the dataset. So, it is a trade-off between time and accuracy. Static implementation of word vectors is fast with good performance while non-static implementation is slow with slightly better performance.
- The small difference between the performance of static and non-static vectors demonstrates that Mikolov's pretrained word2vec model is highly efficient and produces universal word vectors.
- With a sequential network like RNN, the performance decreases with the higher values of dropout especially if the dataset is already preprocessed. Smaller dropout rates are preferred for RNN.

5 Conclusion

In this work, we have done a series of experiments with different RNN architectures, using LSTM and GRU cells, built on the top of word2vec model. We observe that with proper tuning of hyperparameters and attention mechanism the proposed AGRU and ALSTM on vanilla RNN architecture and ABGRU and ABLSTM on bidirectional RNN architecture produce remarkable results. Our results confirm that RNN, being a sequential architecture, is good at the task of sentence classification. Similar to Yoon Kim's observation, our work with static and non-static word vectors obtained from Mikolov's [16] model suggests that these word vectors are universal in nature.

References

1. Hughes, M., Li, I., Kotoulas, S., Suzumura, T.: Medical text classification using convolutional neural networks. Stud. Health Technol. Inform. **235**, 246–50 (2017)
2. Jotikabukkana, P., Sornlertlamvanich, V., Manabu, O., Haruechaiyasak, C.: Effectiveness of social media text classification by utilizing the online news category. In: 2nd International Conference on Advanced Informatics: Concepts, Theory and Applications (ICAICTA), pp. 1–5. IEEE, New York (2015)
3. Schumaker, R.P., Chen, H.: Textual analysis of stock market prediction using breaking financial news: the AZFin text system. ACM Trans. Inf. Syst. (TOIS) **27**(2), 12 (2009)
4. Bahdanau, D., Cho, K., Bengio, Y.: Neural machine translation by jointly learning to align and translate. arXiv preprint arXiv:1409.0473 (2014)
5. Vinyals, O., Toshev, A., Bengio, S., Erhan, D.: Show and tell: a neural image caption generator. In: IEEE Conference on Computer Vision and Pattern Recognition (CVPR), pp. 3156–3164. IEEE, New York (2015)
6. Graves, A., Mohamed, A.R., Hinton, G.: Speech recognition with deep recurrent neural networks. In: 2013 IEEE International Conference on Acoustics, Speech and Signal Processing (ICASSP), pp. 6645–6649. IEEE, New York (2013)
7. Liu, B., Lane, I.: Attention-based recurrent neural network models for joint intent detection and slot filling. arXiv preprint arXiv:1609.01454 (2016)
8. Hochreiter, S.: The vanishing gradient problem during learning recurrent neural nets and problem solutions. Int. J. Uncertainty Fuzziness Knowl.-Based Syst. **6**(02), 107–116 (1998)
9. Hochreiter, S., Schmidhuber, J.: Long short-term memory. Neural Comput. **9**(8), 1735–1780 (1997)
10. Graves, A., Schmidhuber, J.: Offline handwriting recognition with multidimensional recurrent neural networks. In: Advances in Neural Information Processing Systems, pp. 545–552 (2009)
11. Graves, A., Jaitly, N., Mohamed, A.R.: Hybrid speech recognition with deep bidirectional LSTM. In: IEEE Workshop on Automatic Speech Recognition and Understanding (ASRU), pp. 273–278. IEEE, New York (2013)
12. Cho, K., Van Merriënboer, B., Gulcehre, C., Bahdanau, D., Bougares, F., Schwenk, H., Bengio, Y.: Learning phrase representations using RNN encoder-decoder for statistical machine translation. arXiv preprint arXiv:1406.1078 (2014)
13. Chung, J., Gulcehre, C., Cho, K., Bengio, Y.: Empirical evaluation of gated recurrent neural networks on sequence modeling. arXiv preprint arXiv:1412.3555 (2014)
14. Schuster, M., Paliwal, K.K.: Bidirectional recurrent neural networks. IEEE Trans. Signal Process. **45**(11), 2673–2681 (1997)
15. Yu, Z., Ramanarayanan, V., Suendermann-Oeft, D., Wang, X., Zechner, K., Chen, L., Qian, Y.: Using bidirectional LSTM recurrent neural networks to learn high-level abstractions of

sequential features for automated scoring of non-native spontaneous speech. In: IEEE Workshop on Automatic Speech Recognition and Understanding (ASRU), pp. 338–345. IEEE, New York (2015)
16. Mikolov, T., Sutskever, I., Chen, K., Corrado, G.S., Dean, J.: Distributed representations of words and phrases and their compositionality. In: Advances in Neural Information Processing Systems, pp. 3111–3119 (2013)
17. Kim, Y.: Convolutional neural networks for sentence classification. arXiv preprint arXiv:1408.5882 (2014)
18. Lopez, M.M., Kalita, J.: Deep Learning applied to NLP. arXiv preprint arXiv:1703.03091 (2017)
19. Lai, S., Xu, L., Liu, K., Zhao, J.: Recurrent convolutional neural networks for text classification. In: AAAI, pp. 333, 2267–2273 (2015)
20. Liu, P., Qiu, X., Huang, X.: Recurrent neural network for text classification with multi-task learning. arXiv preprint arXiv:1605.05101 (2016)
21. Kowsari, K., Brown, D.E., Heidarysafa, M., Meimandi, K.J., Gerber, M.S., Barnes, L.E.: Hdltex: Hierarchical deep learning for text classification. arXiv preprint arXiv:1709.08267 (2017)
22. Graves, A.: Generating sequences with recurrent neural networks. arXiv preprint arXiv:1308.0850 (2013)
23. Hu, M., Liu, B.: Mining and summarizing customer reviews. In: Proceedings of the Tenth ACM SIGKDD International Conference on Knowledge Discovery and Data Mining, pp. 168–177. ACM, New York (2004)
24. Wiebe, J., Wilson, T., Cardie, C.: Annotating expressions of opinions and emotions in language. Lang. Resour. Eval. **39**(2–3), 165–210 (2005)
25. Pang, B., Lee, L.: Seeing stars: exploiting class relationships for sentiment categorization with respect to rating scales. In: Proceedings of the 43rd Annual Meeting on Association for Computational Linguistics, pp. 115–124. Association for Computational Linguistics (2005)
26. Pang, B., Lee, L.: A sentimental education: sentiment analysis using subjectivity summarization based on minimum cuts. In: Proceedings of the 42nd Annual Meeting on Association for Computational Linguistics, p. 271. Association for Computational Linguistics (2004)
27. Socher, R., Perelygin, A., Wu, J., Chuang, J., Manning, C.D., Ng, A., Potts, C.: Recursive deep models for semantic compositionality over a sentiment treebank. In: Proceedings of the 2013 Conference on Empirical Methods in Natural Language Processing, pp. 1631–1642 (2013)
28. Kingma, D.P., Ba, J.: Adam: A method for stochastic optimization. arXiv preprint arXiv:1412.6980 (2014)

Least Squares Twin Extreme Learning Machine for Pattern Classification

Reshma Rastogi (nee Khemchandani) and Amisha Bharti

Abstract In this paper, we have proposed Least Squares Twin Extreme Learning Machine (LSTELM) for pattern classification. The proposed LSTELM formulation solves Extreme Learning Machine (ELM) problem in twin framework. It owns better generalization ability than standard ELM (Huang et al. in IEEE Trans Syst Man Cybern: Part B (Cybern) 42(2):513–529 (2012) [1]) by finding two nonparallel hyperplanes in ELM random feature space which passes through the origin. Unlike Twin Extreme Learning Machine (TELM) (Wan et al. in Neurocomputing 260:235–244 (2017) [2]), it requires less computation time for training the model as proposed LSTELM can be efficiently solved by solving only two systems of linear equations. The proposed LSTELM formulation combines the benefits of the TWSVM (Jayadeva et al. in IEEE Trans Pattern Anal Mach Intell 29(5):905–910 (2007) [3]) and standard ELM (Huang et al. in IEEE Trans Syst Man Cybern: Part B (Cybern) 42(2):513–529 (2012) [1]) in true sense. Experimental results on several UCI benchmark datasets (Bache et al. in UCI Machine Learning Repository (2013) [4]) show that the proposed LSTELM is well generalizable than standard ELM and take less training time compared to TELM model (Wan et al. in Neurocomputing 260:235–244 (2017) [2]).

Keywords Extreme learning machine · Twin support vector machine
Classification · Least squares

1 Introduction

Extreme Learning Machine (ELM) is one of the emerging machine learning algorithms due to its simple implementation, fast learning rate and better generalization ability [1]. It was originally developed for the Single hidden Layer Feed forward Neural networks (SLFNs) in which input weights and hidden layer biases are randomly

R. Rastogi (nee Khemchandani) · A. Bharti (✉)
South Asian University, New Delhi 110021, Delhi, India
e-mail: contactamishabharti@gmail.com

R. Rastogi (nee Khemchandani)
e-mail: reshma.khemchandani@sau.ac.in

D. Deb et al. (eds.), *Innovations in Infrastructure*, Advances in Intelligent Systems and Computing 757, https://doi.org/10.1007/978-981-13-1966-2_50

selected. The output weights of the network are computed analytical and efficiently by minimizing the training errors and norm of output weights. The minimization of the output weights in ELM makes it consistent with Structural Risk Minimization (SRM) principle [5] and lead to generalizing well on unseen test data. ELM has also been extended to the generalized single hidden layer feed forward networks, which may not be neuron alike [6, 7]. Moreover, several attempts have been made to extend the ELM machine to SVM framework for the classification problems [8]. Frénay et al. [9] showed that the ELM learning approach can be applied to SVM directly by replacing SVM kernels with random ELM kernels. Optimization method-based extreme learning machine (OP-ELM) [10] combine benefits of the ELM and SVM by solving the ELM problem using SVM type optimization problem. They have also shown that SVM's maximal margin property and ELM's minimal norm of output weights are actually consistent. OP-ELM [10] finds a separating hyperplane which passes through the origin of the ELM random feature space.

Wan et al. [2] combines the benefits of the ELM and TWSVM [3] by solving the ELM problem in twin framework in their proposed Twin Extreme Learning Machine (TELM) formulation [2], which passes through the origin. TELM finds a pair of nonparallel hyperplanes in ELM random feature space. Similar to traditional TWSVM, TELM also solves a pair of smaller Quadratic Programming problems (QPPs).

As compared to standard ELM, the OP-ELM [10] and TELM [2] formulations results into better generalization ability by solving SVM-type QPPs in ELM random feature space. However with increase in size, training time complexity increases as it solves QPPs. Where complexity is $O(m^3)$. Where m is no. of constraints. Number of constraints in OP-ELM which is feature of datasets solves dual problem.

Taking motivation from above mentioned challenge, we propose Least Squares Twin Extreme Learning Machine (LSTELM) for pattern classification. The proposed LSTELM modifies the primal QPPs of TELM in least squares sense. The primal problem of LSTELM only contains equalities constraints instead of inequality constraints of TELM [2]. It enables us to obtain solution by solving system of linear equation. Like TELM [2], the proposed LSTELM formulation also owns better generalization ability than standard ELM by finding two hyperplanes in ELM random feature space which passes through the origin. However, unlike TELM [2], it does not take much computation time for training the model. In this way, the proposed LSTELM formulation combines the benefits of ELM and TELM [2] in true sense. Experimental results on various UCI benchmark datasets [4] show the efficacy of the proposed method as compared to standard ELM in terms of better prediction performance and reduced training time complexity.

The rest of the paper is organized as follows. Section 2 briefly describe the existing Twin Extreme Learning Machine [2] formulation. Section 3 contains the proposed Least square Twin Extreme learning machine (LSTELM). Section 4 summarizes experimental results which show the efficacy of the proposed LSTELM over existing TELM [2] and Sect. 5 concludes this paper.

2 Twin Extreme Learning Machine

Twin Extreme Learning Machine [2] is a binary classifier that does classification using two nonparallel hyperplanes instead of a single hyperplane as in the case of OP-ELM [10]. The two nonparallel hyperplanes are obtained by solving two QPPs of smaller size compared to a single large QPP being solved by OP-ELM [10].

Let U and V are the data matrices used to denote the hidden layer output of data points belonging from class 1 to -1, respectively, then the TELM [2] seeks the following two nonparallel hyperplanes in $\Re^n$:

$$\begin{aligned} f_1(x) &:= \beta_1 \cdot h(x) = 0, \\ f_2(x) &:= \beta_2 \cdot h(x) = 0, \end{aligned} \tag{1}$$

such that each hyperplane is close to data points belonging to its own class and far away from the data points of other class. A new data point is assigned to class $+1$ or -1 depending upon its proximity to the two nonparallel hyperplanes.

The idea of linear TELM [2] is to solve the following two QPPs [(2) and (3)] with objective function corresponding to one class and constraints corresponding to the other classes.

$$\begin{aligned} \min_{\beta_1,\xi} \quad & \frac{1}{2}\,||U\beta_1||_2^2 + c_1 e_2^T \xi \\ \text{subject to,} \quad & -V\beta_1 + \xi \geq e_2, \;\; \xi \geq 0, \end{aligned} \tag{2}$$

and

$$\begin{aligned} \min_{\beta_2,\eta} \quad & \frac{1}{2}\,||V\beta_2||_2^2 + c_2 e_1^T \eta \\ \text{subject to,} \quad & U\beta_2 + \eta \geq e_1, \;\; \eta \geq 0, \end{aligned} \tag{3}$$

where ξ and η are error vectors with respect to the training pattern belonging to class -1 and class +1, respectively, $c_1, c_2 > 0$ are trade-off parameters, and e_1 and e_2 denote vectors of ones.

The Wolfe dual of primal problem (2) and (3) is obtained as follows:

$$\begin{aligned} \max_{\alpha} \quad & e_2^T \alpha - \frac{1}{2}\alpha^T V(U^T U + \varepsilon I)^{-1} V^T \alpha \\ \text{subject to,} \quad & 0 \leq \alpha_i \leq c_1, \quad i = \; 1, 2, \ldots, m_2, \end{aligned} \tag{4}$$

and,

$$\begin{aligned} \max_{\gamma} \quad & e_2^T \gamma - \frac{1}{2}\gamma^T U(V^T V + \varepsilon I)^{-1} U^T \gamma \\ \text{subject to,} \quad & 0 \leq \gamma_i \leq c_2, \quad i = \; 1, 2, \ldots, m_2, \end{aligned} \tag{5}$$

respectively. After obtaining the optimal value of Lagrange multipliers α and γ from (4) and (5), respectively, the decision variables β_1 and β_2 are computed as

$$\beta_1 = -(U^T U + \varepsilon I)^{-1} V^T \alpha, \tag{6}$$
$$\text{and} \quad \beta_2 = (V^T V + \varepsilon I)^{-1} U^T \gamma, \tag{7}$$

A new data point $x \in R^n$ is assigned to the class $r(r = +1, -1)$ using

$$f(x) = \arg\left(\min_{r=1,2} d_r(x)\right) = \arg\left(\min_{r=1,2} |\beta_r^T h(x)|\right)$$

TELM [2] was also extended to nonlinear case and similarly we get solution as linear TELM.

3 Proposed Least Squares Twin Extreme Learning Machine

3.1 Linear LSTELM

To extend the ELM [1] in twin framework in true sense, the proposed Least Squares Twin Extreme Learning Machine (LSTELM) solves the primal problem of the TELM [2] in least square sense.

For this, it uses the L2 norm of the slack variables in the place of their L1 norm and replaces the inequalities constraints with equality constraints. The proposed LSTELM modified the primal problem of the TELM [2] is as follows

$$\begin{aligned} &\min_{\beta_1, \xi} \ \frac{1}{2} \, ||U\beta_1||_2^2 + \frac{c_1}{2}\xi^T \xi \\ &\text{subject to,} \quad -V\beta_1 + \xi = e_2, \end{aligned} \tag{8}$$

and

$$\begin{aligned} &\min_{\beta_2, \eta} \ \frac{1}{2} \, ||V\beta_2||_2^2 + \frac{c_2}{2}\eta^T \eta \\ &\text{subject to,} \quad U\beta_2 + \eta = e_1, \end{aligned} \tag{9}$$

where U and V are the hidden layer output matrices of the data points belonging to class +1 and -1, respectively.

Replacing ξ in the objective function of the primal problem (8) leads to the following constrained optimization problem:

$$\min_{\beta_1} \frac{1}{2} \, ||U\beta_1||_2^2 + \frac{c_1}{2}(e_2 + V\beta_1)^T (e_2 + V\beta_1) \tag{10}$$

Setting the gradient with respect to β_1 equal to zero gives

$$(U^T U + c_1 V^T V)\beta_1 + c_1 V^T e_2 = 0$$

form which optimal value of β_1 can be obtained as

$$\beta_1 = -c_1(U^T U + c_1 V^T V)^{-1} V^T e_2, \tag{11}$$

On the similar lines, the optimal value of β_2 in the primal problem (9) can be obtained as

$$\beta_2 = c_2(V^T V + c_2 U^T U)^{-1} U^T e_1, \tag{12}$$

Note that the solution of the optimization problems of the proposed LSTELM only requires the inversion of the two $(n \times n)$ matrices, where n is the numbers of training points with respective classes.

After computing the optimal value of the β_1 and β_2 from (11) and (12), respectively, the separating hyperplanes

$$\beta_1 \cdot h(x) = 0 \text{ and } \beta_2 \cdot h(x) = 0 \tag{13}$$

are obtained.

An unseen test point $x \in \Re^n$ is classified into the classes depending upon its proximity with the two classes:

$$f(x) = \arg\min_{r=1,2} d_r(x) = \arg\min_{r=1,2} |\beta_r^T h(x)| \tag{14}$$

where $|\cdot|$ is the perpendicular distance of the data points x from the hyperplane β.

3.2 Nonlinear LSTELM

The proposed LSTELM can also be extended to nonlinear case. Consider the following two kernel generating surfaces:

$$\begin{aligned} K_{ELM}(x^T \cdot C^T)\mu_1 &= 0. \\ K_{ELM}(x^T \cdot C^T)\mu_2 &= 0. \end{aligned} \tag{15}$$

where $C^T = [A,\ B]^T$ and K_{ELM} represent kernel function. We can calculate the ELM kernel function as

$$\begin{aligned} K_{ELM}(x_j, x_i) = h(x_j) \cdot h(x_i) = [G(w_1, b_1, x_j) \ldots G(w_L, b_L, x_j)]^T \\ \cdot [G(w_1, b_1, x_i) \ldots G(w_L, b_L, x_i)] \quad \forall\, i, j = 1, 2, \ldots, m. \end{aligned} \tag{16}$$

For the surface $K_{ELM}(x^T, C^T)\mu_1 = 0$ the primal problem is given by

$$\min_{\beta_1} \frac{1}{2} ||K_{ELM}(A, C^T)\beta_1||_2^2 + \frac{c_1}{2}||(e_2 + K_{ELM}(B, C^T)\beta_1)||^2, \tag{17}$$

and

$$\min_{\beta_2} \frac{1}{2} ||K_{ELM}(B, C^T)\beta_2||_2^2 + \frac{c_2}{2}||(e_1 + K_{ELM}(A, C^T)\beta_2)||^2. \tag{18}$$

Setting the gradient with respect to β_1 and β_2 to zero will give

$$\beta_1 = -c_1(U^T U + c_1 V^T V)^{-1} V^T e_2, \tag{19}$$

and

$$\beta_2 = c_2(V^T V + c_2 U^T U)^{-1} U^T e_1, \tag{20}$$

where $U = K_{ELM}(A, C^T)$ and $V = K_{ELM}(B, C^T)$, respectively.

Once the solution β_1 and β_2 of the problem (19) and (20) are obtained, a new data points x is assigned to the class $r(r = +1, -1)$ in a manner similar to the linear case.

3.3 LSTELM Algorithm

Algorithm LSTELM : Given a datapoint $X = ((x_i, t_i)|x_i \in \Re^d,\ t_i = (+1, -1),\ i = 1, 2, \ldots, N)$, activation function $G(x)$, L are number of hidden nodes.

Step 1: In ELM [1] networks initially generate random input weight w_i and b_i with L hidden nodes in ELM [1] networks.

Step 2 : Construct input matrices A and B for linear LSTELM,

Step 3 : Then calculate the hidden layer output matrices U and V respectively, For nonlinear LSTELM, kernel functions are defined as $U = [K_{ELM}(A, C')]$ and $V = [K_{ELM}(B, C')]$.

Step 4: Select penalty parameters c_1 and c_2.

Step 5: Calculate output weights β_1 and β_2 using (19) and (20).

Step 6: For a next point x, assign a test data point find proximal distance to each class using (14).

3.4 Comparison ELM and LSTELM

1. LSTELM aims to generate two nonparallel separating hyperplanes similar to TELM [2]. However, ELM [1] finds only one separating hyperplane.
2. Traditional ELM aims to makes training errors equal to zero. However, LSTELM trades off between training error and generalization ability of the classifier.
3. LSTELM solves two smaller systems of linear equations instead of a large system of equation of ELM [1]. Thus, the computational time for LSTELM model is lesser when compared to ELM and TELM.

4 Experimental Results

In this section, we show the effectiveness of the proposed algorithm on UCI binary classification datasets. In order to verify the capabilities of our algorithm, we have compared its performance with ELM and TELM.

4.1 Data Set

We have used six binary classification datasets belonging to UCI repository [4]. We have summarized the characteristics of these binary datasets in Table 1. To achieve more reliable test results, our experiments used the fivefold cross-validation techniques [11].

4.2 Experimental Setting

All of the classification models implemented are simulated in MATLAB 2015 with DELL PRECISION processor with 16 GB of RAM.

Table 1 UCI datasets

Dataset	Feature	# training data	# testing data
Pima	9	614	154
Ionosphere	34	281	70
WPBC	33	158	40
Sonar	61	206	52
Heart	14	216	54
Australian	15	552	138

For the proposed LSTELM, TELM [2] and ELM we need to adjust the number of hidden layer nodes (L). The value of L has been tuned from the set $\{1, 2, 3, \ldots, 1000\}$. Experimental results of LSTELM and TELM which is mentioned in table at the same number of hidden nodes. The value of the parameter c in ELM (Huang et al., [1]) has been chosen from the set $\{2^i : i = -10, -9, \ldots, 9, 10\}$. For the sake of experimental convenience, we have taken $c_1 = c_2$ in the proposed LSTELM and TELM [2]. The value of the parameter c_1 in proposed LSTELM and TELM has been chosen from the set $\{2^i : i = -10, -9, \ldots, 9, 10\}$.

In this section, we shall present the experimental results which will show the efficacy of the proposed LSTELM. We have chosen to compare the proposed LSTELM over standard ELM [1] and TELM [2] on UCI datasets. Each of the dataset has first normalized in the range $[-1, 1]$ and then converted to zero mean dataset.

4.3 Experimental Results

In this subsection, we have compared the accuracy of LSTELM with TELM and ELM and other baselines method in terms of prediction accuracy and computation time.

We first experiment on the binary datasets with linear and *tanh* kernel functions. The experimental results are shown in Tables 2 and 3, respectively, in which the best results are marked in bold.

We can see that LSTELM algorithm achieved much better performance on most tasks compared to TELM [2] and ELM [1] algorithms. Linear kernel function LSTELM perform better than TELM [2] and ELM [1] in three data sets, For *tanh* kernel, LSTELM is better than TELM [2] on six datasets and LSTELM also perform better than ELM on five datasets. LSTELM algorithm performed with linear kernel on Ionosphere, WPBC, and Australian perform better than TELM and ELM, with nonlinear kernel LSTELM algorithm performed on Ionosphere, WPBC, Sonar, Heart, and Australian better than TELM and ELM.

Table 2 Comparison (accuracy ± standard deviation and time(sec)) between LSTELM, TELM and TELM with linear kernel

Dataset	LSTELM		TELM		ELM	Time (s)
	Accuracy	Time (s)	Accuracy	Time (s)	Accuracy	
Pima	73.82 ± 3.10	0.39	78.13 ± 4.75	2.10	**78 ± 5.49**	0.46
Ionosphere	**89.17 ± 1.61**	0.07	87.21 ± 8.01	0.61	85.75 ± 3.37	0.17
WPBC	**79.78 ± 4.5**	0.06	76.85 ± 1.67	0.44	76.75 ± 4.22	0.11
Sonar	75.48 ± 2.60	0.08	76.28 ± 5.07	0.40	**77.41 ± 2.56**	0.08
Heart	82.22 ± 4.41	0.09	**84.44 ± 1.65**	0.50	84.07 ± 5.08	0.44
Australian	**87.53 ± 1.65**	0.32	86.66 ± 1.50	1.80	86.08 ± 1.19	0.34

Table 3 Comparison (accuracy ± standard deviation and time (sec)) between LSTELM, TELM and ELM with nonlinear kernel

Dataset	LSTELM		TELM		ELM	Time (s)
	Accuracy	Time (s)	Accuracy	Time (s)	Accuracy	
Pima	73.05 ± 2.88	0.97	74.99 ± 2.78	2.91	**76.50 ± 3.47**	0.72
Ionosphere	**89.16 ± 5.11**	0.30	87.75 ± 3.19	0.66	82.03 ± 4.85	1.07
WPBC	**81.32 ± 4.85**	0.16	80.78 ± 4.72	0.51	75.74 ± 2.98	0.15
Sonar	**79.3 ± 3.44**	0.12	77.86 ± 4.41	0.50	67.84 ± 3.61	3.55
Heart	**85.18 ± 6.01**	0.12	83.33 ± 5.40	0.56	83.33 ± 6.30	0.47
Australian	**87.53 ± 2.53**	0.35	86.52 ± 1.95	1.06	85.21 ± 2.64	0.80

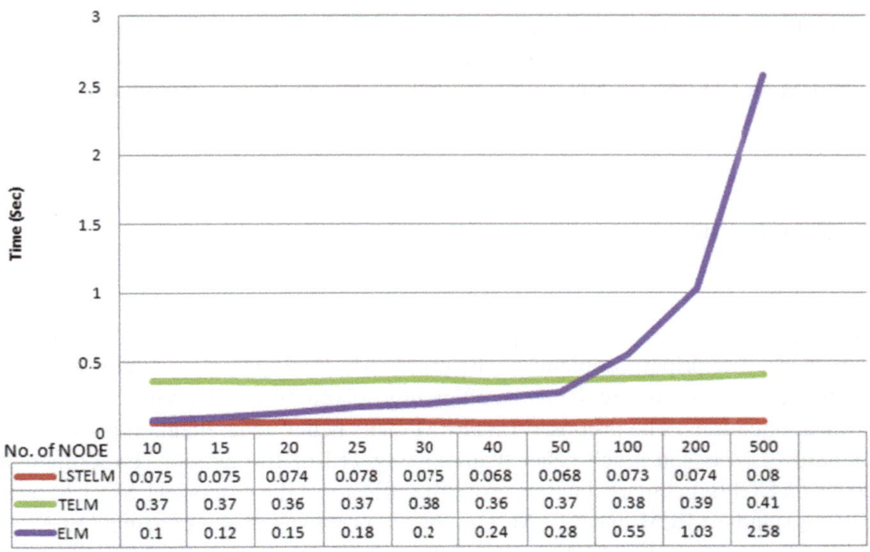

Fig. 1 Time versus no. of nodes comparison graph with Heart UCI benchmark dataset

In order to evaluate the computational efficiency of the proposed algorithm, the training time and accuracy of LSTELM, TELM, and ELM are also listed in (Tables 2 and 3) LSTELM is the most efficient algorithm as its training time is lower than ELM [1] and TELM [2]. It is to be noted that LSTELM and ELM [1] solves least squares problem, where TELM [2] solves QPPs.

Therefore, of TELM computational time is more than ELM [1] and LSTELM. LSTELM solves two small size systems of equations but ELM solve one larger system of equations. Thus, LSTELM is much faster than ELM.

From Fig. 1 this is clear that ELM computation time is dependent on hidden nodes, i.e., if the number of hidden nodes will increase then computation time will also increase, whereas LSTELM and TELM [2] computational time is independent to

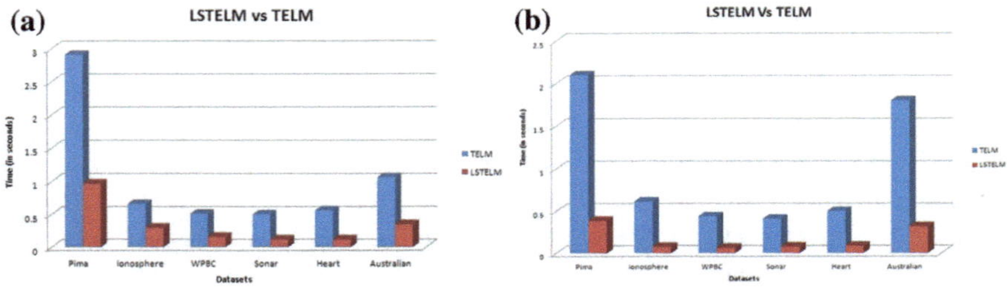

Fig. 2 Training time comparison LSTELM and TELM graph with **a** linear kernel and **b** nonlinear kernel

the hidden nodes. So even if number of node increases then computational time will not increases similarly it holds for TELM. The training time comparison between LSTELM and TELM for both linear as well as nonlinear case is summarized up in Fig. 2.

5 Conclusion

This paper proposed Least Squares Twin Extreme Learning Machine (LSTELM) for binary data classification. The proposed LSTELM solves ELM problem in twin framework. The primal problems of the proposed LSTELM only contains equality constraints only. Therefore, it can be efficiently solved by solving only two systems of linear equations. The proposed LSTELM owns better generalization ability than standard ELM is and much faster than existing TELM formulation. Experimental results on UCI Benchmark datasets prove the efficacy of the proposed LSTELM over existing standard ELM and TELM. One of the problems of the existing TELM as well as the proposed LSTELM is that they are not sparse in nature. It results in an increase in training and testing time. In future, we aim to develop the sparse ELM classification model in twin framework.

References

1. Huang, G.B., Zhou, H., Ding, X., Zhang, R.: Extreme learning machine for regression and multiclass classification. IEEE Trans. Syst. Man Cybern.: Part B (Cybern.) **42**(2), 513–529 (2012)
2. Wan, Y., Song, S., Huang, G., Li, S.: Twin extreme learning machines for pattern classification. Neurocomputing **260**, 235–244 (2017)
3. Jayadeva, Khemchandani, R., Chandra, S.: Twin support vector machines for pattern classification. IEEE Trans. Pattern Anal. Mach. Intell. **29**(5), 905–910 (2007)
4. Bache, K., Lichman, M.: UCI Machine Learning Repository. http://archive.ics.uci.edu/ml (2013)

5. Vapnik, V.N.: An overview of statistical learning theory. IEEE Trans. Neural Networks **10**(5), 988–999 (1999)
6. Huang, G.B., Chen, L.: Convex incremental extreme learning machine. Neurocomputing **70**(16–18), 3056–3062 (2007)
7. Huang, G.B., Chen, L.: Enhanced random search based incremental extreme learning machine. Neurocomputing **71**(16–18), 3460–3468 (2008)
8. Liu, Q., He, Q., Shi, Z.: Extreme support vector machine classifier. In: Pacific-Asia Conference on Knowledge Discovery and Data Mining, pp. 222–233. Springer, Berlin (2008)
9. Frénay, B., Verleysen, M.: Using SVMs with randomised feature spaces: An extreme learning approach. In: ESANN (2010)
10. Huang, G.B., Ding, X., Zhou, H.: Optimization method based extreme learning machine for classification. Neurocomputing **74**(1–3), 155–163 (2010)
11. Kohavi, R.: A study of cross-validation and bootstrap for accuracy estimation and model selection. In: IJCAI, vol. 14,pp. 1137–1145 (1995)

Impact of Power System Stabilizer on Combined ALFC and AVR System

Rumi Rajbongshi, Lalit Chandra Saikia, Arindita Saha, Washima Tasnin and Debdeep Saha

Abstract This study demonstrates the impact of Power System Stabilizer (PSS) on the combined Automatic Load Frequency Control (ALFC) and Automatic Voltage Regulator (AVR) model of two unequal area systems. Both areas of the system have wind, diesel, and reheat thermal plant with proper physical constraints. Fractional-order $PI^{\lambda}DF$ controller is introduced for the system and the performance of the same is compared with some integer-order controllers. The parameters of controllers and PSS are optimized simultaneously using lightning search algorithm. The investigation indicates the superior performance of $PI^{\lambda}DF$ to others. The impact of PSS on system dynamic performance is investigated and infers that the inclusion of PSS damps out the system oscillations, which consequently enhances the system dynamic performance. Also, robust performance of the $PI^{\lambda}DF$ controller and PSS parameters for the considered system are established against changing parameters of generator field.

Keywords Automatic load frequency control · Automatic voltage regulator
Fractional-order controller · Power system stabilizer

R. Rajbongshi (✉) · L. C. Saikia · A. Saha · W. Tasnin · D. Saha
Department of Electrical Engineering, National Institute of Technology
Silchar, Silchar, Assam, India
e-mail: rumi.nits90@gmail.com

L. C. Saikia
e-mail: lcsaikia@yahoo.com

A. Saha
e-mail: sahaarindita.91@gmail.com

W. Tasnin
e-mail: washima.nits@gmail.com

D. Saha
e-mail: saha_debdeep_RS@yahoo.com

D. Deb et al. (eds.), *Innovations in Infrastructure*, Advances in Intelligent Systems and Computing 757, https://doi.org/10.1007/978-981-13-1966-2_51

1 Introduction

In power system, the main concern of power engineers is to supply stable and quality power to the consumers to meet the load demand. Due to mismatch between load demand and generation, the deviation in the frequency, voltage, and tie-line power occurs. Two control loops of generator such as Automatic Load Frequency Control (ALFC) loop and Automatic Voltage Regulator (AVR) loop are responsible to control these deviations. ALFC loop and AVR loop maintain such variations via governor control action and excitation of the generator, respectively. Practically, a poor cross coupling exists between these two because of the quick control response of AVR loop [1, 2]. Therefore, the combined ALFC–AVR system having cross coupling between these two needs to be analyzed.

In the previous literature, various researchers focused their investigations on the separate study of ALFC and AVR problem [3–7]. Authors investigated ALFC problem for isolated wind–diesel system [3] and multi-area thermal system [4, 5]. The analysis on AVR system was carried out in [6, 7]. In recent literature, very limited investigations on combined ALFC and AVR had been done [8–10]. The single-area thermal system for combined ALFC and AVR model [8] had been demonstrated. Chandrakala et al. [9] presented combined model for two-area hydrothermal system having boiler dynamics and Governor Dead Band (GDB). The authors in [10] explored the dish–Stirling solar thermal incorporated multi-area combined system with Generation Rate Constraints (GRC) and GDB. It is noticed that very few investigations had been carried out on application of renewable resources in combined ALFC–AVR system. However, renewable resources such as wind, solar, etc., grab more attention of researchers to maintain the future energy demands. The presence of renewable resources in the system makes the controlling of frequency and voltage more challenging. Moreover, wind–diesel integrated two-area thermal system for combined ALFC–AVR study had not been yet investigated. This calls for further analysis.

The electromechanical oscillations are associated with power system due to the presence of disturbances. These oscillations can be mitigated by applying Power System Stabilizer (PSS) [11]. Kundur et al. [12] designed and analyzed the performance of PSS for Single Machine Infinite Bus (SMIB) system. The authors designed the robust PSS to damp out the oscillations of SMIB system [11, 13]. Mostly the investigation of PSS had been done for SMIB system. Hence, there is a scope of application of PSS in two-area combined system.

The robust secondary controller is required to design for both ALFC and AVR loops. Previously, various classical and Fractional-Order (FO) controllers had been reported in individual study of ALFC and AVR [3–7]. Authors proposed Integral (I) [8], Proportional Integral (PI) [8], Proportional–Integral–Derivative (PID) [9], and FO Integral Double Derivative with Filter [10] controllers in the combined system. The performance of other FO Controller (FOC) like $PI^{\lambda}DF$ was not yet evaluated. Thus, the scope for implementation of the same in combined ALFC–AVR system is still left.

The application of optimization technique for the designing of robust controller plays the vital role. Although numerous optimization techniques exist in the literature, authors implemented simulated annealing [9] and Lightning Search Algorithm (LSA) [10] in combined system to get the optimal controller parameters. Authors in [10] demonstrated the superior performance of LSA than some other algorithms. However, LSA has not yet used for optimization of both PSS as well as FOC parameters of combined system.

Based on the above description, the major objectives of this work are as follows:

(a) To design a two unequal area multi-source combined ALFC–AVR system having GRC and GDB.
(b) To analyze the comparative performance of I, PI, PIDF, and $PI^{\lambda}DF$ controllers for the considered system in (a) for selection of the best controller.
(c) To incorporate PSS and evaluate the impact of PSS in the considered system with suitable controller obtained in (b).
(d) To verify the robust performance of the suitable controller and PSS parameters for the system considered in (c).

2 System Under Investigation

The basic block diagram representation of combined ALFC–AVR system is depicted in Fig. 1. A two unequal area combined ALFC–AVR system having three sources in each area is considered for investigation. In both the areas of the considered system, we have wind, diesel, and reheat thermal units. The capacity ratio for the system is taken as Area1:Area2 = 1:4. The area participation factor (apf) for the system is considered as $apf_{11} = apf_{12} = apf_{21} = apf_{22} = 0.2$ and $apf_{31} = apf_{32} = 0.4$. The PSS is incorporated in both the areas of the system. The Transfer function (T.F.) representation of the considered system with PSS is depicted in Fig. 2. The AVR with cross-coupling coefficients and PSS block are shown separately in Fig. 2 itself. The parameters of the reheat thermal, wind–diesel, and AVR are considered from [3, 5, 9], respectively. To make the system more realistic, the physical constraints of the system such as GDB and GRC for thermal unit are considered here. The values of GRC and GDB are taken as 3%/min and 0.06% (0.036 Hz), respectively [10]. In Area1 of the system, 1% Step Load Perturbation (SLP) is provided. In the considered system, I, PI, PIDF, and $PI^{\lambda}DF$ controllers are applied in ALFC and AVR loop as secondary controller. LSA is implemented with the objective function as integral squared error to get the optimal values of the considered controllers and PSS parameters. The fitness function used in this study can be expressed as (1).

$$J = \int_0^T \left\{(\Delta f_1)^2 + (\Delta V_1)^2 + (\Delta f_2)^2 + (\Delta V_2)^2 (\Delta P_{tie12})^2\right\} dt. \quad (1)$$

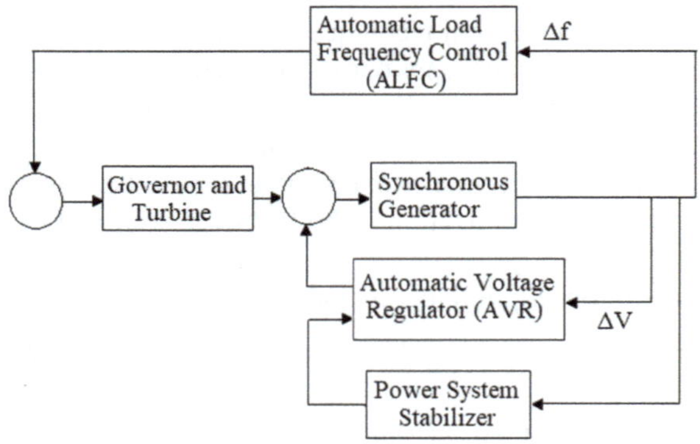

Fig. 1 Block diagram representation of combined ALFC–AVR system with PSS

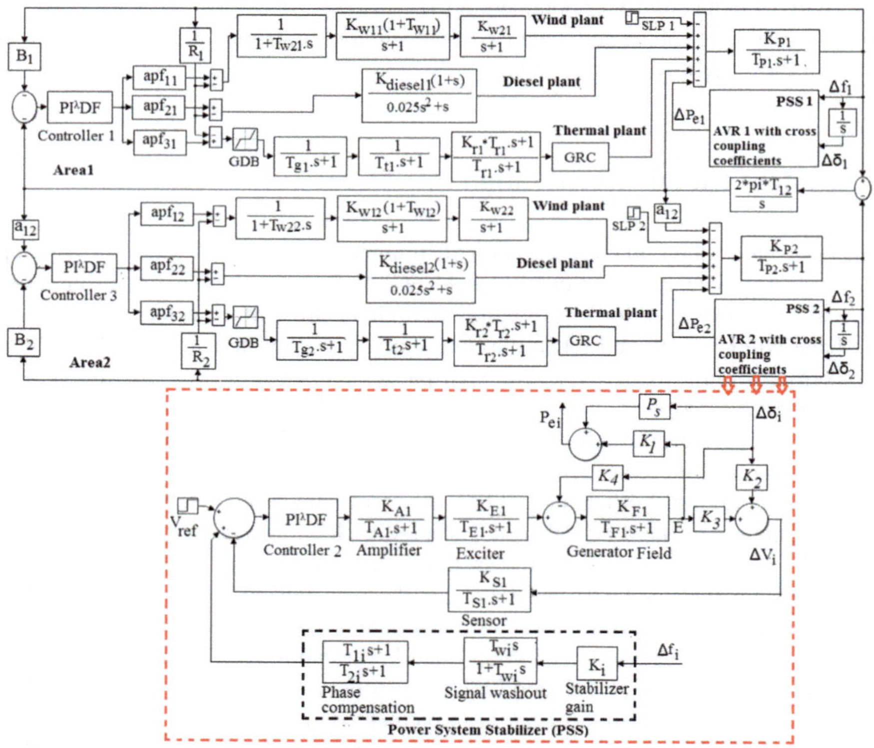

Fig. 2 T.F. depiction of two-area combined ALFC–AVR system having PSS

2.1 ALFC–AVR Combined System

This model controls the voltage and frequency of the system within the predefined values at small disturbance. The change in frequency affects the terminal voltage, since stator emf is directly proportional to frequency (speed). But, AVR provides rapid response than ALFC. Therefore, ALFC loop does not have much effect on AVR loop. However, the controlling action of AVR loop affects the generator stator or internal emf, E, which affects the real power shown in (2) [1, 2]. Hence, AVR loop has considerable effect on ALFC loop.

$$P_e = \frac{|V||E|}{X_S}\sin\delta. \tag{2}$$

In this model, the deviation in real power and terminal voltage with respect to change in rotor angle, $\Delta\delta$ and stator emf, ΔE can be defined in (3) and (4), respectively [1–3].

$$\Delta P_e = P_s\Delta\delta + K_1 E. \tag{3}$$

$$\Delta V = K_2\Delta\delta + K_3 E. \tag{4}$$

Considering the effect of $\Delta\delta$, the generator stator emf can be modified as shown in (5).

$$E = \frac{K_F}{1 + sT_F}(E_e - K_4\Delta\delta). \tag{5}$$

In modeling of combined system, the cross coupling between ALFC and AVR is done using the cross-coupling coefficients like K_1, K_2, K_3, and K_4. The detailed mathematical modeling and explanations of the combined system are referred from [10].

2.2 Power System Stabilizer (PSS)

Power System Stabilizer (PSS) has basically three parts like stabilizer gain, signal washout, and phase compensation, which is depicted in Fig. 1. The gain block measures the total damping, which is introduced by PSS and its value depends on the maximum damping. The signal washout performs like a high-pass filter. Phase compensation with its phase-lead behavior maintains the phase lag of the input to exciter and electrical torque [2]. Thus, PSS with the auxiliary stabilizing signal, which is fed to the excitation system, damped out the generator rotor oscillation [2]. The different parameters of the PSS block are optimized simultaneously along with controller parameters by using LSA. The boundary limits of K_i, T_{wi}, T_{1i}, and T_{2i} are considered as 0–20, 1–10, 0.15–1.5, and 0.015–0.15, respectively [11, 12].

3 Controller and Optimization Technique

Fractional calculus is the foundation of Fractional-Order Controller (FOC) design. The fractional integral can be defined as Riemann–Liouville definition given in (6).

$$ {}_aD_t^{-\alpha} f(t) = \frac{1}{\Gamma(\alpha)} \int_a^t (t-\tau)^{\alpha-1} f(\tau)\mathrm{d}\tau. \tag{6}$$

After taking the Laplace transform, the fractional integral having differential equation can be expressed as transfer function having fractional order of s. The basic concept of FOC is opted from [5].

Nowadays, FOC, with its more control knobs, becomes the favorable option to control the system dynamics. Therefore, in this study, PI$^{\lambda}$DF controller is proposed and transfer function of the same is shown in (7). This controller considers the integer order of derivative term with filter and fractional action of integral.

$$U_i(s) = K_p + \frac{K_I}{s^{\lambda}} + K_d\, s\left(\frac{N}{s+N}\right). \tag{7}$$

The controller parameters $K_p, K_I, K_d, N,$ and λ are optimized by using LSA, which was developed based on the lightning mechanism [14]. The lightning mechanism deals with the propagation of step leader in which the conception of projectiles is the foundation of it. Therefore, the projectiles represent the initial population size of LSA. In LSA, the leader tip energy gives the solution. To obtain optimum solution, the exploration and exploitation can be executed by lead and space projectiles at step leader tip. The tuned values of LSA parameters are taken as population size (=50), maximum iteration (=100), and channel time (=10). The detailed explanations and flowchart of LSA are referred from [14].

4 Results and Analysis

4.1 Performance Analysis of PI$^{\lambda}$DF Controllers

Some controllers, for example, I, PI, PIDF, and PI$^{\lambda}$DF are applied one by one in both ALFC and AVR loops of the considered system. The parameters of each controller have been optimized applying LSA and the obtained optimal values of the same are tabulated in Table 1. The corresponding responses of the controllers are compared in Fig. 3. From the comparative analysis of Fig. 3, it is examined that the responses for PI$^{\lambda}$DF controller have lesser peak overshoot, peak undershoot, and faster settling time than I, PI, and PIDF controllers. The oscillations in the responses for the proposed controller are less as compared to others. Therefore, PI$^{\lambda}$DF controller provides better

Table 1 Optimal values of I, PI, PIDF, and PI$^{\lambda}$DF controller parameters

Controller	Optimal values of controller parameters				
I	$K_{I1}^* = 0.4362$	$K_{I2}^* = 0.2263$	$K_{I3}^* = 0.1346$	$K_{I4}^* = 0.2381$	
PI	$K_{p1}^* = 0.1291$	$K_{I1}^* = 0.1206$	$K_{p2}^* = 0.1002$	$K_{I2}^* = 0.3966$	$K_{p3}^* = 0.0329$
	$K_{I3}^* = 0.0127$	$K_{p4}^* = 0.1000$	$K_{I4}^* = 0.3766$		
	$K_{p1}^* = 0.8655$	$K_{I1}^* = 0.7548$	$K_{d1}^* = 0.3026$	$N_1^* = 83.62$	$K_{p2}^* = 0.1601$
PIDF	$K_{I2}^* = 0.6419$	$K_{d2}^* = 0.2303$	$N_2^* = 91.81$	$K_{p3}^* = 0.9988$	$K_{I3}^* = 0.1028$
	$K_{d3}^* = 0.7535$	$N_3^* = 100$	$K_{p4}^* = 0.2136$	$K_{I4}^* = 0.8086$	$K_{d4}^* = 0.3029$
	$N_4^* = 94.18$				
PI$^{\lambda}$DF	$K_{p1} = 0.9865$	$K_{I1}^* = 0.3599$	$\lambda_1^* = 0.2462$	$K_{d1}^* = 0.1236$	$N_1^* = 97.13$
	$K_{p2} = 0.2031$	$K_{I2}^* = 0.6493$	$\lambda_2^* = 0.9990$	$K_{d2}^* = 0.2768$	$N_2^* = 76.49$
	$K_{p3} = 0.5435$	$K_{I3}^* = 0.6522$	$\lambda_3^* = 0.4972$	$K_{d3}^* = 0.2996$	$N_3^* = 62.23$
	$K_{p4} = 0.1129$	$K_{I4}^* = 0.6854$	$\lambda_4^* = 0.8934$	$K_{d4}^* = 0.2836$	$N_4^* = 86.95$

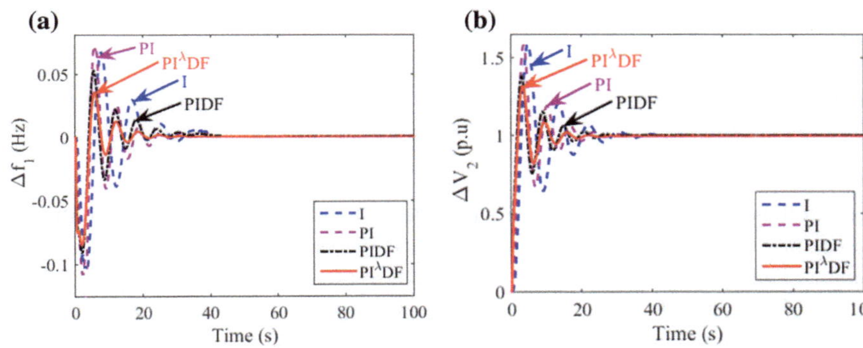

Fig. 3 Comparative dynamic response of different controllers **a** Δf_1 versus time, **b** ΔV_2 versus time

control performance than I, PI, and PIDF controllers. Hence, the analyses of the subsequent sections have been continued with PI$^{\lambda}$DF controller.

4.2 Impact of Power System Stabilizer

In Sect. 4.1, the system performance has been studied without PSS by considering LSA optimized PI$^{\lambda}$DF controller. The controller parameters and the system responses have been shown in Table 1 and Fig. 3, respectively. Here, PSS is incorporated in both

Table 2 Optimal values of controller and PSS parameters for the system

$K_{p1} = 0.4289$	$K^*_{I1} = 0.4548$	$\lambda^*_1 = 0.4144$	$K^*_{d1} = 0.3007$	$N^*_1 = 71.33$
$K_{p2} = 0.2264$	$K^*_{I2} = 0.5773$	$\lambda^*_2 = 0.9933$	$K^*_{d2} = 0.2782$	$N^*_2 = 84.54$
$K_{p3} = 0.1823$	$K^*_{I3} = 0.8537$	$\lambda^*_3 = 0.6840$	$K^*_{d3} = 0.9702$	$N^*_3 = 85.23$
$K_{p4} = 0.2679$	$K^*_{I4} = 0.5674$	$\lambda^*_4 = 0.9594$	$K^*_{d4} = 0.2009$	$N^*_4 = 91.58$
$K^*_1 = 2.291$	$T^*_{w1} = 1.221$	$T^*_{11} = 0.149$	$T^*_{21} = 0.0173$	$K^*_2 = 0.1741$
$T^*_{w2} = 2.008$	$T^*_{12} = 0.1512$	$T^*_{22} = 0.0151$		

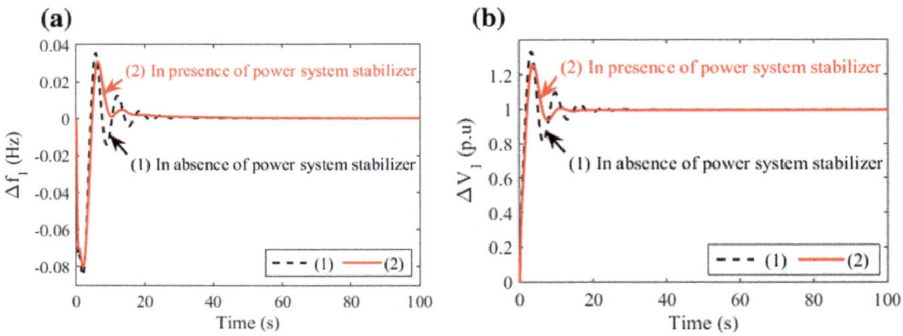

Fig. 4 Comparison of responses of the system in presence and absence of PSS **a** Δf_1 versus time, **b** ΔV_1 versus time

areas of the considered system. The parameters of PSS as well as PI$^\lambda$DF controller are optimized concurrently applying LSA and the optimal values are included in Table 2. The dynamic responses of the system in presence and absence of PSS are compared and depicted in Fig. 4. The critical inspection of this figure leads to a distinct indication that the response of the system with PSS has less peak overshoot, peak undershoot, settling time, and oscillations than that of the system without PSS. The dynamic responses have been improved the in the range of 9–40% with the application of PSS in the system as compared to that of the system in absence of PSS. The analysis reflects that the presence of PSS in the system damped out the oscillations and improved the system dynamics effectively. Thus, it is recommended to integrate PSS in the combined system for enhancing the system performance.

4.3 Sensitivity Analysis

The robustness of PSS and PI$^\lambda$DF controller parameters for the considered system is required to be checked. This can be done by performing the sensitivity analysis for the changed values of gain and time constant of generator filed. The values of the

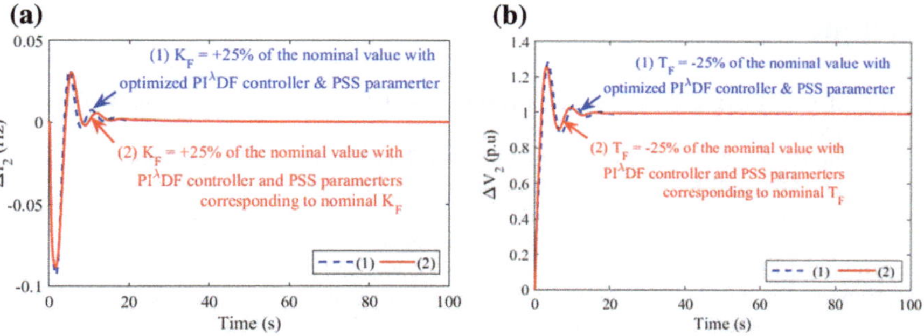

Fig. 5 Dynamic responses for sensitivity analysis at **a** 25% change in K_F in all areas for Δf_2 versus time **b** 25% change in T_F in all areas for ΔV_2 versus time

same are varied as ±25% of the nominal values. The controller and PSS parameters are optimized at each changing condition applying LSA but the values are not shown here. The obtained responses of the system having changes and nominal values are compared in Fig. 5. It is observed that the system with nominal and changed values shows almost the same response. Thus, robustness of the proposed system with PSS and $PI^{\lambda}DF$ controller has been proved for changing parameters of generator filed.

5 Conclusion

A maiden endeavor has been made to study the significance of PSS on the realistic combined ALFC–AVR system with the consideration of physical constraints. A novel FOC, $PI^{\lambda}DF$ is introduced as secondary controller for the system and LSA has been applied to get the optimal values of controllers and PSS parameters. The comparative performance analysis of $PI^{\lambda}DF$ with other classical controllers reveals the superiority of former to others with reduced peak deviation, oscillations, and settling time. Investigation reflects that the integration of PSS in the system reduced the oscillatory behavior of the responses and upgraded the system dynamic performance. Moreover, the proposed system with PSS and $PI^{\lambda}DF$ controller provides the robust performance with the varying AVR parameters. Thus, it can be concluded that the PSS integrated combined system with $PI^{\lambda}DF$ controller enhanced the system performance and maintains the required controlled behavior even at changing generator field parameters.

References

1. Elgerd, O.I.: Electric Energy Systems Theory—An Introduction, 2nd edn. Tata McGraw Hill, New Delhi (2000)
2. Kundur, P.: Power System Stability and Control. Tata McGraw Hill, New York (1994)
3. Das, D., Aditya, S.K., Kothari, D.P.: Dynamics of diesel and wind turbine generators on an isolated power system. Int. J. Electr. Power Energy Syst. **2**, 183–189 (1999)
4. Elgerd, O.I., Fosha, C.E.: Optimum megawatt-frequency control of multi area electric energy systems. IEEE Trans. Power Appar. Syst. **89**(4), 556–563 (1970)
5. Debbarma, S., Saikia, L.C.: Effect of fractional-order controller on automatic generation control of a multi-area thermal system. In: International Conference on Sustainable Energy and Intelligent Systems (SEISCON 2011), Chennai, pp. 223–228 (2011)
6. Chatterjee, S., Mukherjee, V.: PID controller for automatic voltage regulator using teaching–learning based optimization technique. Int. J. Electr. Power Energy Syst. **77**, 418–429 (2016)
7. Tang, Y., et al.: Optimum design of fractional order PID controller for AVR system using chaotic ant swarm. Expert Syst. Appl. **39**(8), 6887–6896 (2012)
8. Rakhshani, E., Rouzbehi, K., Sadeh, S.: A new combined model for simulation of mutual effects between LFC and AVR loops. In: Asia-Pacific Power and Energy Engineering Conference, pp. 1–5 (2009)
9. Chandrakala, K.R.M.V., Balamurugan, S.: Simulated annealing based optimal frequency and terminal voltage control of multi-source multi area system. Int. J. Electr. Power Energy Syst. **78**, 823–829 (2016)
10. Rajbongshi, R., Saikia, L.C.: Combined voltage and frequency control of a multi-area multi-source system incorporating dish-Stirling solar thermal and HVDC link. IET Renew. Power Gener. **12**(3), 323–334 (2018)
11. Sambariya, D.K., Prasad, R.: Robust tuning of power system stabilizer for small signal stability enhancement using meta-heuristic bat algorithm. Int. J. Electr. Power Energy Syst. **61**, 229–238 (2014)
12. Kundur, P., et al.: Application of power system stabilizers for enhancement of overall system stability. IEEE Trans. Power Syst. **4**(2), 614–626 (1989)
13. Khodabakhshian, A., Hemmati, R.: Robust decentralized multi-machine power system stabilizer design using quantitative feedback theory. Int. J. Electr. Power Energy Syst. **41**(1), 112–119 (2012)
14. Shareef, H., Ibrahim, A.A., Mutlag, A.H.: Lightning search algorithm. Appl. Soft Comput. **36**, 315–333 (2015)

Evaporation Ponds: An Effective Measure for Salinity Control

S. S. Khandelwal and S. D. Dhiman

Abstract Evaporation ponds can be effectively used to dispose of saline groundwater in the Limbasi branch canal command area of Mahi Right Bank Canal (MRBC) Project, Gujarat. The present study aims at explaining the concept and design aspects of trapezoidal and circular types of evaporation ponds. Annual volume of water to be disposed through evaporation ponds was worked out as 4.0825 MCM, which was approximately 9% of total groundwater pumped annually and 25% of net annual recharge. Summer months from March to June are considered for effective evaporation. Both the designs are found to yield similar results in terms of cost and volume of disposal of water. Farmers and local authorities are advised to use any one design as per their wish.

Keywords Conjunctive use · Evaporation pond · Groundwater disposal
Limbasi branch canal · Waterlogging

1 Introduction

Irrigation is widely practiced in many parts of arid and semiarid regions of the world to increase agricultural production. Nearly, 80% of the world's developed water resources are consumed by irrigated agriculture [16], while in India 83% of the developed water resources are consumed by irrigated agriculture [10]. The increase in irrigation potential may cause detrimental impacts like waterlogging and salinity [11]. On a global scale, about 80 Mha land is affected to some extent by waterlogging

S. S. Khandelwal (✉)
Department of Civil Engineering, Faculty of Technology, Dharmsinh Desai University,
Nadiad 387001, Gujarat, India
e-mail: sskddu@gmail.com

S. D. Dhiman
Department of Civil Engineering, Birla Vishvakarma Mahavidyalaya,
Vallabh Vidyanagar 388120, Gujarat, India
e-mail: dhimansanjay18@hotmail.com

D. Deb et al. (eds.), *Innovations in Infrastructure*, Advances in Intelligent Systems
and Computing 757, https://doi.org/10.1007/978-981-13-1966-2_52

and salinization [2], while in India the waterlogged areas in Punjab, Haryana, UP, and Rajasthan cover about 10.5 Mha [3]. Andhra Pradesh has maximum waterlogged area (5.83 Mha), while Himachal Pradesh has minimum waterlogged area (0.011 Mha). The waterlogged area in Gujarat is approximately 0.433 Mha as per the report of National Commission for Irrigation [9].

Evaporation ponds can be effectively used to dispose saline groundwater. The construction of on-farm and regional evaporation ponds as one of the options for disposal of saline subsurface drain waters was studied by Tanji et al. [13]. Mickley [8] found evaporation ponds as most appropriate for relatively warm and dry climatic conditions having high evaporation rates, level terrain, and low land costs. Tyagi et al. [14] designed evaporation ponds for disposal of drainage water of lower Ghaggar river basin of Haryana, India. The use of evaporation ponds is reviewed by Ahmad et al. [1] and Tanji et al. [12].

The proximity to the Gulf of Cambay of the Limbasi canal command area (part of MRBC project of Gujarat state, India) with flat topography and insufficient drainage network has led to waterlogging and salinity concerns. To avoid further decline in groundwater quality, pumping of groundwater is needed to maintain the salt balance. The usefulness of evaporation ponds in controlling waterlogging and salinity problems is described in this paper, and two alternative designs (trapezoidal and circular section) are presented.

2 Study Area

Limbasi branch canal command area (Fig. 1) has a culturable command area of 15,764 ha and lies between 22° 31′ 33.19″ N to 22° 36′ 11.79′ N latitude and 72° 32′ 08.63″ E to 72° 48′ 18.69″ E longitude. It has a semiarid climate with the mean rainfall of 785.6 mm. The recorded maximum and minimum temperatures are 46.7 and 4.2 °C, respectively, between 1993 and 2012. In the study area, the depth to water table in pre-monsoon season varies between 0.15 and 16 m, while in post-monsoon season it is varying between 0.1 and 15.40 m. The major crops sown in kharif and rabi season are paddy and wheat, respectively. The elevation of study area from MSL is from 35 to 90 m.

The data on groundwater depth in study area for 36 wells for a period 1992–2010 and for five wells during 2000–2010 are used for analysis. These data were obtained from Gujarat Engineering Research Institute (GERI), Vadodara [5]. The data on groundwater quality, especially, electrical conductivity (EC) of canal water and groundwater were also obtained to calculate the salt load.

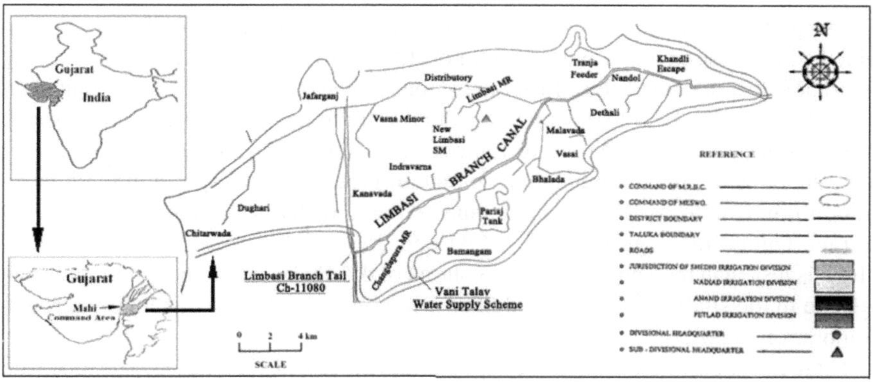

Fig. 1 Index plan of the study area

3 Evaporation Ponds

For a sustainable conjunctive use of fresh canal water and saline groundwater, the water table can be kept below the critical level by surface drains and the excessive salt in groundwater can be removed using evaporation ponds [14]. In the study area, evaporation ponds are best suited for disposal as there are limited surface drainage facilities available. For the present study, two alternative designs of evaporation ponds (trapezoidal and circular section) are discussed.

The volume of the water to be disposed depends upon the quality and quantity of canal water and quality of groundwater. Canal water quality in the study area is good with EC value 0.3 dS/m. The average EC value of the groundwater of the command area is 3.5 dS/m with maximum value of 28.30 dS/m at village Vasai (Well no. 1608) in the year 2001. The salt load can be calculated by the following formula (FAO—http://www.fao.org/docrep/t0667e/t0667e02.htm) [4]:

$$\text{TDS} = 640\ \text{EC for EC between}\ 0.1\ \text{and}\ 5.0\left(\frac{\text{dS}}{\text{m}}\right) \tag{1}$$

$$\text{TDS} = 800\ \text{EC for EC} > 5.0\left(\frac{\text{dS}}{\text{m}}\right), \tag{2}$$

where TDS is in $\left(\frac{\text{mg}}{\text{l}}\right)$ and EC is in $\left(\frac{\text{dS}}{\text{m}}\right)$.

Estimation of the groundwater recharge of the study area shows that there was 51.26 MCM of canal-induced recharge (average of 8 years from 2002–2003 to 2008–2009) in the study area [6]. The salinity of canal water is 0.3 dS/m. From Eq. (1), 1 MCM of water with EC 0.3 dS/m brings a salt load of 192 ton per year. Hence, 51.26 MCM of canal water will bring a total salt of 9842 ton annually in the command area. Assuming that no build up of salinity has occurred in the crop root zone, these salts are passed to the groundwater as deep percolation.

Using proper groundwater pumping strategies, the water table depth can be maintained within desired limits and the groundwater quality can be maintained by keeping the total salt input and output loads the same. The average salinity of groundwater is 3.5 dS/m. Using Eq. (1), volume of groundwater to be disposed annually through evaporation ponds is 4.4 MCM to remove salt load of 9842 ton. Tyagi et al. [14] suggested that the maximum annual volume of groundwater disposal through evaporation ponds is to be 25% of net annual average recharge. In the study area, net average annual recharge is 16.33 MCM and 25% of this value is 4.0825 MCM, which is less than the annual disposal volume of 4.4 MCM (required to maintain salt balance in the study area). Hence, 4.0825 MCM is to be disposed annually through evaporation ponds.

There are approximately 700 wells in the study area of 15764 ha [7]. The tube well density is, therefore, 1 tube well per 22 ha (approx.). Evaporation ponds are generally constructed near the wells to minimize conveyance losses. Hence, total number of evaporation ponds will also be 700. The total volume of groundwater to be disposed annually is 4.0825 MCM. Therefore, the volume of water to be disposed through 1 tube well (for 22 ha) will be 0.005697 MCM (5697 m^3). To prevent seepage losses, polythene sheet is provided on the wetted area. However, 5% unavoidable losses may take place, and hence, the volume of water to be disposed by one evaporation pond will be 5413 m^3.

Most of the annual evaporation takes place during 4 months from March to June. The disposal through evaporation ponds would be more efficient during these months. Average pan evaporation (of 10 years from the year 2000 to 2009) during March, April, May, and June in the area was 7.3, 8.8, 9.2, and 8.0 mm/day, respectively [15]. These values are converted into monthly values (226.3, 264, 285.2, and 248 mm, respectively). For the months of March, April, May, and June, 20% (1082.6 m^3), 25% (1353.25 m^3), 30% (1623.9 m^3), and 25% (1353.25 m^3) of the annual groundwater pumped for disposal (5413 m^3) was found out based on these evaporation values.

3.1 Design of Trapezoidal Section of Evaporation Pond

Assume a trapezoidal cross section with top width "T" and side slope 2H: 1 V. In the plan, consider square section at top and bottom. Let depth of water is 1.5 m. Volume of water stored in the trapezoidal section (when it is full of water) is given by Eq. (3).

$$V = \left(A_1 + A_2 + \sqrt{A_1 * A_2}\right)\left(\frac{1.5}{3}\right), \tag{3}$$

where V = volume of water stored in the trapezoidal section, A_1 = top area (T^2), and A_2 = bottom area $(T–6)^2$. Putting the values of A_1 and A_2 in Eq. (3),

$$V = \left(T^2 + (T-6)^2 + T(T-6)\right)\left(\frac{1.5}{3}\right) = \left(1.5\,T^2 - 9\,T + 18\right). \tag{4}$$

Table 1 Calculations of top width (T) of the trapezoidal section of the evaporation pond during March and April

Description	March	April
Volume of water to be evaporated (m^3)	1082.6	$(1082.6–0.2263A_1)+1353.25$
Volume of water that can be evaporated from the top surface in (m^3)	$0.2263\ A_1$	$0.264\ A_1$
Residual volume of water at the end of month for which storage required (m^3)	$1082.6–0.2263\ A_1$	$2435.85–0.2263\ A_1–0.264\ A_1$
Equating volume of water (Eq. 4) to the volume of residual water	$V = 1082.6–0.2263\ A_1$	$V = 2435.85–0.4903\ A_1$
Top width (T) required (m)	27.58	37.19

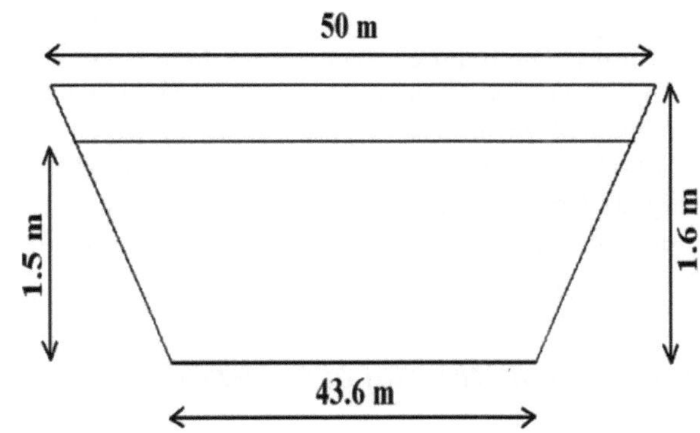

Fig. 2 Evaporation pond of trapezoidal section

The calculations of top width during March and April are shown in Table 1.

Applying the same procedure as above, top width required will be 44.17 and 48.06 m for the months of May and June, respectively. The pond will be designed for a maximum top width of 50 m. Considering square area in plan, the required area (for which land rent is to be calculated) is 2500 m^2. A freeboard of 0.1 m is provided. Hence, total depth of pond is 1.6 m. Corresponding bottom width of the pond will be 43.6 m for a depth of 1.6 m (Fig. 2). Volume of the material to be dug will be 3539.2 m^3 and wetted area (where polythene sheets are to be provided) will be 2576.1 m^2.

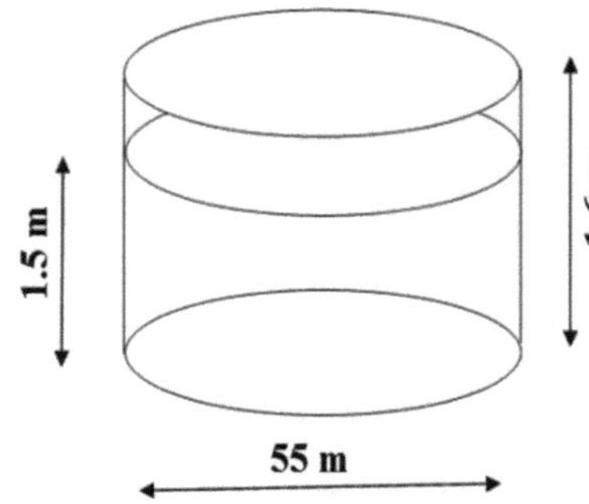

Fig. 3 Evaporation pond of circular section

3.2 Design of Circular Section of Evaporation Pond

For a circular section also, data of Table 1 can be used with only difference that $A_1 = \frac{\pi}{4}D^2$ and $V = 1.5\ A_1$ (where D is a diameter of a circle and depth of water as 1.5 m). The required diameter for March, April, May, and June will be 28.28, 39.47, 47.66, and 52.26 m, respectively. The pond will be designed for a maximum diameter of 55 m. The required area (for which land rent is to be calculated) is 2374.63 m^2. A freeboard of 0.1 m is provided. Hence, total depth of pond is 1.6 m (Fig. 3). Corresponding volume of material to be dug will be 3799.4 m^3 and wetted area (where polythene sheets are to be provided) will be 2651 m^2.

3.3 Cost of Disposal of Groundwater

For calculating cost of disposal of groundwater, life of a pond is assumed as 40 years. The polythene sheet has to be replaced after every 10 years. The Present Worth (PW) concept was applied to calculate replacement cost of polythene sheet. The present worth of an item that has to be replaced after "n" years is given by Eq. (5):

$$PW = \frac{RC(1+r)^n}{(1+i)^n}, \quad (5)$$

where PW = present worth of the future replacement (Rs.), RC = replacement cost (Rs.), n = number of years after which the replacement has to be made, r = inflation rate (7%), and i = interest rate (12%). Cost of earthwork is considered as Rs. 20 per m^3 and cost of polythene sheet as Rs. 10 per m^2. Cost of disposing water was calculated as Rs. 38,180/ha-m for trapezoidal section and Rs. 37,826/ha-m for circular section. The details are given in Table 2.

Table 2 Estimation of cost of water disposal through trapezoidal and circular sections of evaporation pond

Item	Trapezoidal section			Circular section		
	Initial investment (Rs.)	Present Worth (Rs.)	Cost (Rs.)	Initial investment (Rs.)	Present Worth (Rs.)	Cost (Rs.)
Cost of excavating earthwork			70,784			75,988
Initial cost of polythene sheet	25,761			26,509		
Replacement after 10 years		16,316			16,790	
Replacement after 20 years		10,334			10,634	
Replacement after 30 years		6545			6736	
Total capital cost			58,956			60,670
Annualized capital cost			9488			9763
Annual maintenance cost @ 2% of cost of construction			1179			1213
Land rent @ Rs. 40,000/ha/year (50 m × 50 m for trapezoidal section and 55 m diameter for circular section)			10,000			9499
Total annual cost			20,667			20,475
Volume of water disposed (m^3)			5413			5413
Cost of disposing water Rs/m^3(Rs/ha-m)			3.818 (38,180)			3.7826 (37,826)

4 Conclusions

Salinity control in irrigated area requires proper disposal of the groundwater through evaporation ponds as adequate surface drainage facilities are not available in the Limbasi branch canal command area. The volume of groundwater pumped annually is 45 MCM, and 4.0825 MCM is to be disposed through evaporation ponds. This was approximately 9% of total groundwater pumped annually and 25% of net annual recharge (16.33 MCM). Two alternative designs of evaporation ponds (trapezoidal and circular section) are presented. Both the sections are found equally good as far as volume and cost of disposal are concerned. Farmers and local authorities can use any one type of the evaporation ponds discussed here for managing salinity and waterlogging problem prevailing in the command area.

Acknowledgements We thank the officers of WALMI and B. A. College of Agriculture, Anand; MIC and MGVCL, Nadiad and GERI, Vadodara for providing the data for this investigation.

References

1. Ahmed, M., Shayya, W.H., Hoey, D., Mahendran, A., Morris, R., Al-Handaly, J.: Use of evaporation ponds for brine disposal in desalination plants. Desalination **130**(2), 155–168 (2000)
2. Bakker, D.M., Hamilton, G.J., Hetherington, R., Spann, C.: Salinity dynamics and the potential for improvement of waterlogged and saline land in a Mediterranean climate using permanent raised beds. Soil Tillage Res. **110**(1), 8–24 (2010)
3. Chadha, D.K.: Conjunctive use—coping with water logging and salinity. Water Energy Int. **70**(1), 43–47 (2013)
4. Food and Agricultural Organization of the United Nations (FAO): The use of saline waters for crop production - http://www.fao.org/docrep/t0667e/t0667e02.htm
5. Gujarat Engineering Research Institute (GERI): Vadodara: depth to water table below ground level, EC and pH data of Limbasi branch canal command area (soft copy)
6. Khandelwal, S.S., Dhiman, S.D.: Estimating Net Groundwater Recharge in the Limbasi Branch Command Area of Mahi Right Bank Canal Project, Gujarat. J. Indian Water Resour. Soc. **34**(4), 17–27 (2014)
7. Madhya Gujarat Vij Company Limited—(MGVCL).: Details of electric connections given to farmers of the Limbasi branch command area to operate tube wells (2013) (unpublished)
8. Mickley, M.: Environmental considerations for the disposal of desalination concentrates. In: Proceedings of International Desalination Association Conference, Abu Dhabi vol. 7, pp. 351 (1995)
9. National Commission for Irrigation: Ministry of Irrigation and Power. GOI. New Delhi, India (1998)
10. Planning Commission: Report of the Steering Committee on Irrigation for the Tenth Five Year Plan (2002–2007). Government of India: New Delhi, India (2002)
11. Singh, A.: Soil salinization and waterlogging: a threat to environment and agricultural sustainability. Ecol. Ind. **57**, 128–130 (2015)
12. Tanji, K., Davis, D., Hanson, C., Toto, A., Higashi, R., Amrhein, C.: Evaporation ponds as a drain water disposal management option. Irrigat. Drain. Syst. **16**(4), 279–295 (2002)
13. Tanji, K., Grismer, M., Hanson, B.: Subsurface drainage evaporation ponds. Calif. Agric. **39**(9), 10–12 (1985)
14. Tyagi, N.K., Srinivasulu, A., Kumar, A., Tyagi, K.C.: Modeling conjunctive use of water resources: hydraulic and economic optimization. Central Soil Salinity Research Institute, Karnal, India (1995) Report (unpublished)
15. Water and Land Management Institute: Anand.: average climatological data for 10 years (2000–2009) Report (unpublished)
16. Wolff, P., Stein, T.-M.: Efficient and economic use of water in agriculture—possibilities and limits. Nat. Res. Dev. **49**(50), 151–159 (1999)

Effect Friction Stir Welding of Aluminum Alloys AA6061/AA7075: Temperature Measurement, Microstructure, and Mechanical Properties

Pratik S. Godhani, Vivek V. Patel, Jay J. Vora, Nishit D. Chaudhary and Rishab Banka

Abstract Dissimilar aluminum alloys AA6061 and AA7075 were welded by friction stir welding process with the tool traverse feed of 31.5 mm/s, tool rotational speed of 765 rpm, and tool tilt angle of 2° forward. This experiment was done with the help of cylindrical tool pin profile, and its effect on macrostructure, microstructure, and tensile property of welded specimen was observed. Further, temperature measurement at six different positions (three on advancing and three on retreating side) was carried out with the help of *K*-type thermocouple and temperature profile was obtained. The results of defect-free stir zone were obtained and tensile values.

Keywords Friction stir welding (FSW) · AA6061 · AA7075

1 Introduction

Aluminum is very much in use nowadays, because of its better mechanical and metallurgical properties and its promising applications. Aluminum is the most challenging in terms of welding. Welding of aluminum 2xxx and 7xxx series is difficult with conventional welding process [1–4]. These aluminum alloys are generally classified as nonweldable because of the poor solidification microstructure and porosity in the fusion zone. Friction stir welding is derived from friction welding, which was invented in 1991 by Wayne Thomas and his team at The Welding Institute at Cambridge [5]. Friction stir welding is a solid-state welding process as the materials to be welded are fused and not melted and joined through mechanical deformation [6, 7]. In this process, a cylindrical shouldered tool with a profiled probe is rotated and slowly plunges into the joint line between two pieces of plate/sheet materials, which are butted together. As the tool pin rotates and traverses, the frictional heat is gener-

P. S. Godhani (✉) · V. V. Patel · J. J. Vora · N. D. Chaudhary · R. Banka
Pandit Deendayal Petroleum University, Gandhinagar 382007, Gujarat, India
e-mail: pratik27gmc@gmail.com

V. V. Patel
e-mail: profvvp@yahoo.com

D. Deb et al. (eds.), *Innovations in Infrastructure*, Advances in Intelligent Systems and Computing 757, https://doi.org/10.1007/978-981-13-1966-2_53

ated between the shoulder face and the top surface of the plates, and thus the plastic deformation of the material takes place. A lot of research is done on optimization of parameters [8–12] but less light is focussed on the variation of tool pin profile and its effect on various properties. Various parameters are tool rotation speed, tool traverse speed, tool tilt angle, and tool pin profile. In fact, tool pin profile plays a major role in mixing of material and ensuring better quality of welded specimen [13, 14]. The effects of cylinder tool pin profile on mechanical properties, microstructure analysis, and temperature profiles are studied. Even the arrangements of materials on advancing and retreating side play an important role [15, 16]. During dissimilar materials' welding, the softer material is placed on retreating side and the harder material on the advancing side. From the previous studies [15, 16], AA6061 (softer material) is placed on the advancing side, whereas AA7075 (harder material) is placed on the retreating side which is opposite to which is generally seen. Large amount of research work is done in the field of FSW of AA6061 aluminum grade [1, 2, 17] as it has many applications and lesser work is done in the field of FSW of AA7075 [18, 19], and also less focus has been raised in the FSW of AA6061 and AA7075 dissimilar materials joining [12, 15, 16] till now according to the best of the author's knowledge. In the present investigation, the novel approach FSW is done for the dissimilar aluminum grade materials (AA6061 and AA7075) with the help of the cylinder pin profile, and to obtain the variation in the mechanical properties, micro- and macrostructure analysis along with the temperature profiles at six different positions has been carried out.

2 Experimental Setup

The base materials used in the study are rolled plates of AA6061 and AA7075 Al alloys (both are 6 mm thick) in T6 temper condition. The chemical compositions of AA6061 and AA7075 are shown in Table 1, whereas the mechanical (tensile) properties are shown in Table 2. All the plates were cut before welding into dimensions measuring 100 mm long and 50 mm wide. The holes were drilled on the face to fix thermocouple into the plate. The diameter of drilled hole was taken as 2 mm and with different depths such as 1.5, 3, and 4.5 mm. These holes were drilled at every 25 mm distance from the top on the longitudinal axis. The tool pin profile selected is the cylindrical tool pin profile whose dimensions are shown in Fig. 1.

Various tests were performed like tensile test, micro- and macrostructure, etc. Tensile specimen was taken perpendicular to the welded joint with the help of the

Table 1 Chemical composition (%wt)

	Mg	Zn	Cu	Si	Fe	Cr	Al
AA6061	0.11	0.2	0.27	0.56	0.56	0.18	Balance
AA7075	2.5	5.67	1.5	0.35	0.61	0.24	Balance

Table 2 Tensile properties of different samples

	UTS1(MPa)	UTS2 (MPa)	UTS3 (MPa)	Avg UTS (MPa)	Joint efficiency
AA6061	290.18	286.27	288.31	288.25	
Welded specimen	173.52	179.99	177.18	176.89	61.36%

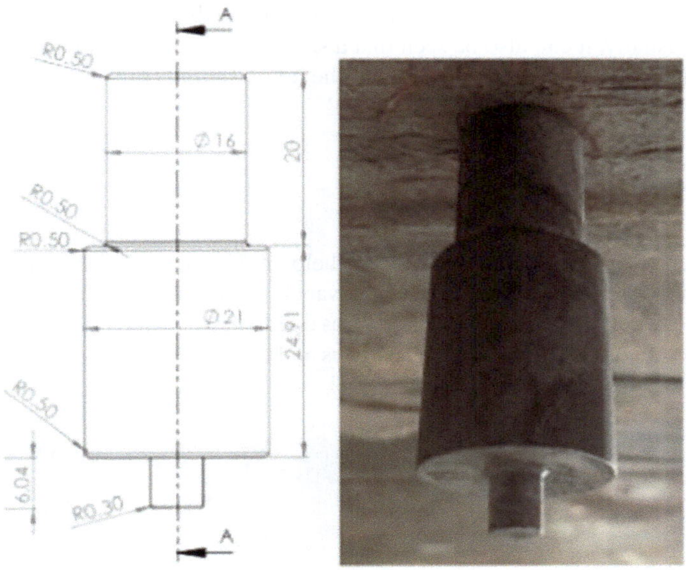

Fig. 1 Dimensions of cylindrical tool pin profile

wire-cut EDM, and the specimen is shown in Fig. 4. Three different samples were taken into consideration and the simple average has been done for the same. For the temperature measurement, *K*-type thermocouple was used and thus the temperature profile was obtained at different positions and different depths. The microstructure specimen was made in the transverse direction, 10 mm away from the exit hole and having 10 mm width. The etchant used was Keller's reagent.

As the base metals are two different aluminum alloys AA6061 and AA7075, the mechanical properties are very different. Thus, it becomes very important to determine the processing conditions for successful solid-state welding. Improper welding parameters lead to improper material flow and improper heat input, which would further lead to pinhole, tunnel defect, piping defect, zigzag lines, kissing bond in friction stir weld joints, and inappropriate onion rings. The tool rotational speed was kept as 765 rpm and the transverse feed as 31.5 mm/min with two degrees forward tilt with no offset provided.

3 Results and Discussion

3.1 Visual Inspection

Initially, the specimen was visually inspected, and the penetration was in full front and back as shown in Figs. 2 and 3 with almost no visual defect. Onion rings were formed on the surface that can be clearly seen, and the distance between the rings is uniform. Visually, it can also be seen that a small amount of aluminum got stuck on the tool pin and that can be removed with the help of filing (Fig. 4).

3.2 Temperature

Temperature profile is obtained with the help of K-type thermocouple, and here there are six different positions (three at advancing side and three at retreating side); position 1 has depth of 1.5 mm, position 2 has depth of 3 mm, and position 3 has depth of 4.5 mm. According to different researches and theoretical knowledge, the temper-

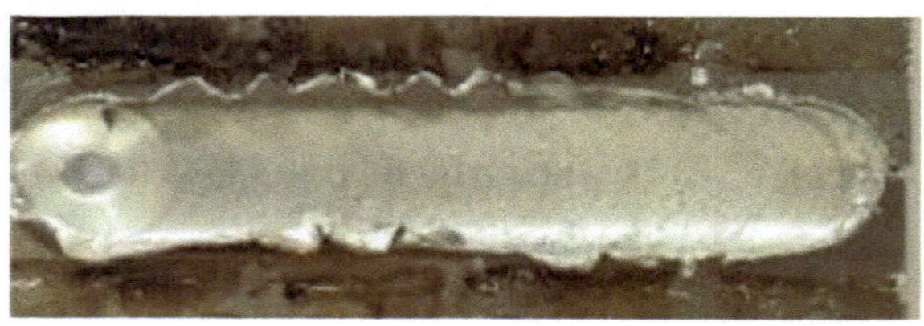

Fig. 2 Top view of welded specimen

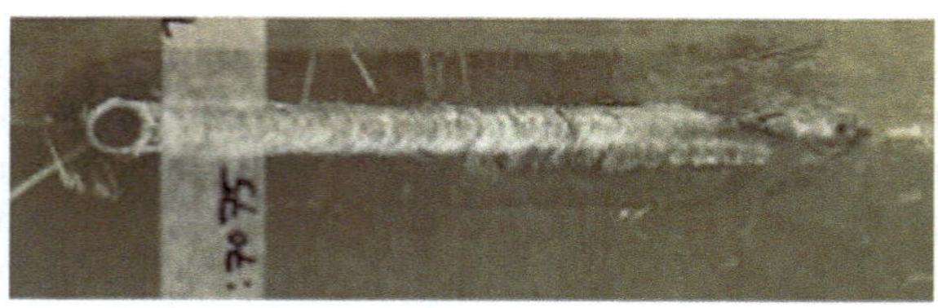

Fig. 3 Bottom view of welded specimen

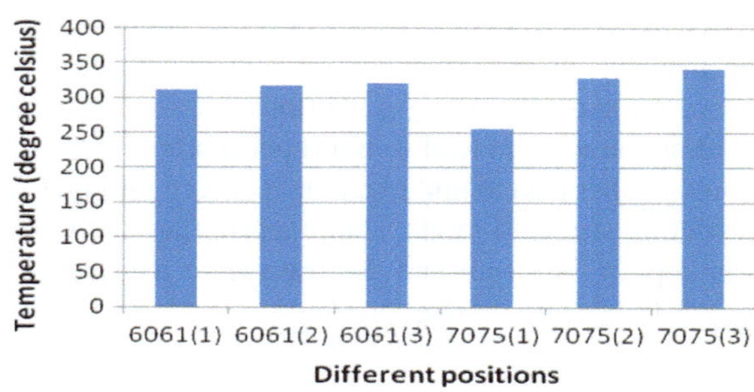

Fig. 4 Temperature at different positions

ature of advancing side would be higher as compared to retreating side, and also the temperature at the center would be highest as this point would be subjected to both preheating (when tool is at position 1) and postheating (when tool is at position 3).

But according to the experiments, we got more or less nearly the same temperatures at advancing and retreating side as there is no large difference in properties of these alloys and also we got maximum temperature at position 3 during the experiments as the tool was allowed to stay in the plunged area at the end of the experiment for some amount of time, and thus the temperature rise took place at this position [19–21].

3.3 Macrostructure and Microstructure

The specimen was treated with the help of Keller's reagent and then observed under optical microscope. The macrostructure of the welded specimen is shown in Fig. 5. It depicts the different alloys AA6061 and AA7075 by the color variation. Additionally, it also shows that the tool pin has been completely penetrated.

The microstructure of the specimen shows the grain size and structure in the various zones. The size of the grains varies in the different zones. The different zones are nugget zone (NZ), thermo-mechanically affected zone (TMAZ), heat-affected zone (HAZ), and base metal (BM). The different regions where the microstructure is obtained are base metal (Fig. 6a AA6061, Fig. 6b AA7075), left interference zone (Fig. 6c, interference between nugget zone and base metal AA6061), the nugget zone (Fig. 6d), and right interference (Fig. 6e, interference between nugget zone and base metal AA7075). The interface zones differentiate the nugget zone from the base metal. The differentiation is done by looking at the grain structures [22–24]. The zone in the middle is called weld pool/stir zone/nugget zone (Fig. 6d).

Size of the grain depends on the various factors such as amount of heat input, mixing of the materials, and the rate of cooling. The size of the grain and grain density decides the strength. The microstructure of the base metal of AA6061 and AA7075 shows that the grain density of AA7075 is high, and hence the strength of it is higher than AA6061. Further, looking at the other microstructures we can see that the grain density in the nugget zone is the highest, and hence the welded specimen has higher strength compared to the base metals, which is one of the big advantages of friction stir welding. Comparing the interfaces on advancing (AA6061)

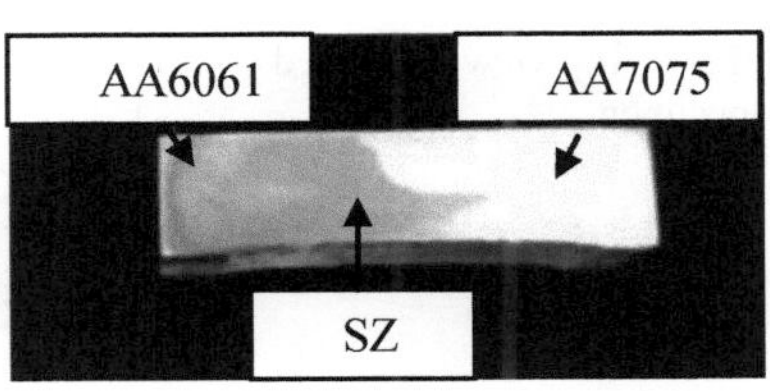

Fig. 5 Macrostructure view

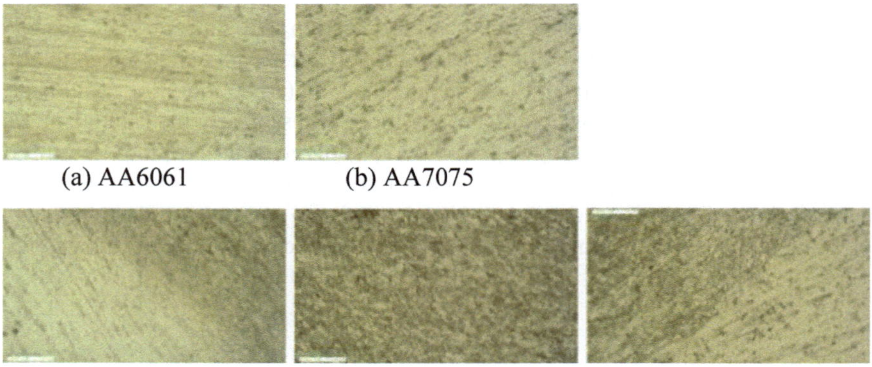

(a) AA6061 (b) AA7075

(c) AA6061 Interface (d) Nugget Zone (e) AA7075 Interface

Fig. 6 Microstructure at different locations

and retreating (7075) side, we can conclude that the failure would occur from the advancing side as the density of grain is less compared to other locations, and this can be proved from the tensile tests that the failure occurred from AA6061 side.

3.4 Tensile Test

The tensile specimen was cut as shown in Fig. 7. The size of the specimen was according to the ASTM E8 standard subsize section. The tensile properties of the AA6061 and the welded specimen are shown in Table 2 along with the joint efficiency. Only the tensile properties of AA6061 are mentioned because the joint efficiency is calculated with respect to weak base metal among the two. Additionally, the fracture took place from the HAZ of the AA6061 side. As reported previously in dissimilar FSW of other Al alloys, the friction stir welded joints usually fractured at locations in HAZ on the weaker material side. The failure locations in HAZ also indicate that the seamless bonding has been achieved between dissimilar AA6061 and AA7075 alloys under all investigated welding conditions. The pictorial view of the tensile specimen after the breakage is shown in Fig. 8. The breakage is in the cup-and-cone form, and this proves that it is ductile failure.

Fig. 7 Top view of welded specimen

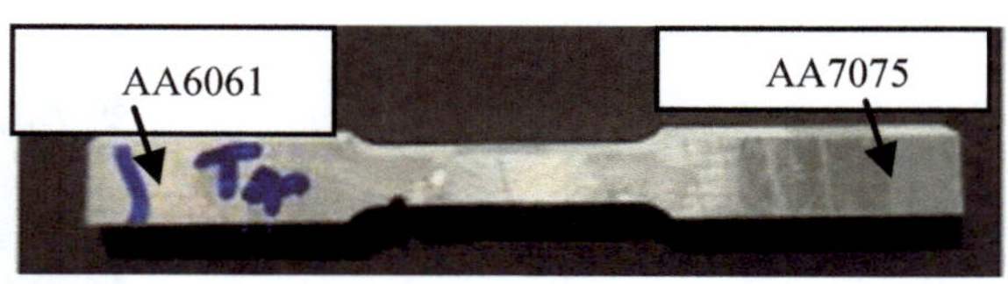

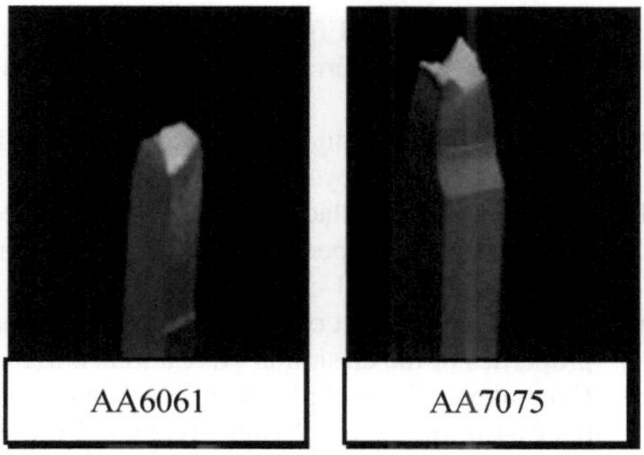

Fig. 8 Welded specimen after failure

4 Conclusion

1. The use of cylindrical tool pin profile produces a successful weld specimen with minimum amount of flash.
2. The temperature is nearly the same on both advancing and retreating sides, whereas the highest temperature is recorded at the third position that is near the exit hole.
3. Further, the grains in the nugget zone are much finer than the base metal which strengthens the weld zone.
4. The failure during tension testing leads to cup-and-cone failure which depicts the ductile behavior of the material.

Acknowledgements The authors' sincerely express their gratitude to Office of Research and Sponsored Program (ORSP/R&D/SRP/2016/GPVP), Pandit Deendayal Petroleum University, Gandhinagar, Gujarat, India for their financial support.

References

1. Leon, J.S., Jayakumar, V.: Investigation of mechanical properties of aluminium 6061 alloy friction stir welding. Am. J. Mechanical Eng. Autom. **1**(1), 6 (2014)
2. Heidarzadeh, A., et al.: Tensile behavior of friction stir welded AA 6061-T4 aluminum alloy joints. Mater. Des. **37**, 166–173 (2012)
3. Patel, V.V., Badheka, V., Kumar, A.: Friction stir processing as a novel technique to achieve superplasticity in aluminum alloys: process variables, variants, and applications. Metallography, Microstruct. Anal. **5**(4), 278–293 (2016)
4. Patel, V.V.: Experimental investigation on friction stir processed AA7075 process variables microstructure and superplasticity (2017)
5. Thomas, W., et al.: Friction stir welding. International patent application no. PCT/GB92102203 and Great Britain patent application, 1991(9125978.8)
6. Midling, O.T., Morley, E.J., Sandvik, A.: Friction stir welding. Google Patents (1998)
7. De Koning, A.U.: Friction stir welding. Google Patents 2002
8. Mishra, R.S., Ma, Z.: Friction stir welding and processing. Mater. Sci. Eng.: R: Rep. **50**(1–2), 1–78 (2005)

9. Cavaliere, P., et al.: Effect of welding parameters on mechanical and microstructural properties of AA6056 joints produced by friction stir welding. J. Mater. Process. Technol. **180**(1–3), 263–270 (2006)
10. Hirata, T., et al.: Influence of friction stir welding parameters on grain size and formability in 5083 aluminum alloy. Mater. Sci. Eng. A **456**(1–2), 344–349 (2007)
11. Cavaliere, P., Squillace, A., Panella, F.: Effect of welding parameters on mechanical and microstructural properties of AA6082 joints produced by friction stir welding. J. Mater. Process. Technol. **200**(1–3):364–372 (2008)
12. Xue, P., et al.: Effect of friction stir welding parameters on the microstructure and mechanical properties of the dissimilar Al–Cu joints. Mater. Sci. Eng. A **528**(13–14), 4683–4689 (2011)
13. Emamikhah, A., et al.: Effect of tool pin profile on friction stir butt welding of high-zinc brass (CuZn40). Int. J. Adv. Manuf. Technol. **71**(1–4), 81–90 (2014)
14. Patel, N., Bhatt, K., Mehta, V.: Influence of tool pin profile and welding parameter on tensile strength of magnesium alloy AZ91 during FSW. Procedia Technol. **23**, 558–565 (2016)
15. Guo, J., et al.: Friction stir welding of dissimilar materials between AA6061 and AA7075 Al alloys effects of process parameters. Mater. Des. **56**, 185–192 (2014)
16. Cole, E., et al.: Weld temperature effects during friction stir welding of dissimilar aluminum alloys 6061-t6 and 7075-t6. Int. J. Adv. Manuf. Technol. **71**(1–4), 643–652 (2014)
17. Lakshminarayanan, A., Balasubramanian, V., Elangovan, K.: Effect of welding processes on tensile properties of AA6061 aluminium alloy joints. Int. J. Adv. Manuf. Technol. **40**(3–4), 286–296 (2009)
18. Rajakumar, S., Muralidharan, C., Balasubramanian, V.: Influence of friction stir welding process and tool parameters on strength properties of AA7075-T 6 aluminium alloy joints. Mater. Des. **32**(2), 535–549 (2011)
19. Patel, V.V. et al.: Effects of Various Cooling Techniques on Grain Refinement of Aluminum 7075-T651 During Friction Stir Processing. In: ASME 2016 International Mechanical Engineering Congress and Exposition. 2016. American Society of Mechanical Engineers
20. Patel, V.V., Badheka, V.J., Kumar, A.: Influence of pin profile on the tool plunge stage in friction stir processing of Al–Zn–Mg–Cu alloy. Trans. Indian Inst. Met. **70**(4), 1151–1158 (2017)
21. Patel, V.V., et al.: Experimental Investigation on Hybrid Friction Stir Processing using compressed air in Aluminum 7075 alloy. Mater. Today: Proc. **4**(9), 10025–10029 (2017)
22. Patel, V.V., Badheka, V., Kumar, A.: Influence of friction stir processed parameters on superplasticity of Al-Zn-Mg-Cu alloy. Mater. Manuf. Processes **31**(12), 1573–1582 (2016)
23. Patel, V.V., Badheka, V., Kumar, A.: Effect of polygonal pin profiles on friction stir processed superplasticity of AA7075 alloy. J. Mater. Process. Technol. **240**, 68–76 (2017)
24. Patel, V.V., Badheka, V.J., Kumar, A.: Effect of velocity index on grain size of friction stir processed Al-Zn-Mg-Cu alloy. Procedia Technol. **23**, 537–542 (2016)

Progressive Collapse Analysis of Two-Dimensional Reinforced Concrete Framed Structure

Eshwar Kuncham and Venkata Dilip Kumar Pasupuleti

Abstract Two-dimensional reinforced concrete frames with the storey of different heights are simulated to the column removal scenario to understand the progressive collapse behaviour by using General Services Administration (GSA 2003) guidelines. Numerical modelling of 4-, 8- and 12-storey structures with different analyses procedures were evaluated such as Linear Static Analysis (LSA), Linear Dynamic Analysis (LDA) and Nonlinear Static Analysis (NLSA) using ETABS 2015. The liner results demonstrate the corner column removal and thus make the structure more vulnerable to progressive collapse when compared to the middle column removal and nonlinear results indicate that the progressive collapse of the structures decreases with increasing the height of the storey.

Keywords Progressive collapse · Linear static analysis
Linear dynamic analysis Nonlinear static analysis

1 Introduction

Major causes of the building collapse are seen when they are subjected to earthquake, flood, cyclone, gas explosion, impact and terrorist attacks, which are termed as extreme loads. These loads affect the performances of structure. Generally, members of the structure are not designed to resist this kind of a typical loading, and thus result in the collapse of structure. When a key structural element fails due to the applied extreme loads, it results in failure of adjacent structure members, which in turn results in further structural failure. This type of failure is referred as progressive collapse of structure. Therefore, the members of the structures should be designed to resist and redistribute the additional applied load. Thus, it is necessary to provide

E. Kuncham (✉) · V. D. K. Pasupuleti
School of Engineering Sciences, Mahindra École Centrale, Hyderabad, India
e-mail: eshwar.research@gmail.com

V. D. K. Pasupuleti
e-mail: venkata.pasupuleti@mechyd.ac.in

D. Deb et al. (eds.), *Innovations in Infrastructure*, Advances in Intelligent Systems and Computing 757, https://doi.org/10.1007/978-981-13-1966-2_54

sufficient redundancy, ductility and continuity, which helps the structure to find alternate paths for load distribution during undesired failure event and thus to reduce the progressive collapse [1].

Earlier cases of the progressive collapse are recorded on 14 July 1902, the 323 foot St Mark's Campanile in Venice, Italy collapsed after its northern load-bearing wall began to separate from the main structure. The concept of progressive collapse comes to the image after the collapse of the 22-storey Ronan Point Apartment Tower in 1968. On 11 September 2001, World Trade Center buildings in New York City collapsed. On 19 January 2017, the Plasco Building in Iran, Tehran that caught fire and this leads to its progressive collapse. Apart from this, many more incidents were recorded across the world [2]. A 7-storey (stlit + 5 + Penthouse) building in Nanakramguda, Hyderabad was collapsed on 8 December 2016 due to the removal of interior column and the effect of controlled blasts carried out at the neighbouring building weakened the foundation of the structure [3].

In the field of structural engineering, the progressive collapse incidents from the past alert the community for better understanding the phenomena of progressive collapse resistance and the failure of structures. Henceforth, the thorough analytical research and experiments result in the development of general procedures and guidelines for the progressive collapse prevention and their implementation of design codes and standards. In parallel, to access the potential of progressive collapse of buildings have been developed by US government agencies such as the General Service Administration (GSA 2003) [4] and the Unified Facilities Criteria (UFC).

2 Literature Review

Marjanishvili [5] investigated the causes of progressive collapse of structure and proposed various analysis procedures with their advantages, disadvantages and limitations. Tsai and Lin [6] have discussed the effectiveness of different structural systems on progressive collapse resistance. Zahrai and Ezoddin [7] have conducted a detailed study on intermediate moment-resistant reinforced concrete frame buildings, and finally proposed the adequate level of reinforcement to be provided so as to develop Alternative Load Paths (ALP). By implementing the following ALP, the loss of an individual member and transfer of the load to the load transfer mechanism of each member to ground can prevent progressive collapse and maintain their serviceability. Petrone et al. [8] have proposed an energy-based method for identifying the elements influencing progressive collapse of structure. Wenga et al. [9] have discussed numerical examples with different loading and column removal scenarios, which are given to validate the suggested damage assessment procedure and the collapse searching algorithms as well as to demonstrate the effectiveness of the proposed modelling approach.

3 Methodology

The aim of GSA guidelines is to help in evaluating the risk of progressive collapse in new and existing buildings. The following analysis has been considered for assessment as shown in Fig. 1a. **Case1**. Analyse for the sudden loss of a column for one floor above ground level situated at the corner of the building. When the corner column is removed, the affected area is called factored area (only 5) and rest of the area is the un-factored area as shown in Fig. 1. **Case2**. Analyse for the sudden loss of a column for one floor above ground level situated in the middle of the building. When the middle column is removed, the factored areas (6 and 7) and rest of the area are the un-factored area. **Case3**. Analyse for the sudden loss of a column for one floor above ground level situated in the interior column of the building. When the interior column is removed, the factored areas (1, 2, 3 and 4) and rest of the area are the un-factored area.

3.1 Analysis Loading

For static analysis purposes, the following vertical load can be applied in the downward direction of the structure. If any column is removed in the static analysis to get progressive behaviour, then the gravity loads increases. Load combination for the factored area of the above-removed column is given by Eq. (1) and the un-factored area is Eq. (2) as shown in Fig. 1b. There is no specific load combination to suggest for dynamic analysis in guidelines, so assume load combination from the static analysis for the un-factored area as shown in Fig. 1c. Because dynamic analysis does not require to increase in any gravity loads.

(a) Above the removed column location

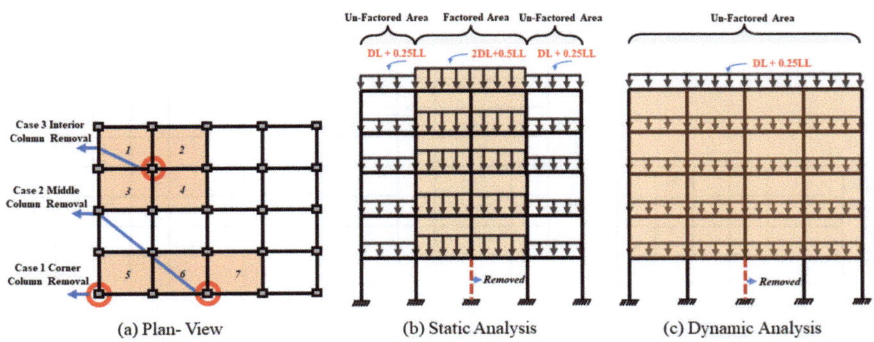

Fig. 1 **a** Location of removal column member, load combinations and load locations for middle column removal of **b** Static analysis and **c** Dynamic analysis

$$\text{Factored Area Load} = 2(\text{DL} + 0.25\ \text{LL}), \quad (1)$$

(b) Adjacent to the removed column location

$$\text{Un-Factored Area Load} = \text{DL} + 0.25\ \text{LL}. \quad (2)$$

4 Modelling

Two-dimensional reinforced concrete framed structures of 4-storey, 8-storey and 12-storey having each floor of height 3 m with typical structural bays of width 5 m were simulated using the finite element method based on ETABS 2015 [10]. Basic model structures considered for this study are shown in Fig. 2.

4.1 Material and Sectional Property

First, structures are analysed and designed as per IS 456-2000 [11] load combination and geometry of columns and beams is shown in Table 1. Assume the thickness of the square slab (5 m × 5 m) as 0.12 m and the thickness of the wall as 0.115 m in all three-frame models. The characteristic compressive strength of concrete (f_{ck}) for the beam is taken as 20 N/mm^2 and for the column is taken as 25 N/mm^2 with a unit weight of concrete as 25 N/mm^3. The yield strength of reinforcing steel (f_y) is 415 N/mm^2.

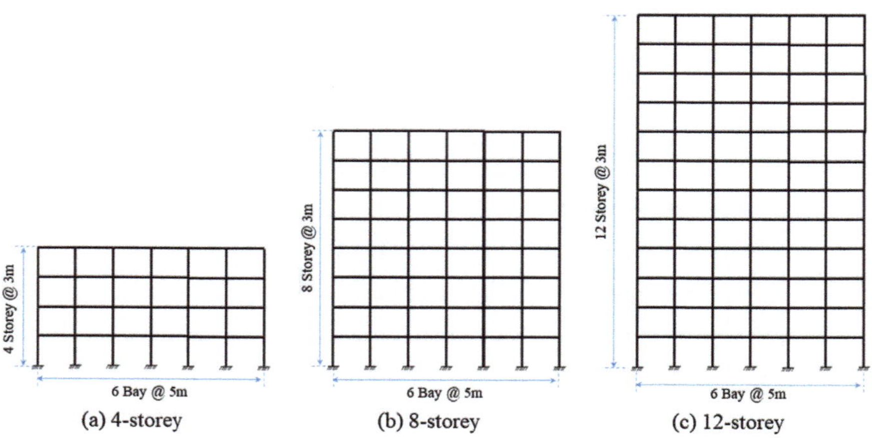

Fig. 2 Elevation of 4-, 8- and 12-storey with a uniform bay width

Table 1 The geometry of sections in the different storeys

No. of Storeys	Beam (mm^2)	Column (mm^2)
4-storey	300 × 230	300 × 300
8-storey	300 × 230	300 × 450
12-storey	300 × 230	300 × 600

Table 2 Intensities of load considered in the storey

Types of load	Top storey		Other storey	
	In 3D frame (kN/m^2)	In 2D frame (kN/m)	In 3D frame (kN/m^2)	In 2D frame (kN/m)
Live load	1.5	2.5	2.5	4.17
Floor load	2	3.33	1.25	2.08
Wall load	2.19	2.19	2.19	6.55
Slab load	3	5	3	5

4.2 Intensities of Loading

Load transfer from top slab to the beam is according to the yield line theory. The amount of load from square slab to the beam is given in Eq. (3).

$$\text{For Square slab} = \frac{Wl_x}{3}. \quad (3)$$

Table 2 shows the load transfer from the assumed three-dimensional (3D) structure to real two-dimensional (2D) structure for the top storey and other storeys which are considered. All the loads have been taken from IS 875-1987 (Part 1, 2 and 5) [12].

5 Analysis and Results

Modal Analysis: A modal analysis calculates the natural frequencies of a structure. The typical structural behaviour depends on its mode shapes which actually depends on the geometry, material and boundary conditions. Figure 3 shows mode shapes of side swinging for all the three cases considered. It can be observed that the time period increases with increase in height of the structure, which is true and obvious. But, the important point to be noted is that the time period increases linearly with increase in storeys. For 8-storey, the time period is double of 4-storey and for 12-storey it is three times more than the 4-storey structure, which is linearly increased.

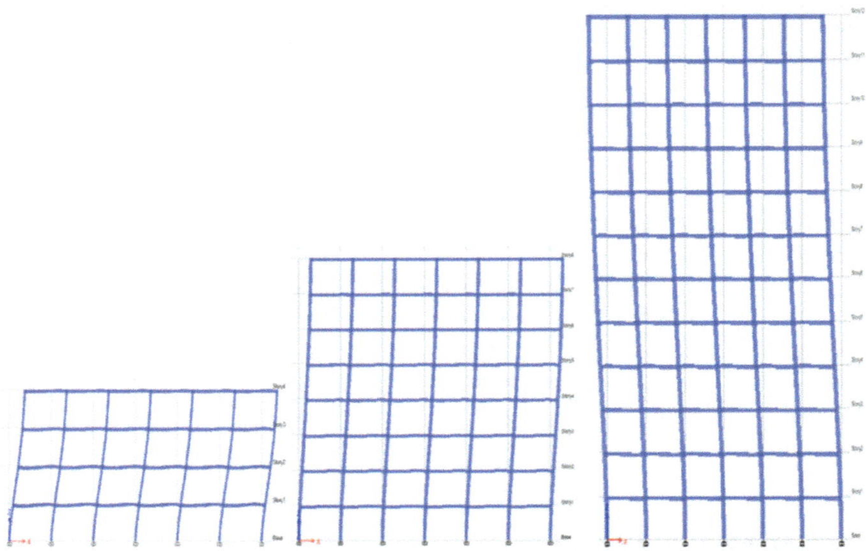

Fig. 3 Fundamental mode shape for 4-, 8- and 12-storey structures

5.1 Linear Static Analysis (LSA)

The structure is analysed and designed for gravity load, which is carried out based on IS 456-2000 load combination. Now, remove a vertical column (C1 or C4) from the position under consideration. Add an extra load for this model with 2 (DL + 0.25 LL) for factor area and 1.2 (DL + 0.5 LL) for un-factored area from Eq. 1. These extra loads are applied on the slab to get the effect of sudden column removal. The load combinations were entered into the program to generate a model.

Liner static results demonstrate that the corner column (C1) removal had more displacement when compared with middle column (C4) removal. And, when compared with the storey for both corner column (C1) removal and middle column (C4) removal, the displacement increases with increase in storeys from 4-storey to 12-storey. But in case of an 8-storey, there is a slight decrease in corner column removal and linear increment in middle column removal with respect to 4- and 12-storey.

5.2 Nonlinear Static Analysis (NLSA)

This analysis is carried out for multiple steps with defining auto hinges for every beam and column at the relative distance by using ASCE 41-13. Similar to LSA, the load combinations are applied to the structure and push-down analysis has been

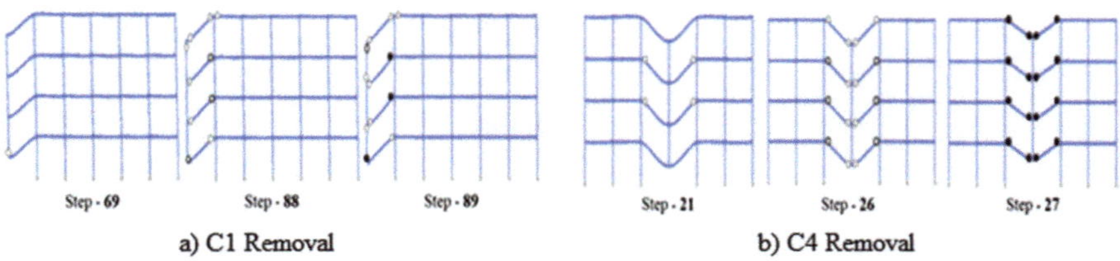

Fig. 4 Formation of hinges after the removal of column C1 and C4 in 4-storey

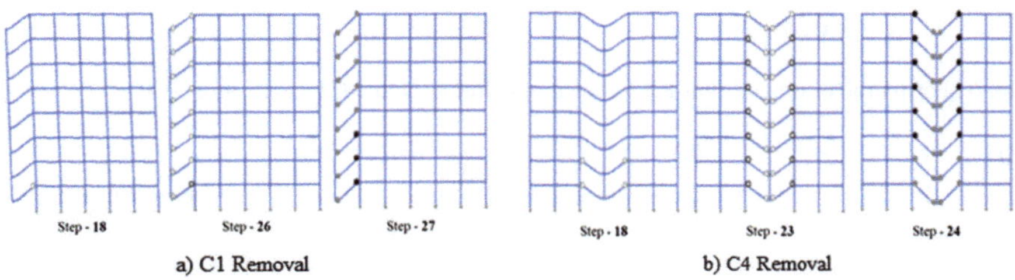

Fig. 5 Formation of hinges after the removal of column C1 and C4 in 8-storey

carried out for all the three cases based on the guidelines. The failure mechanism of elements, it can be observed with different shards in black colours such as immediate occupancy, life safety, collapse prevention and the collapse of the structural member.

Figure 4a shows the hinge formation of structural elements in three particular steps. It indicates first, the joint at which the column is removed has been yielded followed by the yielding of the adjacent column on the second and third floor of column removal. Slowly, the hinges transform causing the failure of the structure. Figure 4b shows in the case of column C4 removal, adjacent joints of the second and third floor have yielded first. Slowly, the yielding starts at the top and bottom floors. In this case, all the adjacent joints get yielded, and first top floors have shown the collapse prevention hinge formation, indicating that the structure has actually experienced a good amount of damage. This indicates the middle column removal is more dangerous than corner column removal. Two main reasons could play a vital role in the failure of complete adjacent joints and just above joints of middle column removal causes more load transfer compared to corner column removal and fast yield formation. Similarly, hinge formation in 8- and 12-storey is shown in Figs. 5 and 6.

5.3 *Linear Dynamic Analysis (LDA)*

This analysis is carried out by using time history method, which consists of a user-defined function such as ramp and nodal function. Ramp function is used for dead load and live load, which has to remain active even if the column is removed with respect to time. The nodal function used for the nodal reaction has to be active when

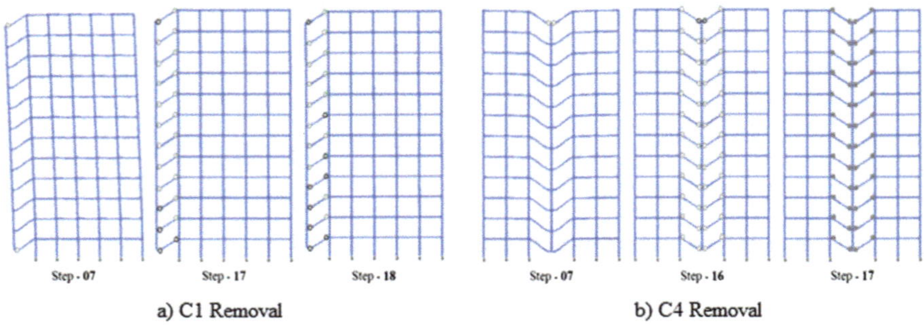

Fig. 6 Formation of hinges after the removal of column C1 and C4 in 12-storey

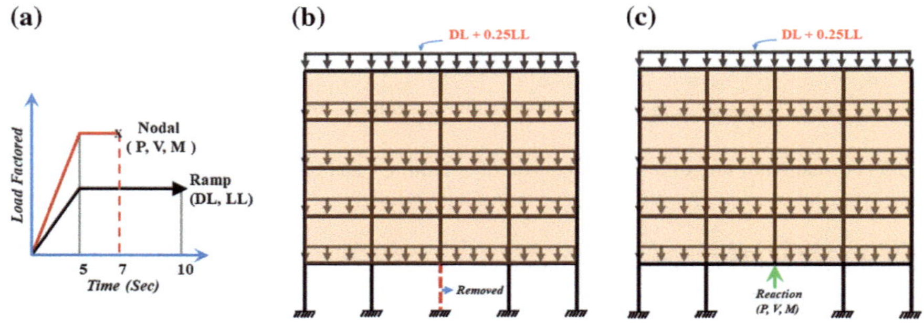

Fig. 7 **a** Application of dynamic load factored for progressive collapse **b** Position of middle column removed **c** Reaction is replaced in middle column position with same load combination

the column is removed, to get the sudden removal of column behaviour with respect to time. Both functions are linearly increasing with respect to time up to 5 s and ramp function remains the same from fifth second to tenth second. The nodal reaction is zero after seventh second, which gives a sudden effect similar to the progressive behaviour of the structure as shown in Fig. 7a.

In LDA, these extra loads are not applied to the structure on the removal location, as the dynamic loading itself would create the effect. To simulate the phenomena of column removal and to carry out LDA, the axial force (P), shear force (V), and bending moment (M) acting on the column are computed before it is removed as per load combination is given in Eq. 2, as shown in Fig. 7b. Then, the column is replaced by point load equivalent to its member force, i.e. nodal reaction as shown in Fig. 7c.

Figure 8a–c shows vibrations of corner column removal. Vibration starts from seventh second because of the sudden removal of column, i.e. making zero nodal reaction and its effect is continued to 15–20 s. LDA that is comparatively lower than this, which is due to the addition of extra loads on to the structure. The 8-storey structure had much more vibration comparative to all three storeys. Figure 8d–f shows the vibrations of middle column removal. There are no vibrations in all three

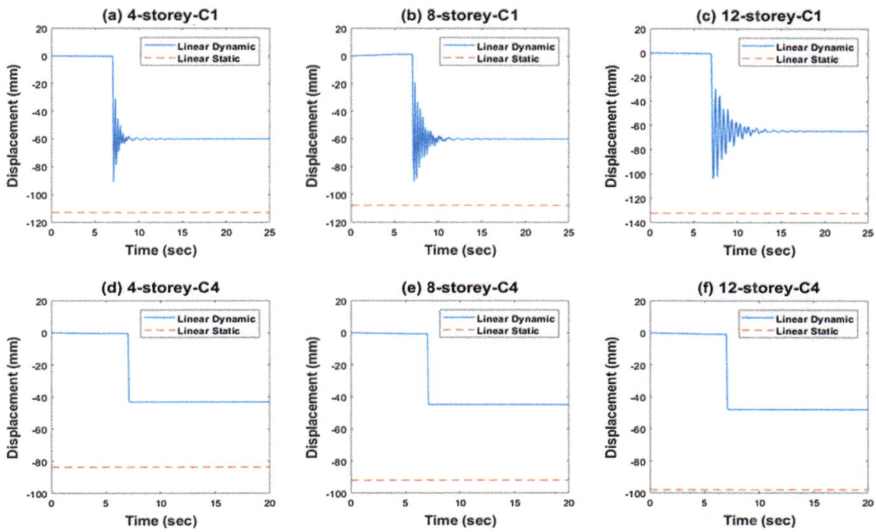

Fig. 8 The response of joints after the removal of C1 and C4 column for three different structures

storeys after seventh second and displacements gradually increase with the increase in the storey number.

6 Conclusions

Progressive collapse analysis was carried out for 4-, 8- and 12-storey reinforced concrete structures. Only two-dimensional numerical modelling has been carried out for proper understanding of the behaviour of the structure.

The obtained results indicate the following points:

- From linear static and linear dynamic analysis, it can be concluded that the removal of the corner column is more vulnerable than the middle column.
- In case of 8-storey there is a slight decrease in corner column removal and linear increment in middle column removal when compared with 4- and 12-storey.
- Middle column does not undergo the vibrations like corner column removal.
- From push-down analysis, 4-storey corner column removal effects in the slight failure of the structure, whereas middle column removal leads to the complete failure. The removal of middle column in the 8-storey structure leads to higher failure than the removal of corner column.
- It clearly indicates that the progressive collapse of the structures decreases with increasing the height of the storey.

The same behaviour can be observed in the real case, which happened in Hyderabad city on 8 December 2016. Even though the study is very fundamental in its nature, the results are very important in understanding the actual behaviour.

References

1. Nimse, R.B., Joshi, D.D., Patel, P.V.: Behaviour of wet precast beam-column connections under progressive collapse scenario: an experimental study. Int. J. Adv. Struct. Eng. **6**, 149–159 (2014)
2. From progressive collapse wikipedia website https://en.wikipedia.org/wiki/Progressive_collapse
3. From times of India epaper website http://epaperbeta.timesofindia.com
4. General Service Administration (GSA): Progressive collapse analysis and design guidelines for new federal office buildings and major modernization projects. General Service Administration, Washington DC (2003)
5. Marjanishvili, S.M.: Progressive analysis procedure for progressive collapse. J. Perform. Constructed Facil. **18**, 79–85 (2004)
6. Tsai, M., Lin, B.: Investigation of progressive collapse resistance and inelastic response for an earthquake-resistant RC building subjected to column failure. Eng. Struct. J. **30**, 3619–3628 (2008)
7. Zahrai, S.M., Ezoddin, A.R.: Numerical study of progressive collapse in intermediate moment resisting reinforced concrete frame due to column removal. Civil Eng. Infrastructures J. **47**(1), 71–88 (2014)
8. Petrone, F., Shan, L., Sashi, K.K.: Modeling of RC frame buildings for progressive collapse analysis. Int. J. Concrete Struct. Mater. **10**, 1–13 (2016)
9. Weng, J., Tan, K.H., Lee, C.K.: Modeling progressive collapse of 2D reinforced concrete frames subject to column removal scenario. Eng. Struct. **141**, 126–143 (2017)
10. CSI, ETABS 2015 V-15.2.2.0: Integrated finite element analysis and design of structures basic analysis reference manual. Computers and Structures Inc
11. Indian Standard 456:2000: Code of practice for plain and reinforced concrete (Fourth Revision)
12. Indian Standard 875–1987 (Part 1, 2 and 5): code of practice for design loads (Other than Earthquake) and for Buildings and Structures. (Second Revision)

Tri-band Semicircular Slit Based Wearable Antenna for Defense Applications

Rama Sanjeeva Reddy and Amulya Kumar

Abstract This article presents a novel tri-band semicircular slit based wearable antenna covering defense and LTE applications. Initially, the proposed antenna is constructed with three adjusted minor semicircular slit cuts on the right top layer of the patch to obtain multiple responses at 2.48, 3.5, and 5 GHz. Simulations with iterative designs are carried using CST Microwave Studio EM simulator and experimentally measured −10 dB return loss-impedance bandwidths are 34, 99, and 115 MHz, respectively, on the FR4 laminate. In addition, a parametric study is carried on to use the same prototype model as a wearable/off-body antenna by choosing *Jeans* substrate with a dielectric constant of 1.7. At last, it is found that the optimized antenna with better radiation patterns preferably at low resonant band and also a stable high gain is obtained over all the operating bands.

Keywords Wearable antenna · Off-body communication · LTE
Semicircular slit · Gain

1 Introduction

Wearable technology is widely popular in recent times for both in civilian and military domains. Significance of this technology introduced in military domain would take time owing to the defense primarily ensuring several security measures, safety of army personnel, and simplified operations [1]. Transforming antenna design to be the type of wearable, one must be focused on selecting suitable material types, both for the nonconductive as well as conductive categories, with a perfect topology. Evolution of wearable clothes to be as laminates is an obvious choice implanting

R. S. Reddy (✉)
Faculty, Department of ECE, BVRIT, Narsapur, Medak (Dt) 502313, Telangana, India
e-mail: sanjeev.reddy@bvrit.ac.in

A. Kumar
Department of ECE, BVRIT, Narsapur, Medak (Dt) 502313, Telangana, India
e-mail: amulyakumar.a@bvrit.ac.in

D. Deb et al. (eds.), *Innovations in Infrastructure*, Advances in Intelligent Systems and Computing 757, https://doi.org/10.1007/978-981-13-1966-2_55

on to the garments [2]. Tsolis et al. explain the textile substrate material property, which varies with the choice of the material and the operating frequency, which are specially designed for emergency rescue workers and space applications [3]. A dual-band antenna using a U-slot-shaped patch implanted using a copper tape assembled with a fleece-type fabric substrate is discussed by Salonen et al. [4]. Printed substrate to textile transition is investigated using FR4 material with planar design operating at a single commercial band of 2.45 GHz [5]. Person wearing the jacket must hold comfort with prior selection of wires and copper tapes, which are conductively flexible. Linear polarized radiation patterns are obtained due to symmetrical strips in the design of CPW-fed slot antennas and found a low-gain value observed at low-frequency band [6]. Li et al. discuss a dual rectangular ring shape with open-ended monopole antenna used on telephone handsets and the peak gain varies between 1.22 and 4.41 dBi [7]. Few other antenna prototypes with L-shaped slits for WLAN operation have been reported in Kim et al. [8]. However, these existing works limit with narrow bandwidths, low efficiency, and have poor radiation patterns. Printed planar monopole antennas are attractive for defense and LTE applications due to the advantages of high-impedance bandwidth and low profile. Few antenna designs existing are covering the significant bands, but it is found that the there is a mismatch impedance at few operating bands [9]. Hertleer et al. show the flexibility of using textile materials for the design of antennas working with dedicated frequency of operations and available electro-textiles such as Flectron, Sheildit, super, and Zelt support the antenna prototype for better improvement of antenna efficiency [10]. Meandered dipole with 30 cm length is realized with a copper adhesive foil material and demonstrated on a worn jacket that operates at a center frequency of 406.05 MHz with a reflection coefficient of −25 dB as discussed by Serra et al. [11].

In this paper, a novel multiband semicircular slit-based planar wearable antenna is designed, simulated, and measured. The antenna consists of three semicircular slit cuts on the right side of the top-layered patch satisfying the impedance characteristics involved in design criteria. The simulated model provides the change in bandwidth at higher resonant modes and generates tri-band operation. The radius R and probe feed positions are managed to obtain the best optimal dimensions from multiple iterations, which enhance the impedance bandwidth of the antenna at lower and higher bands. The designed prototype is evaluated for FR4 substrate (4.4) and Jeans substrate (1.7). It is observed that a stable gain is obtained for the wearable antenna design when compared to the FR4 substrate. Wearable antenna dimensions are optimized and simulated with the help of commercial three-dimensional electromagnetic (EM) CST Microwave Studio software tool. Good agreement is found from the comparison of obtained measured and simulated results.

2 Design and Parametric Analysis

The conventional rectangular shape patch is chosen with dimensions $L_1 \times W_1$. Dual narrowed rectangular sections on top and bottom portions of the patch are modified as shown in Fig. 1a. By varying the dimensions of the patch, an additional resonant mode is obtained accompanied with improvement in bandwidth.

2.1 Design Parameters

The parametric analysis due to the effect of both FR4 and Jeans wearable dielectrics with coaxial probe feed is studied in this section. The resonant frequency along the patch with TM_{11} resonant mode gives broadband response and semicircular slit that only modified the operating frequency of TM_{21} mode. Different iterative shapes with single semicircular cut, two semicircular cuts, and three semicircular cuts are as shown in Fig. 1b, c, and d , respectively. Coaxial probe feed is preferred compared with other feeding techniques because it is easy to fabricate and the feed is placed at preferred location on the patch layer in order to match the impedance value of 50 ohms.

By compensating the probe feed at different points using a trial and error method, it is noticed that gradual improvement in return loss (dB) and impedance bandwidth (Hz) are observed. Finally, optimized feed point is chosen at which better return loss and bandwidth are obtained. The dimensions parameters for the simulated structure are optimized with $L_1 = 38.2$ mm, $W_1 = 28$ mm, $A = 20$ mm, $B = 20$ mm, $C = 4$ mm, $D = 18.2$ mm, $R_1 = R_2 = R_3 = 3.5$ mm, and probe feed location at (5.1, 5.2 mm).

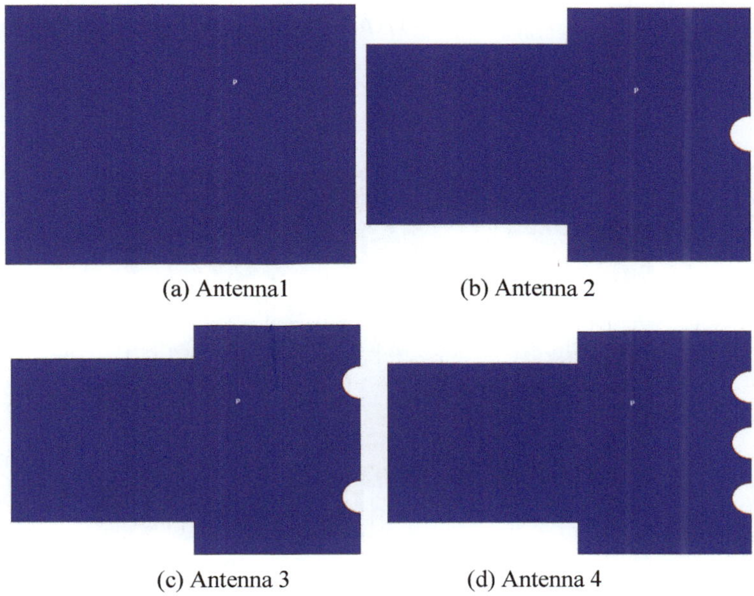

Fig. 1 Design of antenna iterations **a** Antenna1 **b** Antenna2 **c** Antenna3 **d** Antenna4

2.2 Effect of Different Dielectric Constants

The designed prototype is typically simulated on CST 3D simulator considering both FR4 substrate and Jeans dielectric used for wearable antennas as represented in Fig. 2. The significance of antenna with reasonable large bandwidth is designed, which inherently operates the complete ISM band on one end and shifts in resonant mode due to textiles. However, the thickness of the weft and warp yarns in the woven variable dimensions of the textile and presence of the human body are also have significant importance in the proposed antenna. The rectangular shaped geometry with semicircular slits is preferred due to its simple shape, generally, implanted on electro structures that make to determine the ideal choice in designing the prototype model. A small variation in the dimensions of the conductive fabric as a textile antenna would result in a slight shift or change in the antenna characteristics, as represented in Fig. 3. Generally, the e-fabrics hold a very low dielectric constant due to the property of porous materials and the presence of air makes the relative permittivity to unity. The low dielectric constant makes the right choice for designing the wearable antenna for effective propagation in the form of substrates. Impedance bandwidth and antenna efficiency metrics of a rectangular patch antenna is mainly determined by the right choice of substrate dielectric constant value and its thickness.

The relevant critical parameter is the conductivity of the fabric, σ, holding the units as Siemens per meter (S/m) as part of antenna design. Equation (1) gives the relation between the surface resistivity, ρ_s, and the thickness of the fabric

$$\sigma = 1/(\rho_s * t). \tag{1}$$

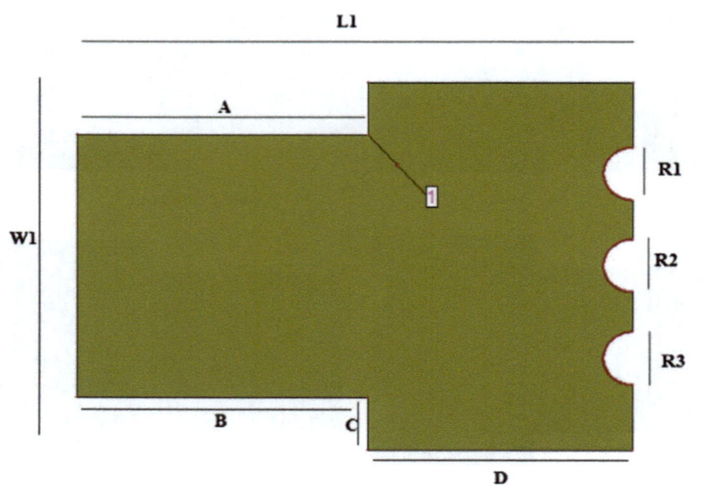

Fig. 2 Proposed simulation structure for dielectric constants 4.4(FR4) and 1.6 (Jeans)

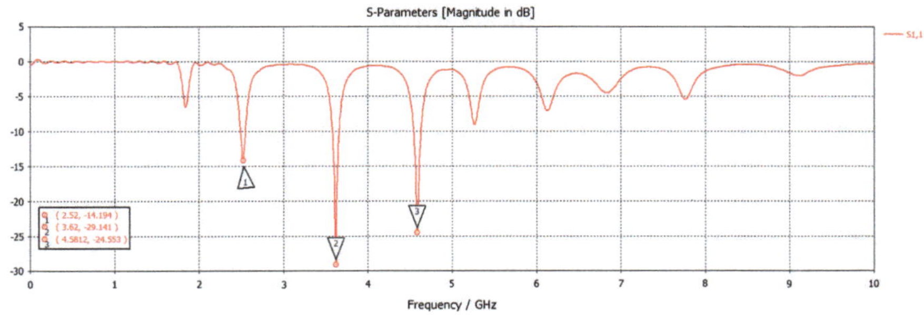

Fig. 3 Return loss characteristic response for dielectric constants 4.4 (FR4)

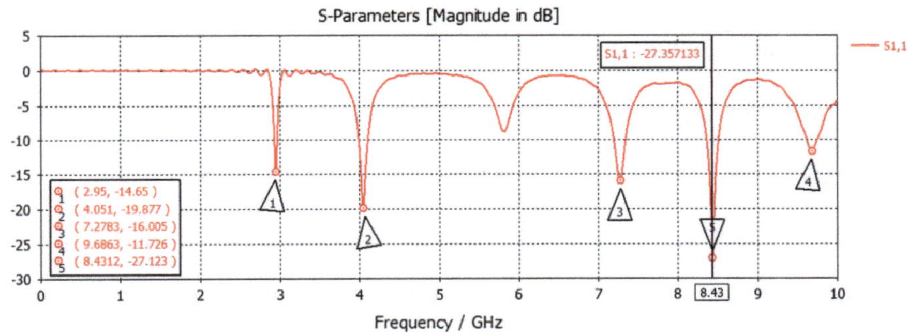

Fig. 4 Return loss characteristic response for textile dielectric constant 1.7 (Jeans)

The ideal choice of conductive fiber for the top and bottom planes is also very important to assure good performance of the antenna, besides, the substrate constant. The next section discusses the return loss characteristics for the proposed design in both planar and wearable antennas.

3 Results and Discussion

Modified rectangular patch of dimensions (38.2 * 28) mm^2, which is deposited on an FR4 high-frequency laminate of relative permittivity ($\varepsilon_r = 4.4$) with the given thickness and the return loss response is characterized as displayed in Fig. 3.

Figure 4 shows the return loss response for the wearable patch antenna design with the dielectric constant of 1.7 (Jeans cloth) with the low-thickness value of 0.8 mm. The return loss characteristics for the best iteration are represented in Fig. 5.

Table 1 indicates the values of different resonant frequencies for both FR4 and Jeans substrate for the proposed antenna design. The major contribution of replacing the design with the textile is to extract the best bandwidth at low-range resonant frequencies and also improving the antenna performance. The VSWR value is main-

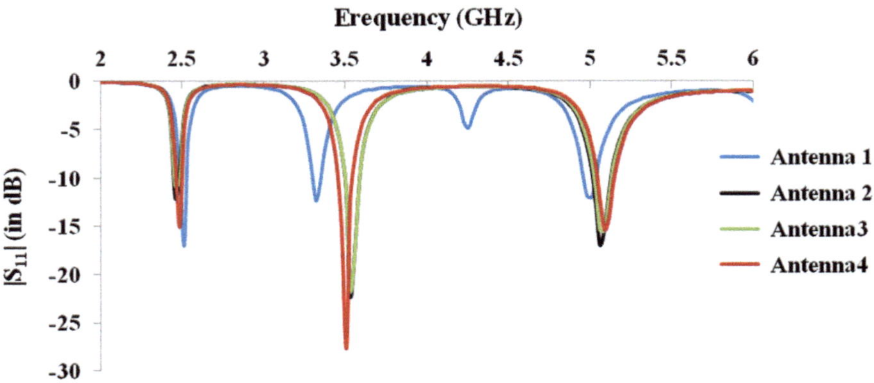

Fig. 5 Representation of best iterative return loss response (Ant 1-4)

Table 1 Comparison of performance metrics for the proposed design for FR4 and Jeans dielectrics

Dielectric constant (ε_r)	Thickness (mm)	Resonating frequencies (GHz)	Impedance bandwidth (MHz)	Gain (dBi)
FR4 (4.4)	3.2	2.48, 3.5, and 5	34, 99, and 115	5.19, 8.25, and 5.78
Jeans(1.7)	0.8	2.95, 4.05, and 7.27	45, 122, and 164	4.32,4.56, and 5

tained less than two for all the obtained resonant frequency bands. The values of impedance bandwidth are also represented in the formulated table. It is also observed that a stable gain of around 5.5 dBi is obtained which is as represented in the table.

Pictorial representation of fabricated antenna along with the measured return loss curve with the help of Keysight E5710C Vector Network Analyzer (VNA) is as shown in Fig. 6. Good performance of the designed wearable patch in terms of high-impedance bandwidth, available stable gain, and better radiation patterns are obtained with Jeans dielectric and the tabulated values display the improved gain at respective resonant bands. The decrease in the thickness values of wearable dielectric, increases the efficiency of the designed antenna. Far-field radiation patterns are represented in Fig. 7 for the lower resonant frequency band (2.95 GHz) for $\phi = 0^{\circ}$ and 90°. The proposed antenna presents a higher gain when compared to the previous works at these bands. Because of slits cut, it is found that the radiation patterns are consistent at different operating bands of the structure.

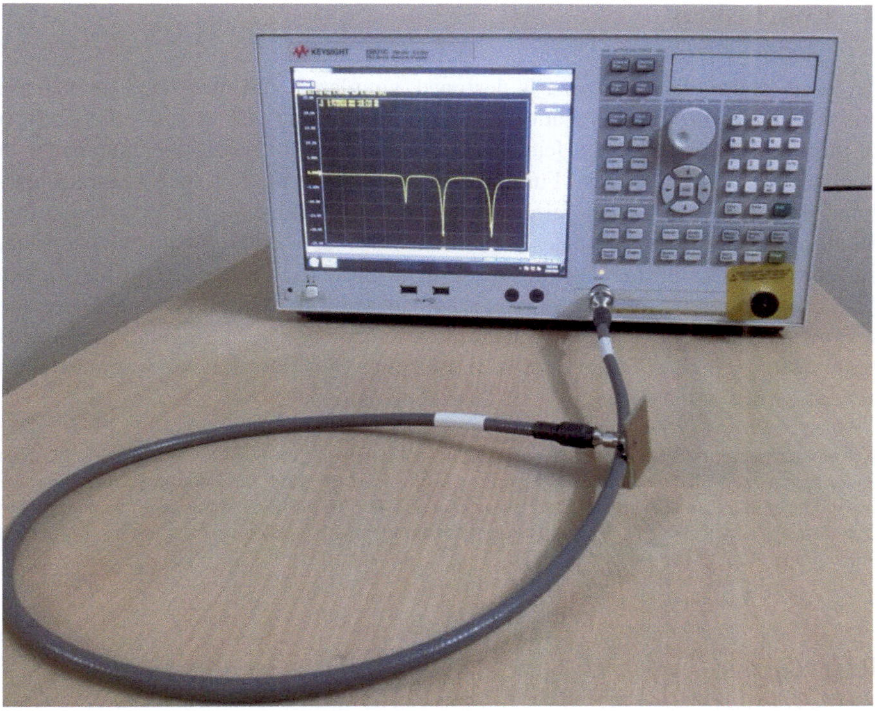

Fig. 6 Fabricated prototype tested using vector network analyzer

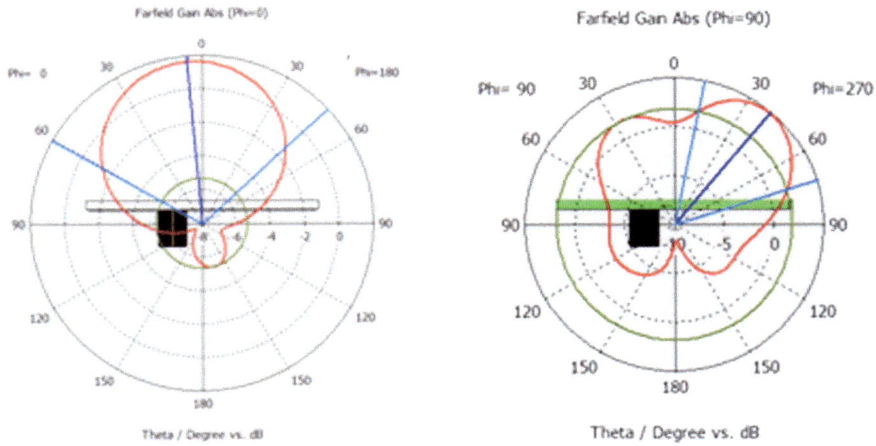

Fig. 7 Radiation patterns for the lower resonant frequency band (2.95 GHz) for $\phi = 0^{\circ}$ and 90°

4 Conclusions

Tri-band textile planar antenna is designed, simulated, and fabricated for defense and ISM band wireless applications. The parametric study is carried with the effect of placing FR4 substrate and Jeans wearable substrate for the design antennas with thickness of 3.2 and 0.8 mm. Further, the performance metrics such as return loss-impedance bandwidths and gain are measured and analyzed. It is also found that the designed prototype resonates at multibands, though with low-substrate constant and low thickness of the wearable substrate as observed for the case of Jeans cloth (1.7). The proposed prototype generates tri-bands and also maintains a stable average gain of 5.5 dBi. Better radiation patterns are displayed at low-operating range of frequencies, which can be effectively used for off-body wireless communication covering most of the short-range defense applications.

Acknowledgements The authors would like to acknowledge the support of SERB/ECR team under Department of Science and Technology with financial grant No. 000585 for carrying this work. The authors are also greatly indebted to Prof. NVSN. Sarma, Dr. IA. Pasha, and colleagues of Department of ECE, BVRIT, Medak for their encouragement, support, and inspiration throughout in the preparation of this research article.

References

1. Hall, P. S., Hao, Y.: Antennas and propagation for body centric communications. In: European Conference on Antennas and Propagation (EuCAP) (2006)
2. Salvado, R., Loss, C., Goncalves, R., Pinho, P.: Textile materials for the design of wearable antennas: a survey. Sensors **12**(11), 15841–15857 (2012)
3. Tsolis, A., Whittow, W.G., Alexandridis, A.A., Vardaxoglou, J.C.: Embroidery and related manufacturing techniques for wearable antennas: challenges and opportunities. Electronics **3**(2), 314–338 (2014)
4. Salonen, P., Rahmat-Samii, Y., Hurme, H., Kivikoski, M.: Dual-band wearable textile antennas. In: Proceedings IEEE International Symposium Antennas Propagation, pp. 463–466 (2004)
5. Nurul, H.M.R., Malek, F., Soh, P.J., Vandenbosch, G.A.E., Volski, V., Ooi, S.L., Adam, I.: Evaluation of a wearable hybrid textile antenna. In: Antennas and Propagation Conference (LAPC). Loughborough, pp. 337–340 (2010)
6. Hu, W., Yin, Y.Z., Fei, P., Yang, X.: Compact tri-band square slot antenna with symmetrical L-strips for WLAN/WiMAX applications. IEEE Antennas Wireless Propag Lett. **10**, 462–465 (2011)
7. Li, B., Yan, Z.-H., Wang, C.: Dual rectangular ring with open-ended CPW-fed monopole antenna for WiMAX/WLAN applications. Progr. In Electromagnetics Res. Lett. **25**, 101–107 (2011)
8. Kim, T.H., Park, D.C.: Compact dual-band antenna with double L slits for WLAN operations. IEEE Antennas Wirel. Propag. Lett. **4**(1), 249–252 (2005)
9. Zhai, H., Ma, Z., Han, Y., Liang, C.: A compact printed antenna for triple-band WLAN/WiMAX applications. IEEE Antennas Wireless Propogation Lett. 12, 65–68 (2013)
10. Hertleer, C., Tronquo, A., Rogier, H., Langenhove, L.: The use of textile materials to design wearable microstrip patch antennas. Text. Res. J. **78**(8), 651–658 (2008)

11. Serra, A.A., Nepa, P., Manara, G.: A wearable multi antenna system on a life jacket for Cospas Sarsat rescue applications. In: Antennas and Propagation (APSURSI), 2011 IEEE International Symposium pp. 1319–1322 (2011)

Experimental Comparison of TIG and Friction Stir Welding Process for AA6063-T6 Aluminum Alloy

Navneet Khanna, Bhavesh Chaudhary, Jay Airao, Gaurav Dak and Vishvesh J. Badheka

Abstract Welding is an essential process for infrastructure development since its wide application in automobile, aerospace, pipelines, power plant, construction, etc. Among welding of different materials, welding of aluminum alloys remains the point of focus for researchers because aluminum alloys are widely used in the architectural, structural, aerospace, and automotive applications. But welding of aluminum is a challenging task as aluminum easily reacts with the atmospheric air and causes serious defects. Again, repeatability of welding depends on its control on welding speed and other processing parameters. The present paper focuses on the defects and strength analysis of AA6063-T6 aluminum alloy welded by two different welding processes: AC TIG welding process and solid-state joining process called as friction stir welding (FSW). The defects and tensile strength for both welding processes were analyzed by using X-ray radiography and universal testing machine, respectively, and the results were compared. The results revealed that FSW can be a better option for welding AA6063-T6 compared to TIG welding.

Keywords AA6063-T6 · TIG welding · Friction stir welding · Defect analysis Tensile strength

1 Introduction

Welding industry in India is heading toward growth since the government has started spending on infrastructure building such as construction and energy. Welding is an essential element in industrial infrastructure, shipbuilding, oil and gas, automotive

N. Khanna (✉) · B. Chaudhary · J. Airao · G. Dak
IITRAM University, Ahmedabad 380026, India
e-mail: navneetkhanna@iitram.ac.in

V. J. Badheka
PDPU, Raysan, Gandhinagar 382421, India
e-mail: vishvesh.badheka@spt.pdpu.ac.in

D. Deb et al. (eds.), *Innovations in Infrastructure*, Advances in Intelligent Systems and Computing 757, https://doi.org/10.1007/978-981-13-1966-2_56

Table 1 Nomenclature

TIG	Tungsten inert gas welding
GTAW	Gas tungsten arc welding
FSW	Friction stir welding
DOE	Design of experiment
AC	Alternating current
DCSP	Direct current straight polarity
DCRN	Direct current reverse polarity
mm/min	Millimeter per minute
L/min	Liter per minute
UTM	Universal testing machine
NSD	No significant defect
UC	Undercut

manufacturing, aerospace, and many more. As the technology continues to advance, many new techniques have been developed for the joining of rare earth element.

Among different materials, aluminum is widely used in many industries due to its excellent mechanical properties. The use of heat-treatable Al alloys in aerospace applications is very popular due to their high strength-to-density ratio. However, these aluminum 6XXX series alloys are difficult to weld by conventional fusion welding technique. Al–Mg–Si alloys belong to 6XXX series as magnesium and silicon are major alloying elements. The addition of these elements improves the mechanical properties. The hardening element in AA 6063 alloy is Mg_2Si. The 6XXX series alloys are sensitive to weld metal cracking. A weld metal with a composition close to that of the parent metal may age-harden naturally or may be artificially aged to achieve a strength approaching, but never matching, that of the aged parent metal. The strength losses in the 6XXX series alloys are less in the naturally aged conditions than in the artificially aged condition [1] (Table 1).

The present paper focuses on the effect of the fusion welding process (TIG) and a solid-state joining process (FSW) on tensile strength and defect formation for an AA6063 alloy.

In TIG welding, an arc is created between nonconsumable tungsten electrode and the base metal. The arc melts the workpieces and joining takes place on solidification. The high temperature induced in fusion welding processes produces several microstructural changes, which are responsible for the alteration of the mechanical properties of the weld material [2]. The peak current and pulse frequency are directly proportional to the tensile properties of the welded joints, and the base current and pulse on time are having an inversely proportional relationship with the tensile strength [3]. The increase in welding current leads to an increase in the depth/width ratio and deteriorates the surface appearance of the welded seams with TiO_2 coating. The welded joints with TiO_2 coating exhibit a deeper weld penetration and larger grain size compared to the welded joint without TiO_2 coating [4]. Excitation current of arc-ultrasonic has a great effect on the pores distribution and tensile property [5].

Table 2 AA6063-T6 material elements and their proportion

Name	Mg	Si	Mn	Cu	Zn	Ti	Cr	Fe
Percentage (%)	0.560	0.425	0.016	0.004	0.001	0.002	0.0	0.105

TIG current and pulse duration are most significant parameters in influencing weld bead width due to increase in heat input to the material [6].

The solid-state joining technique called friction stir welding (FSW) was patented by The Welding Institute (TWI) in 1991. This joining technology uses the advance of a rotating hard steel pin extended by a cylindrical shoulder between two mating workpiece. The frictional heat is generated between the contact surface of the rotating tool and the base metal. This frictional heat is utilized to melt the base metal. The advantage of this process is that there is little or no porosity and no other significant defects are produced as compared to the conventional fusion welding process. However, due to low stability, the hardening precipitates which are responsible for the good mechanical properties of heat-treatable alloys Al alloys (2XXX, 6XXX, and 7XXX series) are shown to be extremely affected by this process [7, 8].

Most of the researchers are focusing on the effect of FSW parameters and tool profiles on tensile properties and investigation of mechanical properties by experimental comparison between FSW and TIG welding for Al alloys [9–16]. In fact, the industrial interest of this study is to evaluate the possible benefits of FSW compared to TIG, considering the lower heat input of the solid-state joining process. The process parameters taken in this study are as per the requirement of the sponsored industry and literature survey. The results have been submitted to the sponsored industry, and the industry is expected to gain improvement in defect-free welds as well as improvement in tensile strength.

2 Experimental Procedure

Rolled plates of AA6063-T6 alloy of 3 mm thickness were taken as workpiece material. The chemical composition of AA6063-T6 is given in Table 2. These plates were fusion welded using TIG (Fig. 1a) and FSW (Fig. 1b). The full factorial design was used for TIG welding. The process parameters and their levels for the TIG welding are shown in Table 3. The FSW experiments were performed on modified traditional milling machine; the rotational speed was 500 rpm and the welding speed was 50 mm/min. The FSW parameters were chosen based on the heat supplied to the weld specimen, which was having the highest tensile strength in TIG welding. In both welding processes, the welding direction coincided with the rolling direction. The defects in each workpiece were analyzed by X-ray radiography. The tensile specimens were taken perpendicular to the welding direction and welding strength was measured on UTM at a rate of 2 mm/min followed by radiography tests.

(a) 

(b) 

Fig. 1 **a** TIG welding of AA6063-T6 alloy; **b** FSW of AA6063-T6 Alloy

Table 3 Process parameters and their levels

Welding parameter	Unit	Level	Factors of level		
			1	2	3
Arc current	A	3	130	140	150
Welding speed	mm/min	3	150	75	50
Gas flow rate			14–15 liter/min		
Torch angle			70–80°		
Filler angle			10–15°		
Filler material			ER5356		
Filler diameter			1.5 mm		
Tungsten diameter			3.2 mm		
Root gap			1.5 mm		

3 Results and Discussion

The defects revealed by the radiography testing for the TIG welding are shown in Table 4.

3.1 Radiography

The X-ray radiography is the most sophisticated nondestructive testing used to determine the internal defects produced in the weld metal, which are not visible to the naked eye. The X-ray radiography technique uses the X-ray source, and the beam of X-rays passes through the object being inspected and develops the latent image on the radiography film. This latent image then passes through various processing techniques such as developing, rinsing, fixing, and washing. All these processes are performed in dark room because these radiography films are light sensitive. The film

Table 4 Radiography of TIG welded specimens

Exp. No.	Specimen No.	Parameters		Defects
		Arc current (A)	Welding speed (mm/min)	
1	1	130	150	Porosity, inclusion
2	5	140	150	Lack of penetration, undercut
3	6	140	75	Porosity
4	2	140	50	Porosity
5	3	130	75	No significant defects
6	4	130	50	Porosity
7	7	150	150	Porosity
8	8	150	75	No significant defects
9	9	150	50	Porosity

darkness (density) will vary with the amount of radiation absorbed by different parts of weld metal, less dense area will appear as darker areas on the film to give the shadow picture of the internal features of the test sample once the film has been processed, and therefore, voids, porosity, slag, cracks, and defects of geometry can all be identified.

Porosity

In the experiment nos. 1, 3, 4, 6, 7, and 9, the porosity (entrapment of gas in the weld metal) produced results in a cavity. This cavity may be localized or aligned because of the following reasons: as aluminum is highly reactive material, it easily reacts with the atmospheric air, in order to prevent it from atmospheric contamination; sufficient supply of shielding gas must be ensured, and the shielding gas creates the layer of inert gases around the weld pool, which prevents the atmospheric air from being entrapped into the weld pool. Also, the quality of the shielding gas plays an important role, and the oxygen in the shielding gas must be within the permissible limit.

Effect of Current and Welding Speed on Porosity

If the arc current increases, it will increase the overall heat supplied to the weld pool and thereby increases the rate of absorption of hydrogen in the molten metal. A similar effect is achieved by reducing the travel speed, i.e., by reducing the travel speed, heat input increases, and it eventually increases the rate of hydrogen absorption in the weld pool.

Lack of Penetration

Lack of penetration fails to attain full penetration specified by design. This is observed in the experiment no. 2 because of the too high welding speed. If the welding speed is too high, proper penetration may not be achieved, which has produced the lack of penetration.

Undercut
Undercut is the notch parallel to weld at weld toe, which is observed in the experiment no. 2. This is because of the too high welding speed or inadequate cleaning.

No Significant Defects
It means the weld is free of defects and the quality of the weld is good. In the experiment nos. 5 and 8, there are no significant defects observed. It depends on the skills of the operator much compared to current and welding speed.

Generally, the porosity is confined to the weld metal. It arises from gas dissolved in the molten weld becoming trapped as it solidifies than forming bubbles in the solidified weld. The main source of the porosity in the weld is the hydrogen. Hydrogen has a high solubility in the molten state but very low solubility in the solid state. Porosity increases because of contamination from the wire and filler rod. Magnesium raises the solubility and reduces absorption of hydrogen by as much as twice at 6% magnesium. During the welding, the moisture decomposes in the arc to give hydrogen resulting in large porosity. The source of contamination may come from the gas hoses as the moisture condenses inside the hose and entrains in the shielded gas. The hydrogen dissolved within the aluminum workpiece also causes the porosity [1].

3.2 Tensile Strength

The ultimate tensile strength of the all the workpieces welded by TIG and FSW was measured on the universal testing machine (Tables 5 and 6, respectively). For the better accuracy, three samples were cut from each welded piece with the help of electric discharge machining. The specimens of the tensile test were made according to the standard ASTM-E8M that is shown in Fig. 2. The plot for the arc current and welding speed vs. tensile strength are shown in Figs. 3 and 4, respectively. During the tensile test, all the specimens were broken from the weld zone. This concludes that the weld zone is the weakest zone among the HAZ and WM. This indicates the acceptance of TIG welding process in terms of tensile strength.

It is found that as the welding speed increased keeping current as constant, the weld bead penetration decreased, which decreased the welding strength. At high welding speed because of the less heat input, the weld bead penetration decreases and so the welding strength. As the arc current increased from 130 to 150 A keeping welding speed as constant, more weld bead penetration occurred because of the more heat input, which increased the tensile strength of the weld bead.

For the lower value of arc current (130 A) and high value of the welding speed (50 mm/min), heat input to the weld bead was more and so more penetration occurred, which leads to higher value of tensile strength. Also, at low welding speed

Table 5 Tensile strength experimental results of TIG welding

Exp. No.	Specimen No.	Current (A)	Welding speed (mm/min)	Ultimate tensile strength (MPa)				Defects
				Trial 1	Trial 2	Trial 3	Avg.	
1	1	130	150	96.03	83.97	134.7	**104.9**	Porosity
2	5	130	75	127.1	127.0	128.8	**127.6**	NSD*
3	6	130	50	126.4	128.4	127.8	**127.5**	Porosity
4	2	140	150	109.9	111.4	110.8	**110.7**	LOP, UC
5	3	140	75	131.7	127.1	135.8	**131.5**	Porosity
6	4	140	50	122.9	123.9	125.0	**124**	Porosity
7	7	150	150	124.1	121.8	124.4	**123.4**	Porosity
8	8	150	75	127.3	119.9	103.4	**116.8**	NSD*
9	9	150	50	128.9	130.1	131.0	**130.0**	Porosity

Table 6 Tensile strength experimental result of FSW

Pin shape	Pin diameter	RPM	Welding speed (mm/min)	Ultimate tensile strength (MPa)				Defects
				Trial 1	Trial 2	Trial 3	Avg.	
Cylindrical	3 mm	1000	50	116.67	116.51	117.73	**116.97**	NSD*

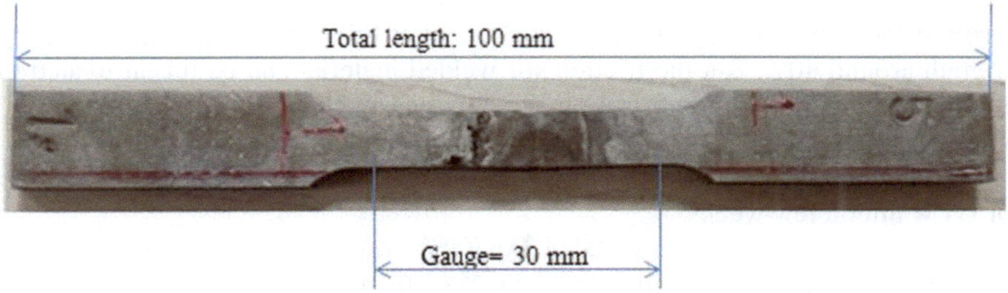

Fig. 2 Tensile specimen

(150 mm/min) and high value of arc current (150 A), the tensile strength obtained was quite good.

The maximum welding strength was obtained at the arc current of 140 A and welding speed of 75 mm/min for third specimen. Though the porosity was present in the third specimen, welding strength obtained was higher because the porosity was negligible that did not affect the tensile strength of the specimen. Table 6 shows the tensile strength obtained for friction stir welded joint. The UTM was carried out immediately after the FSW.

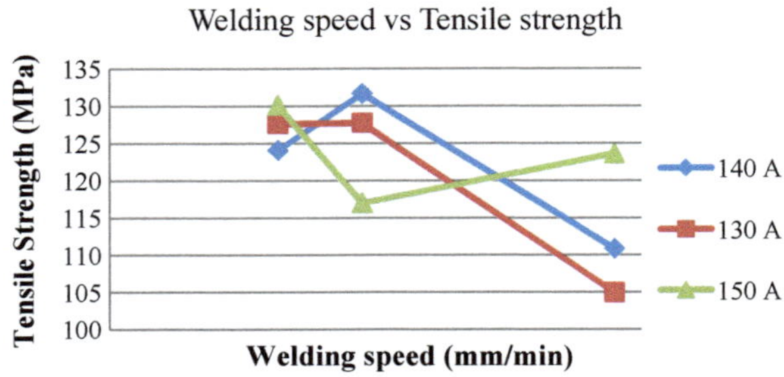

Fig. 3 Plot of welding speed vs. tensile strength for constant arc current

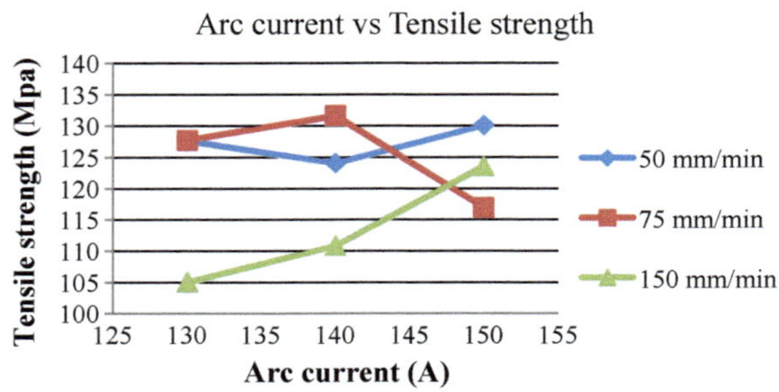

Fig. 4 Plot of arc current vs. tensile strength for constant welding speed

The above tensile strength was compared with the tensile strength of the fifth specimen, which was having no significant defect in TIG welding, and it was found that the tensile strength in FSW was lower than TIG welding. The difference in tensile strength was not significant. Generally, after FSW, there is a loss of welding strength around 40%. But the friction stir welded material can be reheat treated to restore T6 temper condition to get the strength nearly equal to the parent material. Also, the material can be naturally aged to restore some of its strength as well. The natural aging takes 2–3 weeks to restore the strength, and hence, it is preferred to go for UTM after a few weeks.

Also, it can be clearly seen in Table 5 that the strength in TIG welding is higher compared to FSW for the fifth specimen. But, for the same heat input in FSW, this weld was accepted since no significant defect was observed. From the past experiences, many researchers have suggested not to go for UTM immediately after FSW. The reason behind this suggestion was the natural aging, which improves the tensile strength [14, 16].

Strength Versus Porosity
Generally, as the amount of porosity increases in the weld metal, it reduces the tensile strength of the joint. Fine porosity when present in sufficient quantity to contribute a total area comparable to that of large pores also causes a loss of strength. Small pores, if present in sufficient numbers to influence the area loss, will reduce the strength accordingly [1]. But from the radiography films, it was found that the size of porosity obtained in specimens 3, 4, 6, 7, and 9 was negligible, and hence, the porosity did not affect the welding strength too much compared to the welding strength of specimens

5 and 8 with no significant defect. The friction stir welded specimen was also found with no significant defect.

4 Conclusions

The effect of two welding process, i.e., fusion (TIG) and solid-state (FSW) welding processes, on defects and tensile strength of AA6063-T6 alloy was investigated.

- It was observed that the depth of penetration of weld bead increased with increase in current quantity; however, it decreased with an increase in welding speed. The maximum welding strength 131.58 MPa was observed at the arc current of 140 A and welding speed of 75 mm/min for third specimen in TIG welding.
- Tensile strength was observed higher with lower weld speed (50 mm/min) and concluded that lower range of weld speed is suitable for achieving maximum tensile strength; however, tensile strength was observed higher with higher value of current (150 A), which showed that higher the value of current up to certain limit is suitable for achieving maximum tensile strength. Further, the effect of these two parameters was observed on first specimen, which yields the minimum strength of 104.9 MPa because of the high welding speed (150 mm/min) and low value of current (130 A).
- Though there was a porosity observed in third specimen, it yielded the highest tensile strength of all the specimens, which pointed that the effect of small porosities was negligible on welding strength. However, the results of seventh and eighth specimens deviated from the strength pattern observed in other specimens may be because of error in holding specimen in UTM.
- There was no significant difference found in tensile strength of TIG and FSW though the specimen was not allowed for natural aging after FSW. If the specimen would have been tested after natural aging, the tensile strength in FSW would have been more compared to TIG welding according to past experiences.

Acknowledgements The present work was carried out in collaboration with a renowned industry "Power high Engineers Pvt. Ltd." located at Ahmedabad. We are thankful to "**Power high Engineers Pvt. Ltd**." for their support and providing the opportunity to work with them.

References

1. Mathers, G.: The Welding of Aluminum and Its Alloy. CRC Press, New York (2002)
2. Muncaster, P.: Practical TIG (GTA) Welding. Wood head Publishing, Cambridge (1991)
3. Kumar, S.: Influences of pulsed current tungsten inert gas welding parameters on the tensile properties of AA 6061 aluminium alloy. Mater. Des. **28** (2007)
4. Shen, S.: Effects of welding current on properties of A-TIG welded AZ31 magnesium alloy joints with TiO_2 coating. Trans. Nonferrous Metal Soc. China **24** (2014)

5. Zhu, Q.: Effects of arc-ultrasonic on pores distribution and tensile property in TIG welding joints of MGH956 alloy. Fusion Eng. Des. **89** (2014)
6. Raghvendran, G.: Optimization of hybrid laser—TIG welding of 316LN steel using response surface methodology (RSM). Laser Opt. Eng. **94** (2017)
7. Lohwasser, D., Chen, Z.: Friction Stir Welding. CRC press, Woodhouse Publishing Limited (2010)
8. Thomas.: GB Patent Application No. 9125978 (1991)
9. Squillace, A.: A comparison between FSW and TIG welding techniques: modifications of microstructure and pitting corrosion resistance in AA 2024-T3 butt joints. J. Mater. Process. Technol. **152**, 97–105 (2004)
10. Zhen, B.: Comparison of FSW and TIG welded joints in Al-Mg-Mn-Sc-Zr alloy plates. Trans. Nonferrous Met. Soc. China **21**, 1685–1691 (2011)
11. Cabello, L.: Comparison of TIG welded and friction stir welded Al–4.5Mg–0.26Sc alloy. J. Mater. Process. Technol. **197**, 337–43 (2008)
12. Fahimpour, U.: Corrosion behavior of aluminum 6061 alloy joined by friction stir welding and gas tungsten arc welding methods. Mater. Des. **39**, 329–333 (2012)
13. Lakshminarayanan, K.: Effect of welding processes on tensile properties of AA6061 aluminium alloy joints. Int. J. Adv. Manufact. Technol. **40**, 286–296 (2009)
14. Deng, B.: Effects of Sc and Zr on mechanical property and microstructure of tungsten inert gas and friction stir welded aerospace high strength Al–Zn–Mg alloys. Mater. Sci. Eng. A **639**, 500–513 (2015)
15. Zhao, J.: Comparative investigation of tungsten inert gas and friction stir welding characteristics of Al–Mg–Sc alloy plates. Mater. Des. **31**, 306–311 (2010)
16. Wang, Z.: Comparison of fatigue property between friction stir and TIG welds. Materials **15**, 280–288 (2008)

Optimal Design and Simulation of High-Efficiency SnS-Based Solar Cell

Ayush Bakliwal and Saurabh Kumar Pandey

Abstract Silicon-based solar cells are ruling the present photovoltaic market, but the associated high cost is a hindrance for competing with low-cost fossil fuel, coal-based energy sources. In semiconductor technology, efforts persist on sustainable materials. In this paper, tin sulfide is used as absorber layer for device modeling and discussed for an offset at the heterojunction, carrier density at the absorber/buffer layer. Different factors affecting solar cells outcomes are analyzed, mainly physical parameters of absorber and buffer layer. After optimization of different parameters of the cell module, the photo-conversion efficiency under the standard AM 1.5 illumination condition of the SnS-based photovoltaic cell is 24.87%.

Keywords Conversion efficiency · Defect density · Device modeling
Heterojunction · SnS · Solar cell

1 Introduction

In the modern era, with the advancement of technology, research area in the field of thin-film solar cell has found a great potential due to its higher conversion efficiency and lower manufacturing cost [1]. Many attempts have been made over the past century for making solar cells, which are environment friendly. Physical and chemical properties of each material are different which affects the outcome of the device. A keen observation on individual component is essential for designing the device [2, 3].

For achieving higher conversion efficiency, optical and electrical properties of materials should be taken care of. The various properties include optical band gap, a high optical absorption coefficient, a high quantum yield for excited carriers, and a

A. Bakliwal · S. K. Pandey (✉)
Indian Institute of Technology Patna, Patna, Bihar 801103, India
e-mail: saurabh@iitp.ac.in

A. Bakliwal
e-mail: bakliwal.ee14@iitp.ac.in

D. Deb et al. (eds.), *Innovations in Infrastructure*, Advances in Intelligent Systems and Computing 757, https://doi.org/10.1007/978-981-13-1966-2_57

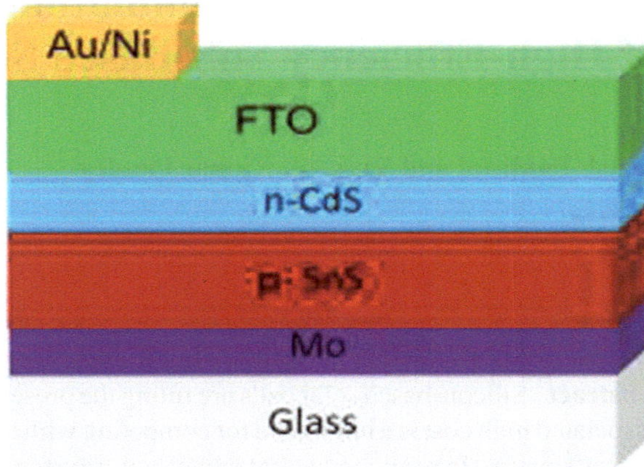

Fig. 1 Device structure of SnS solar cell

long minority carrier diffusion length [4, 5]. Many requirements are to be fulfilled by the absorbing material in the device. SnS is a nontoxic and earth-abundant material used for making solar cells, electronics, and display devices. The various properties of SnS include its direct bandgap of 1.1–1.4 eV and conductivity in both p and n regions [6].

In this paper, we have presented a design approach for the SnS-based solar cell with different constituent layer parameters such as thickness, doping, and alloy composition calibration. The present optimized results give a helpful indication for experimental verification with precise control of process parameters.

2 Device Structure

Different simulating tools are used for analysis of thin-film solar cells. To realize the optimized cell structure, numerical simulations help to understand certain device parameters and processes. We have simulated the SnS solar cell by device simulation software, SCAPS, Solar cell capacitance software to analyze the properties of the hetero-junction solar cells [7]. Maximum seven layers of cells are supported. All device parameters as well as interface study and defect analysis for dark and light can be carried out at particular frequency or spectrum for a range of temperatures. The two layers which have been emphasized in this simulation are n-CdS buffer layer and p-SnS absorber layer. Defect density factor was also taken into consideration during simulating the cell module. By changing the various parameters of materials into SCAPS for different analyses, changes in the efficiency, V_{oc}, $J_{sc,}$ and FF as well as effect of operating temperature are observed. Current transport mechanism of carriers in the heterojunction solar cell is mainly described by a standard system of drift–diffusion Poisson equation with suitable boundary conditions at the interfaces and contacts wherever required (Fig. 1).

Table 1 Defects considered in the SnS absorber during the simulation

Defect type	Custom multilevel, +/0, 0/−
Energetic distribution	Single
Energy level with respect to reference E_t	f0.22; 1.10 g eV
Capture cross section electrons	10^{-15} cm^2
Capture cross section holes	10–15 cm^2
N_t (total) grating uniform	10^{14} cm^{-1}
Radiative recombination coefficient (cm^3/s)	10–14
Auger electron capture coefficient (cm^6/s)	10–29
Auger hole capture coefficient (cm^6/s)	10–29

Table 2 Basic parameters for simulation of SnS-based solar cell

Parameters	SnS (P-Type)	CdS (N-Type)
Thickness	2 mm	50 nm
Bandgap (ev)	1.35 [8]	2.42 [9]
Electron affinity (eV)	3.59 [8]	3.47 [11]
Dielectric permittivity (relative)	12.41 [10]	10 [12]
Electron thermal velocity (cm/s)	10^7	10^7[13]
Hole thermal velocity (cm/s)	10^7	10^7 [14]
Electron mobility (cm^2/Vs)	0.5	350 [15]
Hole mobility (cm^2/Vs)	100 [10]	40 [16]
Absorption constant A (1/cm $eV^{(1/2)}$)	9×10^4 [8]	10^4
Lattice constant/parameters	Orthorhombic (a = 4.33, b = 11.18, c=3.98°A) [10]	Hexagonal (wurtzite) (a = 4.138, c = 6.718 A)

2.1 Defects Considered in the SnS Absorber During the Simulation

Defects are important while designing the solar cell; here, we have considered the recombination factor also, and defect values in the absorber (SnS) are given in Table 1.

The various material parameters for the simulation of device structure are shown in Table 2.

3 Results and Discussion

3.1 Effect of Buffer Layer Thickness and Bandgap Factor

The buffer layer bandgap is changed from 1 to 3.3 eV. It is observed that the efficiency increases for the Eg buffer as shown in Fig. 2a. It is observed that the absorption should be lower for buffer layer for higher efficiency as seen in Fig. 2b. At lower thickness, efficiency is independent of absorption efficiency, and at the high

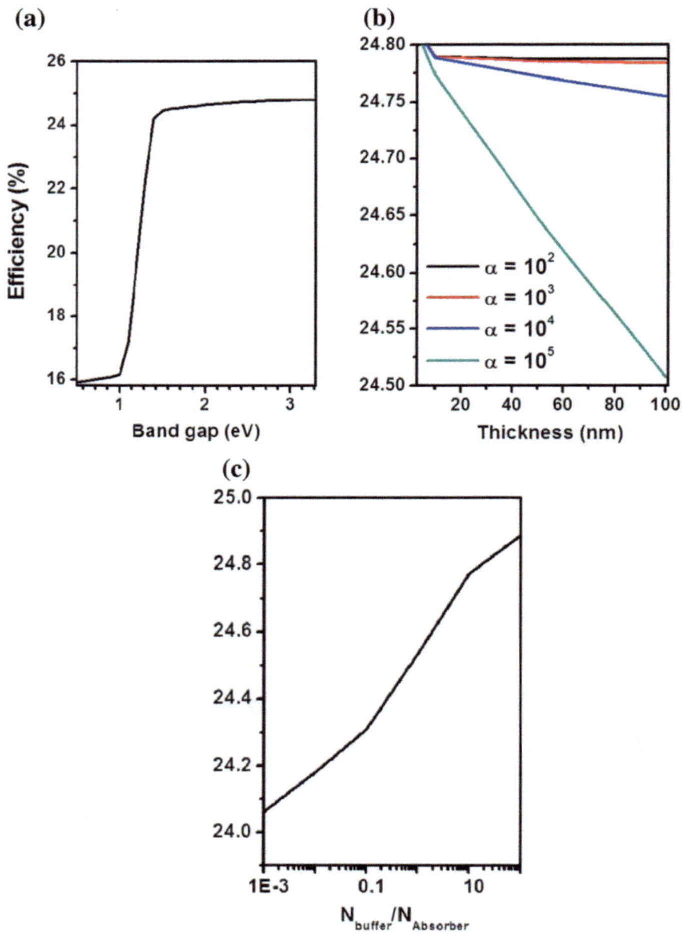

Fig. 2 **a** Cell performance as function of bandgap. **b** Cell performance for different CdS thicknesses. **c** Cell performance due to the ratio of carrier density of the buffer and absorber layer

thickness, the efficiency degrades with high absorption. If the bandgap of the buffer is greater than absorber, then the thickness should be minimum and absorption should be minimum. Figure 2c shows the ratio of doping density in the absorber and the buffer layer. The doping density ratio is plotted from 0.001 to 1000 which is showing a near linear increase in efficiency for the SnS heterojunction. The higher doping in the buffer as compared to the absorber is desirable, essentially to make a pn$^+$ junction. This condition is favorable as symmetric doping density in p and n layers will have excess of electrons populating the interface trap state. Photo-generated electrons in the player crossover to n layer will not recombine with the electron in the trap state as if the interface state is absent for them. Also, the radiative recombination is the maximum, when we have the equal number of the holes and the electrons.

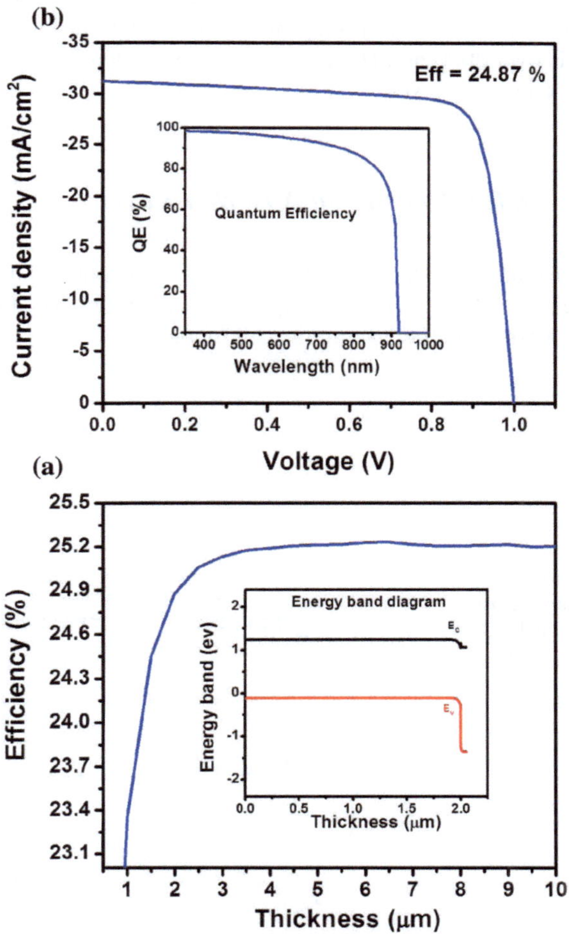

Fig. 3 Effect of SnS thickness on efficiency, V_{OC}, fill factor, J_{SC}

Thus, interface recombination can be lowered by asymmetric doping profile in the pn device.

3.2 *Effect of Absorber Layer Thicknesses*

SnS layer thickness is changed from 500 to 5000 nm for studying the influence of thickness of absorber layer in the cell performance, while material parameters of different layers are kept unchanged. It can be seen from Fig. 3 that both J_{sc} and V_{oc} of the solar cell increase with the increasing SnS layer thickness. As the thickness of absorber layer increases, the number of electron–hole pairs increases, which leads to high radiative recombination rate. Hence, solar cell efficiency increases with the SnS absorber layer thickness, but it has a much slower rate of increment when the layer thickness is over 2000 nm. Hence, most of the incident photons are absorbed by 2000 nm thick absorber layer.

3.3 *Effect of Defect Density on Absorber Layer*

Since the performance of device mainly depends on output parameters. Thus we have studied the defect density and their capture cross sections. The Fig. 4 shows the effect of defect density on efficiency in the cell module.

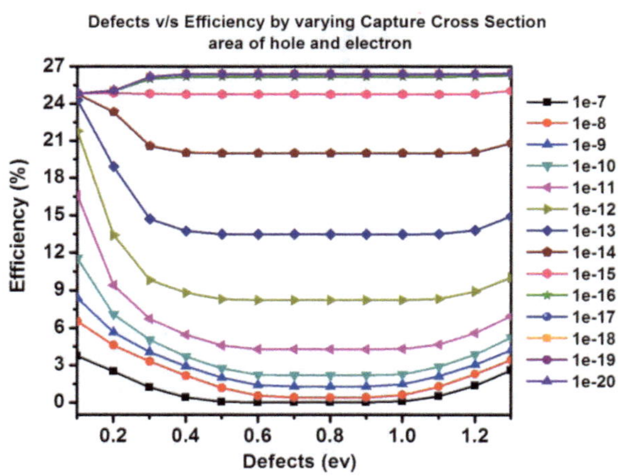

Fig. 4 Defect versus efficiency by capturing cross section area of hole and electron

4 Conclusion

The performance of SnS solar cell is numerically simulated by using device simulation software. The numerical simulations have been done by adjusting parameters such as the thickness of the SnS absorber layer, carrier density, and operating temperature, and analyze their effects on the cell performance. The simulations show that the optimized thickness for the absorbing layer and the n-type should be from 2500 to 4000 nm and 50 nm, respectively. In the range of calculation, the optimal photovoltaic properties have been achieved with an efficiency of 24.87% (with FF = 79.68%, $V_{oc} = 0.9992$ V, and $J_{sc} = 31.244$ mA/cm^2), when the thickness and carrier density of the SnS are 4000 nm and 1×10^{17} cm^{-3}, respectively, and the thickness and the carrier density of the n-type are 50 nm and 1×10^{17} cm^{-3}, respectively. The results above will give some important guidelines for feasibly fabricating higher efficiency SnS-based solar cells.

References

1. Sinsermsuksakul, P., Sun, L., Lee, S.W., Park, H.H., Kim, S.B., Yang, C.: Overcoming efficiency limitations of SnS based solar cells. Adv. Energy Mater **4**, 1400496 (2014)
2. Li, H., Cheng, S., Zhang, J., Huang, W., Zhou, H., Jia, H.: Fabrication of CdS/ZnS hetero junction for photovoltaic applications. World Journal of Condensed Matter Physics **5**, 10–17 (2015)
3. Reddy, K.T.R., Reddy, N.K., Miles, R.W.: Photovoltaic properties of SnS based solar cells. Sol. Energy Mater. Sol. Cells **90**, 3041–3046 (2006)
4. Schneikart, A., Schimper, H.J., Klein, A., Jaegermann, W.: Efficiency limitations of thermally evaporated thin-film SnS solar cells. J. Phys. D Appl. Phys. **46**, 305109 (2013)
5. Kawano, Y., Chantana, J., Minemoto, T.: Impact of growth temperature on the properties of SnS film prepared by thermal evaporation and its photovoltaic performance. Curr. Appl. Phys. **15**, 897–901 (2015)
6. Burton, L.A., Walsh, A.: Band alignment in SnS thin film solar cells, possible origin of the low conversion efficiency. Appl. Phys. Lett. **102**(1–3), 132111 (2013)
7. Burgelman, M., Nollet, P., Degrave, S.: Modelling polycrystalline semiconductor solar cells. Thin Solid Films **361**, 527–532 (2000)
8. Reddy, V.M., Gedi, S., Park, C., Miles, R.W., Ramakrishna Reddy, K.T.: Development of sulphurized SnS thin film solar cells. Curr. Appl. Phys. **15**, 588–598 (2015)
9. Hotje, U., Rose, C., Binnewies, M.: Lattice constants and molar volume in the system ZnS, ZnSe, CdS, CdSe. Solid State Sci. **5**, 1259–1262 (2003)
10. Devika, M., Ramakrishna Reddy, K.T., Koteeswara Reddy, N., Ramesh, K., Gopal, E.S.R., Gunasekhar, K.R.: The effect of substrate surface on the physical properties of SnS films. J. Appl. Phys. **100**, 023518 (2006)
11. Ikuno, T., Suzuki, R., Kitazumi, K., Takahashi, N., Kato, N., Higuchi, K.: SnS thin film solar cells with Zn1−xMgxO buffer layers. Appl. Phys. Lett. **102**, 193901 (2013)
12. Park, H.H., Heasley, R., Sun, L., Steinmann, V., Jaramillo, R., Hartman, K.: Cooptimization of SnS absorber and Zn(O,S) buffer materials for improved solar cells. Prog. Photovolt. Res. Appl **23**, 901–908 (2015)

13. Steinmann, V., Jaramillo, R., Hartman, K., Chakraborty, R., Brandt, R.E., Poindexter, J.R.: 3.88% efficient tin sulfide solar cells using congruent thermal evaporation. Adv. Mater **26**, 7488–7492 (2014)
14. Britt, J, Ferekides, C.: Thin-film CdS/CdTe solar cell with 15.8% efficiency. Appl. Phys. Lett. **62**, 2851–2852 (1993)
15. Yeon, D.H., Mohanty, B.C., Lee, S.M., Cho, Y.S.: Effect of band aligned double absorber layers on photovoltaic characteristics of chemical bath deposited PbS/CdS thin film solar cells. Sci. Rep. **5**, 14353 (2015)
16. Madelung, O. Semiconductors: Data Handbook, 3rd edn. Springer, Berlin (2004)

Author Index

D. Deb et al. (eds.), *Innovations in Infrastructure*, Advances in Intelligent Systems and Computing 757, https://doi.org/10.1007/978-981-13-1966-2

MIX
Papier aus verantwortungsvollen Quellen
Paper from responsible sources
FSC® C105338

If you have any concerns about our products,
you can contact us on
ProductSafety@springernature.com

In case Publisher is established outside the EU,
the EU authorized representative is:
Springer Nature Customer Service Center GmbH
Europaplatz 3, 69115 Heidelberg, Germany

Printed by Libri Plureos GmbH
in Hamburg, Germany